Advanced Principles of Proteomics

Advanced Principles of Proteomics

Editor: Steven Tiff

R CALLISTO REFERENCE

www.callistoreference.com

Callisto Reference,
118-35 Queens Blvd., Suite 400,
Forest Hills, NY 11375, USA

Visit us on the World Wide Web at:
www.callistoreference.com

ISBN: 978-1-64116-157-2 (Hardback)

Cataloging-in-Publication Data

Advanced principles of proteomics / edited by Steven Tiff.
    p. cm.
Includes bibliographical references and index.
ISBN 978-1-64116-157-2
1. Proteomics. 2. Proteins. I. Tiff, Steven.
QP551 .A38 2019
572.6--dc23

Table of Contents

Preface

The proteome is the complete set of proteins that are produced by a living organism or system. It is subject to modification due to changes in requirement and stresses that an organism undergoes. The study of proteins is under the discipline of proteomics. Protein detection with antibodies, antibody-free protein detection, hybrid technologies and high-throughput proteomic technologies are prominent methodologies that are used for the detection and analysis of proteins. The study of proteomics has aided in new drug-discovery and in the understanding of complex plant-insect interactions. This book is a compilation of chapters that discuss the most vital concepts and emerging trends in the field of proteomics. The various advancements are glanced at and their applications as well as ramifications are looked at in detail. The extensive content of this book provides the readers with a thorough understanding of the subject.

The researches compiled throughout the book are authentic and of high quality, combining several disciplines and from very diverse regions from around the world. Drawing on the contributions of many researchers from diverse countries, the book's objective is to provide the readers with the latest achievements in the area of research. This book will surely be a source of knowledge to all interested and researching the field.

In the end, I would like to express my deep sense of gratitude to all the authors for meeting the set deadlines in completing and submitting their research chapters. I would also like to thank the publisher for the support offered to us throughout the course of the book. Finally, I extend my sincere thanks to my family for being a constant source of inspiration and encouragement.

Editor

Comparative proteomics analysis of differential proteins in respond to doxorubicin resistance in myelogenous leukemia cell lines

Shi Qinghong[1], Gao Shen[1], Song Lina[1], Zhao Yueming[1,3], Li Xiaoou[3], Wu Jianlin[4], He Chengyan[1*], Li Hongjun[1] and Zhao Haifeng[2]

Abstract

Background: Chemoresistance remains a significant challenge in chronic myelogenous leukemia (CML) management, which is one of the most critical prognostic factors. Elucidation the molecular mechanisms underlying the resistance to chemoresistance may lead to better clinical outcomes.

Results: In order to identify potential protein targets involved in the drug-resistant phenotype of leukemia, especially the chronic myelogenous leukemia (CML), we used a high-resolution "ultra-zoom" 2DE-based proteomics approach to characterize global protein expression patterns in doxorubicin-resistant myelogenous leukemia cells compared with parental control cells. Ultra-high resolution of 2DE was achieved by using a series of slightly overlapping narrow-range IPG strips during isoelectric focusing (IEF) separation. A total number of 44 proteins with altered abundances were detected and identified by MALDI-TOF or LC-MS/MS. Among these proteins, enolase, aldolase, HSP70 and sorcin were up-regulated in doxorubicin-resistant myelogenous leukemia cell line, whereas HSP27 was down-regulated. Some of the results have been validated by Western blotting. Both enolase and aldolase were first reported to be involved in chemoresistance, suggesting that process of glycolysis in doxorubicin-resistant myelogenous leukemia cells was accelerated to some extent to provide more energy to survive chemical stress. Possible roles of most of the identified proteins in development of chemoresistance in myelogenous leukemia cells were fully discussed. The results presented here could provide clues to further study for elucidating the mechanisms underlying drug resistance in leukemia.

Conclusions: As a whole, under the chemical stress, the doxorubicin-resistant myelogenous leukemia cells may employ various protective strategies to survive. These include: (i) pumping the cytotoxic drug out of the cells by P-glycoprotein, (ii) increased storage of fermentable fuel, (iii) sophisticated cellular protection by molecular chaperones, (iv) improved handling of intracellular calcium, (v) increased glucose utilization via increased rates of glycolysis. In the present study, proteomic analysis of leukemia cells and their drug resistant variants revealed multiple alterations in protein expression. Our results indicate that the development of drug resistance in doxorubicin-resistant myelogenous leukemia cells is a complex phenomenon undergoing several mechanisms.

Background

Doxorubicin, also known as hydroxydaunorubicin, is a drug commonly used in the treatment of a variety of cancers, especially acute myelogenous leukemia (AML) and chronic myelogenous leukemia (CML) [1]. However, resistance to doxorubicin is often observed in patients with leukemia, resulting in failure in chemotherapy. Such chemoresistance is a phenomenon found in many types of cancers including hematological malignancies (leukaemia and lymphoma), many kinds of carcinoma (solid tumors) and soft tissue sarcomas. Once the cancers develop chemoresistance to a distinct drug, they often display low sensitivity to a variety of other chemotherapy drugs and might not respond to these drug therapies, especially in the case of multi-drug resistance in chronic myelogenous leukemia (CML).

Efforts have been made to reveal the molecular mechanisms underlying the development of chemoresistance,

* Correspondence: chengyanhe469@vip.sina.com
[1]Jilin University China-Japan Union Hospital, Changchun 130033, China
Full list of author information is available at the end of the article

particularly multi-drug resistance (MDR) in CML. Early studies had established that drug resistant CML is mediated by P-glycoprotein (Pgp), a protein that functions as a drug efflux pump [2]. Glutathione S transferase (GST) is another important protein found to be associated with MDR [3,4], of which the expression is often observed to be up-regulated in drug resistant cell lines. A recent study indicated that CXCL12 could enhance chemoresistance of K562 cells to doxorubicin by increasing the expression of CXCR4, a seven-transmembrane G-protein-coupled chemokine receptor [5]. Apart from above-mentioned proteins, several other proteins such as sorcin [6,7], survivin [8] and endothelin-1 [9], etc., have been observed to be associated with the development of chemoresistance in CML. Accumulated evidence has shown that these proteins are involved in multiple different pathways and often interact with each other, indicating that the mechanisms mediating drug resistance in CML are multifaceted and still not clearly defined.

Due to the complexity of the changes occurring upon the development of drug resistance, it is of great importance to apply more comprehensive approaches to decipher the codes embedded in drug resistance in leukemia. During the past decade, proteomics has become the powerful tool to perform large scale analysis of complex protein mixtures [10-12]. Two-dimensional gel electrophoresis (2-DE) combined with mass spectrometry (MS) has been one of the most widely used tools in proteomics studies.

Specifically, 2-DE separates proteins according to their isoelectric point (pI) and molecular mass (Mr), both of which are orthogonal parameters of a protein molecule. For separation of a given proteome at protein level, 2-DE has always been the most powerful method to study protein expression and function in living organisms and diseases [13,14]. In a typical 2-DE experiment, immobilized pH gradients (IPGs) with wide pH range (e.g., pH 3–10 or pH 3–11) are used to resolve complex protein mixtures as the first dimensional separation, usually allowing over 1000 protein spots to be visualized on a standard 2-D gel. To achieve better separation in the first dimension of 2-DE, a series of IPG stripes with overlapping narrow pH range such as ultra-zoom IPG stripes (e.g., pH 4.5-5.5 and pH 5.5-6.7) may be used, allowing much more protein spots to be detected with increased resolution [15,16].

In the present study, in order to effectively search for more proteins involved in the development of chemoresistance in CML, we utilized ultra-zoom gels to analyze differences in global protein expression in doxorubicin-resistant myelogenous leukemia cell line K562/A02 and its parental control cell line K562. A series of IPGs with pH range 3.9-5.1, 4.7-5.9, 5.5-6.7 and 6.3-8.3 were used to obtain high resolution 2-DE gels. Up to 44 differentially expressed proteins were identified. The involvement of the identified proteins in the development of drug resistance in leukemia cells was discussed.

Results
Cytotoxicity assay
Cell proliferation assays for both K562/A02 and K562 cultured in the presence or absence of doxorubicin were performed to investigate the drug-resistance characteristics of K562/A02 cells, as well as the viability of K562 cells under chemical stress (Figure 1). Then IC50 values of doxorubicin for both K562/A02 and K562 were calculated. As expected, the IC50 of K562/A02 cells was much greater than that of K562 cells.

2-DE-based proteomics analysis using ultra-zoom IPG strips
The protein expression pattern of the drug resistant K562/A02 cell line was compared to that of the parental K562 cell line using PDQuest software (Version 7.1.1; Bio-Rad). For each cell line, quadruplet of gel images with identical pI range in the IEF separation was assigned for a group using PDQuest software. Thus, two groups, namely one containing the gels of the chemoresistant cell line K562/A02 and one containing the gels of the parental sensitive cell line K562 were summarized to a statistical analysis set. The average quantity of each quadruplet of matched spots in the replicate gels in each group was calculated and the differentially expressed protein spots with p value smaller than 0.05 through Student's t-test were registered. From the statistically differentially expressed protein spots, only those protein spots with quantitative difference between the two groups greater than 2-fold in magnitude were subject to further identification by MS and MS/MS analyses.

Figure 2 shows a group of gel images of the drug resistant cell line K562/A02 (Figure A, C, E and G) and the parental cell line K562 (Figure B, D, F and H), in which a series of ultra-zoom IPG strips with slightly overlapping pH range (pH 3.9-5.1, 4.7-5.9, 5.5-6.7 and 6.3-8.3, 17 cm, Bio-Rad) were used in the first dimensional IEF procedure to greatly increase resolution of IEF. The analysis revealed 44 protein changes between the drug resistant cells and their parental cells, as listed in Table 1. Among the 44 proteins, 30 were up-regulated in drug resistant cells and 14 were down-regulated. Because the IPG strips were slightly overlapped with their neighboring IPG strips, several proteins were detected on more than one gel such as protein 4, 9, 36, 10, 32, 6, 17, 21, 22, etc., of which the robustness of our proteomics analysis was significantly enhanced. For example, detection of protein 9 on gels with IPG strip (pH 3.9-5.1) was confirmed by detection of the same protein on gels with IPG strip (pH 4.7-5.9). All the protein spots were identified by either PMF or MS/MS method alone or the both. Notably, three protein spots (protein

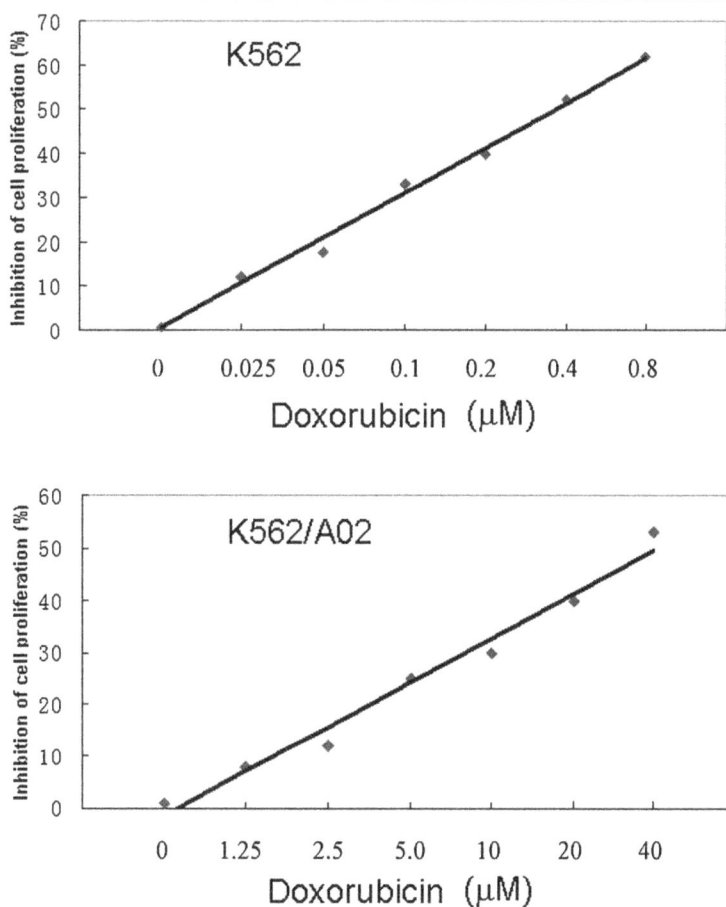

Figure 1 Cell proliferation assays for both K562/A02 and K562 cultured in the presence or absence of doxorubicin.

17) were identified as enolase 1, which were believed to be three isoforms of the same protein. From the comparative proteomics analysis of the doxorubicin-resistant myelogenous leukemia cells and their parental cells, protein 14 on Figure 1 was observed to be highly differentially expressed, which was identified as sorcin, a soluble resistance-related calcium-binding protein. The detailed information regarding to the identification of sorcin by mass spectrometry, as well as the corresponding MS/MS spectra, was shown in Figure 3. Protein spot 43, identified as ATP synthase subunit beta, was found to be down-regulated in K562/A02, of which the result was well consistent with that of a previous study using K562/A02 [17]. The original MS/MS data have been provided in "MS/MS data" file as Additional file 1.

Western blot confirmation of some of results from proteomics analysis

From the identified candidates, HSP70, HSP27 and enolase 1 were selected for western blot analysis as shown in Figure 4. The expression changes of the three selected proteins were consistent with 2-DE results when normalized to actin level. HSP70 and enolase 1 were up-regulated, while HSP27 were down-regulated in K562/A02 cells compared with the parental K562 cells. All experiments were performed for at least three biological replicates. The results of elevated expressions of both HSP70 and enolase 1 were consistent with those from a previous study by Zou L, et al., in which a series of leukaemia-associated antigens in chronic myeloid leukaemia were detected by sera of patients with CML [18]. In addition, We performed Western blotting analysis of both HSP27 and HSP70 for K562 cells exposed to 1 microgram/ml doxorubicin for 2 hrs. In contrast to decreased expression in K562/A02 cells, HSP27 was up-regulated upon drug treatment in K562 cells. For HSP70, elevated expression level was also observed in K562 cells upon drug treatment, of which the fold change was much bigger than that from our proteomics data when expression data of HSP70 in K562/A02 and K562 cells with and without doxorubicin treatment, respectively, were compared.

Discussion

Among the identified proteins with differently expressed abundances, some of the key enzymes involved in

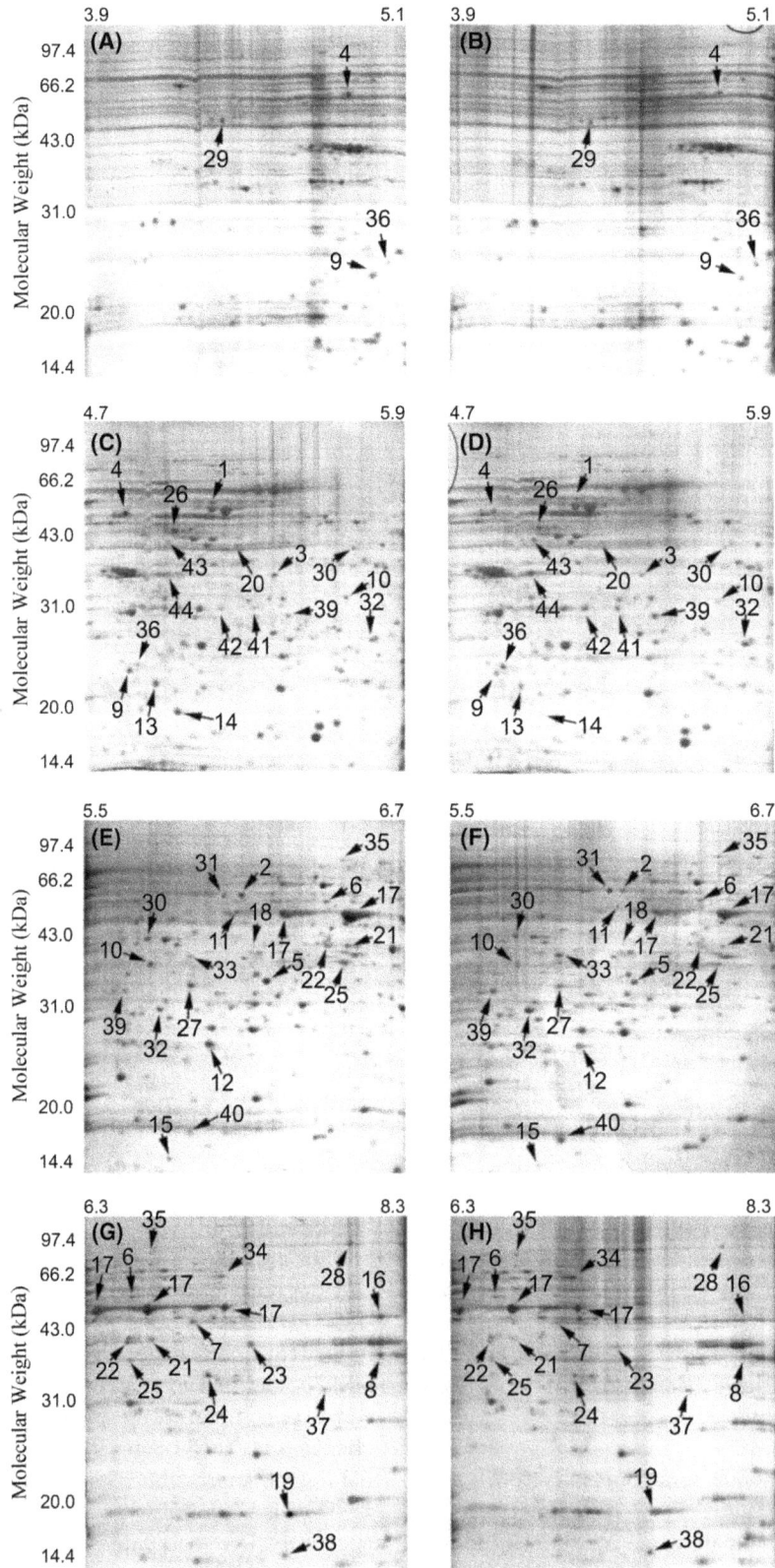

Figure 2 (See legend on next page.)

Figure 2 A group of pairs of protein 2-D maps of doxorubicin-resistant myelogenous leukemia cell line K562/A02 (left) and its parental control cell line K562 (right) with sample loading of 0.6 mg protein each. The isoelectric focusing was carried on 17 cm IPG strips with a pH range of either 3.9-5.1 (**A, B**), 4.7-5.9 (**C, D**), 5.5-6.7 (**E, F**) or 6.3-8.3 (**G, H**). For the second dimension, acrylamide gels of 12.5% were used. The gels were stained by Colloidal Coomassie blue G-250 over night. Numbers associated with the spots on the gel images refer to the identified proteins listed in Table 1.

carbohydrate and energy metabolism were significantly up-regulated, implying that much more energy has to be recruited through glycolysis to help cancer cells survive when exposed to chemical stress. The other interesting finding from the identified protein list is that several kinds of proteins that are involved in intracellular ion regulation were identified, including sorcin, protein S100-A4, annexin A4, calreticulin, chloride intracellular channel protein 1, etc., indicating that metal ions such as calcium might play a pivotal role in the development of chemoresistance.

Proteins involved in carbohydrate and energy metabolism

Early in 1956, Warburg reported that normal cells derive most of their energy though the Krebs cycle (aerobically), whereas cancer cells derive most of their energy from glycolysis (anaerobically). Over the past few years, several studies published in leading scientific journals showed that an increased rate of glycolysis maintained by tumor cells [19] may possibly be due to altered expression of enzymes and as a result, tumor cells burn much more glucose for energy than normal cells [20,21].

In our study, two enzymes involved in glycolysis, namely fructose-bisphosphate aldolase A, fructose-bisphosphate aldolase C, transaldolase and alpha-enolase, were found to be up-regulated in doxorubicin-resistant myelogenous leukemia cells. It was worthy of note that one of the enzymes involved in glycolysis process, namely enolase, was dramatically up-regulated in doxorubicin-resistant myelogenous leukemia cells. Enolases are glycolytic enzymes that interconvert 2-phosphoglycerate to phosphoenolpyruvate. in glycolysis, by which low-energy phosphate ester bond of 2-phosphoglycerate is converted into the high-energy phosphate bond of PEP. In a study by Tu SH and his co-workers [22], alpha-enolase was found to be up-regulated in both tamoxifen-resistant breast cancer and head-and-neck cancer. Additionally, both chemoresistance and invasive ability of these tumors can be dramatically suppressed through knockdown of alpha-enolase expression, implying that up-regulation of enolase 1 may have a protective effect against chemical pressure by augmenting anaerobic metabolism.

The other up-regulated enzyme involved in glycolysis found in doxorubicin-resistant myelogenous leukemia cells, aldolase, splits fructose 1,6-biphosphate into two three-carbon molecules, glyceraldehydes 3-phosphate and dihydroxyacetone phosphate, in glycolysis. Glyceraldehyde 3-phosphate is the only molecule that can be used for the rest of glycolysis. However, the dihydroxyacetone phosphate can be converted to glyceraldehyde 3-phosphate by triose phosphate isomerase. Unfortunately, the evidence of the direct relationship between aldolase and chemoresistance in tumor cells has been poorly documented and the role of aldolase in development of chemoresistance in doxorubicin-resistant myelogenous leukemia cells needs to be further investigated.

Our findings suggest that elevated level of enolase 1 and aldolase C may contribute directly or indirectly to the chemoresistance demonstrated by doxorubicin-resistant myelogenous leukemia cells and the cells need more energy through glycolysis to survive when exposed to chemical stress. Therefore, oncologists should pay much attention on glycolysis as a major target area for the development of new strategies to overcome chemoresistance in cancer patients [23,24].

Proteins involved in intracellular ion regulation

Accumulated evidences have suggested that intracellular calcium may play a role in the development of chemoresistance in some cell lines [25,26]. In our present study, several calcium-binding proteins were identified in doxorubicin-resistant myelogenous leukemia cells, such as sorcin, protein S100-A4, annexin A4, calreticulin, etc., indicating that metal ions such as calcium might play a pivotal role in the development of chemoresistance. A soluble cytosolic calcium-binding protein, sorcin, has been reported to be implicated in development of resistance to cytotoxic chemicals [27,28], although its contribution to multi-drug resistance remains uncertain. Protein S100-A4 belongs to a member of the S100 calcium-binding protein family, and it was observed to be up-regulated in doxorubicin-resistant myelogenous leukemia cells. Similar results were obtained from a recent study by Yang M et al., in which elevated expression level of S100-A8 was detected in drug resistance leukemia cell lines relative to their drug sensitive cell lines. Further studies indicated that S100A8 might contribute to drug resistance in leukemia by regulating autophagy [29]. Interestingly, another member of the S100 calcium-binding protein family, S100P, was reported to contribute to chemosensitivity in gastric cell lines by increasing drug inflow [30]. Our data suggested that S100-A4 may be a novel target for improving leukemia therapy.

Table 1 Summary of differentially expressed proteins

Spot no.	Accession number	Protein name	Function	Identification method	Unique peptides by MS/MS	Theoretical pI/Mw (kDa)	Mascot score	Diffenrential ratio
Proteins with increasing spot intensity								
1	P08107	Heat shock 70 kDa protein 1A/1B	Unfolded protein binding	PMF, MS/MS	8	5.47/70.05	98	2.03 (2.55)*
2	P17987	T-complex protein 1 subunit alpha	Molecular chaperonin	MS/MS	17	5.80/60.34	103	3.91
3	P50453	Serpin B9	Cysteine-type endopeptidase inhibitor activity involved in apoptotic process	PMF, MS/MS	4	5.61/42.40	67	2.34
4	P08670	Vimentin	Double-stranded RNA binding	PMF		5.05/53.65	122	4.84
5	P04083	Annexin A1	Calcium ion binding	PMF		5.85/38.71	132	2.94
6	P20839	Inosine-5′-monophosphate dehydrogenase 1	IMP dehydrogenase activity, metal ion binding	PMF, MS/MS	5	6.43/55.41	116	2.17
7	P12532	Creatine kinase U-type	ATP binding	PMF, MS/MS	3	7.31/43.08	89	3.15
8	P04075	Fructose-bisphosphate aldolase A	Fructose-bisphosphate aldolase activity	PMF		8.30/39.42	142	4.14
9	P52565	Rho GDP-dissociation inhibitor 1	GTPase activator activity	PMF, MS/MS	5	5.01/23.21	127	3.06
10	P07195	L-lactate dehydrogenase B chain	L-lactate dehydrogenase activity	PMF		5.71/36.64	84	2.58
11	V9HW96	Chaperonin containing TCP1, subunit 2 (Beta), isoform CRA_b	Molecular chaperonin	PMF		6.01/57.49	133	2.09
12	P30041	Peroxiredoxin-6	Antioxidant activity	PMF, MS/MS	8	6.00/25.03	93	3.56
13	O00299	Chloride intracellular channel protein 1	Chloride channel activity	PMF, MS/MS	9	5.09/26.92	122	4.19
14	P30626	Sorcin	Calcium channel regulator activity	PMF, MS/MS	4	5.32/21.68	142	50.70
15	P31949	Protein S100-A11	Calcium ion binding	PMF, MS/MS	3	6.65/11.74	89	2.92
16	P00558	Phosphoglycerate kinase 1	Phosphoglycerate kinase activity	PMF		8.30/44.61	91	3.13
17	P06733	Alpha-enolase	Phosphopyruvate hydratase activity	PMF, MS/MS	22	7.01/47.17	123	2.17 (2.93)*
18	Q9NR45	Sialic acid synthase	N-acetylneuraminate synthase activity	PMF		6.29/40.31	97	2.04
19	P62837	Ubiquitin-conjugating enzyme E2 D2	Ubiquitin-protein ligase activity	PMF, MS/MS	4	7.69/16.74	137	2.46
20	P05787	Keratin, type II cytoskeletal 8	Scaffold protein binding	PMF, MS/MS	9	5.52/53.71	152	2.03
21	Q14320	Protein FAM50A	Poly(A) RNA binding	PMF, MS/MS	6	6.39/40.24	104	2.38
22	P09972	Fructose-bisphosphate aldolase C	Fructose-bisphosphate aldolase activity	PMF		6.41/39.46	79	2.11
23	Q08752	Peptidyl-prolyl cis-trans isomerase D	Hsp70 protein binding	PMF, MS/MS	3	6.77/40.76	83	3.17
24	P16422	Epithelial cell adhesion molecule	Protein complex binding	PMF		7.42/34.93	131	2.51
25	P37837	Transaldolase	Monosaccharide binding	PMF, MS/MS	4	6.36/37.54	89	2.04
26	P11926	Ornithine decarboxylase	Ornithine decarboxylase activity	PMF		5.10/51.15	79	3.23
27	P09525	Annexin A4	Calcium ion binding	PMF, MS/MS	2	5.83/35.88	169	4.24
28	P06744	Glucose-6-phosphate isomerase	Glucose-6-phosphate isomerase activity	PMF, MS/MS	4	8.42/63.15	213	3.78

Table 1 Summary of differentially expressed proteins *(Continued)*

29	P27797	Calreticulin	Calcium ion binding	PMF, MS/MS	4	4.29/48.14	228	2.22
30	P05783	Keratin, type I cytoskeletal 18	Scaffold protein binding	PMF, MS/MS	11	5.34/48.06	142	2.13
Proteins with decreasing spot intensity								
31	P30101	Protein disulfide-isomerase A3	Cysteine-type endopeptidase activity	PMF, MS/MS	7	5.98/56.78	103	0.32
32	P28070	Proteasome subunit beta type-4	Threonine-type endopeptidase activity	PMF, MS/MS	2	5.70/29.20	187	0.23
33	Q99LB4	Capping protein (Actin filament), gelsolin-like	Cell projection assembly	PMF		6.47/38.77	109	0.29
34	Q9NNW7	Thioredoxin reductase 2, mitochondrial	Thioredoxin-disulfide reductase activity	PMF		7.24/56.51	93	0.43
35	O95782	AP-2 complex subunit alpha-1	Protein transporter activity	PMF, MS/MS	7	6.63/107.5	167	0.41
36	Q16082	Heat shock protein beta-2	Enzyme activator activity	PMF		5.07/20.23	102	0.38 (0.52)*
37	P42330	Aldo-keto reductase family 1 member C3	Androsterone dehydrogenase activity	PMF		8.06/36.85	146	0.45
38	Q16553	Lymphocyte antigen 6E	Epinephrine secretion	PMF		8.06/13.51	129	0.48
39	P35232	Prohibitin	Sequence-specific DNA binding RNA polymerase II transcription factor activity	PMF, MS/MS	2	5.57/29.80	114	0.28
40	Q92729	Receptor-type tyrosine-protein phosphatase U	Beta-catenin binding	PMF		6.46/16.24	92	0.21
41	P02649	apolipoprotein E	Antioxidant activity	PMF		5.65/36.15	83	0.31
42	Q15181	Inorganic pyrophosphatase	Inorganic diphosphatase activity	PMF, MS/MS	4	5.54/32.66	162	0.25
43	P06576	ATP synthase subunit beta, mitochondrial	Proton-transporting ATP synthase activity, rotational mechanism	PMF, MS/MS	7	5.26/56.56	127	0.41
44	P52597	Heterogeneous nuclear ribonucleoprotein F	RNA binding	PMF, MS/MS	3	5.37/45.67	215	0.46

Note: *The data in the brackets represent the diffenrential ratio obtained from Western blot analysis.

Proteins that function in cell motility or structure

Besides the mechanical role in cell integrity [31,32], cytoskeletal keratins (CKs) also perform important functions in cellular defense mechanism in response to various stresses, including chemical agents. Several studies have shown that both sumoylation and phosphorylation of CKs were raised under cell stress [33,34]. In addition, hyperphosphorylation of CK 8 at Ser-73 was implicated in apoptosis induced by anisomycin or etoposide in cultured HT-29 cells [35]. Tao He, et al., suggested that Ser-73 could be a putative target for the stress-activated MAPK family members [36]. Besides the regulatory role of phosphorylation of CKs in cell signaling pathway under stress, notably, from an energetic point of view, marked overexpression and hyperphosphorylation of CK 8 and CK 18 may support cells in handling several forms of stress that may lead to cell death, because phosphorylation of CKs reduces intracellular ATP levels and may help maintain a phosphate reserve.

Vimentin, another kind of intermediate filament proteins mainly expressed in cells originating from mesenchyme, is thought to play a key role in cell growth, cell cycling and cell resistance to mechanical or chemical stress. In contrast to keratins, vimentin tends to assemble homopolymeric filaments. Belichenko I [37] suggested that the full-length vimentin may be implicated in survival signaling and confers resistance to nuclear apoptosis after photodynamic treatment. Significant elevation of vimentin expression during the process of tumorgenesis was reported in recent years [38,39]. In principal, our data suggest that dramatically increased expression level of vimentin in doxorubicin-resistant myelogenous leukemia cells contributes to the resistance to cytotoxic agent to some extent.

Group of molecular chaperones

It is well known that heat shock and other chemical stresses stimulate synthesis of a specific group of proteins called

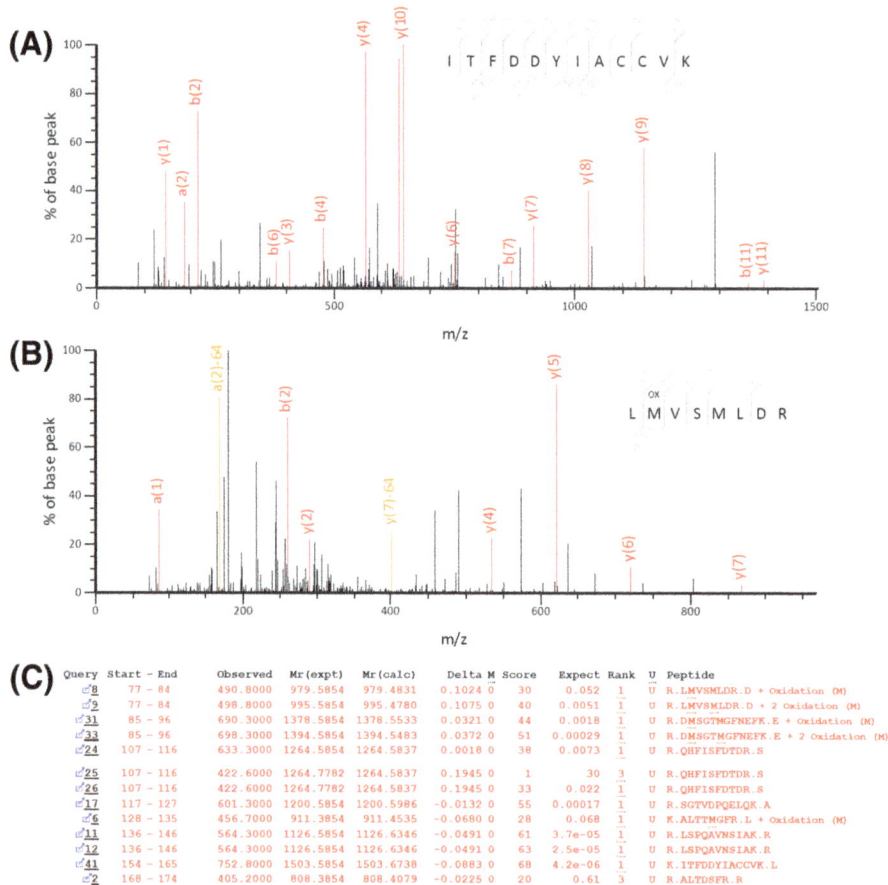

Figure 3 Identification of sorcin by mass spectrometry analysis and database searching. **(A)** MS/MS spectrum of a doubly-charged peak at m/z 752.8. The corresponding peptide is identified as ITFDDYIACCVK (154–165). **(B)** MS/MS spectrum of a doubly-charged peak at m/z 490.8. The corresponding peptide is identified as LMVSMLDR (77–84), in which the methionine is oxidized. **(C)** Identification of sorcin by MS/MS analysis. The data were searched against SwissProt database through Mascot engine.

Query	Start – End	Observed	Mr(expt)	Mr(calc)	Delta M	Score	Expect	Rank	U	Peptide
8	77 – 84	490.8000	979.5854	979.4831	0.1024 0	30	0.052	1	U	R.LMVSMLDR.D + Oxidation (M)
9	77 – 84	498.8000	995.5854	995.4780	0.1075 0	40	0.0051	1	U	R.LMVSMLDR.D + 2 Oxidation (M)
31	85 – 96	690.3000	1378.5854	1378.5533	0.0321 0	44	0.0018	1	U	R.DMSGTMGFNEFK.E + Oxidation (M)
33	85 – 96	698.3000	1394.5854	1394.5483	0.0372 0	51	0.00029	1	U	R.DMSGTMGFNEFK.E + 2 Oxidation (M)
24	107 – 116	633.3000	1264.5854	1264.5837	0.0018 0	38	0.0073	1	U	R.QHFISFDTDR.S
25	107 – 116	422.6000	1264.7782	1264.5837	0.1945 0	1	30	3	U	R.QHFISFDTDR.S
26	107 – 116	422.6000	1264.7782	1264.5837	0.1945 0	33	0.022	1	U	R.QHFISFDTDR.S
17	117 – 127	601.3000	1200.5854	1200.5986	-0.0132 0	55	0.00017	1	U	R.SGTVDPQELQK.A
6	128 – 135	456.7000	911.3854	911.4535	-0.0680 0	28	0.068	1	U	K.ALTTMGFR.L + Oxidation (M)
11	136 – 146	564.3000	1126.5854	1126.6346	-0.0491 0	61	3.7e-05	1	U	R.LSPQAVNSIAK.R
12	136 – 146	564.3000	1126.5854	1126.6346	-0.0491 0	63	2.5e-05	1	U	R.LSPQAVNSIAK.R
41	154 – 165	752.8000	1503.5854	1503.6738	-0.0883 0	68	4.2e-06	1	U	K.ITFDDYIACCVK.L
2	168 – 174	405.2000	808.3854	808.4079	-0.0225 0	20	0.61	3	U	R.ALTDSFR.R

heat shock proteins (HSPs). These proteins are implicated in cellular protection mechanisms and play an important role in protection of proteins from denaturation by stress-inducing agents and the repair or degradation of poly-peptides that have been denatured under stresses in an ATP-dependent manner [40,41]. The increased expression of several HSPs was very common in most chemoresistant variants. However, in the present report, only the HSP70 was found to be up-regulated significantly. An unexpected decrease in the expression of HSP27 was reproducibly observed in the doxorubicin-resistant myelogenous leukemia cells. These results suggest that molecular chaperones, such as HSP27 and HSP70, etc., may execute protective function in systematically different ways in doxorubicin-resistant myelogenous leukemia cells.

To our knowledge, this uncommon expression pattern of HSP27 in doxorubicin-resistant myelogenous leukemia cells has not been previously described. Unlike HSP70, known be implicated in cell proliferation, HSP27 may be involved in cell growth arrest and increased differentiation [42,43]. Several lines of evidence have been clearly indicated that HSP27 is involved in the regulation of actin polymerization [44,45]. In doxorubicin-resistant myelogenous leukemia cells, the decrease of expression level of HSP27 and actin filament (also termed capping protein) revealed in our experiments implied that down-regulation of these proteins may be indirectly implicated in the cell resistance against cytotoxic agents by retarding or reducing differentiation events. In principal, however, the down-regulation of the these proteins in doxorubicin-resistant myelogenous leukemia cells is intriguing and further studies will be required to elucidate the mechanism underlying the reduced expression level of them in this drug resistant cell line.

Proteins that are related to oxidation or reduction

Another interesting finding is the involvement of protective proteins, peroxiredoxins, in the development of drug resistance the chemoresistant cell line K562/A02. Besides playing a role as a peroxidase, these protective

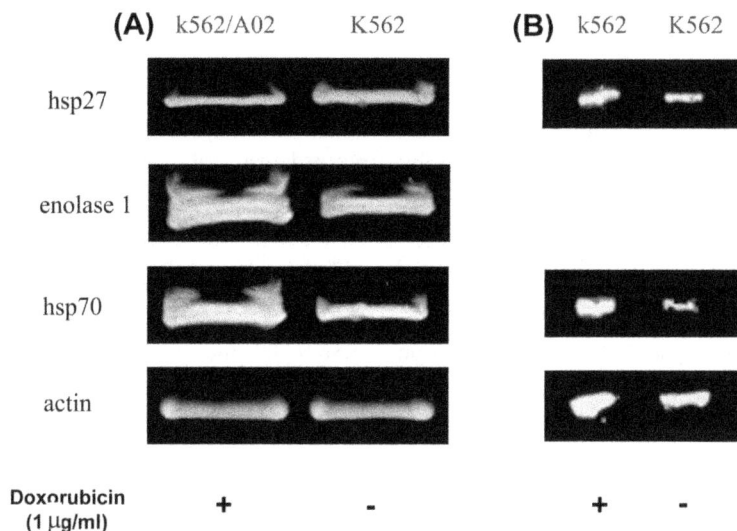

Figure 4 Western blotting analysis of some of differentially expressed proteins. (A) Western blotting analysis of HSP70, HSP27, and enolase 1 between K562 cells and K562/A02 cells. The amount of total protein loaded was 20 μg per lane. Differential expression of the protein of interest between doxorubicin-resistant myelogenous leukemia cell line K562/A02 and its parental control cell line K562 is normalized to actin by probing the same membrane with a monoclonal antibody specific for the protein. **(B)** Western blotting analysis of HSP70 and HSP27 in K562 cells with and without 1 microgram/ml doxorubicin treatment for 2 hrs, respectively.

proteins have been suggested to serve multiple functions involved in various biological processes such as the detoxification of oxidants, cell differentiation and intracellular signaling. Non-selenium glutathione peroxidase (also termed peroxiredoxin 6 or antioxidant protein 2) is a member of family of antioxidant proteins [46]. Unlike other members in the family of anti-oxidative proteins, nonselenium glutathione peroxidase contains only one Cys residue and has been reported to exhibit glutathione peroxidase and phospholipase A2 activities [47]. A body of evidence has accumulated to suggest that protein glutathionylation may be a mechanism of redox regulation of protein functions. A variety of proteins, such as enzymes, transcription factors, and oncogenes, have been reported to be glutathionylated [48,49]. Further studies would lead to a better understanding of biological functions of members of the peroxiredoxin family in doxorubicin-resistant myelogenous leukemia cells.

Protein disulfide isomerase, a primary folding catalyst and chaperone located in the endoplasmic reticulum (ER), is involved in rearrangement of disulfide bridges in proteins to form the correct protein structures. In contrast to HSPs in the cytosol where maintains a more reducing condition than ER, the protein disulphide isomerase turns to act as a chaperone primarily when reduced [50]. However, information on the role of protein disulphide isomerase in doxorubicin-resistant myelogenous leukemia cells is poorly documented and down-regulation of it seems that processes involved in endoplasmic reticulum may be quite different with those in cytosol.

Conclusions
As a whole, under the chemical stress, the doxorubicin-resistant myelogenous leukemia cells may employ various protective strategies to survive. These include: (i) pumping the cytotoxic drug out of the cells by P-glycoprotein, (ii) increased storage of fermentable fuel, (iii) sophisticated cellular protection by molecular chaperones, (iv) improved handling of intracellular calcium, (v) increased glucose utilization via increased rates of glycolysis. In the present study, proteomic analysis of leukemia cells and their drug resistant variants revealed multiple alterations in protein expression. Our results indicate that the development of drug resistance in doxorubicin-resistant myelogenous leukemia cells is a complex phenomenon undergoing several mechanisms.

Methods
Cell culture
The drug-resistant MDR cell line K562/A02 was obtained from the Institute of Hematology, Tianjin, China. It shows resistance not only to doxorubicin but also to some other structurally unrelated lipophilic cytotoxic drugs, including harringtonine, vincristine, amsacrine, etc. The cells were maintained in RPMI 1640 (GIBCO) supplemented with 10% fetal bovine serum (Hyclone) at 37°C in a humidified 5% CO_2 atmosphere. In addition, 1 μg/ml doxorubicin was added into the medium to maintain the drug resistance. The drug-sensitive parental cell line K562 was kept in our laboratory, and cultured in the same condition without doxorubicin.

Cytotoxicity assay

Cytotoxicity of doxorubicin was measured using the MTT (3-[4, 5-dimethylthiazol-2yl]-2, 5-diphenyl tetrazolium bromide) (MTT; Sigma-Aldrich) assay [51]. Briefly, 2×10^4 cells per well were seeded into 96-well culture plates and kept for 24 h. After doxorubicin exposure for 48 hours, the medium was removed, followed by addition of an equal volume of fresh medium containing 0.5 mg/ml MTT. After the cells were incubated with MTT for 4 h at 37°C, the medium was replaced with 200 µl DMSO and kept for 30 min at room temperature. The absorbance was recorded using a microplate reader at absorption wavelength of 570 nm. The cytotoxic effects of drugs were calculated according to the OD values.

Sample preparation for 2-DE

Cells were washed three times with a solution containing 10 mM Tris, 1 mM EDTA and 250 mM sucrose which was adjusted to pH 7.0-7.5, and cell lysates were then prepared using lysis buffer (8 M urea, 4% w/v CHAPS, 50 mM DTT, 25 mM spermine, a cocktail of proteinase inhibitors from Roche). After 60 min of gentle stirring at room temp, the sample was centrifuged at 40 000 g for 60 min. The supernatant was then collected and protein concentration was determined using the DC RC protein assay kit (Bio-Rad), following the manufacturer's instructions. The samples were then aliquoted and stored at –80°C until used for 2-DE.

2-DE

Isoelectric focusing was carried out using Protean IEF Cell (Bio-Rad). Samples containing 0.6 mg for semi-preparative gels, were diluted to 300 µl with rehydration solution (6 M urea, 2% w/v CHAPS, 65 mM DTT, trace bromophenol blue), to which either pH 3.9-5.1, 4.7-5.9, 5.5-6.7 or 6.3-8.3 Bio-lyte was added to final concentration of 0.5% v/v. The samples were accordingly applied to IPG gel strips (pH 3.9-5.1, 4.7-5.9, 5.5-6.7 and 6.3-8.3, 17 cm, Bio-Rad) for 14 h in a passive mode. Proteins absorbed into IPG gel strips were focused for 80 kVh at 20°C. After equilibrated in equilibration solution, gel strips were applied on second-dimensional PAGE with 12.5% polyacrylamide. Separation was then carried out on a Protean II xi electrophoresis system (Bio-Rad) at a current setting of 7.5 mA/gel for the initial 1 h and 15 mA/gel until the bromophenol blue reached the bottom of the gel.

Protein visualization and image analysis

For colloidal Coomassie blue G-250 staining of semi-preparative gels, gels were fixed in 10% TCA for 60 min and rinsed in Milli Q water for 20 min. Gels were then stained in a solution containing 20% methanol, 2% phosphoric acid, 10% ammonium sulfate and 0.1% Coomassie Brilliant Blue G-250 for 24 h. The stained gels were then rinsed in MilliQ water for 20 min to remove any dye residue. Then the stained gels were canned with a high-resolution scanner (Umax 1120) and the gel images were analyzed using PDQuest software (Version 7.1.1; Bio-Rad) according to the protocols provided by the manufacturer. To accurately compare spot quantity between gels, a normalization based on the total density on each gel was applied for each gel and normalized spot intensities were expressed in ppm. Student's t-test was performed on the replicate gels between the chemoresistant cell line K562/A02 and the parental cell line K562. The significantly differentially expressed protein spots ($p < 0.05$) with 2-fold increased or decreased intensity between the chemoresistant variant K562/A02 and the parental cell line K562 were selected and subject to further identification by MALDI-TOF and LC-MS/MS.

MALDI-TOF MS analysis and database searching

Differential spots were excised from semi-preparative gels and transferred to 1.5 ml siliconized Eppendorf tubes. The gel-spots were washed and then destained by 50% ACN until the blue dye turned invisible. After dried in a vacuum centrifuge, the gel-pieces were incubated in the digestion solution containing 50 mmol/L NH4HCO3 and 0.1 g/L TPCK-trypsin for 12 h at 37°C. The resulting peptides were extracted three times by 50 µl aliquots of 5% trifluoroacetic acid in 60% acetonitrile. Combined extracts were concentrated in a Speed Vac to 3–5 µl. The concentrated tryptic peptide mixture was mixed with saturated CHCA matrix solution and vibrated gently. A volume (1 µl) of the mixture containing CHCA matrix was loaded on a 96 × 2 well hydrophobic plastic surface sample plate (Applied Biosystems) and air-dried. The samples were analyzed with Voyager DE STR MALDI TOF Mass Spectrometer (Applied Biosystems). A peptides mixture containing Angiotensin I, ACTH (1–17) and ACTH (18–39) was used as mass standards for external calibration. Monoisotopic peak masses were used to search against the Swiss-Prot database using MASCOT search engine with the following parameters: one missing cleavage, peptide tolerance of 100 ppm, variable methionine oxidation and fixed cysteine carbamidomethylation.

Nano-ESI-MS/MS analysis and database searching

The lyophilized tryptic peptides were redissolved in 0.1% formic acid and then desalted by using ZipTip C18 pipette tips (Millipore, Bedford, MA, USA). The samples were loaded into a nanoelectrospray needle and analyzed on a quadrupole orthogonal acceleration TOF mass spectrometer (QSTAR, Applied Biosystems) equipped with an external nanoelectrospray ion source. Once the full mass spectra (parent ions) of the samples were obtained in TOF MS mode, MS/MS experiments for individual peptide ions with doubly- or triply-charge state were performed in Product ion mode. The data from each sample were searched against the Swiss-Prot database on a local MASCOT server (version 2.1, Matrix Science) using a script embedded in the ProteinPilot 4.5 software (MDS Sciex, South San Francisco, CA, USA). Following parameters were set during database searching: one missed cleavage; peptide tolerance, 0.2 Da; MS/MS tolerance, 0.2 Da; cysteine carbamidomethylation as fixed modification and methionine oxidation as variable modification.

1-DE and western blot analysis

Samples containing equivalent amounts of protein in SDS loading buffer were subjected to electrophoresis in 12% Tris-glycine-SDS polyacrylamide gel using a Mini-Cell system (Bio Rad). After electrophoresis, proteins were electroblotted to polyvinylidene fluoride (PVDF) membranes (Millipore). After blocked with 1.5% nonfat dried milk in TBST (25 mM Tris, pH 7.5, 150 mM NaCl, 0.05% Tween 20, and 0.001% thimerosal) for 1 h at room temperature, membranes were incubated with primary antibody at room temperature for 2 h. Several primary antibodies were chosen: mouse anti-human Hsp70 (Santa Cruz Biotechnology, used at 1:250 dilution), goat anti-human enolase 1 (Santa Cruz Biotechnology, used at 1:225 dilution), mouse anti-human actin (Santa Cruz Biotechnology, used at 1:250 dilution). After washed 10 min in TBST solution, membranes were incubated with properly diluted secondary antibody conjugated with horseradish peroxidase for 1 hr at room temperature. Membranes were washed again and stained with 0.05% diaminobenzidine in Tris–HCl (100 mM, pH 7.5) containing 0.035% hydrogen peroxide.

Abbreviations
CML: Chronic myelogenous leukemia; 2DE: Two dimensional electrophoresis; IEF: Isoelectric focusing; HSP: Heat shock protein; MALDI-TOF: Matrix-assisted laser desorption/ionization-time of flight; AML: Acute myelogenous leukemia; MDR: Multi-drug resistance; Pgp: P-glycoprotein; IPGs: Immobilized pH gradients; MS: Mass spectrometry; CKs: Cytoskeletal keratins; ER: Endoplasmic reticulum.

Competing interests
The authors declare that they have no competing interests.

Authors' contributions
SQ carried out the proteomics studies, participated in the immunoassays and drafted the manuscript. GS participated in the MALDI MS analyses. SL participated in the nano-ESI MS/MS analyses. ZY carried out the immunoassays. LH participated in the design of the study and helped to draft the manuscript. ZH participated in the database searching work. HC conceived of the study, and participated in its design and coordination and helped to draft the manuscript. All authors read and approved the final manuscript.

Acknowledgements
This work was supported by Jilin Province Science and Technology Department (20110713, 20090461, 20110739), Helongjiang Province Science and Technology Department (D201159), Jilin Province Development and Reform Commission (2013c026-5), Changchun Science and Technology Bureau (2011125), Macao Science and Technology Development Fund (014/2012/A).

Author details
[1]Jilin University China-Japan Union Hospital, Changchun 130033, China. [2]Jiamusi University, Jiamusi 154002, China. [3]Tumor Hospital of Jilin Province, Changchun 130021, China. [4]State Key Laboratory for Quality Research in Chinese Medicines, Macau University of Science and Technology, Avenida Wai Long, Taipa, Macau, China.

References
1. Tacar O, Sriamornsak P, Dass CR. Doxorubicin: an update on anticancer molecular action, toxicity and novel drug delivery systems. J Pharm Pharmacol. 2013;65(2):157–70.
2. Turkina AG, Baryshnikov AY, Sedyakhina NP, Folomeshkina SV, Sokolova MA, Choroshco ND, et al. Studies of P-glycoprotein in chronic myelogenous leukaemia patients: expression, activity and correlations with CD34 antigen. Br J Haematol. 1996;92(1):88–96.
3. Schisselbauer JC, Silber R, Papadopoulos E, Abrams K, LaCreta FP, Tew KD. Characterization of glutathione S-transferase expression in lymphocytes from chronic lymphocyticleukemia patients. Cancer Res. 1990;50(12):3562–8.
4. Takanashi M, Morimoto A, Yagi T, Kuriyama K, Kano G, Imamura T, et al. Impact of glutathione S-transferase gene deletion on early relapse in childhood B-precursor acute lymphoblastic leukemia. Haematologica. 2003;88(11):1238–44.
5. Wang Y, Miao H, Li W, Yao J, Sun Y, Li Z, et al. CXCL12/CXCR4 axis confers adriamycin resistance to human chronic myelogenous leukemia and oroxylin A improves the sensitivity of K562/ADM cells. Biochem Pharmacol. 2014;90(3):212–25.
6. Qi J, Liu N, Zhou Y, Tan Y, Cheng Y, Yang C, et al. Overexpression of sorcin in multidrug resistant human leukemia cells and its role in regulating cell apoptosis. Biochem Biophys Res Commun. 2006;349(1):303–9.
7. Yamagishi N, Nakao R, Kondo R, Nishitsuji M, Saito Y, Kuga T, et al. Increased expression of sorcin is associated with multidrug resistance in leukemia cells via up-regulation of MDR1 expression through cAMP response element-binding protein. Biochem Biophys Res Commun. 2014;448(4):430–6.
8. Souza PS, Vasconcelos FC, De Souza Reis FR, Nestal De Moraes G, Maia RC. P-glycoprotein and survivin simultaneously regulate vincristine-induced apoptosis in chronic myeloid leukemia cells. Int J Oncol. 2011;39(4):925–33.
9. Maffei R, Bulgarelli J, Fiorcari S, Martinelli S, Castelli I, Valenti V, et al. Endothelin-1 promotes survival and chemoresistance in chronic lymphocytic leukemia B Cells through ETA receptor. PLoS One. 2014;9(6):e98818.
10. Liu N, Song W, Wang P, Lee K, Chan W, Chen H, et al. Proteomics analysis of differential expression of cellular proteins in response to avian H9N2 virus infection in human cells. Proteomics. 2008;8(9):1851–8.
11. Li Z, He X, Pan C, Liu N. Mass spectrometric analysis of phosphorylation modification in 14-3-3 epsilon protein. Chinese J Anal Chem. 2013;41 (11):1653–8.
12. Liu N, Zhao L, He C, Cai Z. Advances in technologies and biological applications of 18O labeling strategies in LC-MS based proteomics: an updated review. Curr Anal Chem. 2012;8(1):22–34.
13. Li Z, Liu N, Zhang LS, Gong K, Cai Y, Gao W, et al. Proteomic profiling reveals comprehensive insights into adrenergic receptor-mediated hypertrophy in neonatal rat cardiomyocytes. Proteomics Clin Appl. 2009;3(12):1407–21.
14. Liu N, Gong KZ, Cai YB, Li Z. Identification of proteins responding to adrenergic receptor subtype-specific hypertrophy in cardiomyocytes by proteomic approaches. Biochemistry (Mosc). 2011;76(10):1140–6.
15. Hoving S, Gerrits B, Voshol H, Müller D, Roberts RC, van Oostrum J. Preparative two-dimensional gel electrophoresis at alkaline pH using narrow range immobilized pH gradients. Proteomics. 2002;2(2):127–34.
16. Simon WJ, Hall JJ, Suzuki I, Murata N, Slabas AR. Proteomic study of the soluble proteins from the unicellular cyanobacterium Synechocystis sp. PCC6803 using automated matrix-assisted laser desorption/ionization-time of flight peptide mass fingerprinting. Proteomics. 2002;2(12):1735–42.
17. Li RJ, Zhang GS, Chen YH, Zhu JF, Lu QJ, Gong FJ, et al. Down-regulation of mitochondrial ATPase by hypermethylation mechanism in chronic myeloid leukemia is associated with multidrug resistance. Ann Oncol. 2010;21 (7):1506–14.
18. Zou L, Wu Y, Pei L, Zhong D, Gen M, Zhao T, et al. Identification of

leukemia-associated antigens in chronic myeloid leukemia by proteomic analysis. Leuk Res. 2005;29(12):1387–91.

19. Guillaumond F, Leca J, Olivares O, Lavaut MN, Vidal N, Berthezène P, et al. Strengthened glycolysis under hypoxia supports tumor symbiosis and hexosamine biosynthesis in pancreatic adenocarcinoma. Proc Natl Acad Sci U S A. 2013;110(10):3919–24.

20. Song S, Finkel T. GAPDH and the search for alternative energy. Nat Cell Biol. 2007;9(8):869–70.

21. Tsutsumi S, Fukasawa T, Yamauchi H, Kato T, Kigure W, Morita H, et al. Phosphoglucose isomerase enhances colorectal cancer metastasis. Int J Oncol. 2009;35(5):1117–21.

22. Tu SH, Chang CC, Chen CS, Tam KW, Wang YJ, Lee CH, et al. Increased expression of enolase alpha in human breast cancer confers tamoxifen resistance in human breast cancer cells. Breast Cancer Res Treat. 2010;121(3):539–53.

23. Ganapathy-Kanniappan S, Geschwind JF. Tumor glycolysis as a target for cancer therapy: progress and prospects. Mol Cancer. 2013;12:152. doi:10.1186/1476-4598-12-152.

24. Granchi C, Minutolo F. Anticancer agents that counteract tumor glycolysis. ChemMedChem. 2012;7(8):1318–50.

25. Yang H, Zhang Q, He J, Lu W. Regulation of calcium signaling in lung cancer. J Thorac Dis. 2010;2(1):52–6.

26. Al-Bahlani S, Fraser M, Wong AY, Sayan BS, Bergeron R, Melino G, et al. P73 regulates cisplatin-induced apoptosis in ovarian cancer cells via a calcium/calpain-dependent mechanism. Oncogene. 2011;30(41):4219–30.

27. Colotti G, Poser E, Fiorillo A, Genovese I, Chiarini V, Ilari A. Sorcin, a calcium binding protein involved in the multidrug resistance mechanisms in cancer cells. Molecules. 2014;19(9):13976–89.

28. Padar S, van Breemen C, Thomas DW, Uchizono JA, Livesey JC, Rahimian R. Differential regulation of calcium homeostasis in adenocarcinoma cell line A549 and its Taxol-resistant subclone. Br J Pharmacol. 2004;142(2):305–16.

29. Yang M, Zeng P, Kang R, Yu Y, Yang L, Tang D, et al. S100A8 contributes to drug resistance by promoting autophagy in leukemia cells. PLoS One. 2014;9(5):e97242.

30. Zhao X, Bai Z, Wu P, Zhang Z. S100P enhances the chemosensitivity of human gastric cancer cell lines. Cancer Biomark. 2013;13(1):1–10.

31. McLean WH, Lane EB. Intermediate filaments in disease. Curr Opin Cell Biol. 1995;7(1):118–25.

32. Portet S, Arino O, Vassy J, Schoëvaërt D. Organization of the cytokeratin network in an epithelial cell. J Theor Biol. 2003;223(3):313–33.

33. Snider NT, Weerasinghe SV, Iñiguez-Lluhí JA, Herrmann H, Omary MB. Keratin hypersumoylation alters filament dynamics and is a marker for human liver disease and keratin mutation. J Biol Chem. 2011;286(3):2273–84.

34. Pan X, Kane LA, Van Eyk JE, Coulombe PA. Type I keratin 17 protein is phosphorylated on serine 44 by p90 ribosomal protein S6 kinase 1 (RSK1) in a growth- and stress-dependent fashion. J Biol Chem. 2011;286(49):42403–13.

35. Liao J, Ku NO, Omary MB. Stress, apoptosis, and mitosis induce phosphorylation of human keratin 8 at Ser-73 in tissues and cultured cells. J Biol Chem. 1997;272(28):17565–73.

36. He T, Stepulak A, Holmström TH, Omary MB, Eriksson JE. The intermediate filament protein keratin 8 is a novel cytoplasmic substrate for c-Jun N-terminal kinase. J Biol Chem. 2002;277(13):10767–74.

37. Belichenko I, Morishima N, Separovic D. Caspase-resistant vimentin suppresses apoptosis after photodynamic treatment with a silicon phthalocyanine in Jurkat cells. Arch Biochem Biophys. 2001;390(1):57–63.

38. Korsching E, Packeisen J, Liedtke C, Hungermann D, Wülfing P, van Diest PJ, et al. The origin of vimentin expression in invasive breast cancer: epithelial-mesenchymal transition, myoepithelial histogenesis or histogenesis from progenitor cells with bilinear differentiation potential? J Pathol. 2005;206(4):451–7.

39. Chen S, Cai J, Zhang W, Zheng X, Hu S, Lu J, et al. Proteomic identification of differentially expressed proteins associated with the multiple drug resistance in methotrexate-resistant human breast cancer cells. Int J Oncol. 2014;45(1):448–58.

40. Lu WJ, Lee NP, Fatima S, Luk JM. Heat shock proteins in cancer: signaling pathways, tumor markers and molecular targets in liver malignancy. Protein Pept Lett. 2009;16(5):508–16.

41. Calderwood SK, Khaleque MA, Sawyer DB, Ciocca DR. Heat shock proteins in cancer: chaperones of tumorigenesis. Trends Biochem Sci. 2006;31(3):164–72.

42. Venkatakrishnan CD, Dunsmore K, Wong H, Roy S, Sen CK, Wani A, et al. HSP27 regulates p53 transcriptional activity in doxorubicin-treated fibroblasts and cardiac H9c2cells: p21 upregulation and G2/M phase cell cycle arrest. Am J Physiol Heart Circ Physiol. 2008;294(4):H1736–44.

43. Sadeh-Mestechkin D, Epstein Shochet G, Pomeranz M, Fishman A, Drucker L, Biron-Shental T, et al. The effect of heat shock protein 27 on extravillous trophoblast differentiation and on eukaryotic translation initiation factor 4E expression. Mol Hum Reprod. 2014;20(5):422–32.

44. Clarke JP, Mearow KM. Cell stress promotes the association of phosphorylated HspB1 with F-actin. PLoS One. 2013;8(7):e68978.

45. Schneider GB, Hamano H, Cooper LF. In vivo evaluation of hsp27 as an inhibitor of actin polymerization: hsp27 limits actin stress fiber and focal adhesion formation afterheat shock. J Cell Physiol. 1998;177(4):575–84.

46. Chae HZ, Kim IH, Kim K, Rhee SG. Cloning, sequencing, and mutation of thiol-specific antioxidant gene of Saccharomyces cerevisiae. J Biol Chem. 1993;268(22):16815–21.

47. Fisher AB. Peroxiredoxin 6: a bifunctional enzyme with glutathione peroxidase and phospholipase A activities. Antioxid Redox Signal. 2011;15(3):831–44.

48. Su D, Gaffrey MJ, Guo J, Hatchell KE, Chu RK, Clauss TR, et al. Proteomic identification and quantification of S-glutathionylation in mouse macrophages using resin-assisted enrichment and isobaric labeling. Free Radic Biol Med. 2014;67:460–70.

49. Chiang BY, Chou CC, Hsieh FT, Gao S, Lin JC, Lin SH, et al. In vivo tagging and characterization of S-glutathionylated proteins by a chemoenzymatic method. Angew Chem Int Ed Engl. 2012;51(24):5871–5.

50. Sitia R, Molteni SN. Stress, protein (mis)folding, and signaling: the redox connection. Sci STKE. 2004;2004(239):e27.

51. Zhu HJ, Wang JS, Guo QL, Jiang Y, Liu GQ. Reversal of P-glycoprotein mediated multidrug resistance in K562 cell line by a novel synthetic calmodulin inhibitor, E6. Biol Pharm Bull. 2005;28(10):1974–8.

2

Proteomic characterization of adrenal gland embryonic development reveals early initiation of steroid metabolism and reduction of the retinoic acid pathway

Gry H Dihazi[1], Gerhard A Mueller[1], Abdul R Asif[2], Marwa Eltoweissy[1], Johannes T Wessels[1] and Hassan Dihazi[1*]

Abstract

Background: Adrenal glands are essential endocrine organs composed of two embryological distinct tissues. Morphological changes during their development are well described, but less understood with regard to their molecular mechanisms. To identify proteins and pathways, which drive the initial steps of the specification of the endocrine function of the adrenal gland, rat's adrenal glands were isolated at different embryonic days (E): E14, E16, E18, E19 and postnatal day 1 (P1).

Results: The alteration of the proteome during the stages E16, E19 and P1 was investigated by combining two dimensional gel electrophoresis and mass spectrometric analysis. Out of 594 excised protein spots, 464 spots were identified, resulting in 203 non-redundant proteins. The ontogenic classification of the identified proteins according to their molecular function resulted in 10 different categories, whereas the classification of their biological processes resulted in 19 different groups. This gives an insight into the complex mechanisms underlying adrenal gland development. Interestingly, the expression of retinoic acid pathway proteins was decreased during the development of the adrenal gland, suggesting that this pathway is only important at early stages. On the other hand, key proteins of the cholesterol synthesis increased their expression significantly at E19 revealing the initiation of the endocrine specialization of the adrenal glands.

Conclusions: This study presents the first comprehensive wide proteome analysis of three different stages of embryonic adrenal gland development. The identified proteins, which were expressed in early stages of development, will shed light on the molecular mechanisms underlying embryonic development of the adrenal gland.

Keywords: Adrenal gland, Proteomics, Prenatal development

Background

Adrenal glands (AGs) are complex endocrine organs, which are vitally necessary. Each gland is composed of two embryological distinct tissues, the mesodermal cortex and the medulla, arising from the neural crest ectoderm. The adrenal medulla produces catecholamine, such as epinephrine (adrenaline) [1,2], in the classic immediate stress response (fight or flight), whereas the adrenal cortex regulates the long-term stress response by modulating salt,

pH, and glucose homeostasis through the synthesis of corticosteroids, such as cortisol [3,4].

The development of the adrenal gland begins with the migration of neural crest cells toward the coelomic cavity wall and forms a thickening at the medial side of the fetal cortex anlage of the adrenal gland. When the fetal cortex surrounds these cells, they start to differentiate into catecholamine secreting chromaffin cells of the adrenal medulla. Later, further mesenchymal cells detach from the mesothelium and surround the fetal cortex, building the permanent cortex [5,6].

The adrenal cortex, the outer 90% of the gland, synthesizes corticosteroid hormones from cholesterol. Its

* Correspondence: dihazi@med.uni-goettingen.de
[1]Department of Nephrology and Rheumatology, University Medical Center Goettingen, Georg-August University Goettingen, Robert-Koch-Strasse 40, D-37075 Goettingen, Germany
Full list of author information is available at the end of the article

major secretions are cortisol, an important metabolic hormone, androgens such as androstenedione and dehydroepiandrostenedione (DHEA), and aldosterone, a hormone regulating water and electrolyte concentrations [5,6]. The adrenal medulla, the inner 10% of the gland is the main source of the catecholamines epinephrine (adrenaline) and norepinephrine (noradrenaline).

The adrenal gland is an organ, which undergoes major postnatal developmental modifications [7-9]. Recently, a differential proteomic analysis of adrenal gland during postnatal development focusing on regulated proteins, which are involved in steroidogenesis, cell proliferation, and cholesterol synthesis [10] was published.

The underlying molecular mechanisms of the prenatal development of the adrenal gland are still not completely understood. To unravel the protein expression changes that accompany the AG embryonic development, two main approaches might be considered: the analysis of the transcriptome and proteome investigation. Transcriptomic analysis is a powerful method, widely used for comparison of gene expression of different biological samples. Nonetheless, mRNA translation might be influenced by many factors, which can have a profound impact on the amount of protein synthesized. Additionally, post-transcriptional and post-translational events, which play important roles in embryonic development, increase the diversity of proteins that can be synthesized from a

fixed number of genes. Moreover, gene expression changes, which result in alteration in mRNA level, do not necessarily result in protein expression or activity modification. Proteome investigation might help to overcome the limitations of the transcriptomic analysis and deliver important insights into the molecular mechanisms of AG development on the proteome level. For these reasons we decided to use proteomics methods to investigate changes in the protein expression during AG embryonic development. We performed a comparative proteomic analysis of two prenatal stages (embryonic day 16 (E16) and 19 (E19)) as well as of newborn rats (postnatal day 1 (P1)). Here we report some of the molecular modifications identified, and discuss the subjacent functions of the differentially expressed proteins.

Results
Preparation and HE staining of the adrenal glands
To identify and characterize proteins and pathways involved in the development of the adrenal gland, rats were obtained at different stages of embryonic (E14, E16, E18 and E19) and neonatal (P1) development and the AGs were obtained (Figure 1A). The HE staining of the adrenal gland from E19 embryos showed the two distinct tissues of this endocrine organ, the mesodermal cortex, which is the zone of steroid-synthesizing cells and the medulla, which consists of cells arising from the

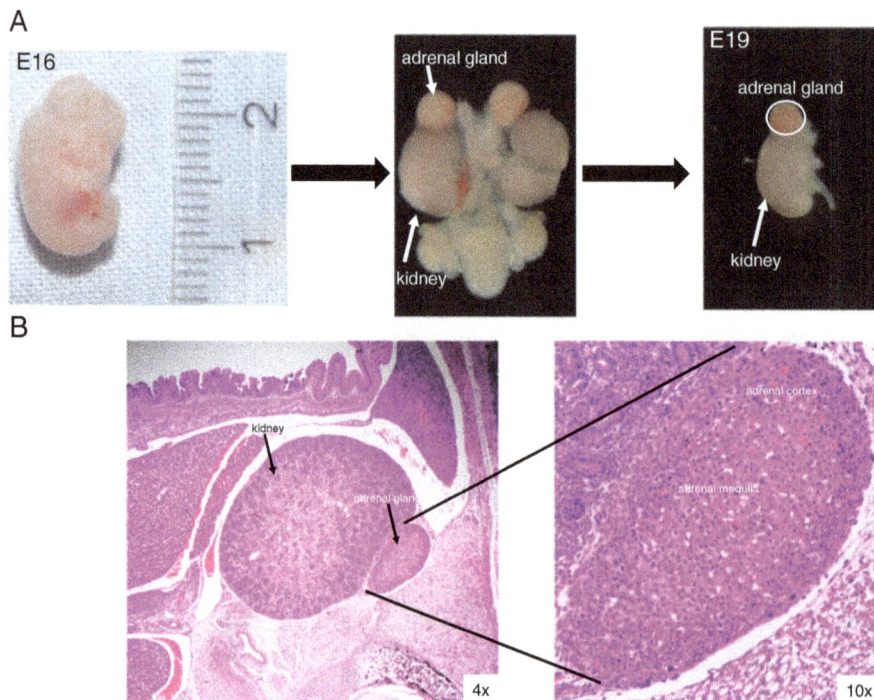

Figure 1 Adrenal gland preparation. A: Embryos were removed from the mother at different days of development. The adrenal glands were excised from the embryos as well as from the newborn rats (P1). **B**: HE staining of a paraffin section of a rat embryo at E19 showing the kidney and the adrenal gland, enclosing the adrenal cortex and the adrenal medulla.

neural crest ectoderm and is responsible for synthesis and secretion of catecholamines (Figure 1B).

Mapping of embryonic adrenal glands proteome

Proteins were extracted and purified from adrenal glands homogenates, and separated by 2-D gel electrophoresis as described in material and methods. The 2-DE analysis was performed using immobilized pI gradients (IPG) (pI 5–8) in the first dimension, and SDS-PAGE in the second dimension. To investigate protein pattern differences between E16 and E19/P1, three independent 2-DE images of each protein extract from three independent biological replicates were selected for comparative and statistical analyses. We assessed the degree of 2-DE technical variation by determining the coefficient of variation (CV) for all spot volumes on a gel for all technical replicates. Protein spot volumes were determined for all matched spots in an experiment set using Delt2D software. We observed only minimal differences in CV values from replicate gels from the same embryonic stage, more than 85% of the matched spots have a CV% < 20%. To investigate the biological variation that can result from the use of different pregnant female animals for AG isolation, CV% was determined for AG 2-DE from same embryonic stage isolated from different pregnant female. Depending on the embryonic stage, between 940 and 1150 protein spots were detected on 2-DE gels. The CV% values were below 10% for 25–31% of the spots and below 20% for 75–86% of the spots.

The spots detected by image analysis in the Flamingo-stained gels are presented in Figure 2A for E19 and in Additional file 1: Figure S1 for E16 (A) and P1 (B). The gels were analyzed by Delta 2D Version 4.3 (Decodon, Braunschweig). The 2-DE gel images of the different stages were overlaid and carefully compared. Evaluation of the 2-DE maps of the three stages revealed high similarity (>95% spots overlapping) in the protein pattern between the late embryonic (E19) and newborn (P1) stages, whereas the adrenal glands from the early embryonic stage (E16) showed different protein patterns than E19 and P1 (Figure 2B and Additional file 1: Figure S1C, D). To obtain an overview of the existing proteins in the AG at the different stages, more than 500 protein spots were excised from the 2-DE gels and further processed for identification. Protein spots, which were present at all three investigated AG stages, were only excised for identification from one of the gels.

The excised spots were in-gel digested with trypsin and processed for mass spectrometric analysis (MS/MS). Proteins were identified by the sequence databases search using Mascot. A total of 464 protein spots were identified resulting in 203 non-redundant proteins (Additional file 2: Table S1). Some of the proteins were identified in multiple spots, as shown in Figure 3.

To gain more information on the biological mechanisms of the identified proteins in embryonic AGs, we combined DAVID bioinformatics with information on the putative function of the protein found in the UniProt and GenBank databases. Thereby, we were able to annotate all 203 gene products. An analysis of the molecular function of the identified proteins based on the Gene Ontology (GO) terms allowed the classification of the proteins into ten different categories (Figure 4A); approximately half of these proteins were classified as binding proteins (48%, 180 proteins) and about one-fourth were involved in catalytic activities (28%, 106 proteins). The rest of the proteins were classified in other categories with less than 5% (20 proteins) per group e.g. transcription regulation, enzyme activity and antioxidant activity. The ontological classification of the identified proteins according to their postulated involvement in biological processes resulted in 19 different categories revealing the complexity of mechanisms that drives embryonic development of adrenal glands (Figure 4B).

Protein expression changes during embryonic development: alteration in cholesterol metabolism and retinoic acid pathway proteins

A total of 168 non-redundant proteins were identified in E16, 190 proteins in E19, and 195 proteins in P1 (Figure 5). Comparative analysis of the protein expression with regards to the embryonic stage showed that 163 proteins were present in all three stages, whereas the other proteins were uniquely found in one or two stages (Figure 5). Several proteins, which were identified in all stages, showed significant expression regulation between E16 and P1. An examination of all regulated proteins revealed 48 (31 non-redundant) proteins for which expression was increased in P1 (Table 1) and 48 proteins (36 non-redundant) for which expression was decreased in stage P1 (Table 2).

The ontological classification of proteins for which expression was significantly altered in AGs during embryonic development, was performed according to their molecular function (Figure 6A, 6B). Although the regulated proteins were divided into two groups; proteins for which expression was increased (Table 1) and proteins for which expression was decreased (Table 2), they partly belong to the same categories of molecular function. Six proteins for which expression was decreased and eight for which expression was increased are involved in lipid binding. Three of the proteins, which expression was increased, are involved in fatty acid binding (Alb, Fabp3, and Fabp6) (Figure 6A), whereas three of the proteins for which expression was decreased, are involved in retinol binding (Rbp1, Crabp1, and Crabp2) (Figure 6B). The expression of some of these was confirmed

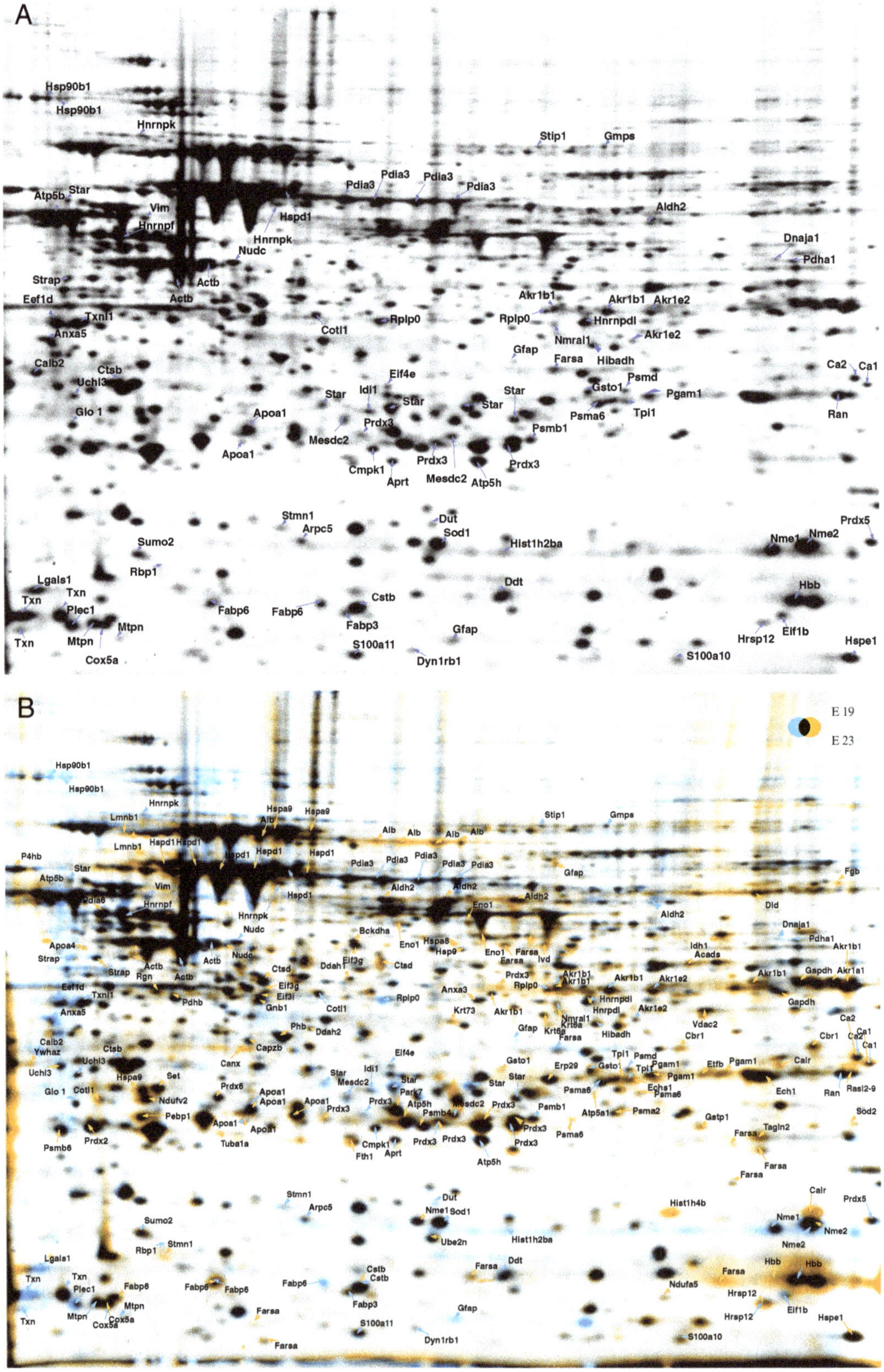

Figure 2 (See legend on next page.)

(See figure on previous page.)
Figure 2 2-DE gel of the adrenal gland. A: 2-DE reference map of proteins extracted from the adrenal gland of an E19 embryo. 150 μg proteins were loaded on an 11-cm IPG strip with a linear pH gradient pI 5–8 for IEF, 12% SDS-polyacrylamide gels were used for the SDS-PAGE. Proteins were stained with Flamingo fluorescent gel stain. Identified spots were assigned a gene name. **B**: An overlay of the 2D gel images of the adrenal glands of E19 embryo and newborn. The identified spots are labeled with the gene names.

by Western blot analysis (6C, D). The expression of proteins, which are involved in transporter activity, especially cholesterol transporter activity, was increased in the adrenal gland during embryonic development (Apoa1, Apoa4, and Star) (Figure 6A, 6C). Among these proteins, the expression of Star (steroidogenic acute regulatory protein, mitochondrial) an essential cholesterol transporting protein, which is expressed exclusively in adrenal cortex, was significantly affected. Star is a protein, which plays a key role in steroid hormone synthesis by enhancing the metabolism of cholesterol into pregnenolone. Star was barely expressed in early stage of adrenal gland development (E16), whereas in later stage (E19) the expression increased significantly. The protein seems to be posttranslationally modified

as evidenced by different spots with similar molecular weights but different pIs (Additional file 3: Figure S2). The proteins involved in the retinoic acid pathway, especially Crabp1, Crapb2, Aldh1a1 and Aldh1a2 were detected in the adrenal medulla shown by immunofluorescence staining of histological sections (Figure 7A, 7B).

Discussion

To date, the growth and function of the adrenal gland have been predominantly investigated in the postnatal and adult stages [10]. Detailed studies of the molecular mechanisms involved in embryonic development of the AGs are still missing. In order to elucidate the molecular mechanisms associated with the embryonic development

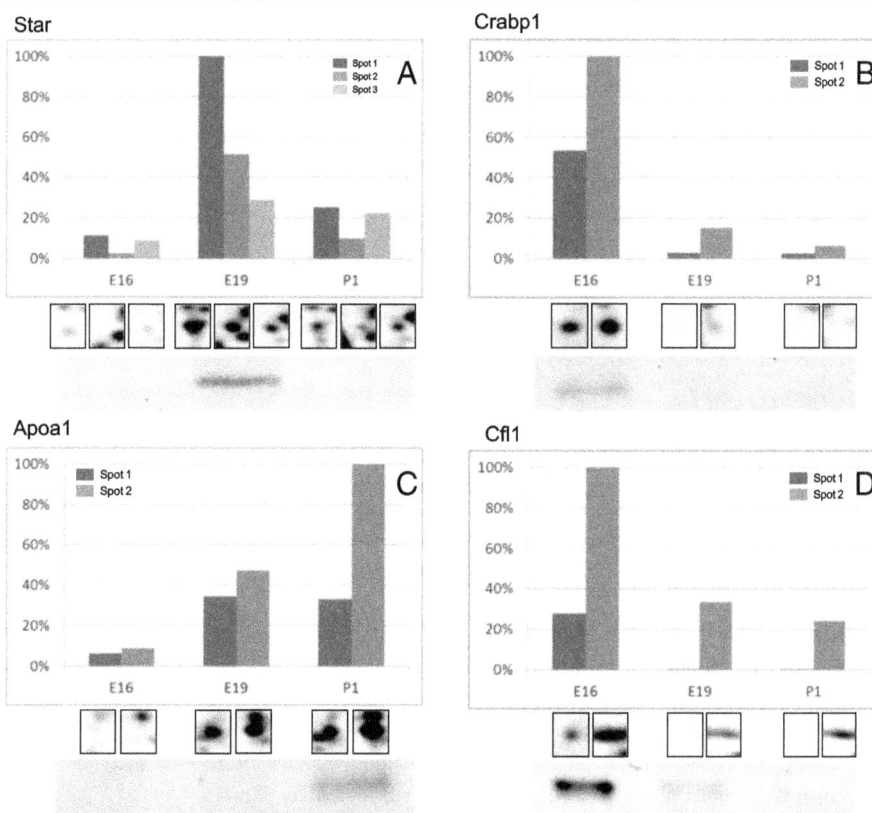

Figure 3 Temporal change in the expression of selected proteins. Graphs represent enlargement of the gel regions of interest showing protein spots found to be differentially expressed in AG from different embryonic stages. The protein expression quantification for selected proteins is given in form of bar diagrams. On the y-axis the relative intensity of spot is given on the x-axis the corresponding embryonic stage from which the AGs were collected. Results are given as the means ± SD of the percentage volume of spot as quantified by Delta 2D software. All the proteins shown, present significant expression changes during AG development (P < 0.05). Three different spots (Spot 1, 2 and 3) were identified as Star **(A)**, two spots (Spot 1 and 2) were identified as Crabp1 **(B)**, two spots (Spot 1 and 2) were identified as Apoa1 **(C)**, and two spots (Spot 1 and 2) were identified as Cfl1 **(D)**. Western blot analyses confirming the protein expression changes are provided.

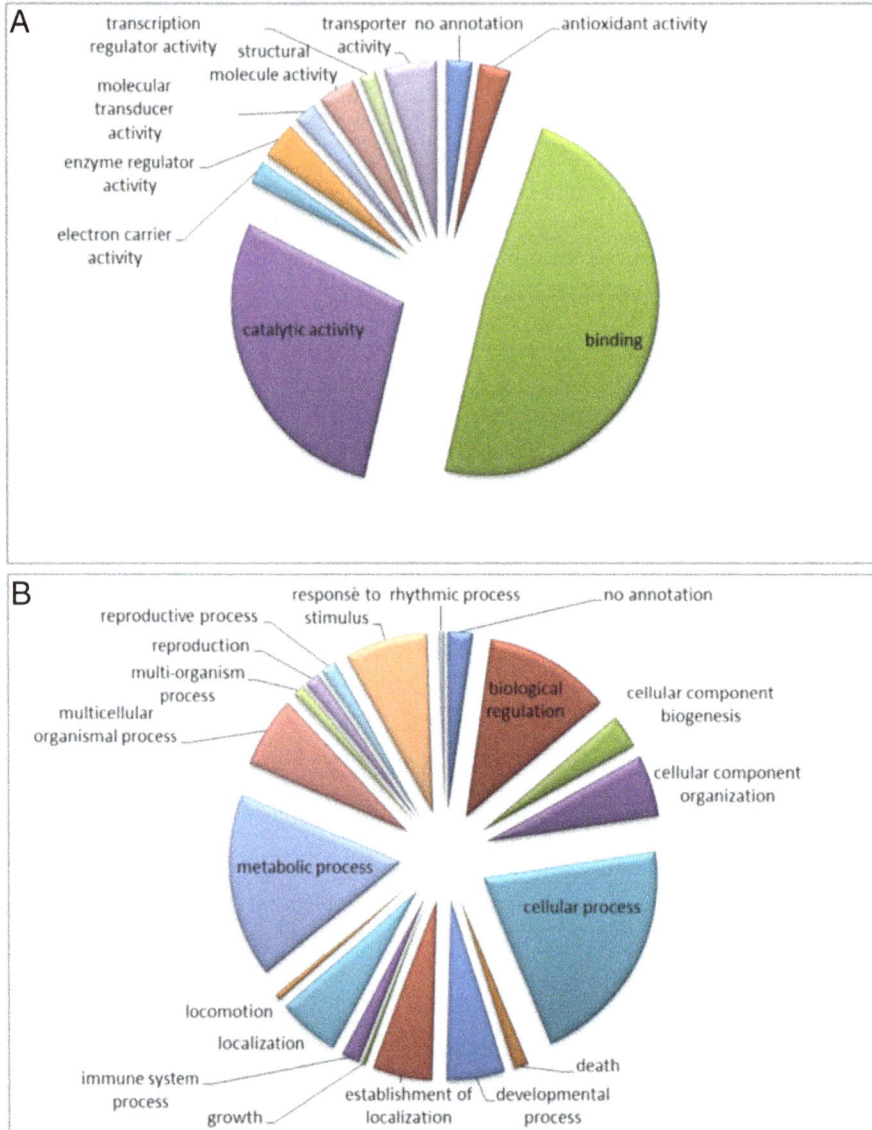

Figure 4 Pie charts of the obtained categories after analysis with DAVID bioinformatics. A: Classification of all identified proteins in the AG according to their molecular function using David Bioinformatics. Most of the proteins were found to be involved in binding functions or catalytic activity. **B**: Distribution of all identified proteins according to their involvement in biological processes.

of the AG, we investigated the proteome changes between early and late embryonic stage of development of the AGs. The data obtained from this study suggest developmental stage specific pathway activation. Taking the biological and functional classification of the proteins, which were differentially expressed during the development into account, we could observe, that expression of the proteins involved in biological processes like response to stress (Ube2n, Gfap, Vcp, Pkm2, Akr1b1, Prdx5, Hspa4), cell cycle process (Phgdh, Stmn1, Tubb5), transport (Khsrp, Crabp1, Crabp2, Clic1, Cfl1, Rbp1, Vcp), gene expression (Fubp1, Hnrpdl, Khsrp), cell differentiation (Cfl1, Phgdh, Stmn1), and embryonic development (Crabp2, Cfl1,

Phgdh) decreases in the course of development (Table 2). In contrast, the expression of proteins, which are involved in lipid metabolic process (Apoa4, Tpi1, Ech1, Apoa1, Star, Prdx6, Fabp3, Atp5a1, Fabp6), steroid metabolic process (Apoa1, Apoa4, Star, Fabp6), response to stress (Star, Phb, Prdx3, Cotl1, Apoa4, Apoa1, Alb, Aldh2, Pebp1, Ctsd, Hspe1, Hspd1), cell differentiation (Actb, Apoa1, Apoa4, Star, Pebp1, Prdx3, Calr), neurogenesis (Actb, Apoa1, Apoa4, Calr, Star), and neuron development/differentiation (Actb, Apoa1, Apoa4) increases in the course of development (Table 1).

The proteomics data revealed interesting aspects in the AG embryonic development. First, the expression of

E16 - 168

E19 - 190

Cfl1   Crabp1
Eefla1   CQO62

Crabp2

1

Calb2   Hist1h2ac
Hist1h2ba

4

163

3

P1 - 195

23

9

Apoa4   Calr
Canx   Cbr1
Cotl1   Ctsd
Fgb   Ivd
Ywhaz

Akr1e2   Alb   Apoa1   Ca1
Ca2   Cox5a   Cstb   Ddah1
Eif1b   Fabp6   Hibadh   Hspe1
Hsrp12   Lgals1   Mesdc2   Mtpn
Plec1   S100a10   S100a11   Star
Sumo2   Txn   Txnl1

Acads	Acat2	Actb	Akr1a1	Akr1b1	Akr1c9	Aldh2	Anxa3	Anxa5	Aprt
Arhgdia	Arpc5	Atic	Atp5a1	Atp5b	Atp5h	Bckdha	Capza1	Capza2	Capzb
Cct2	Cct3	Cct5	Clic1	Cmpk1	Cnn3	Csn1s1	Ddah2	Ddt	Dhfr
Dlat	Dld	Dnaja1	Dut	Dyn1rb1	Echs1	Eef1d	Eif3g	Eif3i	Eif4a3
Eif4e	Eif4h	Eif5a	Eno1	Erp29	Etfb	Fabp3	Fabp5	Fscn1	Fth1
Ftl1	Fubp1	Gapdh	Gfap	Glo1	Gmfb	Gmps	Gnb1	Gnb2	Gnb211
Gpx1	Grpl1	Gsto1	Gstp1	Hbb	Hint1	Hnrnpf	Hnrnph1	Hnrnpk	Hnrpdl
Hsp90b1	Hspa4	Hspa5	Hspa8	Hspa9	Hspb1	Hspd1	Idh1	Idh3a	Idi1
Khsrp	Krt1	Krt10	Krt15	Krt5	Krt6a	Krt73	Lmnb1	Mdh1	Ndufa5
Ndufv2	Nme1	Nme2	Nmral1	Nudc	Nudt5	P4hb	Pafah1b2	Pafah1b3	Park7
Pdha1	Pdhb	Pdia3	Pdia6	Pebp1	Pfdn2	Pfn2	Pgam1	Pgls	Phb
Phgdh	Pkm2	Pnp	Ppid	Prdx2	Prdx3	Prdx4	Prdx5	Prdx6	Psma1
Psma2	Psma3	Psma6	Psmb3	Psmb4	Psmb6	Psmb7	Psmc2	Psmc6	Psmd9
Psme1	Ptpn2	Ran	Rasl2-9	Rbp1	Rgn	Rplp0	Rps12	Sept11	Sept9
Set	Sgta	Sod1	Sod2	St13	Stip1	Stmn1	Strap	Sugt1	Tagln2
Tpi1	Tuba1a	Tubb2c	Tubb5	Tufm	Ube2n	Uchl1	Uchl3	Vcp	Vdac2
Vim	Wdr61	Ywhaq							

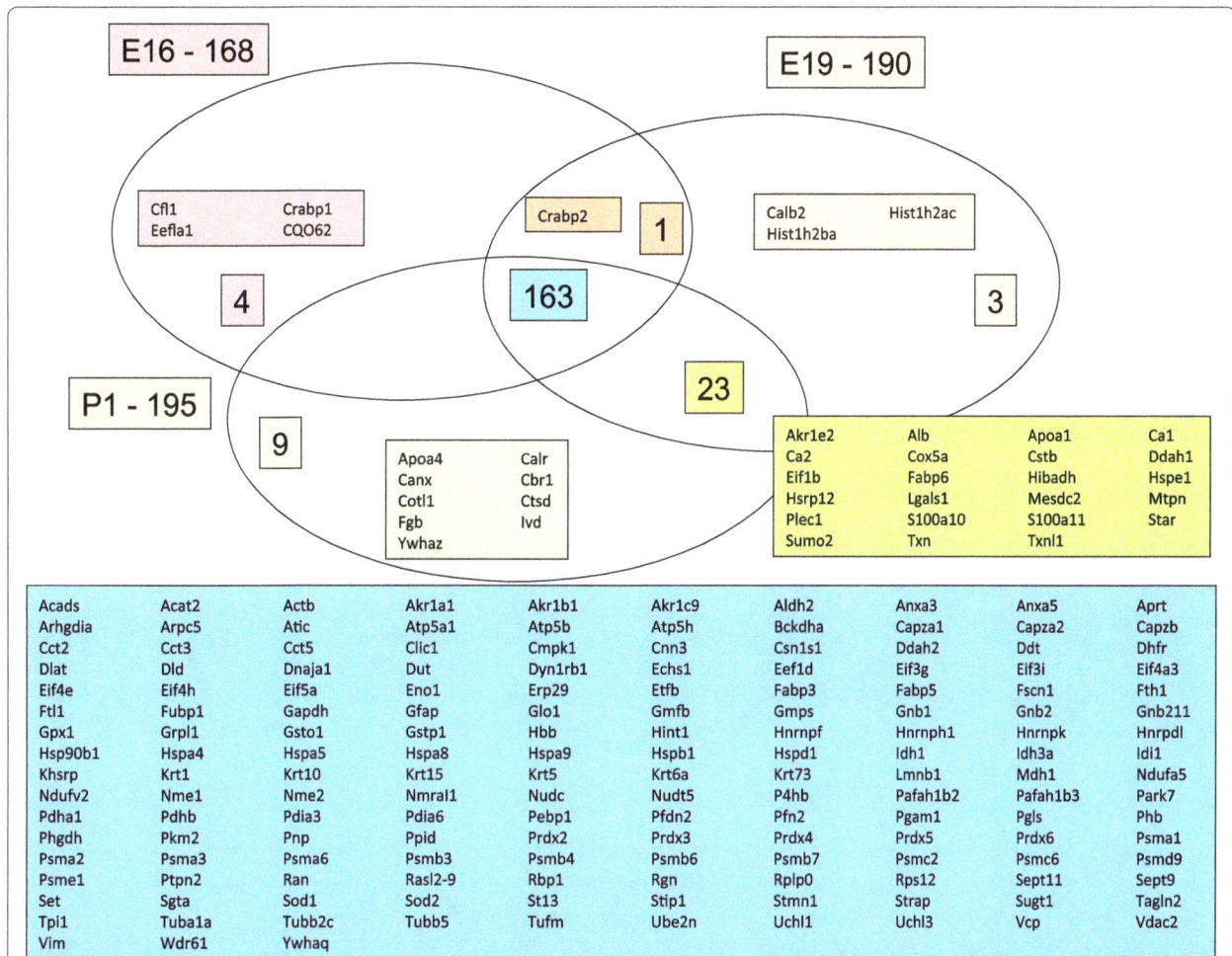

Figure 5 Venn diagram of the distribution of the identified proteins in the adrenal gland at the different days of development. At day E16 a total of 168 proteins were identified, of which 4 were only present at day E16, 1 was present at day E16 and E19 and 163 were present at all three examined developmental days. At day E19 a total of 190 proteins were identified, of which three were only present on the given stage, whereas 23 were present at day E19 and P1. Of the 195 proteins, which were identified in the newborn rat, 9 were only present at this embryonic stage.

key proteins of the RA-pathway decreases in the course of AG development suggesting that this pathway is only important in early stages of development. Second, proteins, which are important in the steroid biosynthetic process or sterol transporter activity, like Apoa1, Apoa4, Star, Fabp3 or Fabp6 were hardly expressed in early stage, whereas their expression increased in the course of development of the adrenal gland. Star (Steroidogenic acute regulatory protein, mitochondrial) stimulates the regulated production of steroid hormones in the adrenal cortex and gonads by facilitating the delivery of cholesterol to the inner mitochondrial membrane [11], which is the rate-limiting step in the production of steroid hormones [12,13]. Production of steroid hormones is one of the main functions of the adrenal cortex. Whereas Star was not detected in the early embryonic stage (E16), four spots were identified as Star in later stages (E19, P1).

These spots may represent four different isoforms of the protein. Three of these were verified by 2-D Western blot analysis. Apoa1, Apoa4 and Fabp6 appeared in late stage of AG development, Apoa1 participates in the reverse transport of cholesterol from tissues to the liver for excretion by promoting cholesterol efflux from tissues and by acting as a cofactor for the lecithin cholesterol acyltransferase (LCAT). Apoa4 may have a role in chylomicrons and VLDL secretion and catabolism. It is also required for efficient activation of lipoprotein lipase by ApoC-II, which is a potent activator of LCAT. Apoa4 is a major component of HDL and chylomicrons. The expression of proteins involved in steroid metabolic process and cholesterol synthesis increased significantly in later stages suggesting that the embryonic development of the adrenal gland is accompanied by coordinated metabolic modifications that facilitate the functional role of the gland.

Table 1 Proteins, which expression increased throughout the embryonic development of the adrenal gland

Increased proteins	E16/P1	p-value	Increased proteins	E16/P1	p-value
Actb	0.25	0.0013	Hnrnpk	0.21	0.0381
Actb	0.29	0.0062	Hnrnpk	0.33	0.0410
Akr1e2	0.31	0.0105	Hrsp12	0.04	0.0013
Alb	0.01	0.0141	Hspa9	0.40	0.0044
Alb	0.06	0.0236	Hspa9	0.40	0.0182
Aldh2	0.30	0.0394	Hspa9	0.41	0.0381
Aldh2	0.52	0.0103	Hspd1	0.40	0.0416
Apoa1	0.06	0.0416	Hspe1	0.04	0.0249
Apoa1	0.17	0.0495	Krt73	0.50	0.0007
Apoa1	0.17	0.0210	Mesdc2	0.19	0.0156
Apoa4	>0.01	0.0150	Pebp1	0.31	0.0031
Atp5a1	0.39	0.0021	Pebp1	0.43	0.0461
Ca2	0.15	0.0043	Phb	0.46	0.0037
Calr	0.04	0.0021	Prdx3	0.01	0.0441
Cmpk1	0.43	0.0071	Prdx3	0.35	0.0134
Cotl1	0.16	0.0092	Prdx3	0.12	0.0023
Cox5a	0.19	0.0115	Prdx3	0.17	0.0372
Cox5a	0.22	0.0003	Prdx3	0.26	0.0501
Ctsd	0.20	0.0046	Star	0.42	0.0019
Eno1	0.17	0.0064	Star	0.52	0.0021
Fabp3	0.42	0.0223	Tpi1	0.46	0.0291
Fabp6	0.14	0.0069	Tpi1	0.33	0.0063
Fabp6	0.30	0.0022	Tuba1a	>0.01	0.0002
Hibadh	0.35	0.0070	Txn	0.10	0.0012

Table 2 Proteins, which expression decreased throughout the embryonic development of the adrenal gland

Decreased proteins	E16/P1	p-value	Decreased proteins	E16/P1	p-value
Akr1b1	6.4	0.0002	Khsrp	2.6	0.0268
Anxa3	2.3	0.0012	Khsrp	3.1	0.0052
Arpc5	2.0	0.0405	Krt6a	2.8	0.0306
Cct3	3.9	0.0052	Krt6a	3.3	0.0188
Cfl1	3.7	0.0061	Krt73	4.1	0.0471
Cfl1	<10	0.0012	Lmnb1	2.2	0.0089
Clic1	2.1	0.0026	Pdia3	2.4	0.0165
Crabp1	<10	0.0003	Pdia3	2.5	0.00401
Crabp1	<10	0.0019	Pdia3	3.6	0.00171
Crabp2	7.3	0.0001	Pdia3	2.0	0.0413
Dhfr	2.2	0.0072	Phgdh	2.4	0.0021
Eef1a1	<10	0.0028	Pkm2	2.3	0.0346
Fscn1	6.5	0.00174	Prdx5	2.6	0.0059
Fscn1	7.3	0.0083	Rbp1	3.8	0.0002
Fubp1	3.1	0.0373	Sept11	2.7	0.0073
Gfap	2.9	0.0367	Stip1	2.1	0.0144
Gmps	3.4	0.0210	Stip1	2.3	0.0185
Gnb1	2.1	0.0209	Stmn1	2.0	0.0027
Gnb211	3.6	0.0196	Stmn1	3.7	0.0003
Hnrnph1	2.1	0.0303	Tubb5	<10	0.0001
Hnrnph1	2.7	0.0415	Ube2n	2.2	0.0221
Hnrpdl	2.2	0.0046	Vcp	2.5	0.0031
Hsp90b1	4.4	0.0275	Vim	2.5	0.0001
Hspa4	3.7	0.0342	Vim	3.7	0.0012

Another important aspect of the embryonic development of AGs seems to be the slowdown of the RA-pathway as revealed by the down-regulation of the key proteins of this pathway in the late prenatal stage. Proteins, which are involved in retinoic acid pathway, like cellular retinoic acid binding proteins (Crabp1, Crabp2) and the retinol binding protein (Rbp1), were found to be expressed in early stages in the AG, but could not be detected in late embryonic stages and new born rats. Rbp1 is binding and transporting retinol in the cell. Once in the cell, retinol is converted into retinoic acid. Crabp1 and Crabp2 are important for the transport of retinoic acid from the cytosol to the nucleus, where it serves as a ligand for nuclear retinoic acid receptors (RARs) that directly regulate gene transcription [14-16] especially of genes that modulate the overall development of the embryo. The down-regulation of the RA-pathway proteins during embryonic development suggests the restricted importance of this pathway in the AG development.

There are several limitations arising from the 2-D electrophoresis as a separation method. The reproducibility of the protein patterns is an issue when using this technique and the range of molecular weight that is resolvable by the method is limited, affecting the detection of very large and very small proteins. Hydrophobic, highly acidic or highly basic proteins are poorly detected resulting in less information on proteome. Moreover, the fact that AGs are constituted from two types of tissue, it might be more informative to resolve the proteome of the two parts separately. This requires additional methods for cell separation or tissue microdissection, which will result in severely reduced amount of usable sample. This will require an even higher number of embryos to be involved in the study and/or more sensitive analytical methods.

Conclusions

In summary, this study provides preliminary proteomic maps of AGs embryonic development and highlights the embryonic stage specific pathway modulation. The RA-pathway seems to be important at the initial steps of the AG development, whereas the molecular changes in later stages revealed an increased importance of the

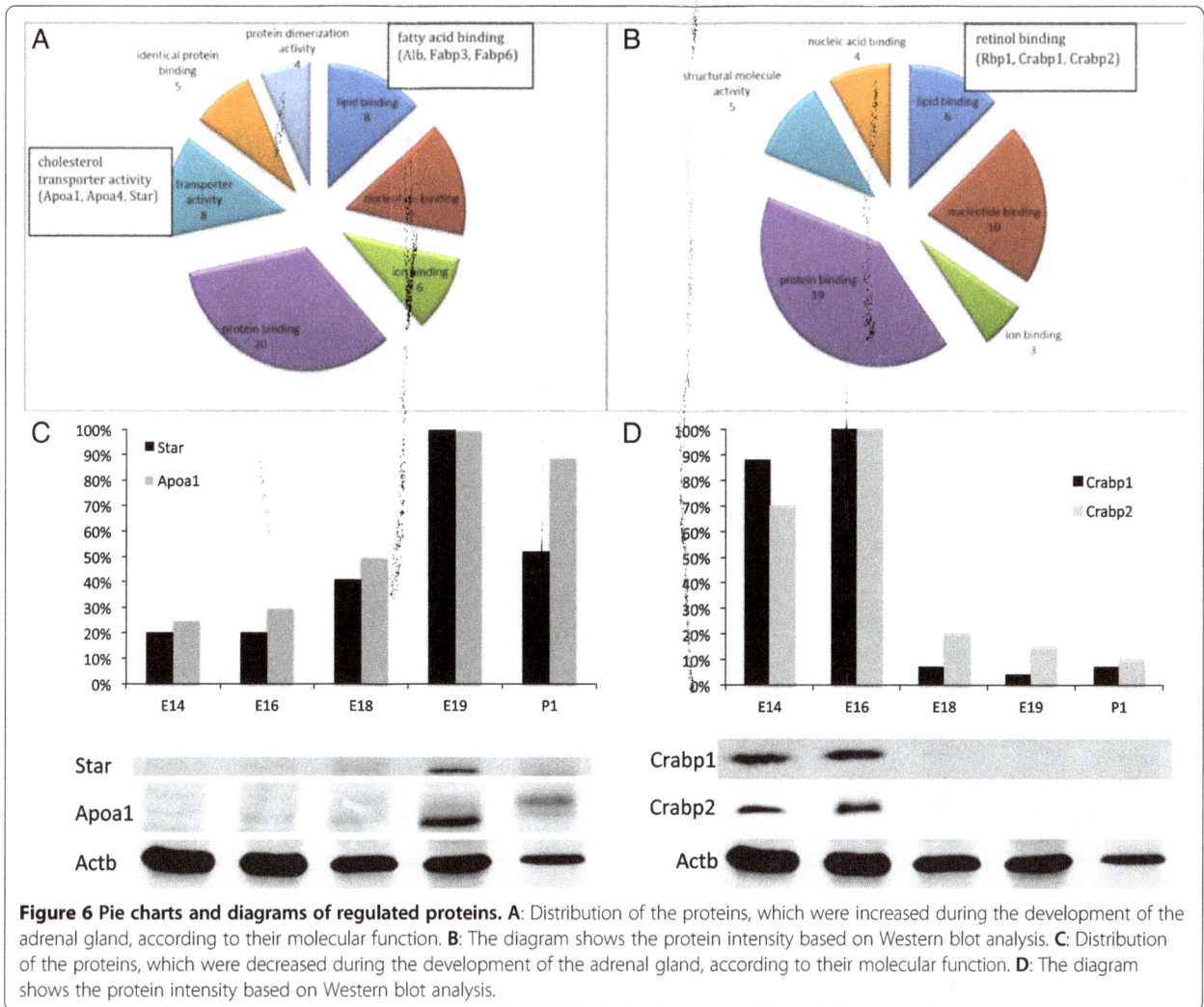

Figure 6 Pie charts and diagrams of regulated proteins. A: Distribution of the proteins, which were increased during the development of the adrenal gland, according to their molecular function. **B**: The diagram shows the protein intensity based on Western blot analysis. **C**: Distribution of the proteins, which were decreased during the development of the adrenal gland, according to their molecular function. **D**: The diagram shows the protein intensity based on Western blot analysis.

regulation of steroid hormones synthesis as a first step towards the endocrine function of this organ. Additional investigations are still needed to elucidate the specific role of the single proteins in the AG development and maturation.

Methods
Animals
Wistar Han rats were kept under 12:12 h cycle of light with *ad libitum* access to food and drink. Pregnant rat females were used to collect the embryos at different embryonic stages: embryonic day 14 (E14), 16 (E16) 18 (E18) and 19 (E19), and newborn (P1). The adrenal glands (AG) were dissected from these embryos as well as from neonatal pups. To prepare the AG protein extracts, 60 AGs were used in the E14, 90 in E16, 50 in E18, 60 in the E19 extracts, and 30 AGs were used for the P1 extract. All experimental procedures were performed according to the German animal care and ethics legislation (NIH standards) and were approved by the local government authorities.

Protein extraction
The protein extraction for 2-D gel electrophoresis was performed as described previously [17]. A single AG, especially from embryonic stage E14 and E16 (200 – 300 μm diameter), will not deliver enough protein for 2-DE analysis. Embryos from the same pregnant rat females have the same genetic background and the AGs from these embryos can be pooled together for proteomic analyses. The AGs from embryos at the same embryonic stage and from the same female (between 14–17 embryos) were pooled, the lysis buffer (9.5 M urea, 2% CHAPS (w/v), 2% ampholytes (w/v), 1% DTT) was added and the samples were vortexed. Thereafter, the samples were incubated for 30 min at 4°C. For removing the cell debris, centrifugation

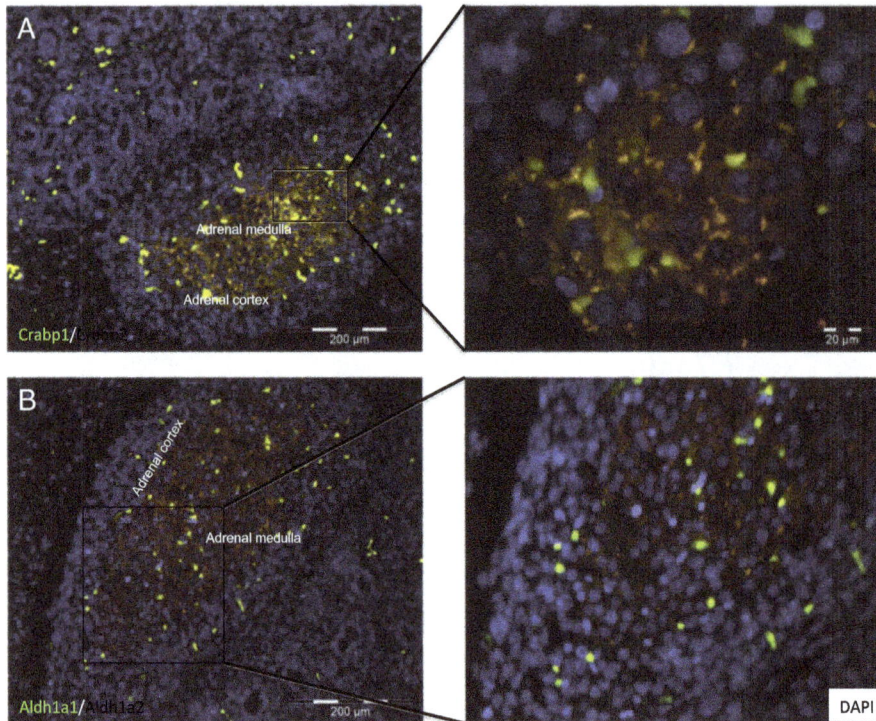

Figure 7 Immunohistological fluorescence staining of the adrenal gland at day E19. The fluorescence staining shows a higher expression of the proteins Crabp1 and Crabp2 **(A)** and Aldh1a1 and Aldh1a2 **(B)** in the adrenal medulla than in the adrenal cortex. The slides were analyzed on an immunofluorescence Zeiss Axiophot microscope (Carl Zeiss, Jena, Germany) using the AnalySIS software (Soft Imaging Systems, Leinfelden, Germany).

was carried out for 30 min at 13,000 ×g and 4°C. The supernatant was recentrifuged at 13,000 ×g and 4°C for an additional 30 min to get maximal purity. The pellet was discarded, and the resulting samples were used immediately or stored at −80°C until use.

Protein precipitation

To reduce the salt contamination and to enrich the proteins, methanol-chloroform-precipitation according to Wessel and Flugge [18] was performed. Briefly, 0.4 ml methanol (100%) was added to 0.1 ml protein sample and mixed together. 0.1 ml chloroform was added to the sample and the mixture was vortexed. Subsequently 0.3 ml water was added and the solution was vortexed and centrifuged at 13,000 ×g for 1 min. The aqueous layer was removed, and another 0.4 ml methanol (100%) was added to the rest of the chloroform and the interphase with the precipitated proteins. The sample was mixed and centrifuged for 2 min at 13,000 ×g and the supernatant was removed. The pellet was vacuum dried and dissolved in lysis buffer.

Total protein concentration was determined using Bio-Rad protein assay (Bio-Rad, Hercules, CA, USA)

according to Bradford [19]. Bovine serum albumin (Sigma, Steinheim, Germany) was used as standard.

2D gel electrophoresis (2-DE)

To assure for high data quality 2-DE, five biological replicates consisting of five pregnant rats (each 14–17 embryos) for every embryonic stage were prepared. For embryonic AGs isolated from embryos collected from the same mother at least three independent experimental replicates from each embryonic stage as well as from newborn pups were performed. IPG strips (11 cm, pI 5–8) were passively rehydrated for 12 h in 185 µl rehydration buffer (8 M urea, 1% CHAPS, 1% DTT, 0.2% ampholytes, and a trace of bromophenol blue) containing 150 µg protein. The IEF step was performed on the PROTEAN$^{(R)}$ IEF Cell (Bio-Rad, Hercules, CA, USA). Temperature-controlled at 20°C, the voltage was set to 500 V for 1 h, increased to 1000 V for 1 h, 2000 V for 1 h and left at 8000 V until a total of 50000 Vhours was reached. Prior to SDS-PAGE, the IPG strips were reduced for 20 min at room temperature in SDS equilibration buffer containing 6 M urea, 30% glycerol, 2% SDS 0.05 M Tris–HCl, and 2% DTT on a rocking table. The strips were subsequently alkylated in the same solution with 2.5% iodoacetamide substituted for DTT, and a trace of bromophenol blue. For the SDS-PAGE 12% BisTris Criterion precast gels (Bio-Rad, Hercules, CA, USA) were used according to manufacturer's instructions. The gels were run at 150 V for 10 min followed by 200 V until the bromophenol blue dye front had reached the bottom of the gel.

Gel staining

For image analysis, 2-DE gels were fixed in a solution containing 50% methanol and 12% acetic acid overnight and fluorescent stained with Flamingo fluorescent gel stain (Bio-Rad, Hercules, CA, USA) for minimum 5 h. Thereafter, gels were scanned at 50 μm resolution on a Fuji FLA-5100 scanner using the Image Reader Software (Fuji). The digitalized images were analyzed using Delta 2D 4.3 (Decodon, Braunschweig, Germany). For protein identification, 2-DE gels were additionally stained with colloidal Coomassie blue, Roti-Blue (Roth, Karlsruhe, Germany) overnight.

Protein identification

Manually excised gel plugs were digested as described previously [20]. After digestion the supernatant was removed and saved, and the additional peptides were extracted with different acetonitrile/trifluoroacetic acid ratio under sonication. All supernatants were pooled together, dried in a vacuum centrifuge, and dissolved in 0.1% formic acid. The mass spectrometric sequencing was performed as described previously [21]. Briefly, the tryptic peptides were subjected to mass spectrometric sequencing using a Q-TOF Ultima Global mass spectrometer (Micromass, Manchester, UK).

Processed data were searched against MSDB and Swiss-Prot databases through Mascot search engine using a peptide mass tolerance of 50 ppm (parts per million) and fragment tolerance of 100 mmu (millimass unit). Protein identifications with at least two peptides sequenced were considered significant.

Bioinformatics

The classification of the identified proteins according to their main known/postulated functions was carried out using DAVID bioinformatics [22,23]. This classification together with the official gene symbol (given in Additional file 2: Table S1) was used to investigate and categorize the gene ontology (GO)-annotations (biological processes and molecular functions).

Histochemistry

Paraffin embedded sections were first heated in an oven at 65°C for 1 h before undergoing several series of washes of xylene and ethanol to deparaffinize/rehydrate. After deparaffinization/rehydration, the slides were immersed in Hematoxylin solution (Merck) for 3 min. The slides were rinsed with water, and counterstained with 0.5% Eosin G-solution (Merck) for 5 min. After 30 s of rinsing in water, the slides were dehydrated in series of increasing ethanol and xylene concentrations. The coverslips were mounted on the slides with Entellan Neu mounting medium (Merck).

Immunohistochemistry

To monitor the expression of the selected proteins in the adrenal gland, indirect immunofluorescence staining of proteins of interest was performed. The deparaffinization of the sections was carried out as described above, thereafter the slides were immersed into a staining dish containing Antigen retrieval solution (18 mM citric acid, 82 mM sodium citrate, pH 6.0) and warmed in a food steamer for 25 min. The slides were allowed to cool down for 20 min before being washed with TBST for 5 min. Inactivation of endogenous peroxidase was performed with 3% hydrogen peroxide for 10 min at 37°C. After three successive washing steps with TBST, sections were blocked with 10% goat serum for 60 min. The incubation with the primary antibodies was carried out overnight at 4°C in a humidified chamber. Molecular Probes Alexa Fluor 647 goat anti-mouse IgG antibody or Alexa Fluor 647 goat anti-rabbit IgG (1:200) were used as secondary antibodies. The incubation was performed at room temperature for 60 min in the dark. The coverslips were mounted on the slides using fluorescence mounting medium with DAPI (Vector Laboratories, Inc., Burlingame, USA). Slides were analyzed on a Zeiss Axiophot microscope (Carl Zeiss, Jena, Germany) using the AnalySIS software (Soft Imaging Systems, Leinfelden, Germany).

Western blot analysis

The validation of the 2-DE data was carried out using Western blot analysis. To assure for the reproducibility of the Western blot analysis, at least three biological and experimental replicates were performed. 40 μg proteins were separated by SDS-PAGE and transferred to Hybond ECL nitrocellulose membrane (GE Healthcare). Immuno-detection was performed according to Towbin et al. [24]. Briefly, membranes were blocked in 5% milk for 2 h at room temperature, followed by overnight incubation at 4°C with diluted specific primary antibody. Mouse monoclonal anti-CRABP1 (1:1000) (abcam), rabbit anti-Cofilin (1:1000) (sigma), mouse monoclonal anti-StAR (1:250) (abcam) and mouse monoclonal anti-ß-actin (1:5000) (sigma) were used as primary antibodies. Molecular Probes Alexa Fluor 647 goat anti-mouse IgG antibody or Alexa Fluor 647 goat anti-rabbit IgG (1:2000) were used as secondary antibodies. Before imaging, the blots were dried in the dark. The blot membranes were scanned at 50 μm resolution on a Fuji FLA-5100 scanner (Fuji Photo) with single laser-emitting excitation light at 635 nm.

2D Western blot analysis

150 μg proteins were separated by isoelectric focusing and SDS-PAGE as described above and transferred to Hybond ECL nitrocellulose membrane (GE Healthcare). Immunodetection was performed as described above. Mouse monoclonal anti-StAR (1:250) (abcam) was used as primary antibodies. Molecular Probes Alexa Fluor 647 goat anti-mouse IgG antibody (1:2000) was used as secondary antibodies. Before imaging, the blots were dried in the dark and scanned as described above.

Statistical analysis

For 2-DE the digitalized images were analyzed; spot matching across gels and normalization were performed using Delta2D 4.3 (Decodon, Braunschweig, Germany). Delta2D computes a 'spot quality' value for every detected spot. This value shows how closely a spot represents the 'ideal' 3D Gaussian bell shape. Based on average spot volume ratio, spots whose relative expression is changed at least 2-fold (increase or decrease) between the compared samples were considered to be significant. To analyze the significance of protein regulation, Student's t-test was performed, and statistical significance was assumed for P values less than 0.01.

All blots were quantified using the ImageJ software. For comparison between two measurements (in the same group) paired t-test was used. Unpaired t-test (for comparing 2 groups) or one-way ANOVA (comparing 3 or more groups) were used. The data were compiled with the software package GraphPad Prism, version 4. The software was used for graphical presentation and statistical analysis. Results are presented as the mean ± SD of at least three independent experiments. Differences were considered statistically significant when p < 0.05.

Additional files

Additional file 1: Figure S1. A: 2D gel of the proteins in the adrenal gland of an E16 embryo and **B**: of a newborn rat. The identified protein spots are labeled with the gene names on the gel. **C**: An overlay of the 2D gels of the adrenal glands of E16 and E19 embryos. **D**: An overlay of the 2D gels of the adrenal glands of E16 embryo and new born. The identified spots are labeled with the gene names. E: HE staining of a paraffin section of adrenal gland and kidney at E19.

Additional file 2: Table S1. List of proteins of non-redundant proteins identified from adrenal glands in all three analysed embryonic stages. Gene name, calculated isoelectric point (CIP) and MS/MS information are given. Additionally protein nominal mass and the number of peptides that were sequenced through MS/MS are also given.

Additional file 3: Figure S2. 2D Western blot of Star in the adrenal gland at E19. Three different spots with the same molecular mass, but with different pI were observed.

Abbreviations

AG: Adrenal gland; E16: Embryonic stage day 16; E19: Embryonic stage day 19; P1: Postnatal day 1; 2-DE: Two-dimensional gel electrophoresis; CV: Coefficient of variation; GO: Gene ontology.

Competing interests

The authors declare that they have no competing interests.

Authors' contributions

GD performed the majority of the experiments in the study, carried out the sample preparation, the 2D electrophoresis, analyzed and interpreted the data and drafted the manuscript; GM conceived of the study, and participated in its design; AA carried out the mass spectrometry analysis, ME contributed in the proteomics data generation; JW contributed in data interpretation and study design; HD conceived of the study, participated in its design, prepared the AG from the embryos, and coordinated and helped to draft the manuscript. All authors read and approved the final manuscript.

Acknowledgments

The authors would like to thank Elke Brunst-Knoblich for her technical assistance.

Author details

[1]Department of Nephrology and Rheumatology, University Medical Center Goettingen, Georg-August University Goettingen, Robert-Koch-Strasse 40, D-37075 Goettingen, Germany. [2]Department of Clinical Chemistry, Georg-August University Goettingen, Robert-Koch-Strasse 40, D-37075 Goettingen, Germany.

References

1. Bürgi U. Normal and pathologic endocrinology of the adrenal glands. Helv Chir Acta. 1989;56:307–14.
2. Mannelli M, Pupilli C, Lanzillotti R, Ianni L, Serio M. Catecholamines and blood pressure regulation. Horm Res. 1990;34:156–60.
3. Neelon FA. Adrenal physiology and pharmacology. Urol Clin North Am. 1977;4:179–92.
4. Roos TB. Steroid synthesis in embryonic and fetal rat adrenal tissue. Endocrinology. 1967;81:716–28.
5. Ishimoto H, Jaffe RB. Development and function of the human fetal adrenal cortex: a key component in the feto-placental unit. Endocr Rev. 2011;32:317–55.
6. Mesiano S, Jaffe RB. Developmental and functional biology of the primate fetal adrenal cortex. Endocr Rev. 1997;18:378–403.
7. Sapolsky RM, Meaney MJ. Maturation of the adrenocortical stress response: neuroendocrine control mechanisms and the stress hyporesponsive period. Brain Res. 1986;396:64–76.
8. Schapiro S. Pituitary ACTH and compensatory adrenal hypertrophy in stress-non-responsive infant rats. Endocrinology. 1962;71:986–9.
9. Walker CD, Sapolsky RM, Meaney MJ, Vale WW, Rivier CL. Increased pituitary sensitivity to glucocorticoid feedback during the stress nonresponsive period in the neonatal rat. Endocrinology. 1986;119:1816–21.
10. Pascual A, Romero-Ruiz A, Lopez-Barneo J. Differential proteomic analysis of adrenal gland during postnatal development. Proteomics. 2009;9:2946–54.
11. Sasaki G, Ishii T, Jeyasuria P, Jo Y, Bahat A, Orly J, et al. Complex role of the mitochondrial targeting signal in the function of steroidogenic acute regulatory protein revealed by bacterial artificial chromosome transgenesis in vivo. Mol Endocrinol Baltim Md. 2008;22:951–64.
12. Crivello JF, Jefcoate CR. Intracellular movement of cholesterol in rat adrenal cells. Kinetics and effects of inhibitors. J Biol Chem. 1980;255:8144–51.
13. Privalle CT, Crivello JF, Jefcoate CR. Regulation of intramitochondrial cholesterol transfer to side-chain cleavage cytochrome P-450 in rat adrenal gland. Proc Natl Acad Sci U S A. 1983;80:702–6.
14. Mangelsdorf DJ, Evans RM. The RXR heterodimers and orphan receptors. Cell. 1995;83:841–50.
15. Mangelsdorf DJ, Kliewer SA, Kakizuka A, Umesono K, Evans RM. Retinoid receptors. Recent Prog Horm Res. 1993;48:99–121.

16. Mangelsdorf DJ, Thummel C, Beato M, Herrlich P, Schütz G, Umesono K, et al. The nuclear receptor superfamily: the second decade. Cell. 1995;83:835–9.

17. Dihazi H, Kessler R, Müller GA, Eschrich K. Lysine 3 acetylation regulates the phosphorylation of yeast 6-phosphofructo-2-kinase under hypo-osmotic stress. Biol Chem. 2005;386:895–900.

18. Wessel D, Flügge UI. A method for the quantitative recovery of protein in dilute solution in the presence of detergents and lipids. Anal Biochem. 1984;138:141–3.

19. Bradford MM. A rapid and sensitive method for the quantitation of microgram quantities of protein utilizing the principle of protein-dye binding. Anal Biochem. 1976;72:248–54.

20. Dihazi H, Kessler R, Eschrich K. High osmolarity glycerol (HOG) pathway-induced phosphorylation and activation of 6-phosphofructo-2-kinase are essential for glycerol accumulation and yeast cell proliferation under hyperosmotic stress. J Biol Chem. 2004;279:23961–8.

21. Dihazi H, Asif AR, Agarwal NK, Doncheva Y, Müller GA. Proteomic analysis of cellular response to osmotic stress in thick ascending limb of Henle's loop (TALH) cells. Mol Cell Proteomics MCP. 2005;4:1445–58.

22. Huang DW, Sherman BT, Lempicki RA. Systematic and integrative analysis of large gene lists using DAVID bioinformatics resources. Nat Protoc. 2009;4:44–57.

23. Huang DW, Sherman BT, Lempicki RA. Bioinformatics enrichment tools: paths toward the comprehensive functional analysis of large gene lists. Nucleic Acids Res. 2009;37:1–13.

24. Towbin H, Staehelin T, Gordon J. Electrophoretic transfer of proteins from polyacrylamide gels to nitrocellulose sheets: procedure and some applications. Proc Natl Acad Sci U S A. 1979;76:4350–4.

A preliminary quantitative proteomic analysis of glioblastoma pseudoprogression

Peng Zhang[1], Zhengguang Guo[2], Yang Zhang[1], Zhixian Gao[1], Nan Ji[1], Danqi Wang[2], Lili Zou[2], Wei Sun[2*] and Liwei Zhang[1*]

Abstract

Backgrounds: Pseudoprogression disease (PsPD) is commonly observed during glioblastoma (GBM) follow-up after adjuvant therapy. Because it is difficult to differentiate PsPD from true early progression of GBM, we have used a quantitative proteomics strategy to identify molecular signatures and develop predictive markers of PsPD.

Results: An initial screening of three PsPD and three GBM patients was performed, and from which 530 proteins with significant fold changes were identified. By conducting biological functional analysis of these proteins, we found evidence that the protein synthesis network and the cellular growth and proliferation network were most significantly affected. Moreover, six of the proteins (HNRNPK, ELAVL1, CDH2, FBLN1, CALU and FGB) involved in the two networks were validated (n = 18) in the same six samples and in twelve additional samples using immunohistochemistry methods and the western blot analysis. The receiver operating characteristic (ROC) curve analysis in distinguishing PsPD patients from GBM patients yielded an area under curve (AUC) value of 0.90 (95% confidence interval (CI), 0.662-0.9880) for CDH2 and.0.92 (95% CI, 0.696-0.995) for CDH2 combined with ELAVL1.

Conclusions: The results of the present study both revealed the biological signatures of PsPD from a proteomics perspective and indicated that CDH2 alone or combined with ELAVL1 could be potential biomarkers with high accuracy in the diagnosis of PsPD.

Keywords: iTRAQ labeling, Pseudoprogression, Quantitative proteomics

Introduction

Glioblastoma (GBM) is one of the most malignant brain tumors. After the postoperative use of radiotherapy for GBM became common, a phenomenon termed pseudo-progression disease (PsPD) was identified [1,2]. With the widely implementation of the Stupp protocol for treating GBM, this phenomenon has been inceasingly reported, with an incidence rate varies among reports (5.5%-64%) [3-6]. PsPD is often misdiagnosed as tumor recurrence and misleads the clinical treatment. However, little is known about why PsPD occurs in a subset of GBM patients and the fundamental biological features of PsPD remain unclear [5;7-10].

From a diagnostic perspective, no single imaging technique, including T1-weighted magnetic resonance imaging (MRI), magnetic resonance spectroscopy (MRS), relative cerebral blood volume (rCBV)-based parametric response mapping and [18]fluorodeoxyglucose (^{18}F-FDG)-positron emission computed tomography (PET), has been adequate for differentiating PsPD from true early tumor progression with high sensitivity and specificity [4,5,11-16]. Moreover, molecular biological studies have failed to uncover biomarkers linked to PsPD for clinical use. Although a multitude of genetic and molecular changes involved in GBM, including O6-methylguanine–DNA methyltransferase (MGMT) promoter methylation, isocitrate dehydrogenase 1 (IDH1) mutation, p53 mutation and Ki-67 expression, have been found to be associated with PsPD, the predictive value of these biomarkers remains debatable [5,8,17-19]. Therefore, except for cases of pathological verification, PsPD is still predominantly diagnosed retrospectively. Thus, there is an urgent need for the exploration of more reliable biochemical markers that can accurately identify PsPD.

* Correspondence: sunwei1018@hotmail.com; zlw.tth@hotmail.com
[2]Core Facility of Instrument, Institute of Basic Medical Sciences, Chinese Academy of Medical Science/School of Basic Medicine, Peking Union Medical College, No. 5 Dongdan Santiao, Dongcheng District, Beijing 100005, China
[1]Department of Neurosurgery, Beijing Tiantan Hospital, Capital Medical University, No. 6 TiantanXili, Dongcheng District, Beijing 100050, China

Proteomic measurements provide a wealth of biological information and several proteomic studies of gliomas have been recently reported [20,21], which demonstrated a possibility to investigate this phenomenon by using proteomics methods. Herein, this present study was designed to identify biological signatures and explore biomarkers for PsPD using differential proteomic techniques (Figure 1).

Results

Identification of proteins with significant fold changes in PsPD versus GBM

In this iTRAQ-labeling proteomic study, by comparing the total proteomes of tissue from PsPDs with the proteomes of tissues from GBMs, we identified 4048 proteins in PsPD and 3846 proteins in GBM (Additional file 1: File s1, Additional file 2: File s2, Additional file 3: File s3 and Additional file 4: File s4). To measure the quantitative correlation between pairwise sample combinations within each group, a Pearson's correlation coefficient (ranged from 0.967 to 0.980) was calculated and showed high biological reproducibility (Additional file 5: Figure s1). To maintain a low false-positive rate of comparative analysis between the groups, an average CV of 0.37 (Additional file 5: Figure s2) was employed to filter out data with poor linearity, corresponding to coverage of more than 80% of the 3390 quantified proteins both in PsPDs and GBMs. Next, a threshold of ≥2-fold and $p < 0.05$ was taken to identify 530 proteins with significant fold changes for further analysis (Figure 2). Among these proteins, 57 proteins were up-regulated and 473 were down-regulated in PsPD compared with GBMs (Additional file 6: File s5 and Additional file 7: File s6).

Interaction networks and functional pathway analysis

Functional pathway analysis was performed for the 530 proteins to better understand the biological features of PsPD. Gene ontology analysis indicated broad distribution of these proteins, with the most frequently represented categories being cellular compartment, molecular function, and biological processes (Figure 3). The results of Ingenuity Pathway Analysis (IPA) analysis indicated that the protein synthesis network and the cellular growth and proliferation network were mostly affected (Figure 4), with a series of cellular functions being significantly inhibited in PsPD compared with GBM (Additional file 5: Figure s3).

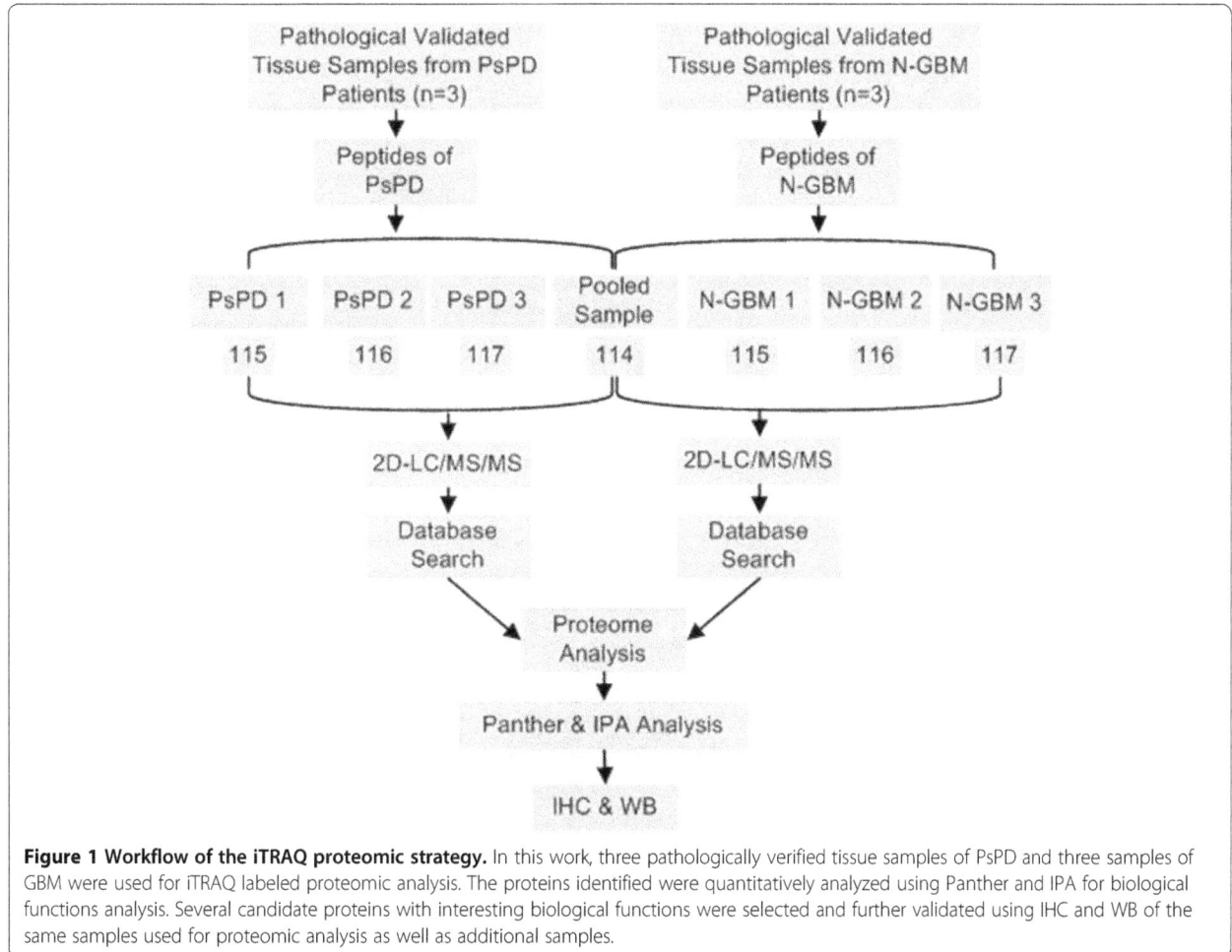

Figure 1 Workflow of the iTRAQ proteomic strategy. In this work, three pathologically verified tissue samples of PsPD and three samples of GBM were used for iTRAQ labeled proteomic analysis. The proteins identified were quantitatively analyzed using Panther and IPA for biological functions analysis. Several candidate proteins with interesting biological functions were selected and further validated using IHC and WB of the same samples used for proteomic analysis as well as additional samples.

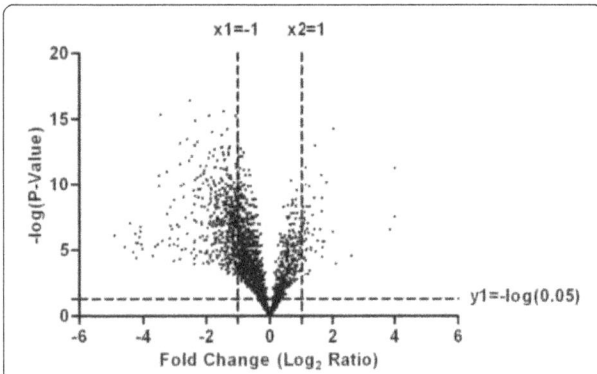

Figure 2 Volcano plots of identified proteins in PsPD vs GBM.
The x-axis of the graph refers to the log transformation of fold
change, whereas the y-axis of the graph refers to the negative log
transformation of the p-value.

For example, the invasion (z-score:-2.575) and proliferation (z-score:-2.886) abilities of tumor cells were significantly downregulated in PsPD compared with GBM. Moreover, the translation (z-score: −2.464), synthesis of protein (z-score: −2.236) and metabolism of protein (z-score:-2.046) were also significantly downregulated in PsPD compared with GBM.

Selection of candidate proteins for validation

Three candidate proteins (HNRNPK, ELAVL1 and CDH2) involved in the two networks and acting as key-point proteins were selected out. In order to explore more promising biomarkers, all secreted proteins with more than 2 fold changes (Additional file 8: File s7) were searched against the protein atlas database (http://www.proteinatlas. org), because the protein atlas database provided the

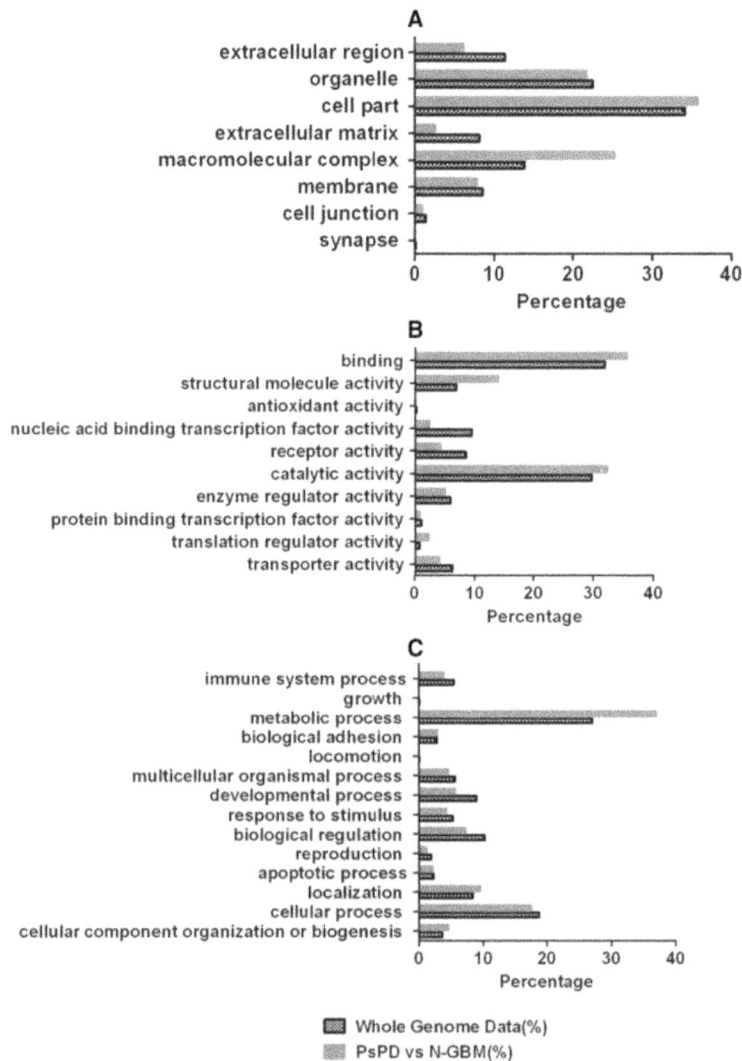

Figure 3 Panther analysis of PsPD vs N-GBM. Graph **A** shows cellular compartment analysis; Graph **B** shows molecular function analysis; and Graph **C** shows biological process analysis.

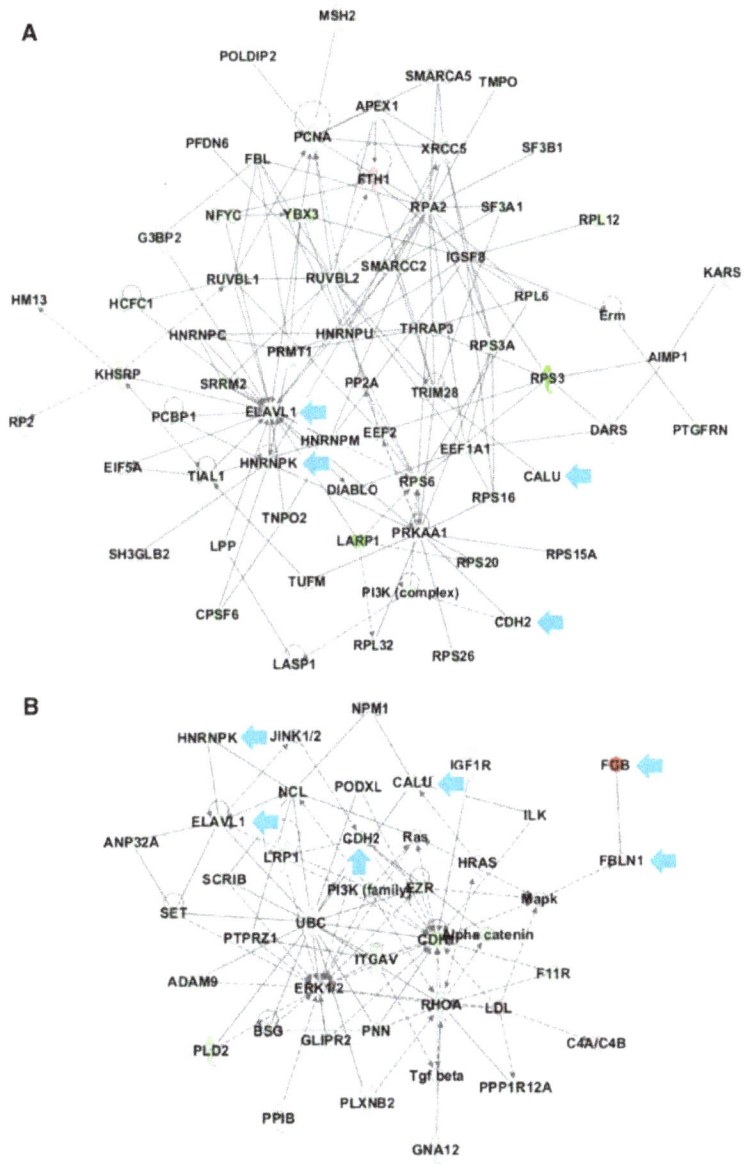

Figure 4 Cellular growth and proliferation network and protein synthesis network from IPA analysis. Graph **A** shows the protein synthesis focused network, and Graph **B** illustratescellular growth and proliferation focused function network. Proteins in red were up-regulated in PsPD compared with N-GBM, and proteins in green were down-regulated in PsPD compared with N-GBM. Proteins pointed by the blue arrow were the selected out candidate proteins used for analysis and further validation.

expression levels of candidate proteins in specific tissues and related antibodies. Proteins with median or high positive expression in glial cell or tissue were chosen for further functional analysis. Three proteins (FBLN1, CALU and FGB) meeting the criteria were selected out. The results of IHC and WB validation of the six proteins were in accordance with the proteomic findings (Figures 5, 6). Moreover, a quantitative analysis of the WB results was performed (Table 1, Figure 6). As shown in the figure, statistically significant differences were found between the groups.

Evaluation of HNRNPK, ELAVL1, CDH2 and FBLN1 as diagnostic markers for PsPD

The WB analysis revealed that HNRNPK, ELAVL1, CDH2 and FBLN1 were of statistical significance and exihibited obvious fold changes between PsPDs and GBMs (Table 1). Furthermore, the area under the ROC curves for ELAVL1, HNRNPK, CDH2 and FBLN1 were 0.86 (p = 0.013), 0.75 (p = 0.077), 0.90 (p = 0.006) and 0.66 (p = 0.258), respectively (Figure 7, Additional file 9: Table S1). A pairwise comparison of ROC curves shows no statistical difference between these four proteins

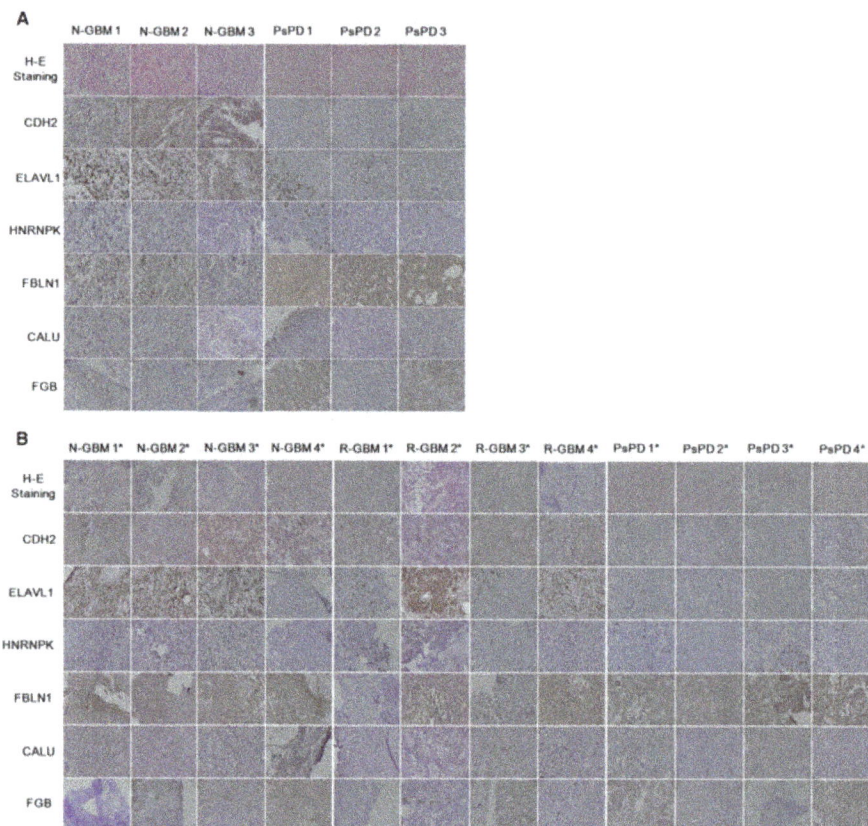

Figure 5 Results of immunohistochemical analysis of CDH2, ELAVL1, HNRNPK, FBLN1, CALU and FGB in tissue samples. Magnification: 200X. Representative images of paraffin-embedded sections of PsPD and GBM tissue that were HE stained and immunostained for CDH2,ELAVL1, HNRNPK, FBLN1, CALU and FGB. Graph **A** shows the validation of these six candidate proteins in the six samples used for proteomic analysis. The first three columns show the validation results in N-GBMs and the second three columns show the results in the PsPDs. Graph **B** shows the validation in an additional twelve samples. The first four column shows the validation results in additionally selected N-GBMs, the second four column shows the results in R-GBMs, and the third four column shows the results in PsPDs. * indicates the twelve additionally selected samples.

(Additional file 9: Table S2). Furthermore, the area under the combined ROC curve for CDH2 and ELAVL1 was 0.92 (P = 0.003), indicating that the diagnostic value of CDH2 alone or combined with ELAVL1 was improved.

Discussion

By using iTRAQ-labeled proteomic analysis and conducting further biological functional analysis of fold-changed proteins, we identified the biological features of PsPD from the perspective of proteomics and explored several candidate proteins to be predictive biomarkers.

Protein metabolism and upstream regulatory mechanisms play fundamental roles

The results of the biological analysis revealed the protein synthesis network to be broadly affected. Based on the data from the present study, the expression level of proteins involved in protein synthesis and upstream regulatory mechanisms, such as RNA post-transcriptional modification, post-translational modification and protein folding are significantly different between PsPDs and

GBMs (Figure 4, Additional file 5: Figure s3, Additional file 7: File s5 and Additional file 8: File s6), indicating these mechanisms may be significantly affected. Two candidate proteins, HNRNPK and ELAVL1, involved in the protein synthesis network were selected and validated.

HnRNPs comprise a large family of proteins with approximately 30 members that share some structural domains. Previous studies have shown that hnRNPs played central roles in several cellular functions, among which HNRNPK was found to play an essential role in cellular proliferation by regulating protein synthesis and is over-expressed in head and neck tumors [22,23]. In recent studies, HNRNPK was also found to play a significant role in the mechanism of DNA damage-related cell cycle arrest under ionizing conditions [24,25], which is similar to the effect of radiotherapy. In the present study, hnRNPs (HNRNPC, HNRNPK, HNRNPM and HNRNP) were found to play roles in the protein synthesis network and were down-regulated in PsPDs compared with GBMs, which may reflect the effect of chemo-radiotherapy treatment in GBM patients.

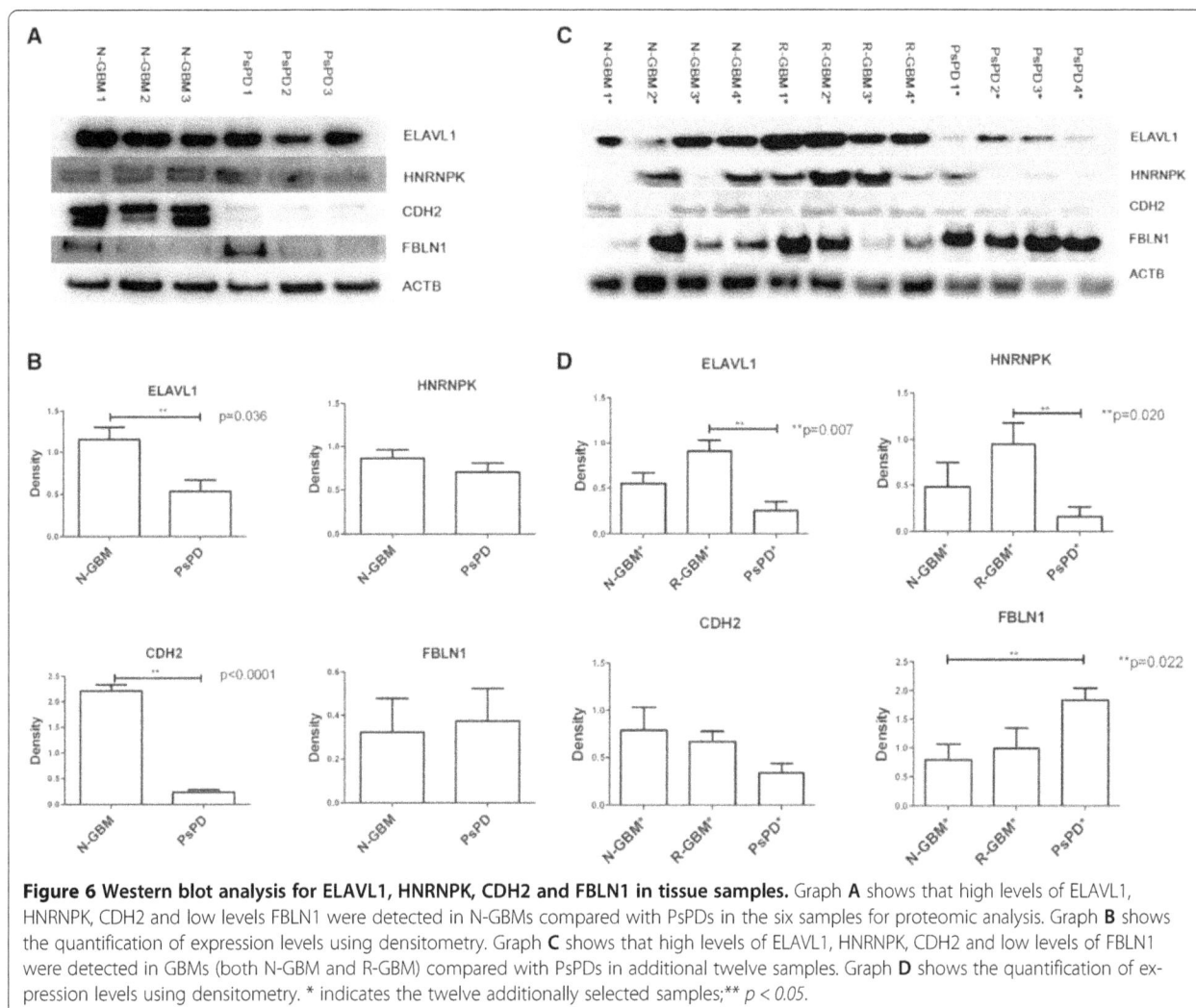

Figure 6 Western blot analysis for ELAVL1, HNRNPK, CDH2 and FBLN1 in tissue samples. Graph **A** shows that high levels of ELAVL1, HNRNPK, CDH2 and low levels FBLN1 were detected in N-GBMs compared with PsPDs in the six samples for proteomic analysis. Graph **B** shows the quantification of expression levels using densitometry. Graph **C** shows that high levels of ELAVL1, HNRNPK, CDH2 and low levels of FBLN1 were detected in GBMs (both N-GBM and R-GBM) compared with PsPDs in additional twelve samples. Graph **D** shows the quantification of expression levels using densitometry. * indicates the twelve additionally selected samples;** $p < 0.05$.

In addition to the hnRNPs, another RNA-binding protein, ELAVL1, was selected. Under hypoxia, ELAVL1 plays a significant role in the regulation of angiogenesis by stabilizing vascular endothelial growth factor A (VEGF-A) mRNA [26,27]. VEGF-A is one of the major mediators of vascular proliferation in astrocytic tumor [28]. Both VEGF and ELAVL1 were identified down-regulated in PsPD compared with GBM, suggesting the possibility of angiogenesis inhibition in PsPD. This result may also help explain how hypoxia is involved in the formation of PsPD, as has been proposed in several studies [18,29].

Table 1 Candidate proteins used for validation and details

Candidate Proteins	Accession Number	iTRAQ		IHC (PsPD vs N-GBM)	WB Quantitative Analysis		
		FC (PsPD vs N-GBM)	P value		PsPD vs N-GBM (MS)	PsPD vs N-GBM#	PsPD vs R-GBM#
HNRNPK	P61978	0.34	**	down-regulation	0.82	0.26	0.14*
ELAVL1	B4DVB8	0.23	**	down-regulation	0.46*	0.33	0.21*
FBLN1	P23142	2.74	**	up-regulation	1.19	1.63*	1.43
CDH2	P19022	0.29	0.05	down-regulation	0.1*	0.35	0.4
CALU	O43852	0.33	0.15	down-regulation	No Validation	No Validation	No Validation
FGB	P02675	16.06	**	up-regulation	No Validation	No Validation	No Validation

Note: The 1st column refers to the candidate proteins used for validation; the 2nd column refers to the corresponding accession number of these candidate proteins; the 3rd column refers to the results of iTRAQ labeled quantitative analysis, FC, fold change, **$p < 0.01$; The 4th column refers to the differential expression levels of these candidate proteins in the immunohistochemistry (IHC) analysis; The 5th column refers to the results of the western blot validation of the 4 candidate proteins, ELAVL1, HNRNPK, CDH2 and FBLN1, #indicates the results in additional samples, *$p < 0.05$.

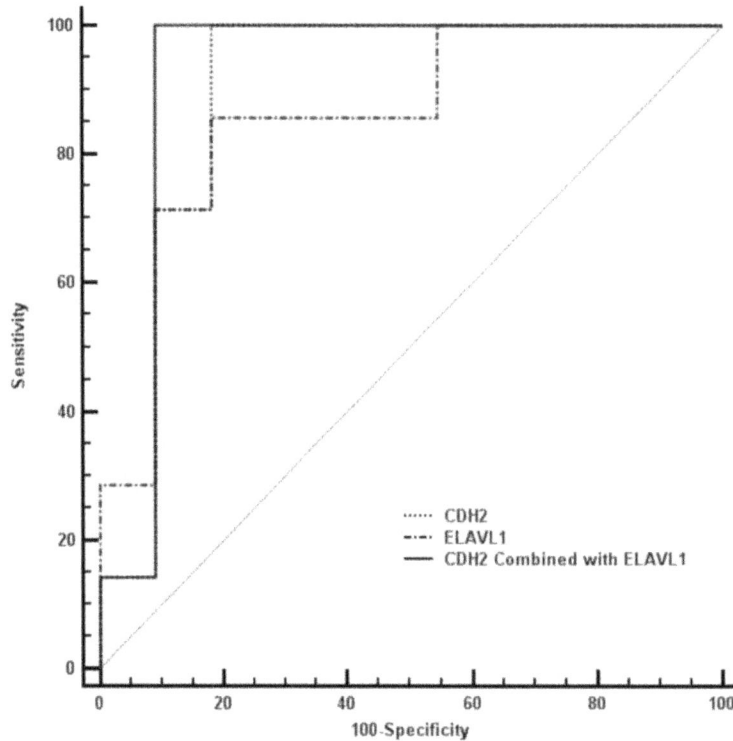

Figure 7 Roc curve of predictive biomarkers. The ROC curve of CDH2, ELAVL1 and the combination of these two candidate proteins was shown in the graph with different lines.

Cellular function interference

Many researchers have proposed that PsPD occurs due to the induction of cell death by radiotherapy and/or chemotherapy of malignant glioma [17,30]. These findings indicate a hypothesis that an underlying relationship between PsPD occurrence and cell death induction by adjuvant therapy may exist [30]. In this present study, the results of biological analysis shows that most of the proteins related to the cellular growth and proliferation functions as well as the invasion and proliferation abilities of tumor cells were down-regulated (Figure 4, Additional file 5: Figure s3, Additional file 6: File s5 and Additional file 7: File s6), demonstrating these functions may have been significantly inhibited. Except for HNRNPK and ELAVL1, another two candidate proteins, CDH2 and CALU, involved in the network of cellular growth and proliferation were selected and validated.

A previous study on brainstem glioma showed that higher expression of CDH2 predicts the progression of malignant tumors and tends to predict a shorter survival time of patients [31]. Other studies also indicated CDH2 may be functionally correlated with tumorigenesis in glioma cells and involved in mediating glioma cell migration [32-34]. In the present study, CDH2 is involved in several cellular functions (Additional file 5: Figure S3, Additional file 9: Table S3) and found to be down-regulated in PsPDs compared with GBMs (Table 1, Figure 4). The results were

in accordance with previous studies and may demonstrate the malignancy changes in PsPD.

Another protein CALU, is a calcium-binding protein located in the endothelium that is involved in protein folding and sorting. This protein was recently found to be highly expressed in normal neural stem cells and GBM stem-like cells compared with the GBM tumor tissue [35]. Additionally, the gene CALU was also observed to be up-regulated in GBM but not in low-grade astrocytoma or oligodendroglioma [36]. These results indicated that the expression levels of CALU may be correlated with tumor cell proliferation ability, which is in accordance with the biological analysis results of this present study.

Validation of secretory proteins as candidate biomarkers

At present, there are no suitable specific biomarkers that can be used to accurately differentiate PsPDs from GBMs. Secretory proteins have the potential to be detected as biomarkers in body fluids. Therefore, we also selected three candidate secretory proteins, CALU (described above), FGB and FBLN1, for validation. The validation results were in accordance with the proteomic findings. It is noteworthy that, previous studies have reported that FBLN1 expression is elevated in breast tumors [37] and ovarian cancer cells [38]. But no details about the roles of FBLN1 in gliomas have been reported previously.

Taken together, the proteomic results as well as the validation results both identified that the expression level of HNRNPK, ELAVL1, CDH2 and FBLN1 in PsPDs were significantly different from GBMs (Figures 5, 6). ROC curves yielded an AUC value of 0.90 (95% CI, 0.662-0.9880) for CDH2 and.0.92 (95% CI, 0.696-0.995) for CDH2 combined with ELAVL1, which indicated that these two proteins could be potential biomarkers with relatively high accuracy in the diagnosis of PsPD.

Conclusion

In summary, our work offers an initial description of the proteins conserved in PsPDs and GBMs as well as novel information on proteins that are differentially expressed between groups. Through biological analysis and validation of the proteomic findings, this present study not only revealed the molecular signatures but also provide novel markers that may help to identify the mechanisms behind and allow the diagnosis of PsPD. However, due to the low number of samples used in the present study, above conclusions were just preliminary results, therefore, it should be careful to use our conclusions. Further verification in additional samples should be helpful and essential to understand the process.

Materials and methods

Sample collection and pathological examination

A set of fresh frozen tissue samples that included PsPD (n = 3) and newly diagnosed GBM (N-GBM, n = 3) was obtained under an Institutional Review Board-approved protocol at the Beijing Tiantan Hospital of Capital Medical University. Consents of clinical data and samples used for the study have been obtained from the patients and their families. PsPD was diagnosed according to the criteria of Macdonald [39] without viable tumor recurrence by pathological verification. The tissue samples were snap-frozen immediately after resection and stored at −80°C. To ensure that the fragments used for proteomic analysis contained a sufficient proportion (at least 80%) of the target tissue, we evaluated each specimen before use. Moreover, twelve additional samples were selected for verification by IHC and WB, including four PsPD, four N-GBM and four recurrent GBM (R-GBM) tissue samples (Additional file 9: Table S4).

ITRAQ sample preparation

First, 80 mg samples from each of the six frozen tissue samples selected for the proteomics screening were rinsed with PBS, and each sample was then mixed with lysis buffer (50 mMTris-HCl, 2.5 M thiourea, 8 M urea, 4% CHAPS, 65 mM DTT) for total protein extraction. The total protein concentration of each sample was determined using the Bio-Rad RC DC Protein Assay.

The proteins from each sample were pooled equally according to the total amount of protein and digested by filter-aided sample preparation combined with a microwave-assisted protein preparation method as previously described [40,41]. The peptides were dried by vacuum centrifugation and stored at −80°C.

The digested PsPD and GBM samples were mixed equally to create the internal standard and labeled by 114 iTRAQ. The three PsPD samples and the three GBM samples, were individually labeled with 115, 116 or 117 iTRAQ according to the manufacturer's protocol (ABsciex).

2D-LC and MS/MS conditions

For offline separation a HPLC from Waters was used, and for online LC/MS/MS analysis a nano-ACQUITYUPLC system from Waters was used. First, the pooled mixture of the labeled samples was fractionated using a high-pH RPLC column from Waters (4.6 mm × 250 mm, C18, 3 μm). For each fraction the injection volume was 8uL. The samples were loaded onto the column in buffer A1 (1‰ aqueous ammonia in water, pH = 10), and eluted by buffer B1 (1‰ aqueous ammonia in 10% water and 90%ACN; pH = 10, flow rate = 1 mL/min) with the gradient of 5–90%for 60 min. The eluted peptides were collected at a rate of one fraction per minute, and pooled into 20 samples. Each sample was analyzed by LC-MS/MS using an RP C18 self-packing capillary LC column (75 μm × 100 mm, 3 μm) and a Triple TOF 5600 mass spectrometer. For Triple TOF 5600 a nano source was used. The MS data were acquired in high sensitivity mode with detailed parameters for Triple TOF 5600 being set as following: ion spray voltage was 2200v, curtain gas was 25, gas 1 was 5, gas 2 was 0, temperature was 150, declustering potential was 100, mass range was 350–1250 for MS and 100–1800 for MS/MS, collision energy was 35, and the resolution of MS and MS/MS was 40000 and 20000. An elution gradient of 5–30% buffer B2 (0.1% formic acid, 99.9% ACN; flow rate, 0.3 μL/min) for 50 min was used for the analysis. Thirty data-dependent MS/MS scans were acquired for every full scan. The normalized collision energy used was 35%, and charge state screening (including precursors with +2 to +4 charge state) and dynamic exclusion (exclusion duration of 15 s) were performed. Analyst TF 1.6 was used to control the instruments.

Database search

The MS/MS spectra were searched against the human subset of the Uniprot database (84910 entries) (http://www.uniprot.org/) using the Mascot software version 2.3.02 (Matrix Science, UK). Trypsin was chosen for cleavage with a maximum number of allowed missed cleavages of two. Carbamidomethylation (C) and iTRAQ 4-plex labels were set as fixed modifications. The searches were performed using a peptide and product ion tolerance of

0.05 Da. Scaffold software was used to further filter the database search results using the decoy database method with the following filter: a 1% false-positive rate at the protein level and two unique peptides per protein. After filtering the results as described above, the peptide abundances in different reporter ion channels of the MS/MS scan were normalized. The protein abundance ratio was based on unique peptide results. Proteins with a fold change ≥ 2 were considered significantly altered.

Bioinformatics analysis
Data filtering was performed according to strict criteria, wherein any missing data values or detection failures were deleted. Pearson's correlation coefficient was calculated to measure the quantitative correlation among the three biological replicates in each group, and the coefficient of variation within groups was set at CV = 0.37 to filter out low-quality data. A Student's t-test was performed between groups, and differences were considered to be significant when $p < 0.05$. Any proteins that satisfied the criteria of a fold change (FC) between groups of ≥2 were selected for bioinformatics analysis using Gene Ontology (GO) and Ingenuity Pathway Analysis (IPA).

GO functional and IPA network analysis
All proteins identified by the two approaches were assigned a gene symbol using the Panther database (http://www.pantherdb.org/). Protein classification was performed based on the functional annotations of the GO project for cellular compartment, molecular functional and biological processed. When more than one assignment was available, all of the functional annotations were considered in the results. Moreover, all of the selected proteins with a significant fold changes were used for pathway analysis using the IPA software (Ingenuity Systems, Mountain View, CA) for network analysis.

Immunohistochemistry and western blot analysis
IHC was performed on the same six tissue samples used for the proteomic analysis and on twelve additional formalin fixed, paraffin embedded tissue samples. The following primary antibodies were used: anti-ELAVL1mouse monoclonal (Santa Cruz), 1:500; anti-HNRNPK mouse monoclonal (Santa Cruz), 1:50;anti-CDH2rabbit monoclonal (Cell Signaling Technology), 1:250; anti-FBLN1 mouse monoclonal (Santa Cruz),1:125; anti-CALU goat polyclonal (Santa Cruz), 1:100; anti-FGB goat polyclonal (Abcam), 1:16000. After deparaffinization and rehydration, antigen retrieval was performed by immersing the slide in antigenretrieval buffer (10 mM sodium citrate, 0.05% Tween 20, pH = 6.0) at 95°C for 5 min using pressure cooker. Endogenous peroxidases were blocked with 0.03% hydrogen peroxide, and nonspecific binding was blocked with 2% fetal calf serum in Tris-buffered saline with 0.1% Triton X-100 (TBST, pH = 7.6).

The sections were then incubated for 1 h at room temperature with primary antibodies followed by peroxidase-labeled polymer conjugate to anti-mouse, anti-rabbit, anti-goat immunoglobulins for 1 h and developed with diaminobenzidine system. The sections were counter stained with the Mayer's hematoxylin and dehydrated, and the image was taken under microscope.

WBs of the same six samples and additional twelve samples was performed to validate the proteomic quantitation of four selected candidate proteins (HNRNPK, ELAVL1, CDH2 and FBLN1). Proteins extracted from GBM or PsPD tissues were resolved by SDS-PAGE (4–20% gradient precast gel; Invitrogen). The protein bands were electro transferred to a PVDF membrane (Millipore, Bedford, MA), blocked with 2% (v/v) BSA in TBST (150 mM NaCl, 20 mM Tris, 0.1% Tween 20, pH = 7.4) for 2 h at room temperature, followed by incubation with primary antibody (anti-ELAVL1, 1:200 (mouse monoclonal, Santa Cruz); anti-HNRNPK,1:3000 (mouse monoclonal, Santa Cruz); anti-CDH2, 1:800 (rabbit monoclonal, Cell Signaling Technology); anti-FBLN1, 1:100, (mouse monoclonal, Santa Cruz)) diluted with 1% BSA in TBST at room temperature for 2 h. After extensive wash with TBST, the membranes were incubated with horseradish peroxidase-conjugated secondary antibody (anti-mouse or anti-rabbit; EarthOX, USA) diluted with 1% BSA in TBST for 90 min at room temperature. The membranes were developed using Immobilon Western chemiluminescent horseradish peroxidase substrate (Millipore). All the selected proteins ELAVL1, HNRNPK, CDH2 and FBLN1 were validated by Western blot analysis with actin as loading control.

Additional files
Additional file 1: File s1. Quantitative Peptide List for PsPD Samples.
Additional file 2: File s2. Quantitative Peptide List for N-GBM Samples.
Additional file 3: File s3. Quantitative Protein List for PsPD Samples.
Additional file 4: File s4. Quantitative Protein List for N-GBM Samples.
Additional file 5: Figure s1. Pearson correlation coefficient plot of each two proteomic runs related to the tissue specimen in each group. The three graphs in the first row of the figure refers to Pearson coefficient of any two samples in PsPD sample group (ranged from 0.974 to 0.980); The three graphs in the second row of the figure refers to the Pearson coefficient of any two samples in GBM sample group (ranged from 0.967 to 0.978).
Additional file 6: File s5. Significantly Fold Changed Proteins between PsPD and N-GBM Samples.
Additional file 7: File s6. Quantitative peptides of differentially expressed proteins between PsPD and N-GBM samples.
Additional file 8: File s7. List of secreted proteins.
Additional file 9: Table S1. Parameters of ROC curve for four proteins.

Abbreviations
[18] F-FDG: [18]fluorodeoxyglucose; CALU: Calumenin; CDH2: N-cadherin; CV: Coefficient of Variance; ELAVL1: Hu-antigen R; FBLN1: Fibulin-1; FC: Fold Change; FGB: Fibrinogen Beta Chain; GBM: Glioblastoma; GO: Gene Ontology; HNRNPK: heterogeneous nuclear ribonucleoprotein K; IDH1: Isocitrate

Dehydrogenase 1; IHC: Immunohistochemistry; IPA: Ingenuity Pathway Analysis; MGMT: O6-methylguanine–DNA methyltransferase; MRI: Magnetic Resonance Imaging; MRS: Magnetic Resonance Spectroscopy; N-GBM: Newly Diagnosed Glioblastoma; Panther: Protein Analysis Through Evolutionary Relationships; PsPD: Pseudoprogression Disease; rCBV: Relative Cerebral Blood Volume; R-GBM: Recurrent Glioblastoma; VEGF: Vascular Endothelial Growth Factor; WB: Western Blot.

Competing interests
The authors declared that they have no competing interests.

Authors' contributions
PZ carried out the sample preparation, proteomic analysis, biological analysis, sample validation using IHC and WB and manuscript drafting. ZG, DW, LZ participated in the 2D-LC analysis of samples and validations using WB. ZG, NJ, WS and LZ participate in the design of the study and the modification of the manuscript. WS and LZ both conceived of the study, and participated in the coordination. All authors read and approved the final manuscript.

Acknowledgements
This work was supported by the National Key Technology Research and Development Program of the Ministry of Science and Technology of China (2013BAI09B03) and Beijing Institute for Brain Disorders (BIBD-PXM2013_014226_07_000084).

References
1. Hoffman WF, Levin VA, Wilson CB. Evaluation of malignant glioma patients during the postirradiation period. J Neurosurg. 1979;50:624–8.
2. De Wit M, De Bruin H, Eijkenboom W, Smitt PS, Van den Bent M. Immediate post-radiotherapy changes in malignant glioma can mimic tumor progression. Neurology. 2004;63:535–7.
3. Brandes AA, Tosoni A, Spagnolli F, Frezza G, Leonardi M, Calbucci F, et al. Disease progression or pseudoprogression after concomitant radiochemotherapy treatment: pitfalls in neurooncology. Neuro-Oncology. 2008;10:361–7.
4. Chaskis C, Neyns B, Michotte A, De Ridder M, Everaert H. Pseudoprogression after radiotherapy with concurrent temozolomide for high-grade glioma: clinical observations and working recommendations. Surg Neurol. 2009;72:423–8.
5. Topkan E, Topuk S, Oymak E, Parlak C, Pehlivan B. Pseudoprogression in patients with glioblastoma multiforme after concurrent radiotherapy and temozolomide. Am J Clin Oncol. 2012;35:284–9.
6. Chamberlain MC. Pseudoprogression in glioblastoma. J Clin Oncol Off J Am Soc Clin Oncol. 2008;26:4359. author reply 4359–60.
7. Van Mieghem E, Wozniak A, Geussens Y, Menten J, De Vleeschouwer S, Van Calenbergh F, et al. Defining pseudoprogression in glioblastoma multiforme. European J Neurol Off J European Federation Neurol Soc. 2013;20:1335–41.
8. Kang HC, Kim CY, Han JH, Choe GY, Kim JH, Kim IA. Pseudoprogression in patients with malignant gliomas treated with concurrent temozolomide and radiotherapy: potential role of p53. J Neuro-Oncol. 2011;102:157–62.
9. Radbruch A, Fladt J, Kickingereder P, Wiestler B, Nowosielski M, Baumer P, et al. Pseudoprogression in patients with glioblastoma: clinical relevance despite low incidence. Neuro Oncol. 2015;17:151–9.
10. Gahramanov S, Muldoon LL, Varallyay CG, Li X, Kraemer DF, Fu R, et al. Pseudoprogression of glioblastoma after chemo- and radiation therapy: diagnosis by using dynamic susceptibility-weighted contrast-enhanced perfusion MR imaging with ferumoxytol versus gadoteridol and correlation with survival. Radiology. 2013;266:842–52.
11. Wen PY, Macdonald DR, Reardon DA, Cloughesy TF, Sorensen AG, Galanis E, et al. Updated response assessment criteria for high-grade gliomas: response assessment in neuro-oncology working group. J Clin Oncol Off J Am Soc Clin Oncol. 2010;28:1963–72.
12. Tsien C, Galbán CJ, Chenevert TL, Johnson TD, Hamstra DA, Sundgren PC, et al. Parametric response map as an imaging biomarker to distinguish progression from pseudoprogression in high-grade glioma. J Clin Oncol. 2010;28:2293–9.
13. Plotkin M, Eisenacher J, Bruhn H, Wurm R, Michel R, Stockhammer F, et al. 123I-IMT SPECT and 1HMR-spectroscopy at 3.0 T in the differential diagnosis of recurrent or residual gliomas: a comparative study. J Neuro-Oncol. 2004;70:49–58.
14. Van Laere K, Ceyssens S, Van Calenbergh F, de Groot T, Menten J, Flamen P, et al. Direct comparison of 18F-FDG and 11C-methionine PET in suspected recurrence of glioma: sensitivity, inter-observer variability and prognostic value. Eur J Nucl Med Mol Imaging. 2005;32:39–51.
15. Terakawa Y, Tsuyuguchi N, Iwai Y, Yamanaka K, Higashiyama S, Takami T, et al. Diagnostic accuracy of 11C-methionine PET for differentiation of recurrent brain tumors from radiation necrosis after radiotherapy. J Nucl Med. 2008;49:694–9.
16. Popperl G, Gotz C, Rachinger W, Gildehaus FJ, Tonn JC, Tatsch K. Value of O-(2-[18F] fluoroethyl)- L-tyrosine PET for the diagnosis of recurrent glioma. Eur J Nucl Med Mol Imaging. 2004;31:1464–70.
17. Brandes AA, Franceschi E, Tosoni A, Blatt V, Pession A, Tallini G, et al. MGMT promoter methylation status can predict the incidence and outcome of pseudoprogression after concomitant radiochemotherapy in newly diagnosed glioblastoma patients. J Clin Oncol. 2008;26:2192–7.
18. Motegi H, Kamoshima Y, Terasaka S, Kobayashi H, Yamaguchi S, Tanino M, et al. IDH1 mutation as a potential novel biomarker for distinguishing pseudoprogression from true progression in patients with glioblastoma treated with temozolomide and radiotherapy. Brain Tumor Pathol. 2013;30:67–72.
19. Pouleau HB, Sadeghi N, Baleriaux D, Melot C, De Witte O, Lefranc F. High levels of cellular proliferation predict pseudoprogression in glioblastoma patients. Int J Oncol. 2012;40:923–8.
20. Turtoi A, Musmeci D, Naccarato AG, Scatena C, Ortenzi V, Kiss R, et al. Sparc-like protein 1 is a new marker of human glioma progression. J Proteome Res. 2012;11:5011–21.
21. Mustafa DA, Dekker LJ, Stingl C, Kremer A, Stoop M, Smitt PAS, et al. A proteome comparison between physiological angiogenesis and angiogenesis in glioblastoma. Mol Cell Proteomics. 2012;11(M111):008466.
22. Lynch M, Chen L, Ravitz MJ, Mehtani S, Korenblat K, Pazin MJ, et al. hnRNP K binds a core polypyrimidine element in the eukaryotic translation initiation factor 4E (eIF4E) promoter, and its regulation of eIF4E contributes to neoplastic transformation. Mol Cell Biol. 2005;25:6436–53.
23. Carpenter B, MacKay C, Alnabulsi A, MacKay M, Telfer C, Melvin WT, et al. The roles of heterogeneous nuclear ribonucleoproteins in tumour development and progression. Biochimica et Biophysica Acta (BBA)-Reviews on. Cancer. 2006;1765:85–100.
24. Moumen A, Magill C, Dry KL, Jackson SP. ATM-dependent phosphorylation of heterogeneous nuclear ribonucleoprotein K promotes p53 transcriptional activation in response to DNA damage. Cell Cycle. 2013;12:698–704.
25. Strozynski J, Heim J, Bunbanjerdsuk S, Wiesmann N, Zografidou L, Becker SK, et al. Proteomic identification of the heterogeneous nuclear ribonucleoprotein K as irradiation responsive protein related to migration. J Proteomics. 2015;113:154–61.
26. Ido K, Nakagawa T, Sakuma T, Takeuchi H, Sato K, Kubota T. Expression of vascular endothelial growth factor-A and mRNA stability factor HuR in human astrocytic tumors. Neuropathol Off J Japanese Soc Neuropathol. 2008;28:604–11.
27. Mukherjee N, Corcoran DL, Nusbaum JD, Reid DW, Georgiev S, Hafner M, et al. Integrative regulatory mapping indicates that the RNA-binding protein HuR couples pre-mRNA processing and mRNA stability. Mol Cell. 2011;43:327–39.
28. Fischer I, Gagner JP, Law M, Newcomb EW, Zagzag D. Angiogenesis in gliomas: biology and molecular pathophysiology. Brain Pathol. 2005;15:297–310.
29. Jensen RL. Brain tumor hypoxia: tumorigenesis, angiogenesis, imaging, pseudoprogression, and as a therapeutic target. J Neuro-Oncol. 2009;92:317–35.
30. da Cruz LC Jr H, Rodriguez I, Domingues RC, Gasparetto EL, Sorensen AG. Pseudoprogression and pseudoresponse: imaging challenges in the assessment of posttreatment glioma. AJNR Am J Neuroradiol. 2011;32:1978–85.
31. Wu W, Tian Y, Wan H, Ma J, Song Y, Wang Y, et al. Expression of beta-catenin and E- and N-cadherin in human brainstem gliomas and clinicopathological correlations. Int J Neurosci. 2013;123:318–23.
32. Kohutek ZA, Redpath GT, Hussaini IM. ADAM-10-mediated N-cadherin cleavage is protein kinase C-α dependent and promotes glioblastoma cell migration. J Neurosci. 2009;29:4605–15.

33. Péglion F, Etienne-Manneville S. N-cadherin expression level as a critical indicator of invasion in non-epithelial tumors. Cell Adhes Migr. 2012;6:327–32.
34. Barami K, Lewis-Tuffin L, Anastasiadis PZ. The role of cadherins and catenins in gliomagenesis. Neurosurg Focus. 2006;21:1–4.
35. Thirant C, Galan-Moya EM, Dubois LG, Pinte S, Chafey P, Broussard C, et al. Differential proteomic analysis of human glioblastoma and neural stem cells reveals HDGF as a novel angiogenic secreted factor. Stem Cells. 2012;30:845–53.
36. Sreekanthreddy P, Srinivasan H, Kumar DM, Nijaguna MB, Sridevi S, Vrinda M, et al. Identification of potential serum biomarkers of glioblastoma: serum osteopontin levels correlate with poor prognosis. Cancer Epidemiol Biomark Prev. 2010;19:1409–22.
37. Greene LM, Twal WO, Duffy MJ, McDermott EW, Hill AD, O'Higgins NJ, et al. Elevated expression and altered processing of fibulin-1 protein in human breast cancer. Br J Cancer. 2003;88:871–8.
38. Roger P, Pujol P, Lucas A, Baldet P, Rochefort H. Increased immunostaining of fibulin-1, an estrogen-regulated protein in the stroma of human ovarian epithelial tumors. Am J Pathol. 1998;153:1579–88.
39. Macdonald DR, Cascino TL, Schold Jr SC, Cairncross JG. Response criteria for phase II studies of supratentorial malignant glioma. J Clin Oncol Off J Am Soc Clin Oncol. 1990;8:1277–80.
40. Wisniewski JR, Zougman A, Nagaraj N, Mann M. Universal sample preparation method for proteome analysis. Nat Methods. 2009;6:359–62.
41. Sun W, Gao S, Wang L, Chen Y, Wu S, Wang X, et al. Microwave-assisted protein preparation and enzymatic digestion in proteomics. Mol Cell Proteomics. 2006;5:769–76.

4

Comparative proteome analysis of abdominal adipose tissues between fat and lean broilers

Chun-Yan Wu[1†], Yuan-Yuan Wu[1,2†], Chun-Dong Liu[1], Yu-Xiang Wang[1], Wei Na[1], Ning Wang[1*] and Hui Li[1*]

†Equal contributors

Abstract

Background: The molecular mechanism underlying broiler fat deposition is still poorly understood.

Method: Currently, we used two-dimensional gel electrophoresis (2DE) to identify differentially expressed proteins in abdominal adipose tissues of birds at 4 week of age derived from Northeast Agricultural University broiler lines divergently selected for abdominal fat content (NEAUHLF).

Results: Thirteen differentially expressed protein spots were screened out and identified by matrix-assisted laser desorption-ionization time-of-flight mass spectrometry (MALDI-TOF-MS). The protein spots were matched to thirteen proteins by searching against the NCBInr database. These identified proteins were apolipoprotein A-I (Apo A-I), cytokeratin otokeratin, ATP synthase subunit alpha, peptidyl-prolyl cis-trans isomerase FKBP4 (PPIase FKBP4), aspartate aminotransferase, carbonic anhydrase II (CA-II), prostaglandin-H2 D-isomerase precursor, fibrinogen alpha chain, lamin-A (LMNA), superoxide dismutase [Mn] (MnSOD), heat shock protein beta-1 (HSPβ1) and two predicted proteins. These differentially expressed proteins are involved mainly in lipid metabolism, amino acid metabolism, signal transduction, energy conversion, antioxidant, and cytoskeleton. Differential expression of Apo A-I, PPIase FKBP4, and cytokeratin otokeratin proteins were further confirmed by Western blot analysis. Quantitative real-time RT-PCR analyses showed that, of these 13 differentially expressed proteins, only PPIase FKBP4 and cytokeratin otokeratin were differentially expressed at mRNA level between the two lines.

Conclusions: Our results have provided further information for understanding the basic genetics control of growth and development of broiler adipose tissue.

Keywords: Abdominal adipose tissue, Broiler, Proteomics, Differentially expressed protein

Background

Great progress has been made in poultry breeding in the past half century. Daily gain and feed conversion have been improved considerably; however, in commercial flocks, the improved productive performance is accompanied by high percentages of body fat content and some other negative effects that bring huge economic losses to the broiler industry [1]. Controlling fat deposition has been one of the major goals in the broiler breeding industry.

Adipose tissue not only serves as a fat storage site, but also as an endocrine organ that plays roles in a wide range of cellular processes including lipid metabolism and glucose homeostasis [2]. In chicken, abdominal adipose tissue is the main tissue of body fat accumulation, accounting for about 22 % of total body fat [3]. To control broiler fat deposition, it is necessary to understand gene expression and its regulation during adipose tissue development. Gene expression profiling of chicken abdominal adipose tissue has been performed, and a number of differentially expressed genes have been identified between fat and lean chickens [4–8]. With the advent of proteomic technologies, comprehensive proteomic

---

* Correspondence: wangning@neau.edu.cn; lihui@neau.edu.cn
†Equal contributors
[1]Key Laboratory of Chicken Genetics and Breeding of Agriculture Ministry, Key Laboratory of Animal Genetics, Breeding and Reproduction of Education Department of Heilongjiang Province, College of Animal Science and Technology, Northeast Agricultural University, Harbin 150030, Heilongjiang, China
Full list of author information is available at the end of the article

approaches have been widely used to identify and relatively quantify proteins.

In the present study, we compared protein expression profiles of abdominal adipose tissues of birds at 4 week of age derived from NEAUHLF, and found 13 differentially expressed proteins between fat and lean broilers. Our findings provide further information for understanding the molecular mechanism of broiler fat deposition.

Methods
Animal and abdominal adipose tissue samples collection
The NEAUHLF [9] was used in the current study. All animal work was conducted on the basis of the guidelines for the care and use of experimental animals established by the Ministry of Science and Technology of the People's Republic of China (Approval number: 2006-398) and approved by the Laboratory Animal Management Committee of Northeast Agricultural University. All broilers were kept in similar environmental conditions and had free access to feed and water. Commercial corn-soybean-based diets, which met all the NRC requirements were provided to the broilers [10].

Six male broilers from the 13th generation of NEAUHLF, three from the lean line and three from the fat line, were used in the present study. The average abdominal fat percentage of the three fat broilers was 9.34 times greater than that of the three lean broilers (Table 1). The broilers were slaughtered at 4 weeks of age, and abdominal adipose tissues were collected, frozen immediately in liquid nitrogen, and stored at -80 °C until further use.

Protein samples preparation for 2DE
Total protein was isolated from the abdominal adipose tissues by Trizol reagent (Invitrogen, Carlsbad, CA, USA) according to the manufacturer's protocol with minor modifications [11]. The samples were then dissolved in lysis buffer containing 8 M urea, 2 M thiourea, 4 % (wt/vol) 3-[(3-cholamidopropyl) dimethylammonio]-1-propanesulfonate, 2 % carrier ampholytes (pH 3 to 10, GE Healthcare, Uppsala, Sweden), 50 mM dithiothreitol (DTT), and 1× protease inhibitor cocktails (Roche

Table 1 Body weight (BW), abdominal fat weigh (AFW) and abdominal fat percentage (AFP) of the fat and lean broilers at 4 weeks of age

Traits	Lean line			Fat line		
	Lean 1	Lean 2	Lean 3	Fat 1	Fat 2	Fat 3
BW (g)	727.50	726.80	741.30	783.50	662.10	853.80
AFW (g)	3.65	3.59	3.85	41.27	30.91	31.44
AFP (%)	0.50	0.49	0.51	5.30	4.60	4.10

Lean 1, Lean 2, Lean 3 were three lean broilers; Fat 1, Fat 2, Fat 3 were three fat broilers

Diagnostics GmbH, Mannheim, Germany). The original protein samples were centrifuged at 20,000 g for 1 h to remove insoluble materials. For 2DE, salt and other small molecular impurities were removed using the 2D Clean-Up Kit (GE Healthcare, Chalfont St Giles, UK). Total protein concentration was determined using a 2D Quant Kit (Amersham Biosciences Corp., Piscataway, NJ, USA) and the protein samples were then stored at -80 °C.

2DE and image analyses
Protein samples were rehydrated at 650 μg/gel in 350 μL of rehydration solution containing 7 M urea, 2 M thiourea, 4 % (wt/vol) 3-[(3-cholamidopropyl) dimethylammonio]-1-propanesulfonate, 50 mM DTT and 0.8 % carrier ampholytes (pH 3 to 10, GE Healthcare). First-dimension electrophoresis was conducted with the IPGphor3 isoelectric focusing system (GE Healthcare) using the dry IPG strips (18 cm pH 3 to 10 nonlinear, GE Healthcare). The program setting was as follows: 50 V for 12 h, 100 V for 1 h, 300 V for 1 h, linear gradient to 1000 V in 2.5 h, linear gradient to 8000 V in 2 h and 8000 V until approximately 60,000 Vh. After the first dimension, the IPG strips were equilibrated in SDS equilibration buffer containing 75 mM Tris-HCl, pH 8.8, 6 M urea, 30 % (v/v) glycerol, 2 % (w/v) SDS, bromophenol blue 1 % (w/v) for 15 min, followed by a second equilibration with 2.5 % iodoacetamide replacing the 2 % DTT for 15 min. After equilibration, the proteins were separated on 12.5 % Trisglycine gels using an Ettan Dalt Six Electrophoresis System (GE Healthcare) at 12 °C. Gels were run at constant power, first with two W/strip for 45 min and then 15 W/strip until the bromophenol blue reached the bottom of the gels. Then, the gels were stained by the blue silver method with coomassie blue brilliant G250 [12]. Finally, six 2-DE gels were obtained.

Protein spot detection, volume calculation, matching, and the patterns were analyzed using Image Master 2D Platinum 6.0 software (GE Healthcare). The parameter used for the quantifications was the % of volume (%VOL: integration of the OD over the feature area (VOL) normalized by the total VOL over the whole image). Differentially expressed protein spots were considered significant if they showed >2-fold relative differences ($P <0.05$, Student's t test) between the fat and lean lines.

Protein identification by MALDI-TOF-MS
After image analyses, the differentially expressed protein spots were selected and excised from the gels. The protein spots were subjected to tryptic proteolysis, and the resultant peptides were analyzed by matrix-assisted laser desorption- ionization time-of-flight mass spectrometry (MALDI-TOF-MS) as described previously [13]. The

resultant peptide mass fingerprint was searched against the NCBInr protein sequence databases using the Mascot search engine [14]. The search parameters were as follows: enzyme search specificity was trypsin for tryptic digest; carbamidomethylation on cysteines was set as fixed modification while methionine oxidation was considered as variable modification; one miscleavage for each peptide was allowed; no restrictions on protein mass and peptide mass tolerance was ±100 ppm. A Mascot score with $P <0.05$ was considered statistically significant [15].

Western blot analysis

The abdominal adipose tissue was homogenized in radio immunoprecipitation assay (RIPA) buffer (1 g/L SDS, 5 g/L sodium deoxycholate, 10 g/L Nonidet P-40, 150 mmol/L NaCl, 50 mmol/L Tris-HCl, pH 8.0), supplemented with protease inhibitors (1 mmol/L phenylmethylsulfonyl fluoride, 0.002 g/L aprotinin and 0.002 g/L leupeptin). Cellular debris and lipids were eliminated by centrifuging the solubilized samples at 13,000 rpm for 60 min. The protein concentration of the samples was determined using a 2D Quant kit.

Protein samples were separated by SDS-PAGE and transferred to an Immun-Blot PVDF membrane (Millipore, Billerica, MA, USA). To block nonspecific binding, the membrane was incubated in blocking buffer (PBS with 5 % nonfat dry milk) for 1 h at room temperature. Membranes were immunoblotted with antibodies against Apo A-I (BIOSS, Beijing, China; 1:500 dilution), PPIase FKBP4 (ProteinTech Group, Chicago, IL, USA; 1: 500 dilution), and cytokeratin otokeratin (ProteinTech Group, Chicago, IL, USA; 1: 500 dilution) for 1 h at room temperature. After washing with PBS with 0.05 % Tween-20 (PBST), the membrane was immunoblotted with goat anti-rabbit IgG conjugated with horseradish peroxidase (1:5000) (ZSGB-BIO, Beijing, China) for 1 h at room temperature. Immunoreactive protein on the membrane was visualized using enhanced chemiluminescence and exposed to X-ray-film (Kodak, New York, NY, USA). β-actin (as the control) was detected first by mouse anti-chicken (β-actin) antibody (Beyotime Institute of Biotechnology, Jiangsu, China) and then by peroxidase-conjugated AffiniPure goat anti-mouse IgG (H + L; ZSGB-Bio). Immunoreactive protein levels were determined semi-quantitatively by densitometric analysis using the UVP system Labworks TM software 3.0 (UVP, Upland, CA, USA). Results were expressed as the relative quantity of Apo A-I/β-actin, PPIase FKBP4/β-actin and cytokeratin otokeratin/β-actin.

Real-time RT-PCR analyses

Total RNA from abdominal adipose tissue was isolated using Trizol reagent. Reverse transcription was performed using 1 μg of total RNA and M-MLV reverse transcriptase (Moloney murine leukemia virus RT, Invitrogen). Reverse transcription conditions for each cDNA amplification were 65 °C for 5 min, 37 °C for 52 min, and 70 °C for 15 min. Real-time RT-PCR was carried out using the 7500 Real-time PCR System (Applied Biosystems) and SYBR Premix Ex Taq (TaKaRa). The primers used for the PCR are listed in Table 2.

Statistical analysis

All results were expressed as mean ± SD and analyzed by student's t-test. Statistical analysis was performed using Prism 5.0 software (GraphPad Software Inc.). $P < 0.05$ was considered to be statistically significant.

Table 2 Primers used for the quantitative real-time RT-PCR analysis

Gene name	Sequence (5'–3')
β-actin	Sense: TCTTGGGTATGGAGTCCTG Antisense: TAGAAGCATTTGCGGTGG
FGA	Sense: GCAGAACAGCATCCAGGAGCAGG Antisense: TCCACCTGGTAATCAAAACTTCTAGCAC
CA2	Sense: CACTGGCACGAGCACTTC Antisense: ACTTCACGCCATCCACAGT
KRT7	Sense: CTGGACGGGTTGTTAAAT Antisense: TCCGCTTCGTAGAGAGAT
GRTP1	Sense: GGCTGCTCCAACGCCCACTT Antisense: CGCAACGCCTTCTGCTCTTT
MnSOD	Sense: CTGACCTGCCCTACGACT Antisense: TGGTATGATTGATATGACCC
LOC429524	Sense: GGAGGAAATGCGGCGCTTAG Antisense: GGCTGGACGAGACGCTGTTGA
ATP5A1	Sense: CAGTTTGGGTCTGATTTGG Antisense: AGCTTAGCTTCCGTCTGG
FKBP4	Sense: TACCTCCCAATGCTACGC Antisense: CCTTCGCCTTTCTTACGG
GOT1	Sense: GCACAGACCCTACTCCAGAC Antisense: AAGCCCTCGGAGACAAAG
LMNA	Sense: GGGGAACTGGCAGGTGAAGC Antisense: CCTCGTCGTCGTCGTTGATG
PTGDS	Sense: CCGAGGTCTTTTGTTTG Antisense: AGGAGGGGGACTTTGATG
HSPβ1	Sense: CAAACACGAGGAGAAACA Antisense: CGTTTATTCAAGGCACTG
Apo A-I	Sense: TCCGCTTCGTAGAGAGATGTGT Antisense: TCAGCGTGTCCAGGTTGTC

β-actin acts as internal control; FGA encodes the fibrinogen alpha chain, CA2 encodes carbonic anhydrase II, Otokeratin encodes the cytokeratin otokeratin protein, GRTP1 was predicted to encode the growth hormone-regulated TBC protein 1 protein, MnSOD encodes the MnSOD protein, LOC429524 was predicted to encode a transcription factor 24-like protein, ATP5A1 encodes the ATP synthase subunit alpha protein, FKBP4 encodes the PPIase FKBP4 protein, GOT1 encodes the aspartate aminotransferase 1 protein, LMNA encodes the lamin-A protein, PTGDS encodes the prostaglandin-H2 D-isomerase precursor protein, HSPβ1 encodes the HSPβ1 protein, Apo A-I encodes the Apo A-I protein

Results

Differentially expressed proteins between the fat and lean lines of broilers

Following staining with coomassie blue brilliant G250, the well-resolved 2DE gels were obtained, which are displayed in Fig. 1. We detected 884 ± 12 well-stained, clearly-delineated protein spots per gel (six gels) using Image Master 2D Platinum 6.0 software (GE Healthcare) and most spots were distributed mainly in range of pH 3-10. Quantitative image analysis of three biological replicates of each line revealed that a total of 13 protein spots showed a more than 2-fold difference ($P < 0.05$) between the fat and lean broilers. Of these, 12 protein spots were up-regulated and 1 protein spot was down-regulated in the lean birds compared to fat birds (Fig. 2a). The magnification of these 13 protein spots were displayed in Fig. 2b. These 13 differentially expressed protein spots were excised, digested in gel with trypsin and identified by MALDI-TOF-MS. All of the 13 protein spots were identified. The names of the identified proteins, their accession number, expression fold changes between the fat and lean broilers, and other information are shown in Table 3.

Western blot analysis

To verify the differential expression of individual proteins between the fat and lean broilers, we performed western blot analysis. Apo A-I, PPIase FKBP4 and cytokeratin otokeratin protein expression were verified using western blot. As shown in Fig. 3, the expression of abdominal adipose tissue Apo A-I, PPIase FKBP4 and cytokeratin otokeratin protein were significantly higher in the lean birds compared with in the fat birds ($P < 0.05$ or $P < 0.01$).

Real-time RT-PCR analysis

We used quantitative real-time RT-PCR to compare the mRNA expression levels of the genes corresponding to 13 differentially expressed proteins between fat and lean birds in abdominal adipose tissues. Surprisingly, the results showed that only two of these transcripts, PPIase FKBP4 and cytokeratin otokeratin, were significantly differentially expressed between fat and lean birds in abdominal adipose tissues ($P < 0.05$). The other 11 differentially expressed proteins were not found to be differentially expressed at the mRNA level between the fat

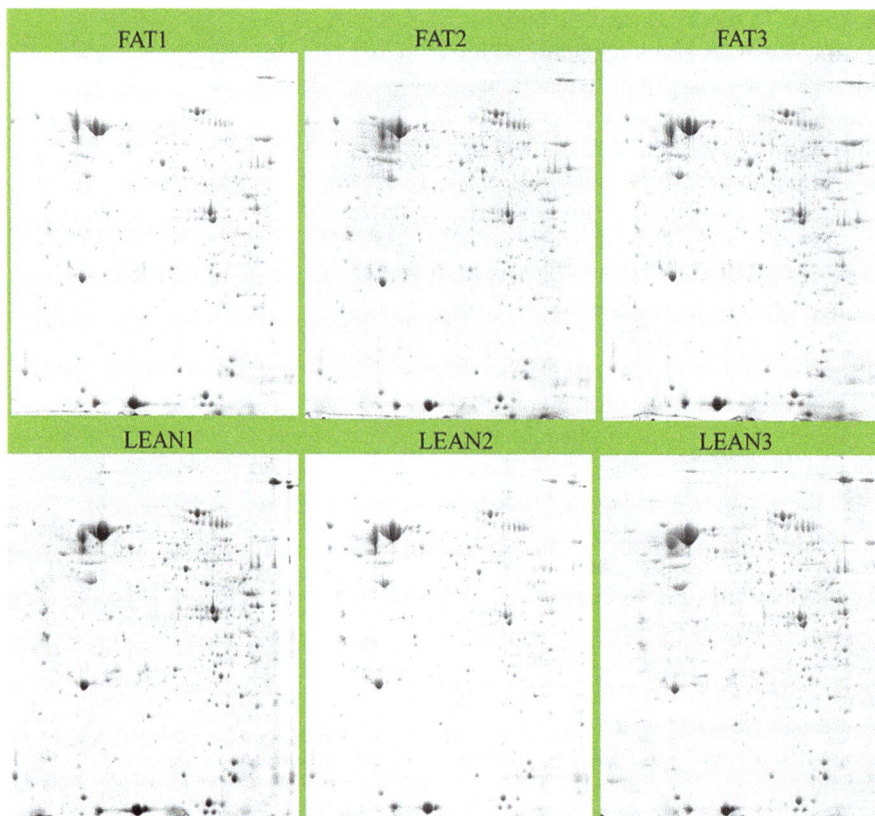

Fig. 1 2DE protein profiles of the 4-week-old abdominal adipose tissues of fat and lean broilers. The top three panels represent three biological replicates of the fat broilers, and the bottom three panels represent three biological replicates of lean broilers. These six gels were analyzed by Image Master 2D Platinum 6.0 software (GE Healthcare), and differentially expressed proteins were identified

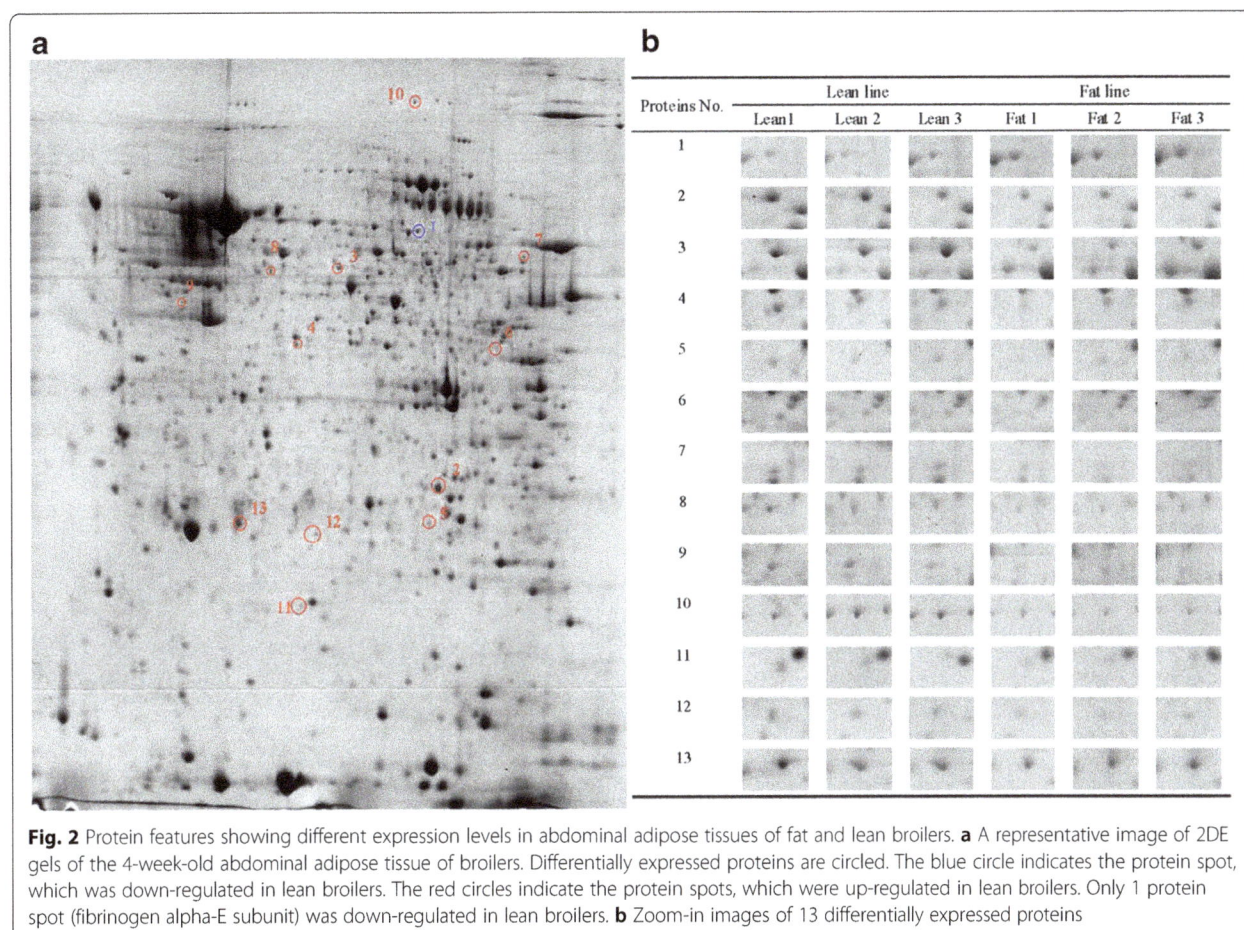

Fig. 2 Protein features showing different expression levels in abdominal adipose tissues of fat and lean broilers. **a** A representative image of 2DE gels of the 4-week-old abdominal adipose tissue of broilers. Differentially expressed proteins are circled. The blue circle indicates the protein spot, which was down-regulated in lean broilers. The red circles indicate the protein spots, which were up-regulated in lean broilers. Only 1 protein spot (fibrinogen alpha-E subunit) was down-regulated in lean broilers. **b** Zoom-in images of 13 differentially expressed proteins

and lean lines (Fig. 4). The results are inconsistent with the results of the proteomic analysis.

Discussion

The NEAUHLF provides an unique experimental model to study growth and development of chicken adipose tissue. In the present study, we identified thirteen differentially expressed proteins in abdominal adipose tissue between fat and lean lines of NEAUHLF at 4 weeks of age. The discovery of these differentially expressed proteins between the NEAUHLF fat and lean broiler lines may provide useful clues for understanding the molecular mechanism of broiler abdominal fat deposition.

Based on the biological process in which they are involved, these differentially expressed proteins could be classified into six categories: lipid metabolism (Apo A-I and prostaglandin-H2 D-isomerase precursor), amino acid metabolism (aspartate aminotransferase), signal transduction (fibrinogen alpha chain and CA-II), energy conversion (ATP synthase subunit alpha), antioxidant (HSPβ1, PPIase FKBP4 and MnSOD), and cytoskeleton (LMNA and cytokeratin otokeratin).

Apo A-I is an important lipid-binding protein, and is the major constituent of high-density lipoprotein (HDL)

cholesterol [16]. It plays important roles in preventing lipid accumulation in tissues and in maintaining cholesterol dynamic balance [17, 18]. Genetic deficiency in Apo A-I has been associated with excessive cholesterol accumulation in human and poultry [19–21]. Association studies showed that a single nucleotide polymorphism (SNP) upstream of the ATG initiation codon of the chicken Apo A-I gene was associated with abdominal fat weight and abdominal fat percentage [22]. Our previous proteomic analysis results showed that Apo A-I protein was down-regulated in the abdominal adipose tissue of fat broilers compared with the lean broilers at 7 weeks of age [8]. In the present study, we also observed that Apo A-I was down-regulated in abdominal adipose tissue of the fat broilers. Taken together, these data suggest that the differential protein expression of Apo A-I in the two divergently selected lines may be partially responsible for the difference in abdominal fat deposition between the two broiler lines.

Prostaglandin-H2 D-isomerase precursor is an essential enzyme in arachidonic acid metabolism. It catalyzes the conversion of prostaglandin-H2 to prostaglandin-D2 [23]. One of the dehydration products of prostaglandin-D2 is 15-deoxy-Δ [12, 14]-prostaglandin J2 (15d-PGJ2)

Table 3 Features of the 13 differentially expressed proteins identified by MALDI-TOF-M

No.	Protein name	Accession number	Fold change[a]	P-value	Variation[b]	Mascot score	MW (kDa)	PI	SC[c] (%)	SL[d]	Biological process
1	Fibrinogen alpha chain	gi|1706798	−3.41	0.025	0.510	204	83.1	5.69	45	S	Signal transduction
2	Carbonic anhydrase II	gi|833606	2.05	0.005	0.595	165	28.8	6.51	68	C	Signal transduction
3	Cytokeratin otokeratin	gi|45384378	2.06	0.002	0.935	264	53.8	5.97	84	NR	Cytoskeleton
4	Predicted: growth Hormone-regulated TBC protein 1	gi|363729049	2.06	0.001	0.273	349	29.5	6.32	62	NR	NR
5	Superoxide dismutase [Mn]	gi|45383702	2.07	0.035	0.468	127	25.1	8.60	48	M	Antioxidant
6	Predicted: transcription factor 24-like	gi|363730972	2.13	0.045	0.855	284	32.9	11.30	71	N	NR
7	ATP synthase subunit alpha	gi|45383566	2.19	0.030	0.595	196	60.1	9.29	48	M	Energy conversion
8	Peptidyl-prolyl cis-trans isomerase FKBP4	gi|57525441	2.20	0.016	0.950	290	51.4	5.34	64	C,N	Antioxidant
9	Aspartate aminotransferase	gi|45384348	2.26	0.003	0.496	164	46.1	8.22	43	C	Amino acid metabolism
10	Lamin-A	gi|45384214	2.30	0.001	1.710	241	73.3	6.50	54	N	Cytoskeleton
11	Prostaglandin-H2 D-isomerase precursor	gi|45383612	2.60	0.010	0.683	155	20.8	6.30	41	C	Lipid metabolism
12	Heat shock protein beta-1	gi|45384222	2.74	0.019	0.518	193	21.6	5.77	37	C	Antioxidant
13	Apolipoprotein AI	gi|227016	2.96	0.039	0.409	253	28.7	5.54	80	S	Lipid metabolism

[a]Fold change: averge relative volume ratio (lean broilers vs. fat broilers)
[b]Variation: Standard deviation
[c]SC: Sequence coverage
[d]SL: Subcellular location. *S* secreted, *C* cytoplasm, *NR* not reported, *M* mitochondrion., N nucleus

[24], which binds directly to peroxisome proliferator-activated receptor γ (PPARγ) and promotes efficient adipocyte differentiation [25]. These findings, combined with the results reported here, suggest that the differential expression of prostaglandin-H2 D-isomerase precursor may contribute to the phenotype difference between the fat and lean broiler lines.

Aspartate aminotransferase 1 is involved in adipocyte glyceroneogenesis, which controls fatty acid homeostasis by promoting glycerol 3-phosphate formation for fatty acid re-esterification when the supply of glucose is reduced [26]. The expression of aspartate aminotransferase 1 is specifically induced by glucose deprivation and rosiglitazone in adipocytes, but it is not directly regulated by PPARγ [27]. In the present study, aspartate aminotransferase 1 was more highly expressed in the lean chickens than in the fat chickens. The differential expression of aspartate aminotransferase 1 may reflect the difference in adipocyte glyceroneogenesis in the two chicken lines.

Fibrinogen alpha chain and CA-II are involved in signal transduction. Fibrinogen alpha chain is an important component of fibrinogen [28]. Kim et al. found that plasma fibrinogen was significantly higher in obese groups compared with in non-obese groups using 2DE and MS, and proposed plasma fibrinogen as a new biomarker of obesity [29]. Fibrinogen is regulated by interleukin-6 (IL-6) in adipose tissue and adipose tissue IL-6 expression was shown to be positively correlated with obesity. IL-6 was found to be the major regulator of fibrinogen, and stimulated fibrinogen synthesis [30]. In high fat diet-induced atherosclerosis in rabbits, high levels of IL-6 and fibrinogen were detected in the plasma [31]. In the present study, we also observed that fibrinogen alpha chain was up-regulated in the abdominal adipose tissue of the fat birds. Taken together, our data suggest fibrinogen alpha chain is involved in broiler fat deposition. Carbonic anhydrases (CAs) are a family of zinc metalloenzymes [32], which is critical to the entire process of fatty acid biosynthesis. A study in human adipose tissue showed that ethoxzolamide, an inhibitor of carbonic anhydrase, significantly reduced the conversion of pyruvate into carbon dioxide, glyceride-glycerol, and fatty acids [33]. In the present study, CA-II was more highly expressed in the lean broilers than in the fat broilers, which is not in agreement with the human study results. This difference may be due to differences in fatty acid synthesis between mammals and birds. In mammals, adipose tissue is one of the major sites for fatty acid synthesis, whereas in birds, adipose tissue is not the major site for fatty acid synthesis, as most of the fatty acid synthesis in birds occurs in the liver.

ATP synthase is responsible for the synthesis of ATP from ADP and inorganic phosphate. Its alpha subunit is essential for its activity and mitochondrial membrane structure [34]. In the present study, ATP synthase subunit alpha was down-regulated in the abdominal adipose

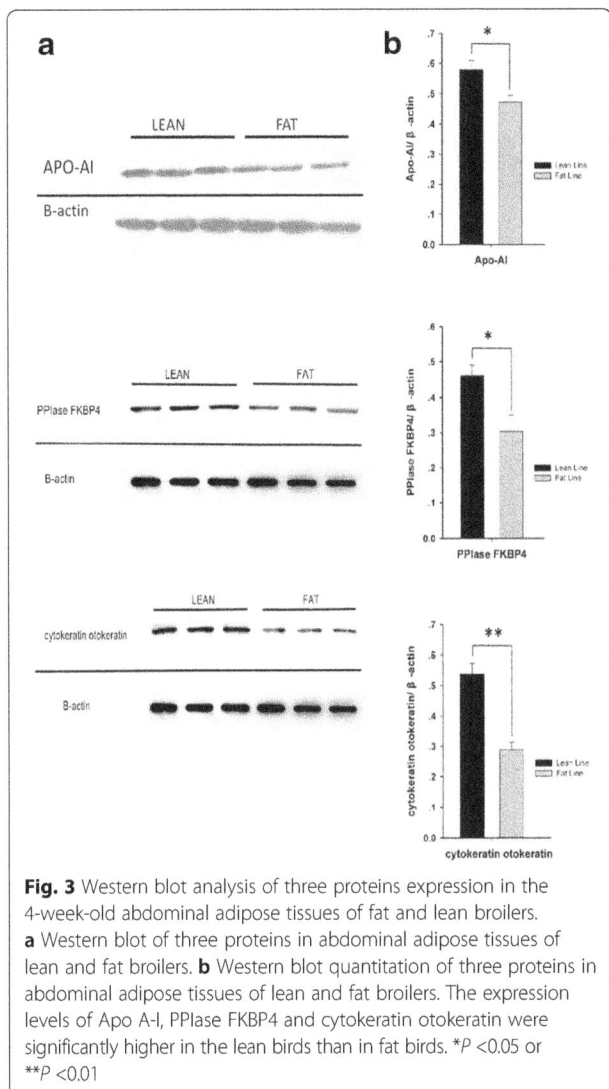

Fig. 3 Western blot analysis of three proteins expression in the 4-week-old abdominal adipose tissues of fat and lean broilers. **a** Western blot of three proteins in abdominal adipose tissues of lean and fat broilers. **b** Western blot quantitation of three proteins in abdominal adipose tissues of lean and fat broilers. The expression levels of Apo A-I, PPIase FKBP4 and cytokeratin otokeratin were significantly higher in the lean birds than in fat birds. *P <0.05 or **P <0.01

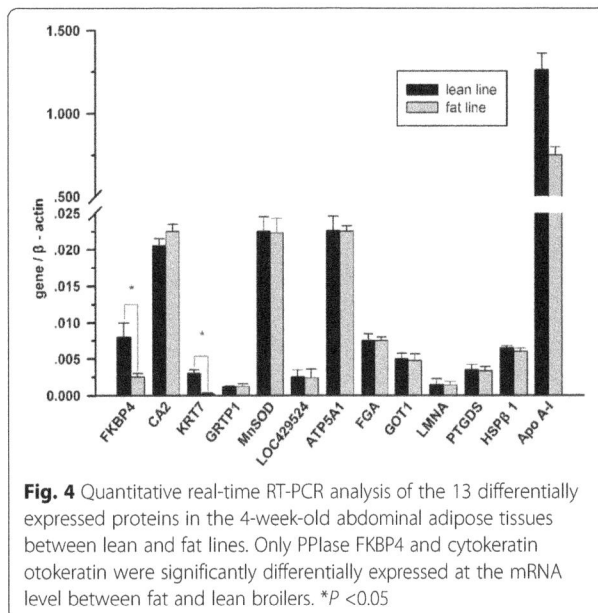

Fig. 4 Quantitative real-time RT-PCR analysis of the 13 differentially expressed proteins in the 4-week-old abdominal adipose tissues between lean and fat lines. Only PPIase FKBP4 and cytokeratin otokeratin were significantly differentially expressed at the mRNA level between fat and lean broilers. *P <0.05

tissue of fat birds compared to lean birds. This differential expression may reflect the difference in energy consumption between the fat and lean chicken lines.

Antioxidant enzymes play important roles in oxidative stress resistance. Animals have a complex network of antioxidant proteins that work together to prevent oxidative damage to cellular components such as proteins and lipids [35]. Antioxidants either prevent reactive species from being formed, or remove them before they damage cell components [36]. Adipogenesis is involved in adipocyte hypertrophy and hyperplasia. In human and mouse, adipocyte hypertrophy is correlated with increased oxidant stress and low-grade inflammation, and both are linked to disturbed cellular redox [37]. In the present study, the antioxidant proteins (HSPβ1, PPIase FKBP4 and MnSOD) were differentially expressed between the fat and lean broilers, suggesting that adipose oxidative stress is different in the two chicken lines.

Two cytoskeleton proteins, LMNA and cytokeratin otokeratin, were found to be differentially expressed in adipose tissue between the fat and lean broiler lines in this study. LMNA plays a role in maintaining nuclear stability and chromatin structure [38]. Mutations in the *LMNA* gene were associated with familial partial lipodystrophy [39]. Further studies have shown that LMNA interacts with the adipocyte differentiation factor, sterol regulatory element-binding protein 1 (SREBP-1), and that the reduced binding of LMNA to SREBP1 may be the cause of the familial partial lipodystrophy [40]. Another cytoskeleton, cytokeratin otokeratin, was first detected in the tegmentum vasculosum in chicken [41]. In the present study, we observed that cytokeratin otokeratin was up-regulated in abdominal adipose tissue of the lean birds compared to the fat birds at 4 weeks of age, consistent with our previous proteomic study of adipose tissue in the same two lines at 7 weeks of age [8]. Taken together, the differential protein expression of LMNA and cytokeratin otokeratin suggests that these two proteins are involved in chicken fat deposition.

In addition to the 11 proteins discussed above, transcription factor 24-like and growth hormone-regulated TBC protein 1 were identified by automated computational analysis. The functions of these two predicted proteins remain to be investigated.

It is noteworthy that in the present study, of these thirteen differentially expressed proteins, only two (PPIase FKBP4 and cytokeratin otokeratin) showed consistent expression results at both the mRNA and protein levels. Poor correlation between protein and mRNA levels has been reported in several genomic, transcriptomic and proteomic studies [42, 43]. There are several possible explanations for this poor correlation. One possible explanation is the complicated and varied post-transcriptional gene regulatory mechanisms, for example, microRNAs can inhibit protein synthesis either by repressing translation or by inducing mRNA degradation [44, 45]. Another possible explanation is that in vivo protein half-lives may differ substantially, and the protein half-lives can vary under different conditions [46]. A third possible explanation is our study's limitation. In the present study, due to the experimental cost, we used a small number ($n = 3$) of biological replicates of the lean and fat broiler lines. Despite the limitation, our study provides the potentially differentially expressed proteins in abdominal adipose tissue between lean and fat broilers.

Conclusions

In the study, we identified 13 proteins that were differentially expressed in abdominal adipose tissue between the fat and lean broiler lines. Of these proteins, one protein (fibrinogen alpha chain) was more highly expressed in fat broilers, while the other 12 proteins were more highly expressed in lean broilers. All or some of these differentially expressed proteins may be responsible for the phenotype difference between the fat and lean broiler lines.

Acknowledgements
The study was supported by the National 863 project of China (No. 2013AA102501), the National Basic Research Program of China (973 Program) (No.2009CB941604), the China Agriculture Research System (No.CARS-42) and the Program for Innovation Research Team at the University of Heilongjiang Province (No. 2010td02).

Funding
The study was supported by the National 863 project of China (No. 2013AA102501), the National Basic Research Program of China (973 Program) (No.2009CB941604), the China Agriculture Research System (No.CARS-42) and the Program for Innovation Research Team at the University of Heilongjiang Province (No. 2010td02). The funders had no role in the design of the study and collection, analysis, and interpretation of data and in writing the manuscript.

Authors' contributions
CYW performed Western blot experiment, participated in data analyses and wrote the manuscript. YYW and CDL carried out the 2DE experiment, participated in MALDI-TOF-MS data analyses, and modified the manuscript. YXW participated in the design of the studies. WN carried out Real-time RT-PCR experiment. HL and NW conceived the study, participated in its design, coordination and helped to draft the manuscript. All authors critically revised the manuscript for important intellectual contents and approved the final manuscript.

Competing interests
The authors declared that they have no competing interests.

Author details
[1]Key Laboratory of Chicken Genetics and Breeding of Agriculture Ministry, Key Laboratory of Animal Genetics, Breeding and Reproduction of Education Department of Heilongjiang Province, College of Animal Science and Technology, Northeast Agricultural University, Harbin 150030, Heilongjiang, China. [2]Weifang Academy of Agricultural Sciences, Weifang 261071, Shandong, China.

References
1. Baéza E, Le Bihan-Duval E. Chicken lines divergent for low or high abdominal fat deposition: a relevant model to study the regulation of energy metabolism. Animal. 2013;7:965–73.
2. Galic S, Oakhill JS, Steinberg GR. Adipose tissue as an endocrine organ. Mol Cell Endocrinol. 2010;316:129–39.
3. Picard FH, Rouvier R, Marche G, Melin JM. Étude de la composition anatomique du poulet. III. – Variabilité de la répartition des parties corporelles dans une souche de type cornish. Genet. Sel. Evol. 1969;1:1–15.
4. Wang H, Li H, Wang Q, Zhang X, Wang S, Wang Y, et al. Profiling of chicken adipose tissue gene expression by genome array. BMC Genomics. 2007;8:193.
5. Wang H, Li H, Wang Q, Wang Y, Han H, Shi H. Microarray analysis of adipose tissue gene expression profiles between two chicken breeds. J Biosci. 2006;31:565–73.
6. Larkina TA, Sazanova AL, Fomichev KA, Olu B, Sazanov AA, Malewski T, et al. Expression profiling of candidate genes for abdominal fat mass in domestic chicken Gallus gallus. Genetika. 2011;47:1140–4.
7. Ji B, Middleton JL, Ernest B, Saxton AM, Lamont SJ, Campagna SR, et al. Molecular and metabolic profiles suggest that increased lipid catabolism in adipose tissue contributes to leanness in domestic chickens. Physiol Genomics. 2014;46:315–27.
8. Wang D, Wang N, Li N, Li H. Identification of differentially expressed proteins in adipose tissue of divergently selected broilers. Poult Sci. 2009;88:2285–92.
9. Guo L, Sun B, Shang Z, Leng L, Wang Y, Wang N, et al. Comparison of adipose tissue cellularity in chicken lines divergently selected for fatness. Poult Sci. 2011;90:2024–34.
10. Berg L, Bearse G. Nutrient requirements of poultry. Poult Sci. 1962;41:1328–35.
11. Young C, Truman P. Proteins isolated with TRIzol are compatible with two-dimensional electrophoresis and mass spectrometry analyses. Anal Biochem. 2012;421:330–2.
12. Candiano G, Bruschi M, Musante L, Santucci L, Ghiggeri GM, Carnemolla B, et al. Blue silver: a very sensitive colloidal Coomassie G-250 staining for proteome analysis. Electrophoresis. 2004;25:1327–33.
13. Hindré T, Didelot S, Le Pennec J-P, Haras D, Dufour A, Vallée-Réhel K. Bacteriocin detection from whole bacteria by matrix-assisted laser desorption ionization-time of flight mass spectrometry. Appl Environ Microbiol. 2003;69:1051–8.
14. Cottrell JS, London U. Probability-based protein identification by searching sequence databases using mass spectrometry data. Electrophoresis. 1999;20:3551–67.
15. Koenig T, Menze BH, Kirchner M, Monigatti F, Parker KC, Patterson T, et al. Robust prediction of the MASCOT score for an improved quality assessment in mass spectrometric proteomics. J Proteome Res. 2008;7:3708–17.
16. Barbaras R, Puchois P, Fruchart J, Ailhaud G. Cholesterol efflux from cultured adipose cells is mediated by LpA I particles but not by LpA I: A II particles. Biochem Biophys Res Commun. 1987;142:63–9.
17. Zannis VI, Chroni A, Krieger M. Role of apoA-I, ABCA1, LCAT, and SR-BI in the biogenesis of HDL. J Mol Med (Berl). 2006;84:276–94.
18. Zannis VI, Fotakis P, Koukos G, Kardassis D, Ehnholm C, Jauhiainen M, et al. HDL biogenesis, remodeling, and catabolism. Handb Exp Pharmacol. 2015;224:53–111.
19. Rosales C, Patel N, Gillard BK, Yelamanchili D, Yang Y, Courtney HS, et al. Apolipoprotein AI deficiency inhibits serum opacity factor activity against plasma high density lipoprotein via a stabilization mechanism. Biochemistry. 2015;54:2295–302.
20. DiDonato JA, Aulak K, Huang Y, Wagner M, Gerstenecker G, Topbas C, et al. Site-specific nitration of apolipoprotein AI at tyrosine 166 is both abundant within human atherosclerotic plaque and dysfunctional. J Biol Chem. 2014;289:10276–92.

21. Kiss RS, Ryan RO, Francis GA. Functional similarities of human and chicken apolipoprotein AI: dependence on secondary and tertiary rather than primary structure. Biochim Biophys Acta. 2001;1531:251–9.

22. Wang Q, Li H, Li N, Leng L, Wang G, Ao J, et al. Polymorphisms of Apo-AI gene associated with growth and body composition traits in chicken. Acta Vet Zootech Sin. 2005;36:751–4.

23. Joo M, Sadikot RT. PGD Synthase and PGD₂ in Immune Resposne. Mediators Inflamm. 2012;2012:503128.

24. Zhu F, Wang P, Kontrogianni-Konstantopoulos A, Konstantopoulos K, Prostaglandin PG. D2 and 15-deoxy-Δ12, 14-PGJ2, but not PGE2, mediate shear-induced chondrocyte apoptosis via protein kinase a-dependent regulation of polo-like kinases. Cell Death Differ. 2010;17:1325–34.

25. Kliewer SA, Lenhard JM, Willson TM, Patel I, Morris DC, Lehmann JM. A prostaglandin J 2 metabolite binds peroxisome proliferator-activated receptor γ and promotes adipocyte differentiation. Cell. 1995;83:813–9.

26. Plee-Gautier E, Aggerbeck M, Beurton F, Antoine Bnd, Grimal H, Barouki R, et al. Identification of an adipocyte-specific negative glucose response region in the cytosolic aspartate aminotransferase gene. Endocrinology. 1998;139:4936–44.

27. Tordjman J, Leroyer S, Chauvet G, Quette J, Chauvet C, Tomkiewicz C, et al. Cytosolic aspartate aminotransferase, a new partner in adipocyte glyceroneogenesis and an atypical target of thiazolidinedione. J Biol Chem. 2007;282:23591–602.

28. Grieninger G. Contribution of the αEC Domain to the Structure and Function of Fibrinogen-420. Ann N Y Acad Sci. 2001;936:44–64.

29. Kim OY, Shin M-J, Moon J, Chung JH. Plasma ceruloplasmin as a biomarker for obesity: a proteomic approach. Clin Biochem. 2011;44:351–6.

30. Lei H, Xu J, Cheng LJ, Guo Q, Deng AM, Li YS. An increase in the cerebral infarction area during fatigue is mediated by il-6 through an induction of fibrinogen synthesis. Clinics (Sao Paulo). 2014;69:426–32.

31. Zhou B, Pan Y, Hu Z, Wang X, Han J, Zhou Q, et al. All-trans-retinoic acid ameliorated high fat diet-induced atherosclerosis in rabbits by inhibiting platelet activation and inflammation. J Biomed Biotechnol. 2012;2012:259693.

32. Supuran CT. Carbonic anhydrases-an overview. Curr Pharm Des. 2008;14:603–14.

33. Mekary RA, Giovannucci E, Willett WC, van Dam RM, Hu FB. Eating patterns and type 2 diabetes risk in men: breakfast omission, eating frequency, and snacking. Am J Clin Nutr. 2012;95:1182–9.

34. Baker LA, Watt IN, Runswick MJ, Walker JE, Rubinstein JL. Arrangement of subunits in intact mammalian mitochondrial ATP synthase determined by cryo-EM. Proc Natl Acad Sci. 2012;109:11675–80.

35. Vertuani S, Angusti A, Manfredini S. The antioxidants and pro-antioxidants network: an overview. Curr Pharm Des. 2004;10:1677–94.

36. Rochette L, Lorin J, Zeller M, Guilland JC, Lorgis L, Cottin Y, et al. Nitric oxide synthase inhibition and oxidative stress in cardiovascular diseases: possible therapeutic targets? Pharmacol Ther. 2013;140:239–57.

37. Guo W, Li Y, Liang W, Wong S, Apovian C, Kirkland JL, et al. Beta-mecaptoethanol suppresses inflammation and induces adipogenic differentiation in 3T3-F442A murine preadipocytes. PLoS One. 2012;7, e40958.

38. Gonzalez-Suarez I, Gonzalo S. Nurturing the genome: A-type lamins preserve genomic stability. Nucleus. 2010;1:129–35.

39. Shackleton S, Lloyd DJ, Jackson SN, Evans R, Niermeijer MF, Singh BM, et al. LMNA, encoding lamin A/C, is mutated in partial lipodystrophy. Nat Genet. 2000;24:153–6.

40. DubandGoulet I, Woerner S, Gasparini S, Attanda W, Kondé E, TellierLebègue C, et al. Subcellular localization of SREBP1 depends on its interaction with the C-terminal region of wild-type and disease related A-type lamins. Exp Cell Res. 2011;317:2800–13.

41. Heller S, Sheane CA, Javed Z, Hudspeth A. Molecular markers for cell types of the inner ear and candidate genes for hearing disorders. Proc Natl Acad Sci. 1998;95:11400–5.

42. Tian Q, Stepaniants SB, Mao M, Weng L, Feetham MC, Doyle MJ, et al. Integrated genomic and proteomic analyses of gene expression in mammalian cells. Mol Cell Proteomics. 2004;3:960–9.

43. Schwanhäusser B, Busse D, Li N, Dittmar G, Schuchhardt J, Wolf J, et al. Global quantification of mammalian gene expression control. Nature. 2011;473:337–42.

44. Eulalio A, Huntzinger E, Izaurralde E. Getting to the root of miRNA-mediated gene silencing. Cell. 2008;132:9–14.

45. Filipowicz W, Bhattacharyya SN, Sonenberg N. Mechanisms of post-transcriptional regulation by microRNAs: are the answers in sight? Nat Rev Genet. 2008;9:102–14.

46. Glickman MH, Ciechanover A. The ubiquitin-proteasome proteolytic pathway: destruction for the sake of construction. Physiol Rev. 2002;82:373–428.

5

Comparative proteomic analyses of Asian cotton ovules with attached fibers in the early stages of fiber elongation process

Bing Zhang🆔, Shao-Jun Du, Jue Hu, Di Miao and Jin-Yuan Liu*

Abstract

Background: Plenty of proteomic studies were performed to characterize the allotetraploid upland cotton fiber elongation process, whereas little is known about the elongating diploid cotton fiber proteome.

Methods: In this study, we used a two-dimensional electrophoresis-based comparative proteomic approach to profile dynamic proteomes of diploid Asian cotton ovules with attached fibers in the early stages of fiber elongation process. One-way ANOVA and Student-Newman-Keuls test were used to find the differentially displayed protein (DDP) spots.

Results: A total of 55 protein spots were found having different abundance ranging from 1 to 9 days post-anthesis (DPA) in a two-day interval. These 55 DDP spots were all successfully identified using high-resolution mass spectrometric analyses. Gene ontology analyses revealed that proteoforms involved in energy/carbohydrate metabolism, redox homeostasis, and protein metabolism are the most abundant. In addition, orthologues of the 13 DDP spots were also found in differential proteome of allotetraploid elongating cotton fibers, suggesting their possible essential roles in fiber elongation process.

Conclusions: Our results not only revealed the dynamic proteome change of diploid Asian cotton fiber and ovule during early stages of fiber elongation process but also provided valuable resource for future studies on the molecular mechanism how the polyploidization improves the trait of fiber length.

Keywords: Asian cotton, Ovule, Fiber elongation, Differentially displayed protein spots, Polyploidization

Abbreviations: 2-DE, Two-dimensional electrophoresis; ChSOD, Chloroplast Cu/Zn superoxide dismutase; DDP, Differentially displayed protein; DPA, Days post-anthesis; EST, Expressed sequence tag; HSP70, Heat shock protein 70; MALDI TOF, Matrix-assisted laser desorption/ionization time of flight; PMF, Peptide mass fingerprinting

Background

As a major source of natural fiber in the world, cotton has been widely used for clothing, papermaking, and other purposes for thousands of years [1]. Apart from its economic value, cotton is also known as an excellent model system for studying cell differentiation, cell elongation and cellulose biosynthesis [2]. All cotton plants belong to the genus *Gossypium* in the family Malvaceae. Among the around 50 known *Gossypium* species, only four cultivated species, *G. arboreum*, *G. herbaceum*, *G. hirsutum*, and *G. barbadense*, can produce spinnable fibers. The first two cultivated species are diploids (AA) and the last two are allotetraploids (AADD) originating by a polyploidization event involving *G. herbaceum* and a diploid hairiness species resembling *G. raimondii* (DD) [3]. Interestingly, the fiber length of the two allotetraploids species are longer than the two diploids species. Considering that the progenitor DD genome species are fiber deficient, the long-fiber phenotype of allotetraploid cotton species must be formed in the polyploidization process through some unknown complex mechanism [4–6]. Elucidation of the molecular mechanism of fiber elongation in the diploid cotton and further comparison of the differences of fiber elongation between the diploid

---

* Correspondence: liujy@mail.tsinghua.edu.cn
Laboratory of Plant Molecular Biology, Center for Plant Biology, School of Life Sciences, Tsinghua University, Beijing 100084, People's Republic of China

and allotetraploid cotton could help us understand how the polyploidization improves the traits of fiber length [7].

With the availability of high-quality EST sequences in public databases, high-throughput proteomic analyses of cotton fiber elongation process were successfully performed in advance of the cotton genome sequencing project. For example, a proteomic study of the *ligon* lintless mutant and wild-type upland cotton fibers identified 81 differentially displayed protein (DDP) spots at 14 days post-anthesis (DPA), suggesting that proteins involved in protein folding and stabilization are important for fiber elongation [8]. Another study that compared the proteomic profiles of wild-type and fuzzless-lintless mutant upland cotton (*G. hirsutum* cultivar Xu142) fibers at 10 DPA identified 104 DDP spots, providing evidence that pectin synthesis is imperative for fiber elongation [9]. In another 2-DE-based comparative proteomic analysis, a total of 235 protein spots were found having different abundance during the entire elongation process at five distinct time points: 5, 10, 15, 20 and 25 DPA of upland cotton (*G. hirsutum* cultivar CRI 35) [10]. Further Kyoto Encyclopedia of Genes and Genomes (KEGG) Orthology-based Annotation System (KOBAS) analyses based on the identified 235 DDPs indicated that glycolysis is the most significantly regulated biochemical pathway during the fiber elongation process [11]. With the availability of cotton genome sequences, post-genomic proteomics studies further shed many new insights into cotton fiber initiation and elongation processes [12–16]. Hu et al. successfully compared the total proteome of two allotetraploid cotton species (*G. hirsutum* and *G. barbadense*) and their diploid parents at 10 and 20 DPA, providing the first evidence that the two allopolyploid species have achieved superficially similar modern fiber phenotypes through different evolutionary routes [17]. However, the mechanisms of which proteins are essential for fiber elongation and how polyploidization increases the fiber length of the allotetraploid upland cotton are poorly understood. Furthermore, because most of these proteomic studies focused on the midst and late stage of fiber elongation process, the early events occur in the fiber elongation process are still poorly characterized at the proteome level.

In the present study, we reported the first comparative proteomic analyses of fiber and ovule proteome of diploid Asian cotton (*G. arboreum* cultivar DPL971) in the early stages of elongation process (1-9 DPA) and the identification of 55 DDPs of developing diploid cotton ovules with attached fibers. Through comparing the dynamic proteome of Asian cotton ovules and attached fibers with differential proteome of upland cotton fibers, we clarified 13 possible essential proteins required for fiber elongation and four important proteins whose increased abundance are correlated with improved fiber length.

Results and discussions

Dynamic proteomes of diploid Asian cotton ovules and fibers in the early stages of fiber elongation process

Fiber length measurement indicated that the diploid Asian cotton (*G. arboreum* cultivar DPL971) shows a gradual increase of fiber length during the early stages of fiber elongation process as the allotetraploid upland cotton (*G. hirsutum* cultivar CRI35) does (Additional file 1: Figure S1). The fastest fiber elongation rate was observed at 5 DPA in both two cultivars. Since then, the elongation rate gradually slowed down. Interestingly, the fiber length of cultivar DPL971 is always shorter than that of cultivar CRI35 (Additional file 1: Figure S1), supporting the notion that polyploidization improves the fiber length trait.

To characterize the dynamic proteome change of cotton ovules and fibers in the early stages of fiber elongation process, total protein extracted from ovules with attached fibers of Asian cotton cultivar DPL971 at five stages (1, 3, 5, 7 and 9 DPA) were separated by 2-DE (Fig. 1a). Approximately 1800 stained protein spots could be reproducibly detected on each 2-D gel, and most of these protein spots showed no significant variance between any two stages. Only 55 protein spots were statistically analyzed being dynamic with their abundance has a greater than two-fold variation (*p*-value < 0.05) during the fiber elongation process (protein abundance at 3-9 DPA in comparison

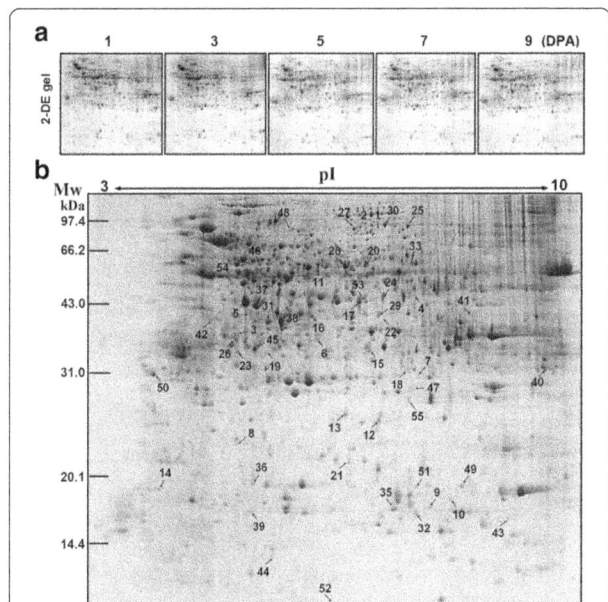

Fig. 1 2-DE map of the 55 identified differentially displayed protein spots. **a** Representative 2-DE gel of total proteins extracted from diploid Asian cotton ovules with attached fibers at five time points. **b** 2-DE map of total proteins extracted from 5 DPA diploid Asian cotton ovules with attached fibers (IPG strip pH 3–10 NL, 12.5 % SDS-PAGE, blue silver staining). 55 differentially displayed protein (DDP) spots are all indicated on the 2-DE map

with that at 1 DPA) using one-way ANOVA and Student-Newman-Keuls tests (Fig. 1b and Additional file 2: Table S1). The 55 distinct spots were manually excised from the 2-DE gels and digested with trypsin. Finally, 55 spots selected for mass spectrometry analysis were all successfully identified and represented 55 distinct proteoforms [18] and 52 unique proteins (Additional file 3: Table S2 and Additional file 4).

The 55 DDP spots were clustered to two types according to their expression patterns (Fig. 2). The type A class contains 16 DDP spots, all of which have a higher protein abundance at 1 and 3 DPA, whereas the remaining 39 DDP spots were all grouped in the type B class having a higher protein abundance between 5 and 9 DPA. Two proteins, chloroplast Cu/Zn superoxide dismutase and heat shock protein 70, belongs to type A and type B

class, respectively. Western blot analysis indicated that expression of chloroplast Cu/Zn superoxide dismutase is gradually decreasing during 1–9 DPA, whereas protein abundance of the two isoforms of heat shock protein 70 are both gradually increasing during the same period (Fig. 3a). These results are in full agreement with the protein abundance variance revealed by 2-DE and mass spectrometry identification (Fig. 3b), confirming the correctness of our analyses. Interesting, initiation of cotton fiber from the ovule epidermis occurs from -3 DPA to 3 DPA, whereas the fast fiber elongation process generally starts at 5 DPA and ends at 25 DPA [2–4]. Thus, the two types of DDP spots correspond exactly to proteins preferentially expressed in the fiber initiation and elongation process, respectively.

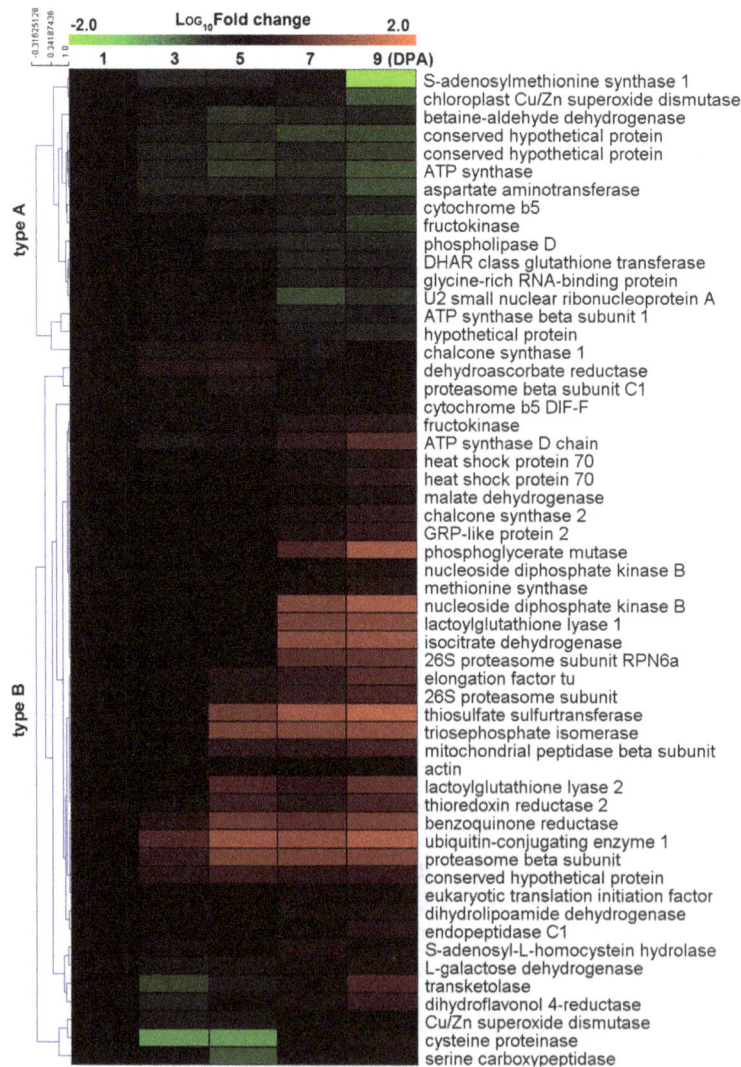

Fig. 2 Protein abundance variance analysis of the 55 differentially displayed protein spots. Complete linkage hierarchical cluster analysis of the 55 DDPs at five time points was performed by comparing the average fold change of the DDPs. Protein names are indicated at the right. Type A and B stands for the proteins preferentially expressed in initiation and elongation process, respectively

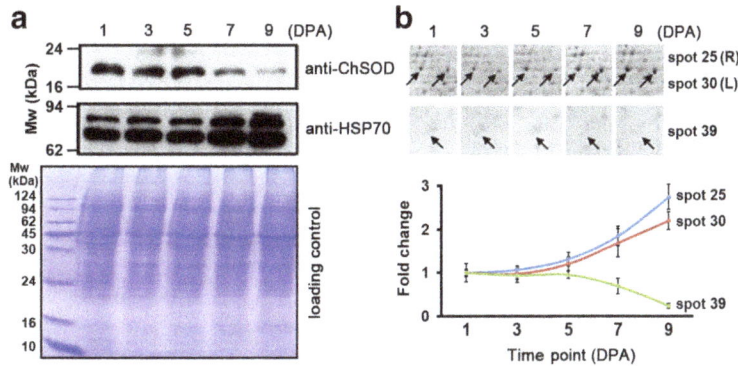

Fig. 3 Comparison of expression profiles of two selected proteins. **a** Western blot assay was performed using peptide-specific antibodies (detailed in Materials and Methods) against chloroplast Cu/Zn superoxide dismutase (anti-ChSOD) and heat shock protein 70 (anti-HSP70). The experiments were repeated for three times using different ovule attached with fiber samples with similar results. The 2-DE image and fold change of chloroplast Cu/Zn superoxide dismutase (spot No.39) and heat shock protein 70 (spot No.25 and 30) were shown in (**b**) based on their Vol% values

Functional analyses of the differentially displayed proteins of diploid Asian cotton ovules and fibers in the early stages of fiber elongation process

AgriGO analysis indicated that the 55 DDPs could be divided into nine functional groups (Fig. 4a). Of these groups, energy/carbohydrate metabolism, including both glycolysis and the TCA cycle (17 DDPs), redox homeostasis (11 DDPs), and protein degradation (9 DDPs) contains the largest number of DDPs. This result is understandable because energy supplement is prerequisite for rapid fiber cell elongation [2, 4], whereas fast cell growth, such as the cotton fiber elongation process, needs an intracellular oxidation-reduction equilibrium [19] and a fast protein turnover rate [2, 8]. In addition, protein synthesis, amino acid and flavonoid metabolism also contained more than three DDPs, suggesting their possible important roles in cotton fiber elongation process [12, 14]. Moreover, different DDPs of each functional groups have unique protein abundance variance (Fig. 2), suggesting their expression were specifically regulated in developing Asian cotton ovules and fibers.

Many of the 55 DDPs were known important proteins required for proper fiber initiation and elongation. For example, DDP No. 31 and 39 were identified as two different Cu/Zn superoxide dismutase, which was previously characterized to be important for cotton fiber development [20]. Similarly, DDP No. 14, 40 and 54 were all identified as ATP synthase subunits, which was previously proved playing vital roles in cotton fiber elongation process [21]. The activity of malate dehydrogenase (DDP No.34) is variant among cotton cultivars with differing fiber traits [22], whereas chalcone synthase (DDP No.24 and 53) and dihydroflavonol 4-reductase (DDP No.16), two important enzymes involved in flavonoid metabolism, were reported being related to cotton fiber quality [23, 24].

Furthermore, although the function of some DDPs are unclear in cotton, investigation of the gene homologs in model plant Arabidopsis also implied their important functions in ovule and fiber development. For example, DDP No. 06 was identified as a thiosulfate sulfurtransferase, which plays important roles in embryo and seed development in *Arabidopsis thaliana* [25]. DDP No. 33 was identified as a dihydrolipoamide dehydrogenase, the E3 subunit of pyruvate dehydrogenase complex. In *Arabidopsis thaliana*, mutations of plastid pyruvate dehydrogenase complex will lead to an early embryo lethal phenotype, suggesting the important function of pyruvate dehydrogenase complex in embryo development [26]. During fiber development process, the fiber cells are always attached with the ovules which provide the fiber cells essential water, carbon source and mineral nutrients [27]. Identification of the important proteins required for embryo/ovule development in this study could give us the clue how the cotton plants delicately regulate the complex sink and source relationship of ovule and fiber to promote the fast elongation of fiber cells. This important information couldn't be obtained through studies only focusing on fibers.

Comparative analyses of the differentially displayed proteins of diploid Asian cotton ovules and fibers with allotetraploid upland cotton fibers

Blast search indicated that 13 of the 55 DDPs of diploid Asian cotton ovules with attached fibers in early stages of fiber elongation process were also identified in the comparative proteomic analyses of the elongating allotetraploid upland cotton fiber cells [10, 11], suggesting these 13 DDPs are required for fiber elongation (Fig. 4b). Moreover, nine of the 13 DDPs were found having more copies in allotetraploid upland cotton fiber cells than in diploid Asian cotton fiber/ovule cells (Fig. 4c). Notably, the different copies of the selected DDPs in allotetraploid upland cotton all have the same protein sequence, suggesting these different proteoforms are differentially post-translationally modified. For example, four DDPs of

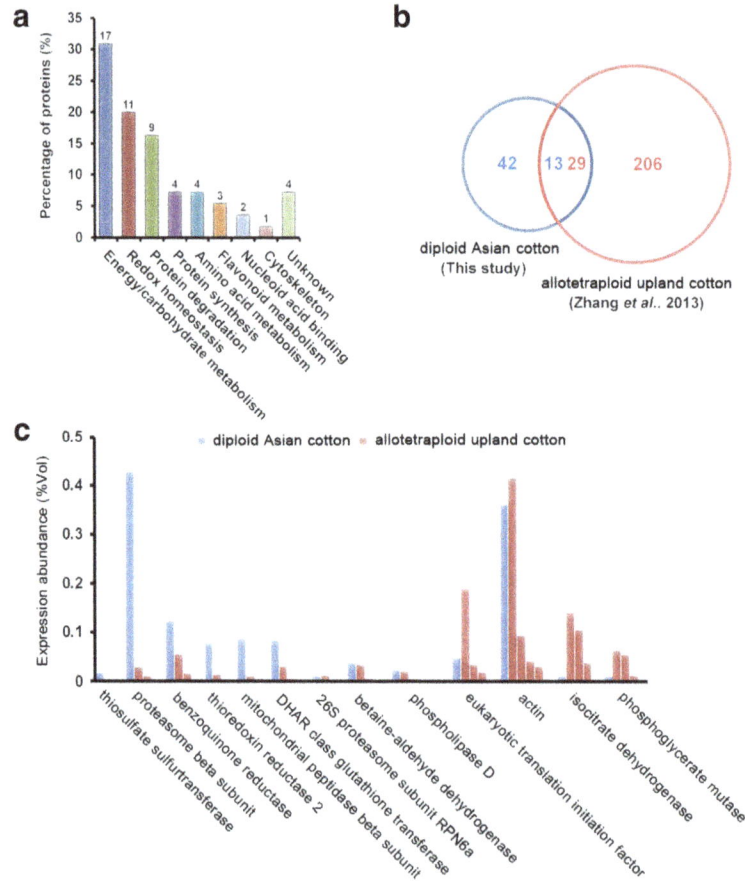

Fig. 4 Functional analysis of the 55 differentially displayed protein spots. **a** Functional classification of the 55 DDP spots. The number of identities in each group are shown on the top. **b** Venn diagram analysis of the 55 DDP spots in this study and 235 DDP spots of elongating allotetraploid cotton fibers indicated that 13 DDPs in diploid Asian cotton have different number of orthologues (total number: 29) in allotetraploid upland cotton. **c** Comparison of the protein abundance at 5 DPA between the 13 DDPs in diploid Asian cotton and 29 homologous DDPs in allotetraploid upland cotton. The nine DDPs having more than one homologous DDPs were compared with all the DDPs of allotetraploid upland cotton in the same group

phospholipase D alpha protein were identified in allotetraploid fiber with the same NCBI accession number whereas only one DDP was identified in diploid fiber/ovule.

Comparison of the protein abundance between the 13 DDPs in diploid Asian cotton and the homologous 29 DDPs in allotetraploid upland cotton at 5 DPA, the same time point of fiber elongation process, further revealed that protein abundance of four proteins, including translation initiation factor, actin, isocitrate dehydrogenase and phosphoglycerate mutase, were all increased in allotetraploid upland cotton (Fig. 4c). Moreover, DDP number of the four proteins were also increased in allotetraploid upland cotton. These results strongly suggested that activity of the four proteins might be selectively up-regulated in the polyploidization process, implying the four proteins are important for improving the fiber length trait. In agreement with this suspicion, fiber length of the transgenic cottons expressing the

actin gene is significantly longer than that of wild-type cotton plants [28].

Conclusions

In summary, we reported the first comparative proteomics study of diploid Asian cotton ovules with attached fibers in the early stages of fiber elongation process. Combined with the proteome dataset of elongating allotetraploid upland cotton fibers, our study provides a reference list of essential proteins supporting the fast cotton fiber elongation process. Information of these essential proteins including protein expression level and MS identification data provide a valuable resource for future functional studies.

Methods
Plant materials
Asian cotton (*G. arboreum* cultivar DPL971) and upland cotton (*G. hirsutum* cultivar CRI35) was grown in a

standard agronomic field during the period from April to September in Beijing. The ages of the ovules with attached fibers selected for total protein extraction and fiber length measurement were 1, 3, 5, 7 and 9 DPA. All of the collected samples were frozen in liquid nitrogen and then stored at −80 °C for protein extraction.

Measurements of fiber length

Ovules were detached from the fresh cotton bolls and boiled at 95 °C water bath for 10 min. Aggregated fibers were combed in water, observed and photographed under a SZX12 anatomy microscope (Olympus, Japan) equipped with a DP70 digital camera system. Five ovules were used to represent each developmental stage whereas length of 10 fibers were recorded for each ovule.

Protein extraction

Total protein was extracted using modified Tris-phenol method as described [29]. About 1 g of frozen cotton ovules with attached fibers was ground with 10 % PVPP (w/w) and 10 % quartz sand (w/w) in liquid nitrogen using a mortar and pestle. The powder was suspended completely in twenty milliliters of ice-cold acetone (adding 2 % β-mercaptoethanol) and centrifuged at 12,000 g for 15 min at 4 °C to wash away impurities, and this step was repeated twice. The freeze-dried powder was homogenized in 5 mL of extraction buffer containing 50 mM Tris–HCl, pH 8.6, 2 % SDS, 2 % (w/w) β-mercaptoethanol, 1 mM PMSF, and then an equal volume of Tris saturated phenol (pH 8.0) was added. The mixture was vortexed thoroughly for 5 min, and the phenol phase was collected and precipitated with 5 volume of 0.1 M ammonium acetate in methanol at −20 °C for 30 min. After centrifuging at 12, 000 g for 15 min, the collected protein pellets were washed three times with cold 0.1 M ammonium acetate in methanol, and then washed three times with cold 80 % acetone in water. The lyophilized pellets were dissolved in rehydration buffer (7 M urea, 2 M thiourea, 4 % CHAPS, 1 % IPG buffer, 20 mM DTT) and centrifuged at 12 000 g for 15 min to remove insoluble materials. The concentration of extracted proteins was quantified using Bradford method with biotechnology grade BSA protein as a quantification standard [30]. The proteins underwent 2-DE immediately or were stored at −80 °C.

2-DE and image analyses

2-DE was performed according to the manufacturer's instruction (2-DE Manual, GE Healthcare). 1 mg protein mixed with rehydration buffer (7 M urea, 2 M thiourea, 4 % CHAPS, 1 % IPG buffer, 20 mM DTT) in a total volume of 1 mL, was loaded onto a nonlinear IPG Drystrip (pH 3-10, 24 cm, GE Healthcare, Piscataway, USA). The strips were hydrated in the rehydration buffer for 18 h at room temperature. Then isoelectric focusing was performed on an Ettan IPGphor isoelectric focusing system (GE Healthcare, Uppsala, Sweden) under the following conditions: 100 V for 40 min, 500 V for 40 min, 1000 V for 1 h, 4000 V for 2 h, and 8000 V for 8 h until total voltage hours of 75,000 was achieved. Before SDS-PAGE analysis, strips were incubated for 2 × 15 min in equilibration buffer (6 M urea, 50 mM pH 8.8 Tris–HCl, 30 % (v/v) glycerol, 2 % (w/v) SDS, a trace of bromophenol blue). One percent DTT (w/v) was added to the above for the first 15 min and 2.5 % iodoacetamide (w/v) instead for the second 15 min. After equilibration, strips were placed on top of a vertical 12.5 % SDS-polyacrylamide self-cast gel and electrophoresis was performed at 4 °C and 5 W/ gel for 45 min, and then 17 W/gel for 5 h until the dye front reached the bottom of gels. For calculation of molecular weight (MW) of the 2-DE protein spots, a filter paper piece pre-loaded with a protein marker (14.4– 97.4 kDa) were placed along with the equilibrated strip on top of the gels. After electrophoresis, 2-D gels were stained by Colloidal Coomassie Blue G-250 [31]. The 2-D gels were scanned at 600 dpi resolution using a UMAX PowerLook 2100XL scanner (Willich, Germany) with following parameters: scan mode, transparent; color, grey; calibration, auto calibration. Image analysis was performed with ImageMaster Platinum software (version 6.0) (GE Healthcare). Proteins extracted from three different samples of each time point were analyzed by 2-DE and triplicates were applied to each protein sample, thus a total of 45 CBB-stained 2-D gel images were obtained. The spots were quantified using the % volume criterion. The match analysis was done in automatic mode using the following detection parameters: Smooth 4, Saliency 300, and Min Area 29, and further manual editing was performed to correct the mismatched and unmatched spots. The relative volume of each spot was assumed to represent its expression abundance. A significant difference was defined by the criterion p-value < 0.05 when analyzing parallel spots between groups with one-way ANOVA and Student-Newman-Keuls test using the SPSS 16.0 statistical software (IBM).

Protein identification by MALDI-TOF/TOF

Protein spots were manually excised from the 2-D gels and digested with trypsin. Briefly, the excised gel pieces were destained with 100 mL of 25 mM NH_4HCO_3 in 50 % ACN until the stain faded sufficiently. After that, the gel pieces were washed twice in 100 % ACN for 10 min and then dried under vacuum for 15 min. Proteins in the gel pieces were digested in 25 mM NH_4HCO_3, 10 ng/mL trypsin overnight at 37C. The digestion solutions were used in the subsequent MS analysis. The 4800 MALDI TOF/TOF™ Analyzer (Applied Biosystems, Framingham, USA) was used for protein identification. The MS spectra were acquired in the positive ion reflector mode, with a mass range from 800 to 4000 Da. The peaks with S/N > 20 were selected for PMF analysis. The MS/MS analysis was performed with the 10 strongest peaks of the MS spectra, and the precursor ions were accelerated at a voltage of 8 kV. The MS/MS spectra were accumulated for at least 2000 laser shots. All of the MS and MS/MS data were analyzed in the combined mode using Global Proteome Server Explorer (software version 3.5, Applied Biosystems) to interface with the Mascot 2.2 search engine (Matrix Science) against a *G. arboreum* peptide sequence database downloaded from Cotton Genome Project website (http://cgp.genomics.org.cn, including 41,331 protein sequences, 14,876,209 residues) [32]. Searching parameters were set as follows: $S/N \geq 3.0$; fixed modification, carbamidomethyl (Cys); variable modification, oxidation (Met); maximum number of missing cleavages, 1; MS tolerance, ± 0.1 Da; MS/MS tolerance, ± 0.5 Da. The identities with the highest score were subsequently analyzed using blast tools against the Uniprot database (http://www.uniprot.org/) to obtain the gene ontology (GO) annotation.

Antibody preparation and western blot

The specific antibodies against selected proteins, chloroplast Cu/Zn superoxide dismutase (ChSOD) and heat shock protein 70 (HSP70), were prepared by the immunisation of rabbits with synthesised protein-specific peptides (MAAHIFTTTPSHL for ChSOD and RSFRRRDIAASPL for HSP70) mixed with Freund adjuvant by Beijing Protein Innovation (Beijing, China). Affinity purifications were further performed to enrich antibodies to ensure the titers against synthesised protein-specific peptides larger than 1: 6400 by ELISA. Goat anti-rabbit antibodies that were conjugated to horseradish peroxidase were used as secondary antibodies. For western blotting, 20 μg of proteins for each sample were denatured in 6 × SDS-PAGE sampling buffer by boiling in a water bath for 5 min, then separated by 12 % SDS-PAGE along with a pre-stained protein marker (10–124 kDa) and transferred to a PVDF membrane. The signals were revealed with Lumi-Light western blotting substrate (Roche, Mannheim, Germany) and captured on autoradiographic films (Kodak, New York, USA). Films were scanned using UMAX PowerLook 2100XL scanner and densitometric analysis of the bands was performed using ImageMaster Platinum software. The experiments were repeated for three times using different ovule attached with fiber samples.

Bioinformatics analyses of the differentially displayed protein spots

A complete linkage hierarchical cluster analysis of the differentially displayed protein (DDP) spots was performed through a comparison of the average fold change

value of each spots (comparison to 1 DPA) using MeV 4.7 software package. Functional classification of the DDP spots were performed using agriGO toolkits (http://bioinfo.cau.edu.cn/agriGO/) in hypergeometric testing mode. Amino acid sequences of the DDP spots were BLAST searched against with known DDPs of allotetraploid elongating cotton fibers to retrieve the most homologous allotetraploid cotton DDPs with an E-value = 0.

Additional files

Additional file 1: Figure S1. Fiber length of two cotton species in the early stages of fiber elongation process. (TIF 44 kb)

Additional file 2: Table S1. Relative abundance of the 55 DDP spots. (XLSX 19 kb)

Additional file 3: Table S2. Information of the identified 55 DDP spots. (XLSX 16 kb)

Additional file 4: Annotated PMF spectra for 55 DDP spots. (PPT 728 kb)

Acknowledgements
The authors want to thank the five anonymous reviewers for providing constructive comments on the manuscript.

Funding
This work was supported by grants from the State Key Basic Research and Development Plan (2010CB126003), the National Natural Science Foundation of China (90608016), the National Transgenic Animals and Plants Research Project (2011ZX08009-003) and State Key Laboratory of Cotton Biology Open Fund (CB2015A01). The funding bodies do not have a role in the design of the study and collection, analysis, and interpretation of data and in writing the manuscript.

Authors' contributions
JL designed the study and helped to draft the manuscript; BZ conceived the study, participated in its design and drafted the manuscript; SD and JH performed 2-DE; DM carried out MS analysis and performed protein identification. All authors read and approved the final manuscript.

Competing interests
The authors declare that they have no competing interests.

References
1. Mansoor S, Paterson AH. Genomes for jeans: cotton genomics for engineering superior fiber. Trends Biotechnol. 2012;30:521–7.
2. Kim HJ, Triplett BA. Cotton fiber growth *in planta* and *in vitro*. Models for plant cell elongation and cell wall biogenesis. Plant Physiol. 2001;127:1361–6.
3. Qin YM, Zhu YX. How cotton fibers elongate: a tale of linear cell-growth mode. Curr Opin Plant Biol. 2011;14:106–11.
4. Arpat AB, Waugh M, Sullivan JP, Gonzales M, Frisch D, Main D, et al. Functional genomics of cell elongation in developing cotton fibers. Plant Mol Biol. 2004;54:911–29.
5. Hu G, Koh J, Yoo MJ, Grupp K, Chen S, Wendel JF. Proteomic profiling of developing cotton fibers from wild and domesticated *Gossypium barbadense*. New Phytol. 2013;200:570–82.
6. Hu G, Koh J, Yoo MJ, Pathak D, Chen S, Wendel JF. Proteomics profiling of fiber development and domestication in upland cotton (*Gossypium hirsutum* L.). Planta. 2014;240:1237–51.
7. Xu Z, Yu JZ, Cho J, Yu J, Kohel RJ, Percy RG. Polyploidization altered gene functions in cotton (*Gossypium* spp.). PLoS One. 2010;5:e14351.
8. Zhao PM, Wang LL, Han LB, Wang J, Yao Y, Wang HY, et al. Proteomic identification of differentially expressed proteins in the ligon lintless mutant of upland cotton (*Gossypium hirsutum* L.). J Proteome Res. 2010;9:1076–87.

9. Pang CY, Wang H, Pang Y, Xu C, Jiao Y, Qin YM, et al. Comparative proteomics indicates that biosynthesis of pectic precursors is important for cotton fiber and Arabidopsis root hair elongation. Mol Cell Proteomics. 2010;9:2019–33.

10. Yang YW, Bian SM, Yao Y, Liu JY. Comparative proteomic analysis provides new insights into the fiber elongating process in cotton. J Proteome Res. 2008;7:4623–37.

11. Zhang B, Yang YW, Zhang Y, Liu JY. A high-confidence reference dataset of differentially expressed proteins in elongating cotton fiber cells. Proteomics. 2013;13:1159–63.

12. Liu K, Han M, Zhang C, Yao L, Sun J, Zhang T. Comparative proteomic analysis reveals the mechanisms governing cotton fiber differentiation and initiation. J Proteomics. 2012;75:845–56.

13. Kumar S, Kumar K, Pandey P, Rajamani V, Padmalatha KV, Dhandapani G, et al. Glycoproteome of elongating cotton fiber cells. Mol Cell Proteomics. 2013;12:3677–89.

14. Wang XC, Li Q, Jin X, Xiao GH, Liu GJ, Liu NJ, et al. Quantitative proteomics and transcriptomics reveal key metabolic processes associated with cotton fiber initiation. J Proteomics. 2015;114:16–27.

15. Mujahid H, Pendarvis K, Reddy JS, Nallamilli BRR, Reddy KR, Nanduri B, et al. Comparative proteomic analysis of cotton fiber development and protein extraction method comparison in late stage fibers. Proteomes. 2016;4:7.

16. Jin X, Wang L, He L, Feng W, Wang X. Two-dimensional gel electrophoresis-based analysis provides global insights into the cotton ovule and fiber proteomes. Sci China Life Sci. 2016;59:154–63.

17. Hu G, Koh J, Yoo MJ, Chen S, Wendel JF. Gene-expression novelty in allopolyploid cotton: a proteomic perspective. Genetics. 2015;200:91–104.

18. Smith LM, Kelleher NL. Consortium for top down proteomics. Proteoform: a single term describing protein complexity. Nat Methods. 2013;10:186–7.

19. Li HB, Qin YM, Pang Y, Song WQ, Mei WQ, Zhu YX. A cotton ascorbate peroxidase is involved in hydrogen peroxide homeostasis during fibre cell development. New Phytol. 2007;175:462–71.

20. Kim HJ, Kato N, Kim S, Triplett B. Cu/Zn superoxide dismutases in developing cotton fibers: evidence for an extracellular form. Planta. 2008;228:281–92.

21. Pang Y, Wang H, Song WQ, Zhu YX. The cotton ATP synthase δ1 subunit is required to maintain a higher ATP/ADP ratio that facilitates rapid fibre cell elongation. Plant Biol. 2010;12:903–9.

22. Kloth RH. Variability of malate dehydrogenase among cotton cultivars with differing fiber traits. Crop Sci. 1992;32:617–21.

23. Al-Ghazi Y, Bourot S, Arioli T, Dennis ES, Llewellyn DJ. Transcript profiling during fiber development identifies pathways in secondary metabolism and cell wall structure that may contribute to cotton fiber quality. Plant Cell Physiol. 2009;50:1364–81.

24. Xiao YH, Zhang ZS, Yin MH, Luo M, Li XB, Hou L, et al. Cotton flavonoid structural genes related to the pigmentation in brown fibers. Biochem Biophys Res Commun. 2007;358:73–8.

25. Mao G, Wang R, Guan Y, Liu Y, Zhang S. Sulfurtransferases 1 and 2 play essential roles in embryo and seed development in Arabidopsis thaliana. J Biol Chem. 2011;286:7548–57.

26. Lin M, Behal R, Oliver DJ. Disruption of plE2, the gene for the E2 subunit of the plastid pyruvate dehydrogenase complex, in Arabidopsis causes an early embryo lethal phenotype. Plant Mol Biol. 2003;52:865–72.

27. Ruan YL, Llewellyn DJ, Furbank RT. The control of single-celled cotton fiber elongation by developmentally reversible gating of plasmodesmata and coordinated expression of sucrose and K+ transporters and expansin. Plant Cell. 2001;13:47–60.

28. Li XB, Fan XP, Wang XL, Cai L, Yang WC. The cotton ACTIN1 gene is functionally expressed in fibers and participates in fiber elongation. Plant Cell. 2005;17:859–75.

29. Yao Y, Yang YW, Liu JY. An efficient protein preparation for proteomic analysis of developing cotton fibers by 2-DE. Electrophoresis. 2006;27:4559–69.

30. Bradford MM. A rapid and sensitive method for the quantitation of microgram quantities of protein utilizing the principle of protein-dye binding. Anal Biochem. 1976;72:248–54.

31. Candiano G, Bruschi M, Musante L, Santucci L, Ghiggeri GM, Carnemolla B, et al. Blue silver: a very sensitive colloidal coomassie G-250 staining for proteome analysis. Electrophoresis. 2004;25:1327–33.

32. Li F, Fan G, Wang K, Sun F, Yuan Y, Song G, et al. Genome sequence of the cultivated cotton Gossypium arboreum. Nat Genet. 2014;46:567–72.

6

Proteome profiling reveals insights into cold-tolerant growth in sea buckthorn

Caiyun He[1], Guori Gao[1], Jianguo Zhang[1,2]*, Aiguo Duan[1] and Hongmei Luo[3]

Abstract

Background: Low temperature is one of the crucial environmental factors limiting the productivity and distribution of plants. Sea buckthorn (*Hippophae rhamnoides* L.), a well recognized multipurpose plant species, live successfully in in cold desert regions. But their molecular mechanisms underlying cold tolerance are not well understood.

Methods: Physiological and biochemical responses to low-temperature stress were studied in seedlings of sea buckthorn. Differentially expressed protein spots were analyzed using multiplexing fluorescent two-dimensional fluorescence difference gel electrophoresis (2D-DIGE) coupled with matrix-assisted laser desorption/ionization (MALDI) time-of-flight/time-of-flight (TOF/TOF) mass spectrometry (MS), the concentration of phytohormone was measured using enzyme-linked immunosorbent assay, and a spectrophotometric assay was used to measure enzymatic reactions.

Results: With the increase of cold stress intensity, the photosynthesis rate, transpiration rate, stomatal conductance in leaves and contents of abscisic acid (ABA) and indole acetic acid (IAA) in roots decreased significantly; however, water-use efficiency, ABA and zeatin riboside in leaves increased significantly, while cell membrane permeability, malondialdehyde and IAA in leaves increased at 7 d and then decreased at 14 d. DIGE and MS/MS analysis identified 32 of 39 differentially expressed protein spots under low-temperature stress, and their functions were mainly involved in metabolism, photosynthesis, signal transduction, antioxidative systems and post-translational modification.

Conclusion: The changed protein abundance and corresponding physiological–biochemical response shed light on the molecular mechanisms related to cold tolerance in cold-tolerant plants and provide key candidate proteins for genetic improvement of plants.

Keywords: Antioxidative systems, DIGE, *Hippophae rhamnoides*, Low temperature, Photosynthesis, Post-translational modification

Background

Plants being sessile have developed intricate adaptation mechanisms to stresses. Among abiotic environmental stresses, low temperature is one of the crucial factors limiting plant productivity and distribution [1–3]. Therefore, identifying the reasons behind cold adaptation in cold-tolerant plants could be of central importance to gaining a deeper understanding of the mechanisms involved, and be beneficial for plant growth and production [4, 5].

Sea buckthorn (*Hippophae rhamnoides*) is well adapted to extreme temperatures ranging from −43 to 40 °C [6], and it is distributed in cold regions of Russia, China, Finland, Sweden and many other countries of Asia, Europe and North America [7], which provided an ideal material to study mechanisms related to low-temperature (LT) stress tolerance [2]. Although there is a considerable wealth of studies on origin, distribution, ecology and nutrient and medicinal values in *Hippophae* [8–15], the physiological and biochemical basis of tolerance and mechanisms

* Correspondence: zhangjg@caf.ac.cn
[1]State Key Laboratory of Tree Genetics and Breeding, Key Laboratory of Tree Breeding and Cultivation of the State Forestry Administration, Research Institute of Forestry, Chinese Academy of Forestry, Beijing, People's Republic of China
[2]Collaborative Innovation Center of Sustainable Forestry in Southern China, Nanjing Forestry University, Nanjing, People's Republic of China
Full list of author information is available at the end of the article

of abiotic stress response, especially low-temperature or cold response, are not well understood [2, 16, 17].

Due to the direct roles of proteins in plant stress responses, profound changes in proteome composition can be observed during plant acclimation to stress. Mass spectrometry (MS)-based proteomics has become an essential tool in unraveling possible relationships between protein abundance and plant stress acclimation [18, 19]. The present study discusses the proteome-wide protein responses to low-temperature stress of *H. rhamnoides* cv. 'Chuisk', a widely cultivated hybrid of *H. rhamnoides* subsp. *mongolica* and *H. rhamnoides* subsp. *rhamnoides* in northeast China, which has excellent cold resistance and good characteristics, including large berries, high content of oil and high production. Using physiological, biochemical and comparative proteomic analyses, we hope to provide insights into cold adaptation mechanisms in this cold-tolerant species.

Results

Physiological and biochemical responses during LT stress

Under the LT treatments 7 d (T1) and 14 d (T2), the values of superoxide dismutase (SOD), glutathione reductase (GR) and zeatin riboside in root [ZR(R)] showed no significant changes, but leaf area (LA) and gibberellins (GA₃) did. The decreases in stomatal conductance (Cond) and abscisic acid in root [ABA(R)] were significant with the extension of LT treatment time (p-value $\leq$ 0.05). There were significant increases in water-use efficiency (WUE), ZR (L) (L indicated leaf) and ABA (L) for T1. Gradual decreases in photosynthesis rate (Pn), transpiration rate (Tr) and indole acetic acid [IAA(R)] were observed among both controls and treatments. Cell membrane permeability (CMP), malondialdehyde (MDA) and IAA(L) showed non-consistent changes (increase at T1 and decrease at T2) under LT stress (Fig. 1).

Variations in LT responsive proteins

The DeCyder image analysis of fluorescent images detected 1466 ± 35 protein spots (Fig. 2, Additional file 1: Figure S1). One-way analysis of variance (ANOVA) showed that 39 different protein spots were significantly affected by LT stress (p-value $\leq$ 0.05) (Fig. 2), of which 32 proteins were identified using the MS/MS method: eight exhibited a gradual increase (A-type), five exhibited a gradual decrease (B-type), 11 exhibited an increase at T1 and decrease at T2 (C-type) and eight exhibited a decrease at T1 and increase at T2 (D-type) (Additional file 2: Table S1).

Gene ontology (GO) and pathway enrichment analysis of differentially expressed protein spots

Ninety-seven GO enrichment terms were obtained using GOEAST [20]. The enriched biological process ontology

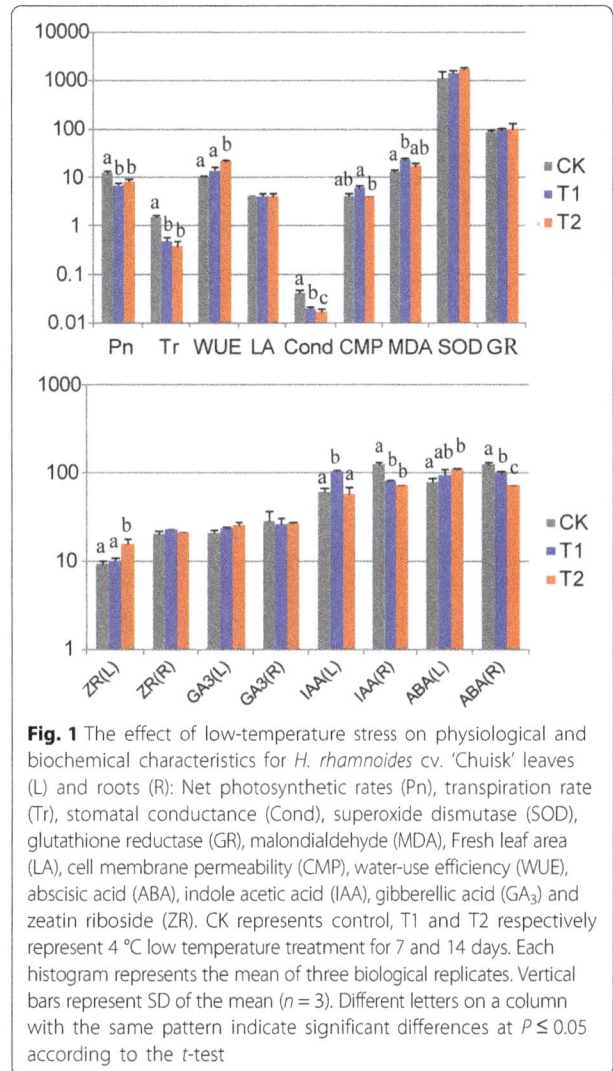

Fig. 1 The effect of low-temperature stress on physiological and biochemical characteristics for *H. rhamnoides* cv. 'Chuisk' leaves (L) and roots (R): Net photosynthetic rates (Pn), transpiration rate (Tr), stomatal conductance (Cond), superoxide dismutase (SOD), glutathione reductase (GR), malondialdehyde (MDA), Fresh leaf area (LA), cell membrane permeability (CMP), water-use efficiency (WUE), abscisic acid (ABA), indole acetic acid (IAA), gibberellic acid (GA₃) and zeatin riboside (ZR). CK represents control, T1 and T2 respectively represent 4 °C low temperature treatment for 7 and 14 days. Each histogram represents the mean of three biological replicates. Vertical bars represent SD of the mean ($n = 3$). Different letters on a column with the same pattern indicate significant differences at $P \leq 0.05$ according to the t-test

included the metabolic process (nitrate assimilation and nucleotide metabolic, oxidoreduction coenzyme metabolic and carotenoid biosynthetic processes), biological regulation (regulation of protein dephosphorylation), response to stimulus (responses to cold, glucose and fructose) and localization (mitochondrial transport) (Fig. 3 and Additional file 2: Table S2).

KEGG pathway analysis [21] showed that 18 terms including genetic information processing (folding, sorting and degradation, and translation) and metabolism (amino acid, carbohydrate and energy metabolisms) were enriched under LT stress (Additional file 2: Table S3). Of the 32 identified protein spots, 14 (43.75 %) proteins were classified to metabolism pathways including energy metabolism (14 protein spots), carbohydrate metabolism (12 spots) and amino acid metabolism (spots 486 and 693, C-type; and spot 437, D-type). Energy metabolism was enriched in carbon

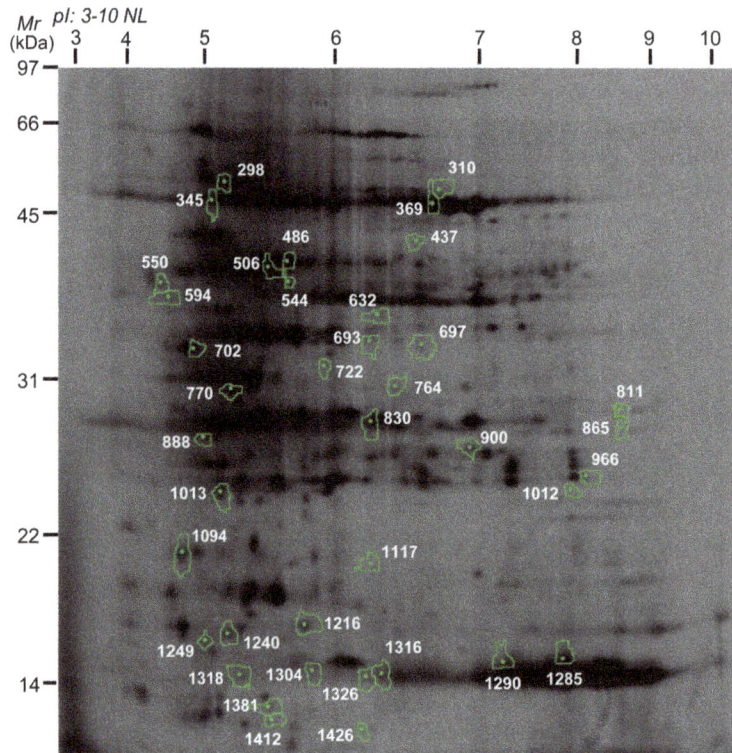

Fig. 2 Representative fluorescent dyes-stained two-dimensional gel of leaf proteins in response to low-temperature stress in *H. rhamnoides* cv. 'Chuisk' seedlings. The relative M_r (on the left) and the pI (on the top) are given. The white numbers represent 39 differentially expressed protein spots

fixation in photosynthetic organisms including ribulose-bisphosphate carboxylase large chain (rbcL; 10 homologous/isoform protein spots), photosynthesis involving oxygen-evolving enhancer protein 1 (spot 770, A-type), photosystem I reaction center subunit II (spot 966, C-type) and nitrogen metabolism (spot 486, C-type, glutamine synthetase, glnA). Carbohydrate metabolism was enriched in glyoxylate and dicarboxylate metabolism (11 protein spots), and citrate cycle and pyruvate metabolism (spot 693, C-type, Nodule-enhanced malate

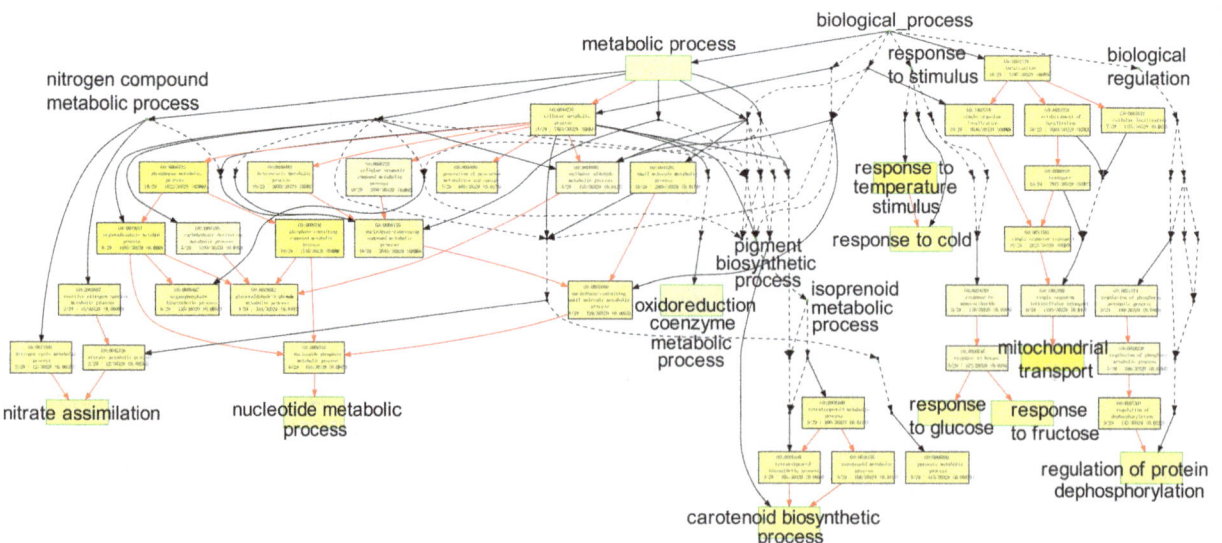

Fig. 3 Biological process enrichment clusters for 37 low-temperature stress proteins with *Arabidopsis thaliana* homologues using GOEAST

dehydrogenase). Four of 32 proteins belonged to the pathway of genetic information processing (spots 298 and 345, A-type, TCP-1/cpn60 chaperonin family protein; spot 1117, D-type, 18.5-kDa class I heat shock protein; and spot 722, D-type, unnamed protein) (Additional file 2: Tables S1 and S3).

mRNA expression of differentially expressed protein spots between different treatments

To confirm the reliability of the MALDI-TOF MS/MS and 2D-DIGE in detecting *H. rhamnoides*, we used RealTime PCR to compare the expression levels of identified protein genes. Ten genes were randomly selected from the differentially expressed protein spots in the three groups. Similar expression patterns between protein and mRNA were observed in six of the ten detected genes (Fig. 4). Of which, two were up-regulated (Spots 298 and 770), one was down-regulated (Spot 1285), one was up-regulated under D1 stress and down-regulated under D2 stress (Spot 966), and two was down-regulated under D1 stress and up-regulated under D2 stress (Spots 1094 and 1117).

Discussion

Endogenous hormones and antioxidant enzymes play roles in physiological and biochemical responses of sea buckthorn under LT stress

Compared with GA$_3$ and ZR, ABA and IAA responded more actively. The LT stress induced the decline of ABA content in sea buckthorn roots, and then accumulation in leaves, which led to stomatal closure and restrained aboveground Pn, Tr and Cond. Meanwhile, the complementary/equilibrium of growth-promoting hormones such as ZR and IAA possibly contributed to maintaining plant growth, while the increase of ABA in leaves also showed growth inhibition under LT stress [22]. The changed protein abundance of probable LRR receptor-like serine/threonine-protein kinase RKF3-like (spot 1426, C-type) involved in hormone-mediated (ABA and GA) signaling pathways, and hypothetical protein (spot 830, C-type) participating in IAA biosynthetic/metabolic process, further showed that ABA, GA and IAA played an important role in responding to LT stress in sea buckthorn, as found in many other cold-stress studies [23–27]. These confirmed that many hormonal proteins

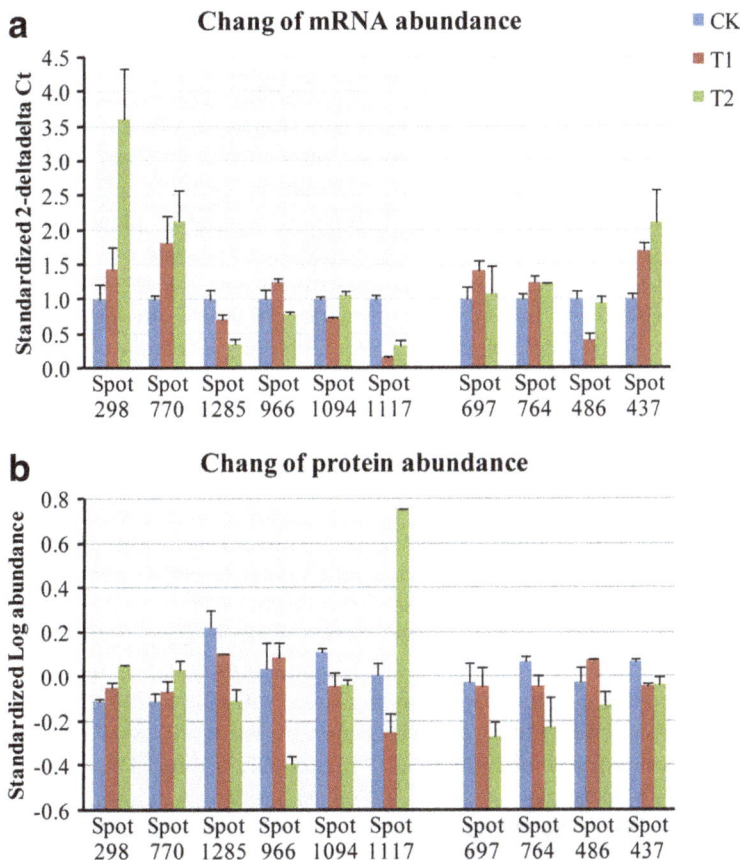

Fig. 4 Quantitative RT-PCR of mRNA changes. **a** and **b** showed expression abundance of mRNA and protein we detected between RT-PCR and 2D-DIGE, respectively. Gray margins showed the agreement in the six genes we detected between RT-PCR and 2D-DIGE

contributed to the changes in ABA, GA and IAA content under LT stress and so induced the hormone signaling pathways responding to LT stress in sea buckthorn.

In addition, the CMP and MDA increase under LT stress suggested that they induced cell damage and membrane lipid peroxidation. Some protein spots of antioxidant enzymes, such as zinc-binding dehydrogenase family oxidoreductase (spot 544, A-type), peroxiredoxin family protein (Prx, spot 888, D-type), dehydroascorbate reductase (DHAR, spot 900, C-type) and glnA (spot 486, C-type), changed dynamically in protein abundance under LT stress, suggesting that they had taken part in changes of enzymatic activity or antioxidants (i.e. SOD and GR). Many proteins involved in antioxidative systems were also induced by LT stress in *H. rhamnoides* [2] and *Arabidopsis* [28]. This study hint at cooperation between hormone and redox signaling pathways in responding to LT stress, and this confers enormous regulatory potential because it allows plants to adapt to changing and often challenging stress conditions [29].

Proteins related to metabolism, signal transduction and post-translational modification contribute to LT stress responses in sea buckthorn

In forest tree species, many genes taking part in metabolic processes were differentially expressed under most abiotic stress conditions [30–34]. Pn significantly decreased under LT stress, and expression of the rbcLs and oxygen-evolving enhancer proteins in sea buckthorn greatly changed consistent with a number of previous reports [35–37]. This suggested negative effects on photosynthetic efficiency and photosynthesis rate. Many studies have shown that LT triggers a cellular signal transduction pathway leading to molecular and metabolic changes [26, 38–40]. This study found glutamine synthetase (spot 486, C-type) was differently expressed during LT stress, suggesting that cold stress stimuli initiated the signal transmission processes in *H. rhamnoides*. During cold acclimation, plants accumulate specific types of solutes, such as soluble sugars (i.e. glucose and fructose) [41] and carotenoids [42, 43] to enhance freezing tolerance. The different expression of Prx (spot 888, D-type) and glnA (spot 486, C-type) under LT stress suggested the accumulation of soluble sugars in *H. rhamnoides*. Meanwhile, two proteins (spots 544 and 594, A-type) involving carotenoid biosynthetic process were enriched in the up-regulated differently expressed protein spots, suggesting an important role of carotenoid in responding to LT stress in sea buckthorn.

It is noteworthy that two identified isoforms had multiple spots located at different positions on the gel with different molecular weight, *pI* or both. For example, spots 298 and 345 were both identified as TCP-1/cpn60 chaperonin family protein and spots 369, 550, 764, 811, 865, 1012, 1285, 1290, 1316 and 1326 were rbcL homologs. These isoforms might represent post-translationally modified forms of the same protein, such as phosphorylated and glycosylated forms under LT stress, hinting at the post-translational regulation of cold acclimation responsein *H. rhamnoides* [44, 45].

According to the findings, we proposed some scenarios involved in LT stress tolerance of *H. rhamnoides*. During the growth season, LT abiotic stress triggered signal transduction, stimulated genetic information processing (such as protein degradation and synthesis, and changes in enzymatic activity and hormone content), promoted cellular processes (cell damage, membrane lipid peroxidation and stomatal opening/closing) and regulated necessary metabolism pathways (especially the energy and carbohydrate metabolism mostly from photosynthesis). This then induced physiological (such as Pn, Tr, Cond, CMP and WUE) and biochemical changes (such as activity of SOD and GR, and contents of MDA and endogenous hormones), eventually leading to growth retardation of *H. rhamnoides* (Fig. 5). Additionally, post-translational modification (PTM), such as dephosphorylation (Fig. 3) might also play important roles in cold stress responses and adaptation in sea buckthorn [46].

Conclusion

The responses of sea buckthorn to cold stress were complex and involved numerous physiological, molecular and cellular adaptations. The proteins involved in photosynthesis, plant hormone signal transduction, antioxidative systems and PTM played vital roles. Furthermore, the 2D-PAGE profile of sea buckthorn under LT stress provided useful candidate proteins for genetic improvement and understanding the general ability of plants to respond to a wide range of external environmental stresses.

Methods

Plant materials and treatments

Root turions of *H. rhamnoides* cv. 'Chuisk' were planted in 18 cm × 22 cm × 26 cm plastic pots containing 6 kg of a mixture of turfy soil/clay soil/sand (5:3:2). The plants were grown in a greenhouse under 25/20 °C of day/night temperatures, relative humidity of 60–70 % and CO_2 concentration of 380 ± 10 µmol mol^{-1}. After three months, the plants were divided into three groups/treatments (CK, T1 and T2), each with three biological replicates, applied as follows. CK was an unstressed control, T1 and T2 were 4 °C LT treatment, in which soil water content was maintained at about 30 % (i.e. 75–85 % of field moisture capacity) by regular watering during the whole experiment (i.e. from 0 to 14 d). Samples taken at

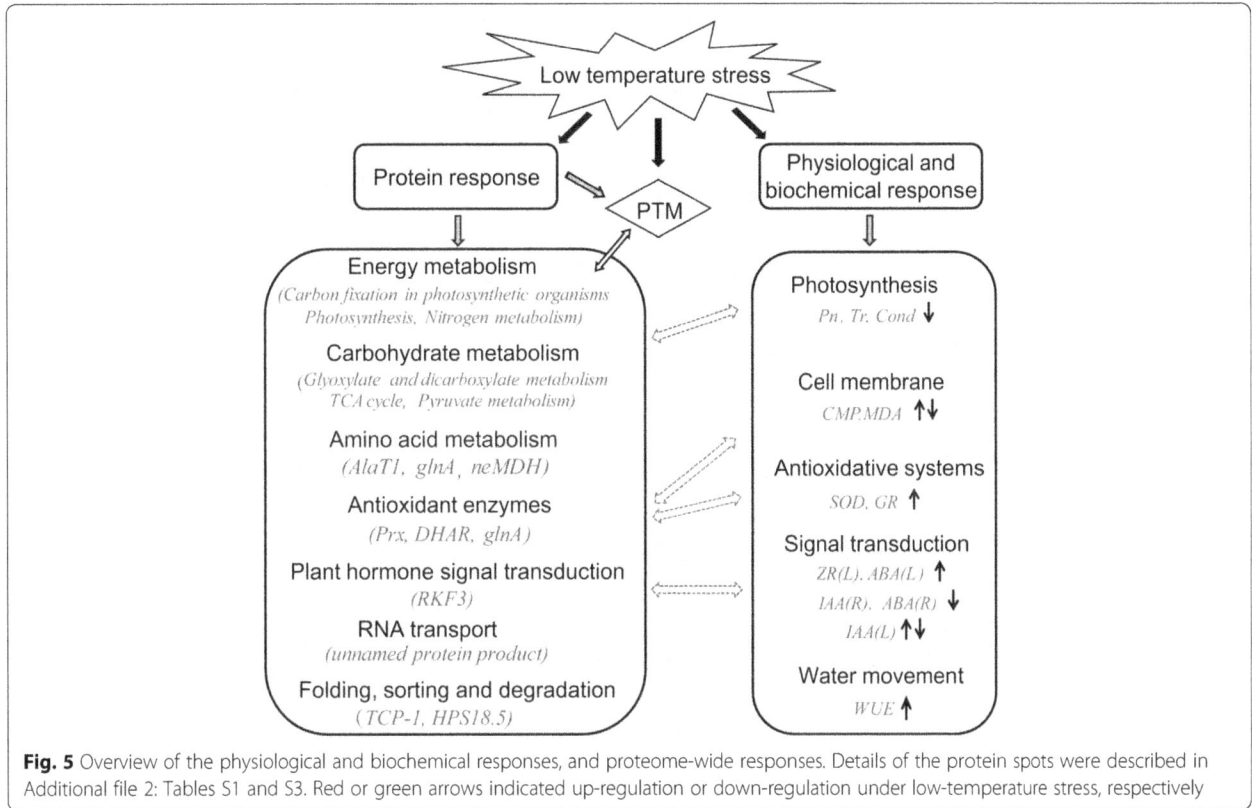

Fig. 5 Overview of the physiological and biochemical responses, and proteome-wide responses. Details of the protein spots were described in Additional file 2: Tables S1 and S3. Red or green arrows indicated up-regulation or down-regulation under low-temperature stress, respectively

7 (T1) and 14 d (T2) were snap-frozen in liquid nitrogen, and stored at −80 °C for further studies. Soil water content was checked every 2 d using a moisture probe type HH2 Meter (Delta-T Devices Ltd., Cambridge, UK). For this, the plants were cultured separately in a 3.5 m × 2.2 m × 3.2 m AGC-2 Growth Cabinet (Zhejiang University Electric Equipment Factory, Hangzhou, China), whose parameters could be regulated strictly to be consistent with the greenhouse conditions.

Measurement of physiological and biochemical parameters

Three fully expanded leaves in the middle portion of each stem were selected for measurement with a LI-6400 Portable Photosynthesis System (LI-COR Inc., Lincoln, NE, USA). Every measurement was repeated five times. The changes in Pn $(\mu mol \cdot m^{-2} \cdot s^{-1})$, Tr $(\mu mol \cdot m^{-2} \cdot s^{-1})$ and Cond $(mol \cdot m^{-2} \cdot s^{-1})$ are shown in Fig. 1. The activities of major antioxidant enzymes (SOD and GR) $(U \cdot g^{-1})$, MDA $(\mu mol \cdot g^{-1})$ content were analyzed according to the method of Gao [47]. Molar conductances, used for CMP (%) were measured using a digital conductivity meter DDS-12DW Microprocessor (LIDA Instrument Factory, Shanghai, China). Fresh LA (cm^2) was measured using a leaf area meter (CI-202, CID Inc., WA, USA). The content $(ng \cdot g^{-1})$ of four kinds of phytohormones (ABA, IAA, GA$_3$ and ZR) were

respectively measured in leaves and roots using enzyme-linked immunosorbent assay method. Each measure had three biological replicates.

Protein extraction and quantification

Of frozen leaves, 2 g was homogenized in a pre-chilled mortar in 25 ml of cold acetone (−20 °C) containing 10 % (V/V) trichloroacetic acid and 65 mM dithiothreitol. The mixture was kept at −20 °C for 1 h, and then centrifuged at 12,000 g for 45 min. The supernatant was discarded and the pellet precipitated with 25 ml of cold acetone. After 1 h at −20 °C, the mixture was centrifuged at 12,000 g for 45 min, and the white pellet was lyophilized and stored at −80 °C. The protein pellet was dissolved in DIGE lysis buffer (7 M urea, 2 M thiourea, 4 % 3-[(3-chol-amidopropyl) dimethylammonio]-1-propane sulfonate and 0.2 % IPG buffer). The mixture was shaken and centrifuged at 12,000 g for 1 h. The supernatant was collected and protein concentrations were determined using the Bradford method with the protein assay reagent (Bio-Rad Laboratories, Hercules, CA, USA).

Two-dimensional difference gel electrophoresis and imaging analysis

Nine protein samples (50 μg, pH 8.0–9.0) labeled with Cy3 and Cy5 dyes (labeled with Cy2 dye for use as internal standard for normalization) were mixed with a strip re-hydration and then separated by 2D-polyacrylamide gel electrophoresis (PAGE). Total volume was adjusted to 50 μL, mixed 1:1 with DIGE lysis buffer and incubated for 30 min on ice in the dark. The reaction was then quenched with 10 mM lysine and additionally incubated for 10 min [48, 49]. Isoelectrofocusing was carried out with the IPGphor system (Amersham Pharmacia Biotech, Uppsala, Sweden) using Immobiline DryStrip gels (13 cm) with nonlinear pH gradients (pH 3–10). The IPGphor system was then programmed as follows: 12 h at 30 V, 1 h at 500 V, 1 h at 1000 V, 8 h at 8000 V and 4 h at 500 V. After IEF, the gel strips were incubated with equilibration buffer, and the strips were then transferred onto vertical 12.5 % (w/v) sodium dodecyl sulfate (SDS)-PAGE gels. The SDS-PAGE was performed on a Hoefer SE600 system (Amersham Pharmacia Biotech) with 15 mA/gel for 30 min and then 30 mA/gel until the bromophenol blue dye reached the bottom (about 0.5 cm) of the gels. The gels were scanned using an UMax Powerlook 2110XL laser scanner (Amersham Pharmacia Biotech), and the images analyzed using DeCyderTM 2D software (Amersham Pharmacia Biotech). Intra-gel spot detection, quantification and inter-gel matching and quantification were performed using Differential In-gel Analysis and Biological Variation Analysis modules respectively. To test for significant differences in expression of protein spots among pairwise groups, one-way analysis of variance (ANOVA) was performed at p-value ≤ 0.05. The differentially expressed protein spots were filtered based on an average volume ratio of twofold with significance at p-value ≤ 0.05. Each treatment had three biological replicates.

In-gel trypsin digestion and MS analysis

These gels were fixed and stained with colloidal Coomassie brilliant blue (CBB). Proteins of interest, as defined by the DeCyder analysis TM 2D software, were excised from the CBB-stained gels for an in-gel trypsin digestion procedure [50]. The resolved peptides were analyzed on matrix-assisted laser desorption/ionization (MALDI) time-of-flight/time-of-flight (TOF/TOF) based on an MALDI TOF/TOF system (ultrafleXtreme, Bruker Daltonics, MA, USA). Protein identification was performed using the Mascot search engine version 2.1.03 (Matrix Science, London, UK) with a mass tolerance of 0.05 Da permitted for intact peptide masses and fragmented ions, and allowance for one missed cleavage in trypsin digests. Gln- > pyro-Glu (N-term Q), Oxidation (M) and Deamidated (NQ) as the potential variable modifications, Carbamidomethyl (C) as fixed modifications. The search that was performed using MASCOT v2.1.03 software (Matrix Science, London, UK) and the following settings: Uniprot_plant_database, one missed

cleavage, fixed modifications of carbamidomethyl, variable modifications of oxidation, peptide tolerance 100 ppm, fragment mass tolerance 0.5 Da, peptide charge 1+. Only peptides with MS/MS ion scores significantly ($P < 0.05$) exceeding the MASCOT identity or extensive homology threshold were reported. Proteins were annotated using Gene Ontology (GO), Kyoto Encyclopedia of Genes and Genomes (KEGG) pathways. GO and pathway enrichment were analyzed using Gene Ontology Enrichment Analysis Software Toolkit (GOEAST) [20] and KEGG Orthology Based Annotation System (KOBAS) [21] Software Toolkit, respectively.

Relative quantification of mRNA expression by RealTime PCR

Total RNAs from *H. rhamnoides* leaves in three groups were used for quantitative PCR analysis. Briefly, the first cDNA strands were obtained using a Thermo First cDNA Synthesis Kit, and were then subjected to quantification with 18S rRNA as an internal control using a standard SYBR Green PCR kit (Bio-Rad, Hercules, CA) on the StepOnePLUS RealTime PCR Detection System (Applied Biosystems, Foster City, CA). Quantitative PCR was then performed using the following conditions: 95 °C for 10 min; 40 cycles of 95 °C for 20 s and a 60 °C annealing temperature for 30 s; 95 °C for 15 s; 60 °C for 30 s; and 95 °C for 15 s. The primers for all 10 genes are listed in Online Additional file 2: Table S4. All reactions were performed in triplicate for each sample. Gene expression was quantified relative to 18S rRNA expression using the comparative cycle threshold (ΔCT) method. Differences in gene expression between the CK and LT stresses (T1 and T2) were detected by using the t-test.

Additional files

Additional file 1: Figure S1. The 1D- and 2D- DIGE maps of H. rhamnoides leaves under control and low-temperature stress. (JPG 635 kb)

Additional file 2: Table S1A. The identified protein summary of H. rhamnoides cv. 'Chuisk' responding to low-temperature stress using MALDI-TOF-TOF. Table S1B. The identified protein detail of H. rhamnoides cv. 'Chuisk' responding to low-temperature stress using MALDI-TOF-TOF. Table S2. GO enrichment terms under low-temperature stress with Arabidopsis thaliana homologues using GOEAST. Table S3. KEGG pathways of protein spots under low-temperature stress. Table S4. Primers used for detecting the gene expression by reverse real-time PCR. (XLS 171 kb)

Acknowledgements

We acknowledge Shanghai Applied Protein Technology and Beijing Genomics Institute at Shenzhen for its assistance in original data processing. We would like to thank many other people for helping with the sampling. This study was supported by the Special Fund for Forest Scientific Research in the Public Welfare (201504103) and National Natural Science Foundation of China (31470616).

Funding
This study was supported by the Special Fund for Forest Scientific Research in the Public Welfare (201504103) and National Natural Science Foundation of China (31470616).

Authors' contributions
CH carried out the experiments, analyzed the data and drafted the manuscript. AD. and HL. collected samples. GG participated in the RT-PCR analysis. JZ. conceived of the study. All authors read and approved the final manuscript.

Competing interests
The authors declare that they have no competing interests.

Author details
[1]State Key Laboratory of Tree Genetics and Breeding, Key Laboratory of Tree Breeding and Cultivation of the State Forestry Administration, Research Institute of Forestry, Chinese Academy of Forestry, Beijing, People's Republic of China. [2]Collaborative Innovation Center of Sustainable Forestry in Southern China, Nanjing Forestry University, Nanjing, People's Republic of China. [3]Experimental Center of Desert Forestry, Chinese Academy of Forestry, InnerMonglia, People's Republic of China.

References
1. Theocharis A, Clément C, Barka EA. Physiological and molecular changes in plants grown at low temperatures. Planta. 2012;235:1091–105.
2. Gupta R, Deswal R. Low temperature stress modulated secretome analysis and purification of antifreeze protein from Hippophae rhamnoides, a Himalayan wonder plant. J Proteome Res. 2012;11:2684–96.
3. Chinnusamy V, Zhu J, Zhu J-K. Cold stress regulation of gene expression in plants. Trends Plant Sci. 2007;12:444–51.
4. Lawlor DW. Genetic engineering to improve plant performance under drought: physiological evaluation of achievements, limitations, and possibilities. J Exp Bot. 2013;64:83–108.
5. Mickelbart MV, Hasegawa PM, Bailey-Serres J. Genetic mechanisms of abiotic stress tolerance that translate to crop yield stability. Nature Rev Genet. 2015;16:237–51.
6. Ruan CJ, Teixeira da Silva J, Jin H, Li H, Li DQ. Research and biotechnology in sea buckthorn (Hippophae spp.). Med Aromat Plant Sci Biotechnol. 2007;1:47–60.
7. Chaudhary S, Sharma PC. DeepSAGE based differential gene expression analysis under cold and freeze stress in seabuckthorn (Hippophae rhamnoides L.). PLoS One. 2015;10:e0121982.
8. Arimboor R, Arumughan C. HPLC-DAD-MS/MS profiling of antioxidant flavonoid glycosides in sea buckthorn (Hippophae rhamnoides L.) seeds. Int J Food Sci Nutr. 2012;63:730–8.
9. Michel T, Destandau E, Floch GL, Lucchesi ME, Elfakir C. Antimicrobial, antioxidant and phytochemical investigations of sea buckthorn (Hippophaë rhamnoides L.) leaf, stem, root and seed. Planta Med. 2012;131:754–60.
10. Dalija S, Daina K, Silvija R, Inta K. The effect of processing on the composition of sea buckthorn juice. J Fruit Ornam Plant Res. 2006;14:257–64.
11. Fatima T, Snyder CL, Schroeder WR, Cram D, Datla R, Wishart D, Weselake RJ, Krishna P. Fatty acid composition of developing sea buckthorn (Hippophae rhamnoides L.) berry and the transcriptome of the mature seed. PLoS One. 2012;7:e34099.
12. Bartish, Jeppsson, Nybom. Population genetic structure in the dioecious pioneer plant species Hippophae rhamnoides investigated by RAPD markers. Mol Ecol. 1999;8:791–802.
13. Sun K, Chen X, Ma R, Li C, Wang Q, Ge S. Molecular phylogenetics of Hippophae L. (Elaeagnaceae) based on the internal transcribed spacer (ITS) sequences of nrDNA. Plant Syst Evol. 2002;235:121–34.
14. Kallio H, Yang B, Peippo P. Effects of different origins and harvesting time on vitamin C, tocopherols, and tocotrienols in sea buckthorn (Hippophae rhamnoides) berries. J Agric Food Chem. 2002;50:6136–42.
15. Jia DR, Abbott RJ, Liu TL, Mao KS, Bartish IV, Liu JQ. Out of the Qinghai–Tibet Plateau: evidence for the origin and dispersal of Eurasian temperate plants from a phylogeographic study of Hippophaë rhamnoides (Elaeagnaceae). New Phytologist. 2012;194:1123–33.
16. Ghangal R, Raghuvanshi S, Sharma PC. Expressed sequence tag based identification and expression analysis of some cold inducible elements in seabuckthorn (Hippophae rhamnoides L.). Plant Physiol Biochem. 2012; 51:123–8.
17. Li C, Yang Y, Junttila O, Palva ET. Sexual differences in cold acclimation and freezing tolerance development in sea buckthorn (Hippophae rhamnoides L.) ecotypes. Plant Sci. 2005;168:1365–70.
18. Xu G, Li C, Yao Y. Proteomics analysis of drought stress-responsive proteins in Hippophae rhamnoides L. Plant Mol Biol Report. 2009;27:153–61.
19. Zhang M, Lv D, Ge P, Bian Y, Chen G, Zhu G, Li X, Yan Y. Phosphoproteome analysis reveals new drought response and defense mechanisms of seedling leaves in bread wheat (Triticum aestivum L.). J Proteomics. 2014;109:290–308.
20. Jin ZP, Wen XF, Zhang JK, Gu YK, Jin SY. Analysis for affect factors of drought stress on production of different kinds of Seabuckthorn sources. The Global Seabuckthorn Research and Development. 2008;2:33–8.
21. Xie C, Mao X, Huang J, Ding Y, Wu J, Dong S, Kong L, Gao G, Li C-Y, Wei L. KOBAS 2.0: a web server for annotation and identification of enriched pathways and diseases. Nucleic Acids Res. 2011;39:W316–22.
22. Huang X, Chen MH, Yang LT, Li YR, Wu JM. Effects of exogenous abscisic acid on cell membrane and endogenous hormone contents in leaves of sugarcane seedlings under cold stress. Sugar Tech. 2015;17:59–64.
23. Sreenivasulu N, Harshavardhan VT, Govind G, Seiler C, Kohli A. Contrapuntal role of ABA: does it mediate stress tolerance or plant growth retardation under long-term drought stress? Gene. 2012;506:265–73.
24. Miao ZY, Xu W, Li DF, Hu XN, Liu JX, Zhang RX, Tong ZY, Dong JL, Su Z, Zhang LW. De novo transcriptome analysis of Medicago falcata reveals novel insights about the mechanisms underlying abiotic stress-responsive pathway. BMC Genomics. 2015;16:818.
25. Maruyama K, Urano K, Yoshiwara K, Morishita Y, Sakurai N, Suzuki H, Kojima M, Sakakibara H, Shibata D, Saito K. Integrated analysis of the effects of cold and dehydration on rice metabolites, phytohormones, and gene transcripts. Plant Physiol. 2014;164:1759–71.
26. Colebrook EH, Thomas SG, Phillips AL, Hedden P. The role of gibberellin signalling in plant responses to abiotic stress. J Exp Biol. 2014;217:67–75.
27. Shibasaki K, Uemura M, Tsurumi S, Rahman A. Auxin response in Arabidopsis under cold stress: underlying molecular mechanisms. Plant Cell. 2009;21:3823–38.
28. Goulas E, Schubert M, Kieselbach T, Kleczkowski LA, Gardeström P, Schröder W, Hurry V. The chloroplast lumen and stromal proteomes of Arabidopsis thaliana show differential sensitivity to short- and long-term exposure to low temperature. Plant J. 2006;47:720–34.
29. Bartoli CG, Casalongué CA, Simontacchi M, Marquez-Garcia B, Foyer CH. Interactions between hormone and redox signalling pathways in the control of growth and cross tolerance to stress. Environ Exp Bot. 2013;94:73–88.
30. Kieffer P, Schröder P, Dommes J, Hoffmann L, Renaut J, Hausman J-F. Proteomic and enzymatic response of poplar to cadmium stress. J Proteomics. 2009;72:379–96.
31. He CY, Zhang JG, Duan AG, Zheng SX, Sun HG, Fu LH. Proteins responding to drought and high-temperature stress in Populus × euramericana cv. '74/76'. Trees. 2008;22:803–13.
32. Ferreira S, Hjernø K, Larsen M, Wingsle G, Larsen P, Fey S, Roepstorff P, Pais MS. Proteome profiling of Populus euphratica Oliv. upon heat stress. Ann Bot. 2006;98:361–77.
33. Costa P, Bahrman N, Frigerio J-M, Kremer A, Plomion C. Water-deficit-responsive proteins in maritime pine. Plant Mol Biol. 1998;38:587–96.
34. Echevarría-Zomeño S, Ariza D, Jorge I, Lenz C, Del Campo A, Jorrín JV, Navarro RM. Changes in the protein profile of Quercus ilex leaves in response to drought stress and recovery. J Plant Physiol. 2009;166:233–45.
35. Evers D, Legay S, Lamoureux D, Hausman J, Hoffmann L, Renaut J. Towards a synthetic view of potato cold and salt stress response by transcriptomic and proteomic analyses. Plant Mol Biol. 2012;78:503–14.
36. Gharechahi J, Alizadeh H, Naghavi MR, Sharifi G. A proteomic analysis to identify cold acclimation associated proteins in wild wheat (Triticum urartu L.). Mol Biol Rep. 2014;41:3897–905.
37. Rocco M, Arena S, Renzone G, Scippa GS, Lomaglio T, Verrillo F, Scaloni A, Marra M. Proteomic analysis of temperature stress-responsive proteins in Arabidopsis thaliana rosette leaves. Mol Biosyst. 2013;9:1257–67.
38. Wang XC, Zhao QY, Ma CL, Zhang ZH, Cao HL, Kong YM, Yue C, Hao XY, Chen L, Ma JQ. Global transcriptome profiles of Camellia sinensis during cold acclimation. BMC Genomics. 2013;14:415.
39. Yang QS, Gao J, He WD, Dou TX, Ding LJ, Wu JH, Li CY, Peng XX, Zhang S, Yi GJ. Comparative transcriptomics analysis reveals difference of key gene

expression between banana and plantain in response to cold stress.
BMC Genomics. 2015;16:1.

40. Sergeant K, Kieffer P, Dommes J, Hausman J-F, Renaut J. Proteomic changes in leaves of poplar exposed to both cadmium and low-temperature. Environ Exp Bot. 2014;106:112–23.

41. Xu J, Li Y, Sun J, Du L, Zhang Y, Yu Q, Liu X. Comparative physiological and proteomic response to abrupt low temperature stress between two winter wheat cultivars differing in low temperature tolerance. Plant Biology. 2013;15:292–303.

42. Bano S, Aslam M, Saleem M, Basra S, Aziz K. Evaluation of maize accessions under low temperature stress at early growth stages. J Anim Plant Sci. 2015;25:392–400.

43. Wu YQ, Wei W, Pang XY, Wang XF, Zhang HL, Dong B, Xing YP, Li XG, Wang MY. Comparative transcriptome profiling of a desert evergreen shrub, *Ammopiptanthus mongolicus*, in response to drought and cold stresses. BMC Genomics. 2014;15:1.

44. Kashyap P, Deswal R. CBF-dependent cold stress signaling relevant post translational modifications. In: Stress Signaling in Plants: Genomics and Proteomics Perspective, Volume 1. New York: Springer; 2013. p. 105–122.

45. Barrero-Gil J, Salinas J. Post-translational regulation of cold acclimation response. Plant Sci. 2013;205:48–54.

46. Ichimura K, Mizoguchi T, Yoshida R, Yuasa T, Shinozaki K. Protein phosphorylation and dephosphorylation in environmental stress responses in plants. Adv Bot Res. 2000;32:355–77.

47. Gao JF. Plant physiology laboratory technology. Xi'an: World Publishing Corporation; 2000. p. 135–49.

48. García-Sevillano MA, Abril N, Fernández-Cisnal R, García-Barrera T, Pueyo C, López-Barea J, Gómez-Ariza JL. Functional genomics and metabolomics reveal the toxicological effects of cadmium in *Mus musculus* mice. Metabolomics. 2015;11:1432-50.

49. Yohannes E, Chang J, Christ GJ, Davies KP, Chance MR. Proteomics analysis identifies molecular targets related to diabetes mellitus-associated bladder dysfunction. Mol Cell Proteomics. 2008;7:1270–85.

50. Li BY, Cheng M, Gao HQ, Ma YB, Xu L, Li XH, Li XL, You BA. Back-regulation of six oxidative stress proteins with grape seed proanthocyanidin extracts in rat diabetic nephropathy. J Cell Biochem. 2008;104:668–79.

iTRAQ-based proteomic analysis of myofibrillar contents and relevant synthesis and proteolytic proteins in soleus muscle of hibernating Daurian ground squirrels (*Spermophilus dauricus*)

Hui Chang[1,2†], Shan-Feng Jiang[1†], Kai Dang[1], Hui-Ping Wang[1], Shen-Hui Xu[1] and Yun-Fang Gao[1,2*]

Abstract

Background: Daurian ground squirrels (*Spermophilus dauricus*) deviate from significant increase of protein catabolism and loss of myofibrillar contents during long period of hibernation inactivity.

Methods: Here we use iTRAQ based quantitative analysis to examine proteomic changes in the soleus of squirrels in pre-hibernation, hibernation and post-hibernation states. The total proteolysis rate of soleus was measured by the release of the essential amino acid tyrosine from isolated muscles. Immunofluorescent analysis was used to determine muscle fiber cross-sectional area. Western blot was used for the validation of the quantitative proteomic analysis.

Results: The proteomic responses to hibernation had a 0.4- to 0.8-fold decrease in the myofibrillar contractile protein levels of myosin-3, myosin-13 and actin, but a 2.1-fold increase in myosin-2 compared to pre-hibernation group. Regulatory proteins such as troponin C and tropomodulin-1 were 1.4-fold up-regulated and 0.7-fold down-regulated, respectively, in hibernation compared to pre-hibernation group. Moreover, 10 proteins with proteolytic function in hibernation, which was less than 14 proteins in the post-hibernation group, were up-regulated relative to the pre-hibernation group. The total proteolysis rates of soleus in hibernation and post-hibernation groups were significantly inhibited as compared with pre-hibernation group.

Conclusion: These findings suggest that the myofibrillar remodeling and partial suppression of myofibrillar proteolysis were likely responsible for preventing skeletal muscle atrophy during prolonged disuse in hibernation. This is the first study where the myofibrillar contents and relevant synthesis and proteolytic proteins in slow soleus was discussed based on proteomic investigation performed on wild Daurian ground squirrels. Our results lay the foundation for further research in preventing disuse-induced skeletal muscle atrophy in mammals.

Keywords: Daurian ground squirrel, Proteomic, Hibernation, Disuse atrophy, Myofibrillar protein, Synthesis and proteolysis

---

* Correspondence: gaoyunf@nwu.edu.cn
†Equal contributors
[1]Key Laboratory of Resource Biology and Biotechnology in Western China
(College of Life Sciences, Northwest University), Ministry of Education, Xi'an
710069, People's Republic of China
[2]Shaanxi Key Laboratory for Animal Conservation, Northwest University, Xi'an
710069, People's Republic of China

Background

In mammals, skeletal muscle accounts for more than 40 % of the mass of a given individual and provides critical functions in metabolism, energy expenditure, physical strength, and locomotor activity [1]. The skeletal muscle atrophy in response to disuse occurs during bed rest or spaceflight associated with the loss of muscle mass and the decline in muscle strength and power [2], muscular activity and cross-sectional area of muscle fiber [3]. The soleus muscle (SOL), which is predominantly composed of slow twitch fibers, is a postural muscle and more sensitive to disuse than fast-twitch muscles (extensor digitorum longus) and hybrid muscles (gastrocnemius) [4–6]. Disuse atrophy results in reduced protein content and a net loss of contractile proteins [7]. A proteomic study on rat soleus muscle after 3-week hindlimb unloading indicates that proteomic responses to disuse had a 0.2- to 0.6-fold decrease in the protein levels of myosin light chain 1 (MLC1), α-actin, tropomyosin β-chain, and troponins T [8]. Moreover, a number of results obtained showed that atrophic changes during a space flight or under head-down bed-rest are accompanied by decrease of total muscle protein [9] and myofibril proteins degradation [10]. Accordingly, disuse atrophy is supposed to be the result of shift of protein synthesis/proteolysis balance towards protein degradation increase [11], although many details remain unknown.

However, the skeletal muscle of hibernators appears to deviate from significant atrophy even after experiencing from extended disuse over three to four months, even six months in the cold North. It has been demonstrated that the muscle-fiber number and cross-sectional area were unchanged in gastrocnemius and biceps femoris of hibernating black bears (*Ursus americanus*), while protein concentration decreased in both muscles during the hibernation period, suggesting only marginal muscle atrophy [12]. In addition, it has also been reported that hibernating ground squirrels (*Citellus undulatus* and *Spermophilus dauricus*) have an evolutionarily determined adaptive mechanism of preservation or increase of slow fibers ratio [13], as the most economic and energetically advantageous, with proteins typical of them, whereas hindlimb unloading of non-hibernators (such as mouse) leads to activation of proteolysis and destruction of myofibrillar integrity, which contributes to considerable atrophy of soleus fibers [14]. Our previous research showed that SOL muscle mass to total body mass ratios (mg/g) were significantly higher in hibernating Daurian ground squirrels compared with that of rats after 14 days of hindlimb suspension, mirroring an effect protecting against disuse atrophy [15]. Daurian ground squirrels (*Spermophilus dauricus*) are obligatory hibernating mammals. They are found across a wide range of latitudes, from steppe and semi-desert and other arid regions of northern China. Hibernation of Daurian ground squirrels provides a useful model to study mechanisms that increases skeletal muscle resilience against atrophy and dysfunction after extended periods of disuse [16].

The study on measuring skeletal muscle protein metabolism of bears suggests that protein synthesis and breakdown are both lower in winter compared to summer but are equal during both early and late hibernation periods, indicating that bears are in protein balance during hibernation [17]. Which plays a predominant role in the maintenance of skeletal muscle homeostasis involved in the mechanism of protecting from muscle atrophy during prolonged disuse in hibernation, the increase of protein synthesis or the decrease of protein degradation? It is noteworthy that the protein biosynthesis category by overexpressed genes exhibits a highly significant enrichment in skeletal muscle of hibernating black bears (*Ursus americanus*) [18]. However, serum- and glucocorticoid-inducible kinase 1 (SGK1) can regulate muscle mass maintenance via downregulation of proteolysis and autophagy during hibernation in 13-lined ground squirrels (*Ictidomys tridecemlineatus*) [19], which is consistent with our previous report demonstrating that the inhibition of calpain activity and consequently calpain-mediated protein degradation by highly elevated calpastatin protein expression levels may be an important mechanism for preventing muscle protein loss during hibernation [15]. Recently, our group reported that the stable expression of atrogin-1 and MuRF1 may facilitate to prevent SOL [20] and extensor digitorum longus [21] muscle atrophy during hibernation. Although more and more regulatory factors involved in the protein metabolism of skeletal muscle during hibernation were found, the detailed mechanisms of protein synthesis and breakdown in hibernation are far from being elucidated.

To our knowledge we yet understand little about myofibrillar contents and relevant synthesis and proteolytic proteins in soleus muscle of hibernating ground squirrels. It is likely that novel mechanisms are involved but are not yet identified. Proteomics approaches are effective at identifying new protein signaling networks. Herein, we conducted isobaric tags for relative and absolute quantitation (iTRAQ) proteomics experiments in order to discover the hibernation-specific skeletal muscle proteomic changes. The aims of the present study were (1) to identify differentially expressed proteins among pre-hibernation, 60-d-hibernation and post-hibernation Daurian ground squirrels; (2) to explore the myofibrillar protein metabolism mechanism underlying the observed anti-atrophy effects in SOL of hibernating ground squirrels with the special ecological environment physiological adaptation.

Methods

Acquisition of animals

Acquisition and use of animals were approved by the Laboratory Animal Care Committee of the China Ministry of Health. As described previously by our laboratory [15], nine male Daurian ground squirrels were obtained from the Weinan region in the Shaanxi province of China and kept in the laboratory for three to four months after collection for acclimation purposes before they were split into groups. The animal colony room was maintained at a temperature range of 18–20 °C, and lighting was changed daily to coincide with local sunrise and sunset. Animals were given wood chips. Squirrels were provided with water and rodent food blocks, and supplemented with fresh fruit and vegetables. In November, three groups were placed in a cold room hibernaculum at (4–6 °C) (2 L: 22D dark). The dates of entering torpor were determined by putting sawdust on the back of each animal. Daily observations were made during the experimental period. Animals were matched for body mass and were randomly assigned to 3 groups: pre-hibernation: Control (no hibernation) animals investigated in late-autumn, about 30–40 d before hibernation; 60-d hibernation: Animals after two months of hibernation; post-hibernation: Animals two days after arousal from 112 ± 14 days of hibernation, with the SOL muscle collected 48 h after arousal. We used 3 animals per time point, then the 3 samples combined before iTRAQ labeling. In order to minimize the impact of individual differences and avoid interference from other factors, we choose animals with the same sex (male), the similar body (350 ± 20 g) weight and frequency of inter-bout arousal in hibernation for one sample. The SOL muscle was in a disuse state in the hibernation, but was not in a disuse state in the pre-hibernation and post-hibernation group.

Immunofluorescent analysis

Ten-μm thick frozen muscle cross-sections were cut from the mid-belly of muscle at −20 °C with a cryostat (Leica, Wetzlar, CM1850, Germany), and stored at −80° Cfor further staining. Immunofluorescent analysis was used to determine muscle fiber cross-sectional area. After fixing in 4 % paraformaldehyde for 30 min, sections were permeabilized in 0.1 % Triton X-100/PBS for 30 min, blocked with 1 % bovine serum albumin (BSA) in PBS for 60 min at room temperature, and then incubated with the anti-laminin rabbit polyclonal antibody solution (1:50; Santa Cruz, CA, USA) at 4 °C overnight. The slides were rinsed twice in PBS and incubated with TRITC-labeled goat anti-rabbit IgG for 60 min and also counterstained with DAPI (0.5 μg/ml) for 30 min. Images were visualized using a confocal laser scanning microscope (Olympus, Osaka, Japan) at an objective magnification of 40× and were counted on at least 3 different fields or 600 cells of each sample.

Muscle collection and protein preparation

All animal procedures were approved by the Northwest University Ethics Committee. Unless otherwise indicated, all chemicals were purchased from Sigma-Aldrich. Animals were anesthetized with 90 mg/kg sodium pentobarbital i.p. After SOL muscles in ground squirrels were excised from both legs, body mass and wet mass of SOL were recorded. At the end of surgical intervention, the animals were sacrificed by an overdose injection of sodium pentobarbital. Then the SOL muscle samples were ground into powder in liquid nitrogen, extracted with Lysis buffer (7 M Urea, 2 M Thiourea, 4 % CHAPS, 40 mM Tris–HCl, pH 8.5) containing 1 mM PMSF and 2 mM EDTA. After 5 min, 10 mM DTT (final concentration) was added to the samples. The suspension was sonicated at 200 W for 15 min and then centrifuged at 4 °C, 30,000 g for 15 min. The supernatant was mixed well with 5× volume of chilled acetone containing 10 % (v/v) TCA and incubated at −20 °C overnight. After centrifugation at 4 °C, 30,000 g, the supernatant was discarded. The precipitate was washed with chilled acetone three times. The pellet was air-dried and dissolved in Lysis buffer (7 M urea, 2 M thiourea, 4 % NP40, 20 mM Tris–HCl, pH 8.0-8.5). The suspension was sonicated at 200 W for 15 min and centrifuged at 4 °C, 30,000 g for 15 min. The supernatant was transferred to another tube. To reduce disulfide bonds in proteins of the supernatant, 10 mM DTT (final concentration) was added and incubated at 56 °C for 1 h. Subsequently, 55 mM IAM (final concentration) was added to block the cysteines, incubated for 1 h in the darkroom. The supernatant was mixed well with 5× volume of chilled acetone for 2 h at −20 °C to precipitate proteins. After centrifugation at 4 °C, 30,000 g, the supernatant was discarded, and the pellet was air-dried for 5 min, dissolved in 500 μL 0.5 M TEAB (Applied Biosystems, Milan, Italy), and sonicated at 200 W for 15 min. Finally, samples were centrifuged at 4 °C, 30,000 g for 15 min. The supernatant was transferred to a new tube and kept at −80 °C for further analysis.

iTRAQ Labeling and SCX fractionation

iTRAQ analysis was implemented at Beijing Genomics Institute (BGI, Shenzhen, China). Total protein (100 μg) was taken out of each sample solution and then the protein was digested with Trypsin Gold (Promega, Madison, WI, USA) with the ratio of protein/trypsin (30/1) at 37 °C for 16 h. After trypsin digestion, peptides were dried by vacuum centrifugation. Peptides were reconstituted in 0.5 M TEAB and processed according to the manufacture's protocol for 8-plex iTRAQ reagent

(Applied Biosystems). Briefly, one unit of iTRAQ reagent was thawed and reconstituted in 24 μL isopropanol. SOL muscle samples were labeled with the iTRAQ tags as follow: pre-hibernation (tag 113), 60-day hibernation (tag 114) and post-hibernation (tag 116). The peptides were labeled with the isobaric tags, incubated at room temperature for 2 h. The labeled peptide mixtures were then pooled and dried by vacuum centrifugation.

SCX chromatography was performed with a LC-20AB HPLC Pump system (Shimadzu, Kyoto, Japan). The iTRAQ-labeled peptide mixtures were reconstituted with 4 mL buffer A (25 mM NaH_2PO_4 in 25 % ACN, pH 2.7) and loaded onto a 4.6×250 mm Ultremex SCX column containing 5-μm particles (Phenomenex). The peptides were eluted at a flow rate of 1 mL/min with a gradient of buffer A for 10 min, 5–60 % buffer B (25 mM NaH_2PO_4, 1 M KCl in 25 % ACN, pH 2.7) for 27 min, 60–100 % buffer B for 1 min. The system was then maintained at 100 % buffer B for 1 min before equilibrating with buffer A for 10 min prior to the next injection. Elution was monitored by measuring the absorbance at 214 nm, and fractions were collected every 1 min. The eluted peptides were pooled into 20 fractions, desalted with a Strata X C18 column (Phenomenex) and vacuum-dried.

LC-ESI-MS/MS analysis based on Triple TOF 5600

Each fraction was resuspended in buffer A (5 % ACN, 0.1 % FA) and centrifuged at 20,000 g for 10 min, the final concentration of peptide was about 0.5 μg/μL on average. 10 μL supernatant was loaded on a LC-20 AD nano HPLC(Shimadzu, Kyoto, Japan) by the autosampler onto a 2 cm C18 trap column. Then, the peptides were eluted onto a 10 cm analytical C18 column (inner diameter 75 μm) packed in-house. The samples were loaded at 8 μL/min for 4 min, then the 35 min gradient was run at 300 nL/min starting from 2 to 35 % buffer B (95 % ACN, 0.1 % FA), followed by 5 min linear gradient to 60 %, then followed by 2 min linear gradient to 80 %, and maintenance at 80 % buffer B for 4 min, and finally return to 5 % in 1 min. Data acquisition was performed with a Triple TOF 5600 System (AB SCIEX, Concord, ON) fitted with a Nanospray IIIsource (AB SCIEX, Concord, ON) and a pulled quartz tip as the emitter (New Objectives, Woburn, MA). Data was acquired using an ion spray voltage of 2.5 kV, curtain gas of 30 psi, nebulizer gas of 15 psi, and an interface heater temperature of 150 °C. The MS was operated with a RP of greater than or equal to 30,000 FWHM for TOF MS scans. For IDA, survey scans were acquired in 250 ms and as many as 30 product ion scans were collected if exceeding a threshold of 120 counts per second (counts/s) and with a 2+ to 5+ charge-state. Total cycle time was fixed to 3.3 s. Q2 transmission window was 100 Da for 100 %. Four time bins were summed for each scan at a pulser frequency value of 11 kHz through monitoring of the 40 GHz multichannel TDC detector with four-anode channel detect ion. A sweeping collision energy setting of 35 ± 5 eV coupled with iTRAQ adjust rolling collision energy was applied to all precursor ions for collision induced dissociation. Dynamic exclusion was set for 1/2 of peak width (15 s), and then the precursor was refreshed off the exclusion list.

Data analysis

Raw data files acquired from the Orbitrap were converted into MGF files using Proteome Discoverer 1.2 (PD 1.2, Thermo), [5600 msconverter] and the MGF file were searched. Proteins identification was performed by using Mascot search engine (Matrix Science, London, UK, version 2.3.02) against database containing 28,942 sequences (up to date 2014-3-11). For protein identification, a mass tolerance of 0.05 Da (ppm) was permitted for intact peptide masses and 0.1 Da for fragmented ions, with allowance for one missed cleavages in the trypsin digests. Gln- > pyro-Glu (N-term Q), Oxidation (M), Deamidated (NQ) as the potential variable modifications, and Carbamidomethyl (C), iTRAQ8plex (N-term), iTRAQ8plex (K) as fixed modifications. The charge states of peptides were set to +2 and +3. Specifically, an automatic decoy data base search was performed in Mascot by choosing the decoy checkbox in which a random sequence of database is generated and tested for raw spectra as well as the real database. To reduce the probability of false peptide identification, only peptides with significance scores ($\geq$20) at the 99 % confidence interval by a Mascot probability analysis greater than"identity" were counted as identified. And each confident protein identification involves at least one unique peptide. For protein quantitation, it was required that a protein contains at least two unique peptides. The quantitative protein ratios were weighted and normalized by the median ratio in Mascot. We only used ratios with p-values < 0.05, and only fold changes of >1.2 was considered as significant.

Function method description

Functional annotations of the proteins were conducted using Blast2GO program against the non-redundant protein database (NR; NCBI). The KEGG database (http://www.genome.jp/kegg/) and the COG database (http://www.ncbi.nlm.nih.gov/COG/) were used to classify and group these identified proteins.

Pathway analysis

KEGG PATHWAY is a database resource and a collection of manually drawn pathway maps [22, 23] representing our knowledge on the molecular interaction and reaction networks between the identified differentially expressed proteins in 60-d hibernation and post-hibernation ground squirrels groups compared to the pre-hibernation groups. Molecules are represented as nodes, and the biological relationship between two nodesis represented as an edge (line).

Western blotting
To confirm the reliability of iTRAQ based quantitative analysis, protein samples used for iTRAQ were further examined by western blot which were undertaken as previously described [24]. Briefly, total protein was extracted from the SOL muscle of ground squirrels and solubilized in a sample buffer (100 mM Tris, pH 6.8, 5 % 2-β-mercaptoethanol, 5 % glycerol, 4 % SDS, and bromophenol blue), with muscle protein extracts resolved by SDS-PAGE using Laemmli gels (10 % gel with an acrylamide/bisacrylamide ratio of 37.5 : 1 for EEF-2; and 12 % gel with an acrylamide/bisacrylamide ratio of 29 : 1 for RPS8, proteasome 20S α5, CAPNS1, actin, and troponin C. After electrophoresis, the proteins were electrically transferred to PVDF membranes (0.45 μm pore size) using a Bio-Rad semi-dry transfer apparatus. The blotted membranes were blocked with 1 % BSA in Tris-buffered saline (TBS; 150 mM NaCl, 50 mM Tris–HCl, pH 7.5) and incubated with rabbit anti-Proteasome 20S α5, rabbit anti-Actin, rabbit anti-Troponin C, rabbit anti-EEF2 and rabbit anti-RPS8 (1:1000, Abcam, Cambridge, MA, USA) and rabbit anti-CAPNS1 (1:1000, Sigma, St. Louis, MO, USA) in TBS containing 0.1 % BSA at 4 °C overnight. After washing 3 times, the membranes were then incubated with HRP-conjugated anti-rabbit secondary antibodies (Pierce Chemical Co., Dallas, USA) for 1 h at room temperature. After another washing, immunoblots were visualized using enhanced chemiluminescence (ECL) reagents (Thermo, Rockford, USA.) according to the manufacturer's protocols. Quantification analysis of blots was performed with the NIH Image J software.

Total protein breakdown in incubated muscles
Rates of protein degradation was determined by the release of tyrosine in incubated soleus muscles as previously described [25]. As tyrosine is present in all proteins, its release reflects total protein breakdown. Briefly, the ground squirrels were anesthetized and the soleus muscles were gently dissected and excised with intact tendons. Immediately after weighing, muscles were secured with custom plastic clips at approximately resting length in order to better maintain their energy level and protein balance. Muscles were incubated for 30 min in oxygenated (O_2 : CO_2 = 95 : 5) Krebs-Henseleit buffer (pH 7.4) containing 5 mM glucose and 0.15 mM pyruvate. After preincubation, one muscle was homogenized

in 3 % perchloric acid for determination of tissue levels of free tyrosine. The contralateral muscle was then blotted and transferred to a new incubation well containing 3 mL Krebs-Henseleit/glucose/pyruvate buffer and 0.5 mM cycloheximide, which inhibited reincorporation of amino acids by protein synthesis. Three hours after incubation, muscles were removed, blotted dry, and frozen in liquid nitrogen. Then the muscle and medium concentrations of free tyrosine were measured by high performance liquid chromatography (Agilent HPLC system, Column: Agilent TC-C18, 5 μm, 4.6 × 250 mm, USA), which was equipped with a binary pump (Agilent G1312A) and a fluorescence detector (Agilent G1321A). Rates of protein degradation are given as nmol tyrosine per gram of muscle wet weight per 3 h.

Statistical analyses
A one-way ANOVA with Fisher's LSD post hoc test was used to determine group differences, and the ANOVA–Dunnett's T3 method was used when no homogeneity was detected. SPSS 19.0 was used for all statistical tests. Statistical significance was accepted for all tests at $P < 0.05$.

Results
Body weight, soleus muscle wet weight and muscle fiber cross-sectional area (CSA)
There was a steady decline in mean body weight from 352 ± 24 g in pre-hibernation to 263 ± 8 g after 60 days of hibernation, and there was even a sharp body weight decrease to 222 ± 11 g in post-hibernation. In other words, the ground squirrels lost 25 and 40 % of the body weight after hibernation for 60 days and post-hibernation, respectively (Fig. 1a). However, the SOL muscle wet weights were decreased slightly in 60-d or post-hibernation ground squirrels in comparison with the pre-hibernation group (Fig. 1b). Meanwhile, the CSA of SOL muscle also showed no significant decrease in 60-d or post-hibernation groups as compared with the pre-hibernation group (Fig. 1c and d).

Proteomics analysis
In the present study, an iTRAQ-based quantitative proteomics approach in combination with LC-ESI-MS/MS was applied to investigate differentially expressed proteins in the SOL of pre-hibernation, 60-d hibernation and post-hibernation ground squirrels. Proteomics analysis identified 11,897 peptides mapped to 2059 proteins. iTRAQ ratio of > 1.20 and < 0.83 (P-value < 0.05) was used to define proteins that are significantly up-regulated or down-regulated, respectively. With this filter, we identified 170 and 333 differentially regulated proteins in the 60-d and post-hibernation groups relative to the pre-hibernation ground squirrels, respectively.

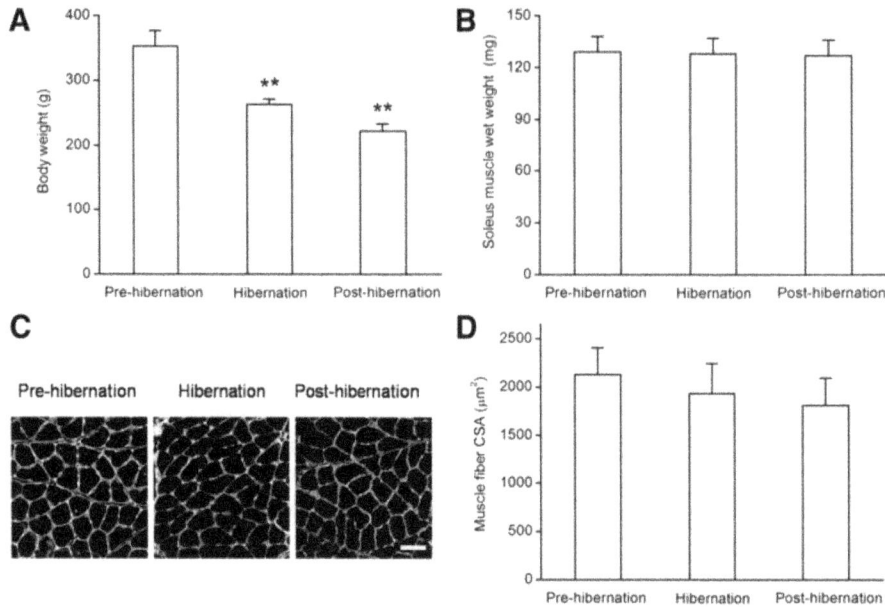

Fig. 1 Body weight (**a**) and soleus muscle wet weight (**b**) of Daurian ground squirrels in pre-hibernation, 60-d and post-hibernation groups. Representative images (**c**) and bar graph (**d**) showing the changes in cross-sectional area of soleus muscle from 3 groups. White represents the anti-laminin stain of interstitial tissue counterstained with DAPI (*blue*) for nuclei identifcation. Scale bar = 50 μm. Values are means ± SEM, $n = 6$ in each group. **$P < 0.01$ versus pre-hibernation group

Besides, 273 differentially regulated proteins were identified in the post-hibernation group compared to the 60-d hibernation ground squirrels. These proteins were subjected to gene-ontology enrichment. Among the three groups, the differences between 60-d hibernation and pre-hibernation were smaller (96 vs 248 up-regulated proteins and 74 vs 85 down-regulated proteins) as compared with the differences between post-hibernation and pre-hibernation ground squirrels. Moreover, 216 up-regulated proteins and 57 down-regulated were identified in the post-hibernation group compared to the 60-d hibernation ground squirrels. Specifically, there were 248 up-regulated expressed proteins between post-hibernation and pre-hibernation ground squirrels, almost one third of the total number (776) of differentially expressed proteins among three ground squirrels groups.

Gene ontology (GO) classification of differentially expressed proteins

To elucidate the biological significance of the 776 differentially modified proteins, we performed GO analysis and categorized these proteins according to their molecular function and biological process using the KEGG database (http://www.genome.jp/kegg/) and the COG database (http://www.ncbi.nlm.nih.gov/COG/). Of the 776 proteins, 123 were selected from these analyses and separated into 3 categories according to their molecular function: protein synthesis, protein proteolysis and structural constituent of muscle. Figure 2 showed the

number of significantly ($P < 0.05$) differentially expressed proteins in the 3 categories. Proteins were annotated according to the gene ontology (GO) classification (Tables 1, 2 and 3).

Pathway analysis

Ribosomes are the sites of protein synthesis, the increased expression of these proteins may have improved protein synthesis in hibernating ground squirrels. Of the proteins identified by iTRAQ that were differentially expressed between 60-d hibernation group and pre-

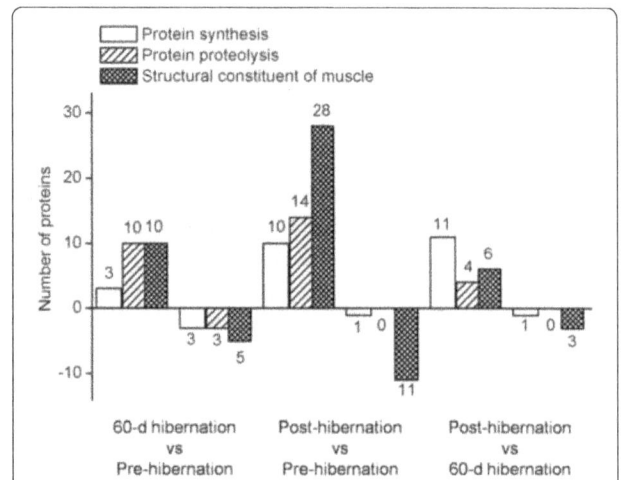

Fig. 2 GO classification of differentially expressed proteins

Table 1 Summary table showing significantly up-regulated or down-regulated proteins in SOL of Daurian ground squirrels identified by iTRAQ

Acc no. (NCBI)	Prot name [Ictidomys tridecemlineatus]	MW (Da)	Peptide	Score	60-d hibernation vs pre-hibernation		Post-hibernation vs pre-hibernation		Post-hibernation vs 60-hibernation	
					Fold change	Sig	Fold change	Sig	Fold change	Sig
Up-regulated proteins										
gi|532099039	elongation factor 1-beta isoform *X2*	32189	6	175	1.348	*	1.202	*		
gi|532060958	60S ribosomal protein L39-like	9445	1	58	1.791	*				
gi|532064086	polyadenylate-binding protein 4 isoform X1	87231	7	188	1.508	*				
gi|532076481	ATP-dependent RNA helicase DDX3X	60127	11	481			1.283	*		
gi|532064280	40S ribosomal protein S8	34224	6	177			1.349	*		
gi|532090722	eukaryotic initiation factor 4A-II isoform X1	51896	9	769			1.232	*		
gi|532100826	60S ribosomal protein L38 isoform *X2*	13441	2	71			1.679	*		
gi|532073691	60S ribosomal protein L30-like	17542	2	66			1.366	*		
gi|532092237	60S ribosomal protein L10-like isoform X1	32972	1	68			1.276	*	1.419	*
gi|532092821	elongation factor 2	116340	23	1556			1.227	*		
gi|532114042	60S acidic ribosomal protein P2 isoform X3	15038	5	699			1.455	*		
gi|532060529	elongation factor 1-delta isoform X4	37312	8	983			1.476	*	1.292	*
gi|532064519	60S ribosomal protein L11	13027	1	53					1.329	*
gi|532085355	40S ribosomal protein S25	22308	2	114					1.228	*
gi|532098257	60S ribosomal protein L22	14812	2	160					1.364	*
gi|532071999	60S ribosomal protein L23	19837	5	246					1.345	*
gi|532082343	60S ribosomal protein L27a	27522	2	86					1.214	*
gi|532110684	40S ribosomal protein S4, X isoform	38325	3	192					1.244	*
gi|532082140	40S ribosomal protein S13	22992	4	72					1.238	*
gi|532094470	60S ribosomal protein L26	24853	1	34					1.225	*
gi|532075928	40S ribosomal protein S15a	18594	3	117					1.223	*
Down-regulated proteins										
gi|532085355	40S ribosomal protein S25	22308	2	114	0.758	*				
gi|532098257	60S ribosomal protein L22	14812	2	160	0.723	*				
gi|532083400	40S ribosomal protein S7	33087	3	90	0.735	*				
gi|532066199	60S ribosomal protein L10a	25228	2	105			0.74	*		
gi|532096351	28S ribosomal protein S25	24779	1	51					0.588	*

Acc no. (Accession number), Prot name (protein name), MW (Molecular mass), Peptide (number of peptides matched), Score (Mowse score), Sig (*P-value < 0.05 was considered statistically significant)

hibernation group, 5 enriched in the ribosomal assembly pathways (60S ribosomal protein L39-like, 40S ribosomal protein S6, 40S ribosomal protein S25, 60S ribosomal protein L22, 40S ribosomal protein S7, Fig. 3), and 8 of the proteins that were differentially expressed between post-hibernation group and pre-hibernation group enriched in this category (60S ribosomal protein L30-like, 60S acidic ribosomal protein P2 isoform X3, 60S ribosomal protein L10a, 40S ribosomal protein S13, 60S ribosomal protein L10-like isoform X1, ubiquitin-60S ribosomal protein L40-like, 40S ribosomal protein S8, 60S ribosomal protein L38 isoform *X2*, Fig. 4). Meanwhile, 12 up-regulated proteins were enriched in the ribosomal assembly pathways between post-hibernation group and 60-d hibernation group (60S ribosomal protein L23, 60S ribosomal protein L26, 40S ribosomal protein S4, 60S ribosomal protein L11, 40S ribosomal protein S15a, 60S ribosomal protein L27a, 60S acidic ribosomal protein P2 isoform X3, 40S ribosomal protein S13, 60S ribosomal protein L10-like isoform X1, 40S ribosomal protein S25, 60S ribosomal protein L22, 40S ribosomal protein S7, Fig. 5).

Numerous proteins were also involved in protein degradation. Proteasome is a multicatalytic proteinase complex with a highly ordered ring-shaped 20S core structure and distributed throughout eukaryotic cells at a high concentration and cleave peptides in an ATP-dependent process [26]. We found that 8 up-regulated

Table 2 Summary table showing significantly up-regulated or down-regulated proteins with protein proteolysis function in SOL of Daurian ground squirrels identified by iTRAQ

Acc no. (NCBI)	Prot name [Ictidomys tridecemlineatus]	MW (Da)	Peptide	Score	60-d hibernation vs pre-hibernation		Post-hibernation vs pre-hibernation		Post-hibernation vs 60-d hibernation	
					Fold change	Sig	Fold change	Sig	Fold change	Sig
Up-regulated proteins										
gi|532073955	proteasome subunit alpha type-6	28470	8	480	1.23	*	1.248	*		
gi|532105099	proteasome subunit alpha type-5	30824	7	614	1.226	*	1.226	*		
gi|59800155	26S protease regulatory subunit 10B	53861	7	589	1.281	*	1.264	*		
gi|532070252	26S proteasome non-ATPase regulatory subunit 5 isoform X1	63125	2	46	1.517	*				
gi|532091148	26S proteasome non-ATPase regulatory subunit 4	48541	5	133	1.488	*				
gi|532103645	proteasome subunit beta type-6 isoform X1	27252	3	115	1.295	*				
gi|532073989	proteasome subunit alpha type-3 isoform X1	34727	5	462	1.209	*				
gi|532066291	cytosol aminopeptidase	66612	11	731	1.202	*				
gi|532101732	26S protease regulatory subunit 6B isoform X1	56577	4	375	1.246	*				
gi|532088665	26S protease regulatory subunit 6A	60713	9	638	1.32	*				
gi|532080894	calpastatin	113441	7	179			1.226	*		
gi|532073188	endoplasmin	117277	14	399			1.316	*		
gi|532076563	ubiquitin-like modifier-activating enzyme 1 isoform X2	135211	19	1304			1.242	*		
gi|532069763	xaa-Pro aminopeptidase 1-like isoform X1	88255	5	128			1.485	*		
gi|532062055	proteasome subunit alpha type-4	37964	6	247			1.289	*		
gi|532114683	calpain small subunit 1	31666	4	257			1.301	*	1.488	*
gi|532105581	ubiquitin carboxyl-terminal hydrolase 5 isoform X1	112519	10	369			1.221	*		
gi|532082471	acylamino-acid-releasing enzyme	91596	7	160			1.567	*		
gi|532088974	prolyl endopeptidase	95929	11	396			1.385	*		
gi|532078165	ubiquitin carboxyl-terminal hydrolase isozyme L3	30900	4	212			1.591	*	1.552	*
gi|532097874	ubiquitin fusion degradation protein 1 homolog isoform X1	41760	5	206			1.419	*		
gi|532116708	proteasome subunit beta type-1	30778	7	506					1.283	*
gi|532085114	26S proteasome non-ATPase regulatory subunit 12 isoform X1	66713	4	67					1.457	*
Down-regulated proteins										
gi|532064840	ubiquinone biosynthesis protein COQ9, mitochondrial	38864	6	713	0.811	*				
gi|532078642	poly(rC)-binding protein 2 isoform X1	54651	5	470	0.732	*				
gi|532086828	cullin-5	114587	6	160	0.774	*				

Acc no. (Accession number), Prot name (protein name), MW (Molecular mass), Peptide (number of peptides matched), Score (Mowse score), Sig (*P-value < 0.05 was considered statistically significant)

proteins enriched in the proteasome pathway between 60-d hibernation group and pre-hibernation group (26S proteasome non-ATPase regulatory subunit 4, 26S protease regulatory subunit 10B, 26S protease regulatory subunit 6A, 26S protease regulatory subunit 6B isoform X1, proteasome subunit alpha type-6, proteasome subunit alpha type-5, proteasome subunit alpha type-3 isoform X1, proteasome subunit beta type-6 isoform X1, Fig. 6), 4 up-regulated proteins enriched in the proteasome pathway between post-hibernation group and pre-hibernation group (26S protease regulatory subunit 10B, proteasome subunit alpha type-6, proteasome subunit alpha type-4, proteasome subunit alpha type-5, Fig. 7), and 2 up-regulated proteins enriched in this

Table 3 Summary table showing significantly up-regulated or down-regulated proteins of structural constituent of SOL muscle in Daurian ground squirrels identified by iTRAQ

Acc no. (NCBI)	Prot name [Ictidomys tridecemlineatus]	MW (Da)	Peptide	Score	60-d hibernation vs pre-hibernation		Post-hibernation vs pre-hibernation		Post-hibernation vs 60-d hibernation	
					Fold change	Sig	Fold change	Sig	Fold change	Sig
Up-regulated proteins										
gi|532098453	small muscular protein isoform X2	12381	3	504	1.238	*	1.392	*		
gi|532095619	cofilin-1	26628	7	639	1.239	*	1.535	*		
gi|532065022	tubulin polymerization-promoting protein family member 3	27600	2	147	1.607	*	1.481	*		
gi|532085001	myosin light chain 4 isoform X2	27241	6	968	1.434	*	1.579	*		
gi|532097561	myelin basic protein	44089	2	69	1.456	*				
gi|532069819	nebulin-related-anchoring protein	245551	22	744	1.211	*				
gi|532094579	myosin-2	244418	102	66551	2.135	*				
gi|532087098	PDZ and LIM domain protein 3 isoform X1	45822	12	3700	1.304	*				
gi|532062716	troponin C	22795	11	5693	1.358	*				
gi|532080240	vimentin	54132	26	3859	1.211	*			1.828	*
gi|532089364	myosin-binding protein H	60653	11	835			1.696	*		
gi|532081847	zyxin	72401	2	97			1.371	*		
gi|532111896	tubulin alpha-8 chain isoform X1	60805	15	2177			1.355	*		
gi|532063429	F-actin-capping protein subunit beta isoform X2	36610	11	763			1.219	*		
gi|532074211	nidogen-2 isoform X1	165553	7	164			1.467	*		
gi|532076843	myozenin-3	30065	6	801			1.401	*		
gi|532100104	myosin light polypeptide 6 isoform X1	20132	8	2096			1.576	*		
gi|532099239	moesin-like	87289	14	565			1.248	*		
gi|532103138	myosin regulatory light chain 12B	24691	6	374			1.771	*		
gi|532081115	talin-1	321070	21	790			1.288	*	1.302	*
gi|532088445	myosin regulatory light polypeptide 9 isoform X1	24375	5	323			1.68	*		
gi|532095126	PDZ and LIM domain protein 7	60050	6	361			1.653	*		
gi|532111349	ankyrin-1 isoform X1	242338	33	1740			2.69	*		
gi|532098546	tubulin beta-5 chain-like	46713	12	1245			1.367	*		
gi|532061621	annexin A2	49151	21	2021			1.502	*		
gi|532068053	myocilin	59868	4	159			2.837	*		
gi|532100974	neurofilament heavy polypeptide	158649	7	679			1.55	*		
gi|532110489	neurofilament medium polypeptide	123496	7	571			1.677	*		
gi|532089731	tubulin alpha-4A chain	56718	18	3110			1.396	*		
gi|532089733	tubulin alpha-1D chain-like	57019	18	3316			2.103	*		
gi|532063781	protein 4.1 isoform X1	120337	9	505			2.552	*		
gi|532111734	alpha-actinin-4 isoform X1	125310	18	2664			1.228	*		
gi|532072736	beta-adducin isoform X1	100809	3	84			1.759	*		
gi|532083652	tubulin beta-2A chain isoform X1	56576	21	2455			1.637	*	1.687	*
gi|532081219	tropomodulin-1	49085	4	141					1.509	*
gi|532102653	tubulin beta-4B chain isoform X1	55122	22	3594					1.25	*
gi|532062201	fibrillin-1	340347	14	616					1.685	*
Down-regulated proteins										
gi|532094519	myosin-3	287961	49	33292	0.587	*	0.399	*	0.758	*

Table 3 Summary table showing significantly up-regulated or down-regulated proteins of structural constituent of SOL muscle in Daurian ground squirrels identified by iTRAQ *(Continued)*

Acc no.	Prot name	MW	Peptide	Score		Sig		Sig		Sig
gi|532085261	transgelin isoform X2	28099	11	691	0.784	*				
gi|532062201	fibrillin-1	340347	14	616	0.534	*				
gi|532072041	telethonin	21240	8	817	0.811	*				
gi|532081219	tropomodulin-1	49085	4	141	0.701	*				
gi|532103167	actin	48418	23	58752			0.823	*	0.806	*
gi|532059726	xin actin-binding repeat-containing protein 2-like	476702	6	84			0.774	*		
gi|532069819	nebulin-related-anchoring protein	245551	22	744			0.747	*		
gi|532094510	myosin-13	285781	37	26764			0.478	*		
gi|532077683	kelch-like protein 41	85237	21	1666			0.772	*		
gi|532097391	supervillin	300575	12	302			0.789	*		
gi|532112765	cytoplasmic dynein 1 heavy chain 1	620997	16	383			0.807	*		
gi|532113339	obscurin, partial	143269	19	783			0.829	*		
gi|532112836	flotillin-1	56752	3	161			0.806	*		
gi|532062845	filamin-B	333949	10	1568			0.743	*		
gi|532094579	myosin-2	244418	102	66551					0.482	*

Acc no. (Accession number), Prot name (protein name), MW (Molecular mass), Peptide (number of peptides matched), Score (Mowse score), Sig (*P-value < 0.05 was considered statistically significant)

category between post-hibernation group and 60-d -hibernation group (26S proteasome non-ATPase regulatory subunit 12 isoform X1, proteasome subunit beta type-1, Fig. 8).

Validation of the quantitative proteomic analysis by Western blotting

Six proteins (EEF-2 and RPS8 associated with proterin synthesis, proteasome 20S α5 and CANPS1 associated with proterin proteolysis, actin and troponin C associated with myofibrillar contents) with the marked differences in expression determined by iTRAQ based quantitative analysis were selected to be verified by western blot analysis. As shown in Fig. 9, EEF-2, RPS8, proteasome 20S α5 and CNPNS1 were up-regulated ($P < 0.05$) in post-hibernation group as compared with pre-hibernation group, and proteasome 20S α5 and troponin C were up-regulated ($P < 0.05$) in hibernation group as compared with pre-hibernation group. However, actin was down-regulated in post-hibernation group as compared with pre-hibernation group ($P < 0.05$), which is consistent with the findings in iTRAQ analysis.

Determination of protein proteolytic rate

To determine whether the anti-muscle atrophy in the soleus of hibernating ground squirrels was a result of decreased proteolysis, we measured protein degradation by the release of the essential amino acid tyrosine. The rates of protein degradation decreased by 73 % ($P < 0.001$) in hibernation group as compared with the pre-hibernation group. Although the proteolytic rate in post-

hibernation squirrels increased approximately 200 % ($P < 0.001$) as compared with the hibernation group, it was still 20 % ($P < 0.01$) lower than the pre-hibernation group (Fig. 10).

Discussion

In present study, we collected the SOL muscle from 60-d and post-hibernation group with 2 months or more than 3 months of hibernation inactivity, which was long enough to cause significant atrophy in SOL muscle of non-hibernators [15]. The SOL muscle wet weight decreased less than the body weight in hibernating ground squirrels, which suggested an anti-atrophy effect during hibernation. Proteomic analysis was performed to investigated the protective proteins changes in soleus muscle of ground squirrels.

Protective remodeling of myofibrillar proteins in preventing atrophy in SOL of Daurian ground squirrels during hibernation inactivity

The cytoplasm of a myofiber contains a regular array of contractile units (sarcomeres) comprised of actin-containing thin filaments and myosin-containing thick filaments that, along with additional regulatory and structural proteins, are arranged longitudinally as myofibrils [27]. With iTRAQ approach, we identified myosin-2 significantly up-regulated while myosin-3 significantly down-regulated in the 60-d hibernation group compared to the pre-hibernation group, meanwhile, myosin-13 significantly down-regulated in the post-hibernation group compared to the pre-hibernation group. Moreover, actin

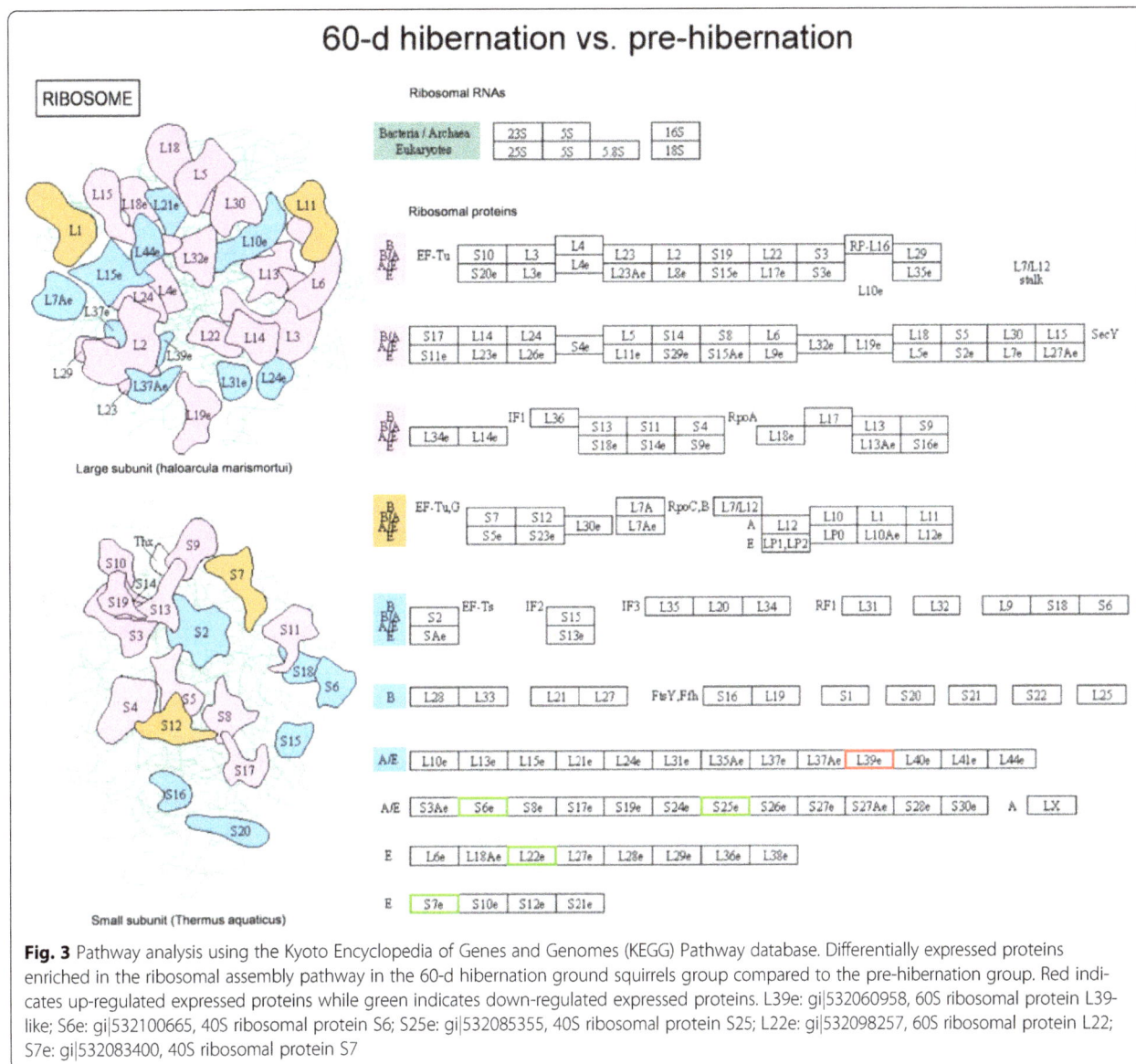

Fig. 3 Pathway analysis using the Kyoto Encyclopedia of Genes and Genomes (KEGG) Pathway database. Differentially expressed proteins enriched in the ribosomal assembly pathway in the 60-d hibernation ground squirrels group compared to the pre-hibernation group. Red indicates up-regulated expressed proteins while green indicates down-regulated expressed proteins. L39e: gi|532060958, 60S ribosomal protein L39-like; S6e: gi|532100665, 40S ribosomal protein S6; S25e: gi|532085355, 40S ribosomal protein S25; L22e: gi|532098257, 60S ribosomal protein L22; S7e: gi|532083400, 40S ribosomal protein S7

exhibited significantly down-regulated expression in the post-hibernation group compared to the pre-hibernation group (Table 3). The ratio of actin to myosin is one of the muscular atrophy hallmarks [28]. After 60-d hibernation inactivity, the ratio of actin/myosin filaments was likely to remain steady, however, a significant loss of myofibrillar proteins occured in post-hibernation group. Inconsistent with our results, the level of one isoform of actin was significantly higher ($P < 0.05$) in hibernation group than in summer active group in pectoral or biceps brachii muscles of the bat *Murina leucogaster* [28], which might be due to that the pectoral or biceps brachii muscles are involved in flight. But compared with the non-hibernator, lost of myofibrillar proteins in Daurian ground squirrel in hibernation is limited. For example, after 3 weeks of hindlimb unloading, levels of contractile

proteins decreased by 40–70 %, and the ratio of actin/myosin filaments decreased by 31 % [8]. Reduction in muscle quality caused by alterations in myofilament contractile proteins (myosin and actin) may scale up from the molecular to the single fiber and tissue level to impact muscle performance [29]. Thus, up-regulation of myosin-2 is one of the most important mechanisms to maintain the integrity of the SOL muscle fiber in the 60-d hibernation ground squirrels. Because the function of myosin-3 and myosin-13 in hibernation is unknown, the relationship between the decrease of these two myosin subtypes and skeletal muscle function is not clarified.

Other regulatory proteins also showed significant changes in hibernation. Troponin C was found significantly up-regulated while tropomodulin-1 down-regulated in the 60-d hibernation group compared to the pre-hibernation

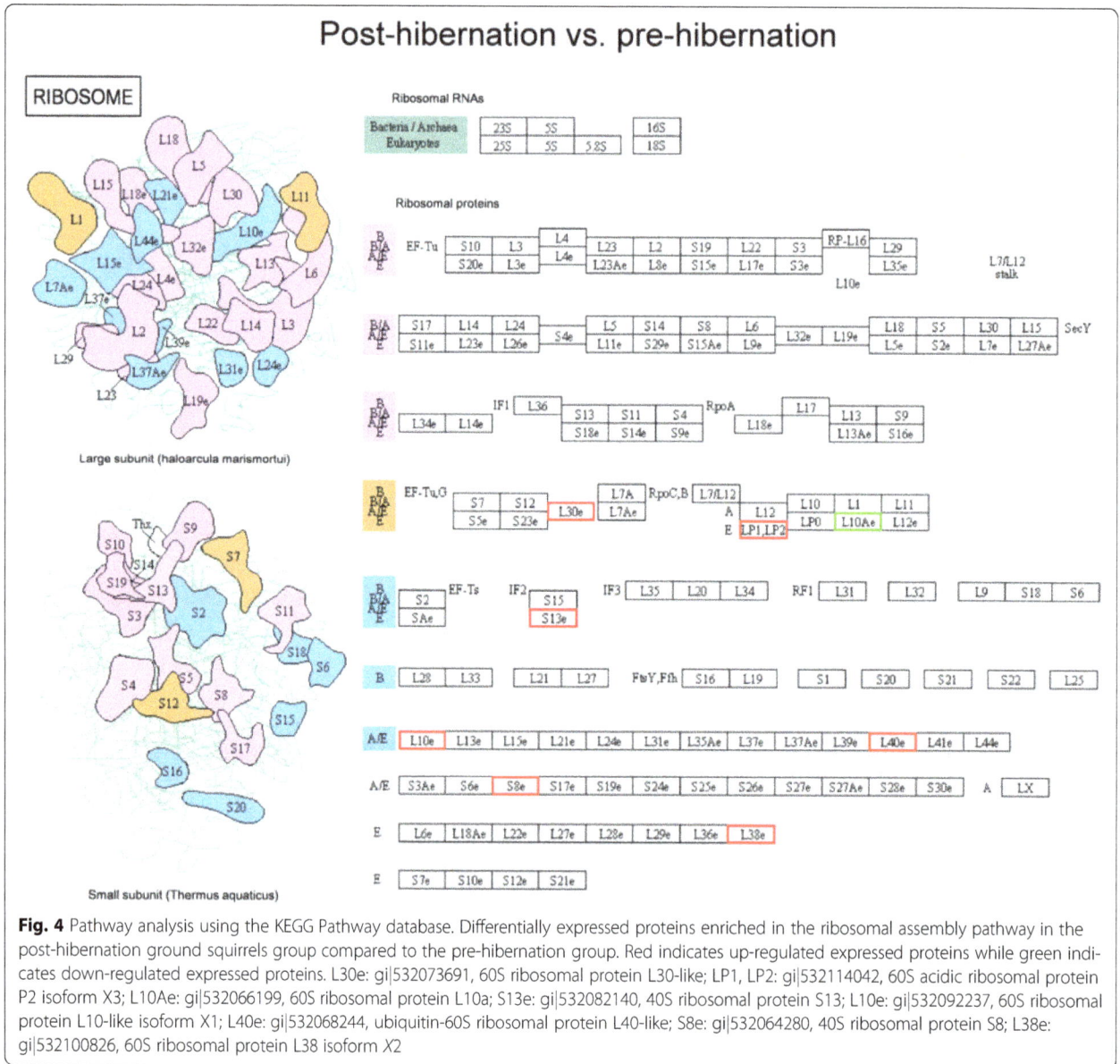

Fig. 4 Pathway analysis using the KEGG Pathway database. Differentially expressed proteins enriched in the ribosomal assembly pathway in the post-hibernation ground squirrels group compared to the pre-hibernation group. Red indicates up-regulated expressed proteins while green indicates down-regulated expressed proteins. L30e: gi|532073691, 60S ribosomal protein L30-like; LP1, LP2: gi|532114042, 60S acidic ribosomal protein P2 isoform X3; L10Ae: gi|532066199, 60S ribosomal protein L10a; S13e: gi|532082140, 40S ribosomal protein S13; L10e: gi|532092237, 60S ribosomal protein L10-like isoform X1; L40e: gi|532068244, ubiquitin-60S ribosomal protein L40-like; S8e: gi|532064280, 40S ribosomal protein S8; L38e: gi|532100826, 60S ribosomal protein L38 isoform X2

group. However, both troponin C and tropomodulin-1 was unaltered in the post-hibernation group. The binding of Ca^{2+} to troponin C induces a series of conformational changes in troponin complex and sarcomeric actin thin filament to activate cross bridge cycling between myosin and actin and muscle contraction [30]. Troponin C increased by 59 % in soleus of human bed rest study [31], which indicated that troponin C in SOL of hibernation changed similarly as in disuse. Another regulatory protein tropomodulin, which is the only protein known to cap the pointed end of actin filaments, plays an important role in actin-driven processes by controlling the addition and dissociation of actin subunits at filament ends [32]. Calpain-mediated proteolysis of tropomodulin isoforms leads to thin filament elongation in dystrophic skeletal muscle [33]. It appears reasonable to assume that the changes of tropomodulin might be an adaptive factor for inhibiting the contractile activity during hibernation. Moreover, over-expression of tropomodulin-1 in mouse hearts results in degenerating myofibrils [34]. Therefore, we assume that the down-regulated tropomodulin-1 in 60-d hibernation might be a crucial component for regulating the length of actin-containing thin filament in soleus during hibernation.

Sarcomeric structural proteins such as α-actinin, titin, tropomyosin and desmin were not detected proteomic variations in present study, which supported our previous findings that the stable expression of atrogin-1 and MuRF1 may facilitate to prevent SOL atrophy via controlling ubiquitination of muscle proteins during hibernation [20, 21]. However, evidence showed that long-term disuse

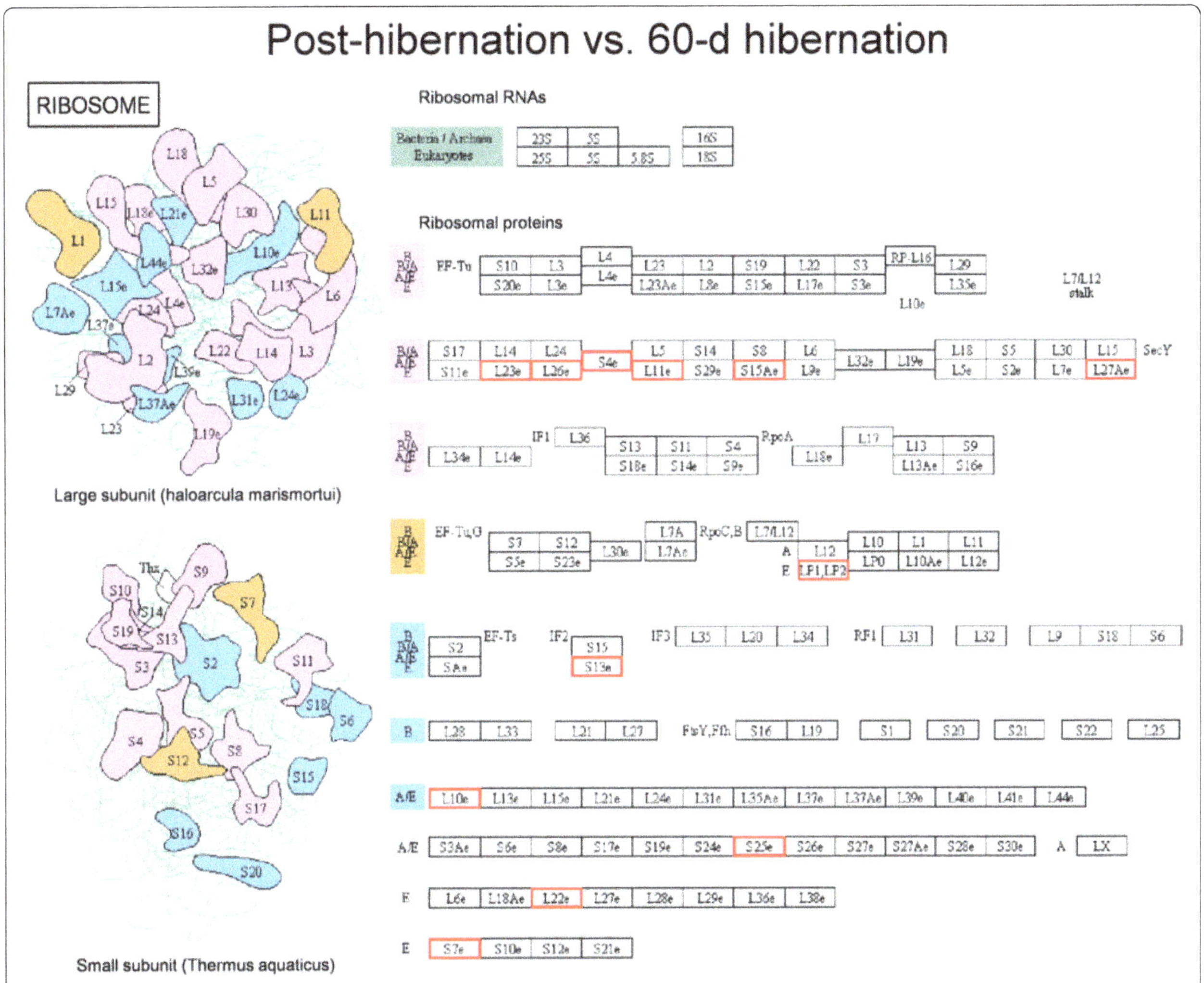

Fig. 5 Pathway analysis using the KEGG Pathway database. Differentially expressed proteins enriched in the ribosomal assembly pathway in the post-hibernation ground squirrels group compared to the 60-d hibernation group. Red indicates up-regulated expressed proteins. L23e: gi|532071999, 60S ribosomal protein L23; L26e: gi|532094470, 60S ribosomal protein L26; S4e: gi|532110684, 40S ribosomal protein S4; L11e: gi|532064519, 60S ribosomal protein L11; S15Ae: gi|532075928, 40S ribosomal protein S15a; L27Ae: gi|532082343, 60S ribosomal protein L27a; LP1,LP2: gi|532114042, 60S acidic ribosomal protein P2 isoform X3; S13e: gi|532082140, 40S ribosomal protein S13; L10e: gi|532092237, 60S ribosomal protein L10-like isoform X1; S25e: gi|532085355, 40S ribosomal protein S25; L22e: gi|532098257, 60S ribosomal protein L22; S7e: gi|532083400, 40S ribosomal protein S7

causes preferential loss of the giant sarcomere protein titin results in altered muscle function via abnormal sarcomeric organization [35]. The level of the T2 fragments of titin was observed decreased in the skeletal muscles of hibernating brown bear (*Ursidae, Mammalia*) [36]. Absence of α-actinin-3 resulted in reduced atrophic response and altered adaptation to disuse [37]. Desmin and titin globally reduced in hibernating myocardium suggested a qualitative cardiomyocyte remodeling [38]. Obviously, homeostasis of most sarcomeric structural proteins is an important mechanism against disuse atrophy in ground squirrels.

In conclusion, these results suggested that myofibrillar protective remodeling marked by dysregulated contractile proteins (myosins and actin) and regulatory proteins (troponin and tropomodulin), and maintain of most sarcomeric structural proteins is a major factor in protecting atrophy in SOL of Daurian ground squirrels during hibernation.

The role of protein synthesis and proteolysis in preventing atrophy in SOL of Daurian ground squirrels during hibernation

Protein balance in skeletal muscles is a delicate interplay between protein synthesis and degradation [39]. High degradation and low synthesis of the proteins is known to cause significant loss of myofibrillar contents in most non-hibernators, including humans, prolonged disuse of skeletal muscle, as seen in bed rest, hindlimb suspension

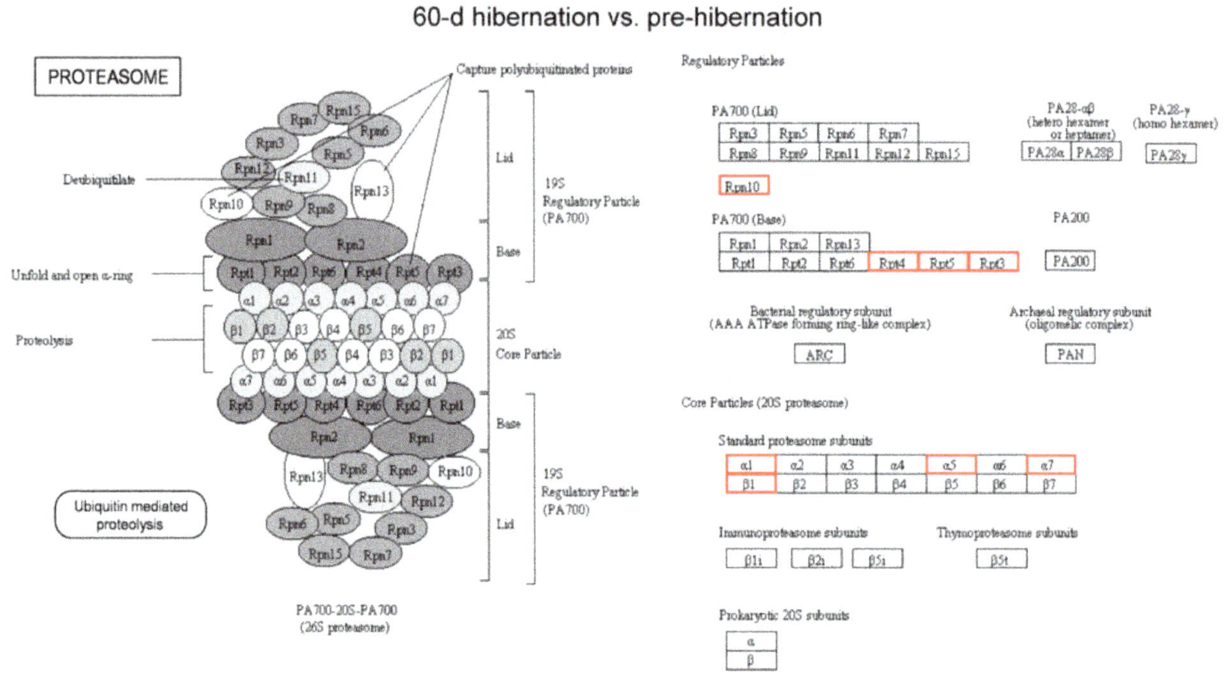

Fig. 6 Pathway analysis using the KEGG Pathway database. Differentially expressed proteins enriched in the proteasome pathway in the 60-d hibernation ground squirrels group compared to the pre-hibernation group. Red indicates up-regulated expressed proteins. Rpn10: gi|532091148, 26S proteasome non-ATPase regulatory subunit 4; Rpt4: gi|59800155, 26S protease regulatory subunit 10B; Rpt5: gi|532088665, 26S protease regulatory subunit 6A; Rpt3: gi|532101732, 26S protease regulatory subunit 6B isoform X1; α1: gi|532073955, proteasome subunit alpha type-6; α5: gi|532105099, proteasome subunit alpha type-5; α7: gi|532073989, proteasome subunit alpha type-3 isoform X1. β1: gi|532103645, proteasome subunit beta type-6 isoform X1

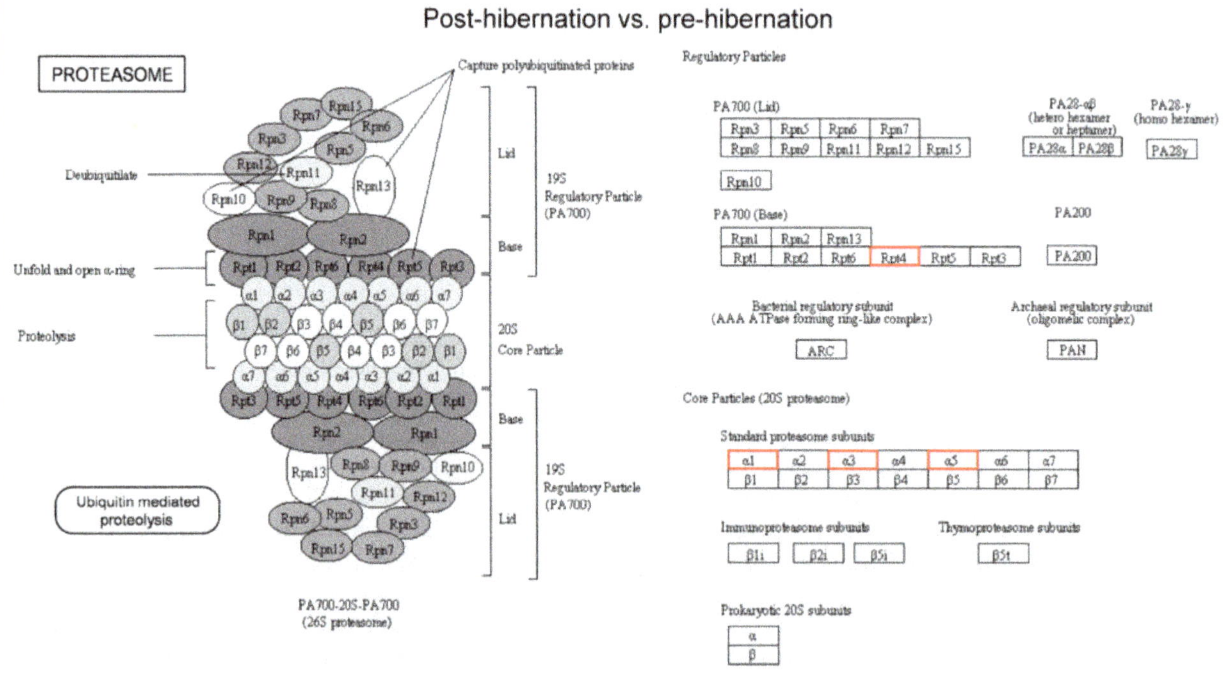

Fig. 7 Pathway analysis using the KEGG Pathway database. Differentially expressed proteins enriched in the proteasome pathway in the post-hibernation ground squirrels group compared to the pre-hibernation group. Red indicates up-regulated expressed proteins. Rpt4: gi|59800155, 26S protease regulatory subunit 10B; α1: gi|532073955, proteasome subunit alpha type-6; α3: gi|532062055, proteasome subunit alpha type-4; α5: gi|532105099, proteasome subunit alpha type-5

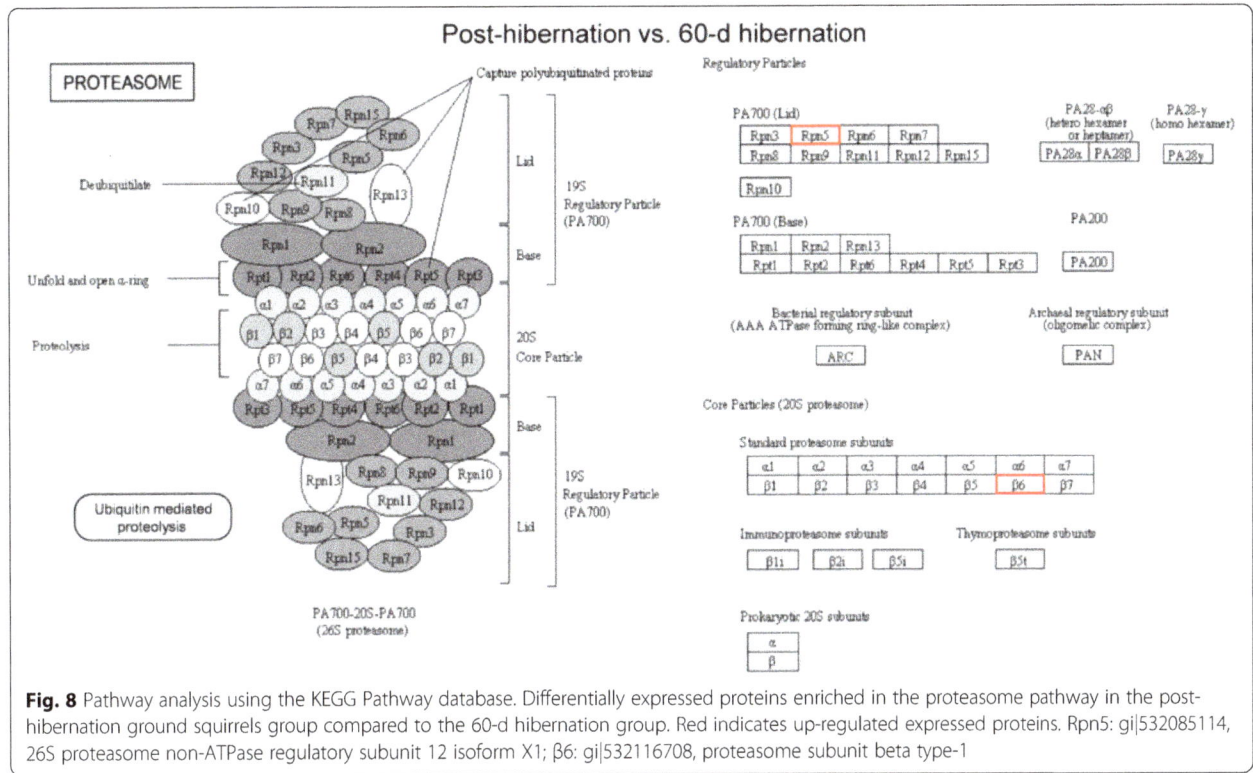

Fig. 8 Pathway analysis using the KEGG Pathway database. Differentially expressed proteins enriched in the proteasome pathway in the post-hibernation ground squirrels group compared to the 60-d hibernation group. Red indicates up-regulated expressed proteins. Rpn5: gi|532085114, 26S proteasome non-ATPase regulatory subunit 12 isoform X1; β6: gi|532116708, proteasome subunit beta type-1

or spaceflight [40]. This study found that the number of up-regulated proteins was much higher in post-hibernation (248) than that of the proteins in 60-d hibernation group (96). Moreover, most of up-regulated proteins were mainly involved in protein binding, catalytic activity, transporter activity, enzyme regulator activity and other metabolic processes, which might be due to that the ground squirrels are recovering and likely produce these types of proteins rapidly upon resuming feeding and activity in 2 days awake after post-hibernation. These results indicated that protein synthesis was even strengthened in late hibernation period, which related to the regulation of contractile function. Our results agree well with the previous findings that the increase of protein synthesis to preserve and augment muscle mass in late winter was observed through direct measurements of protein synthesis by the in vivo SUnSET technique [41], which concurs with the finding here of accelerated synthesis in the later hibernation time point.

Here, we showed that there were 3 and 10 proteins with protein synthesis function significantly up-regulated in the 60-d hibernation group and post-hibernation group, respectively, relative to the pre-hibernation group (Fig. 2). Specifically, 60S ribosomal protein L39-like which was a structural constituent of ribosome participated in RNA binding and translation bioprocess, and polyadenylate-binding protein 4 isoform X1, which was

a RNA-binding protein and locates in cytoplasmic stress granule and nucleus participated in RNA catabolic process and translation, were both significantly increased in the 60-d hibernation group compared to the pre-hibernation group. Besides, all of the up-regulated proteins were involved in translation, 7 of which were structural constituent of ribosome. In addition, elongation factor 1-beta isoform $X2$ which participates in translational elongation biological process was found continuously up-regulated in 60-d and post-hibernation groups relative to the pre-hibernation ground squirrels (Table 1).

In fact, down-regulated 40S ribosomal protein subunits in the 60-d hibernation ground squirrels, which bind to mRNA and modulate of the initiation phase of mRNA translation [42], suggested that the initiation of translation in protein synthesis was inhibited to a certain extent in hibernation. Consistent with our study, previous report demonstrated that mice homozygous for translation elongation factor 1A (eEF1A) deletion in muscle corresponds precisely to the onset of the wasted phenotype, characterized by muscle atrophy [43]. In addition, another serine-threonine kinase, serum- and glucocorticoid-regulated kinase 1 (SGK1), was upregulated during hibernation in 13-lined ground squirrel (*Ictidomys tridecemlineatus*) and contributed to protection from loss of muscle mass via an increased protein synthesis [44]. Hibernation factors including RMF

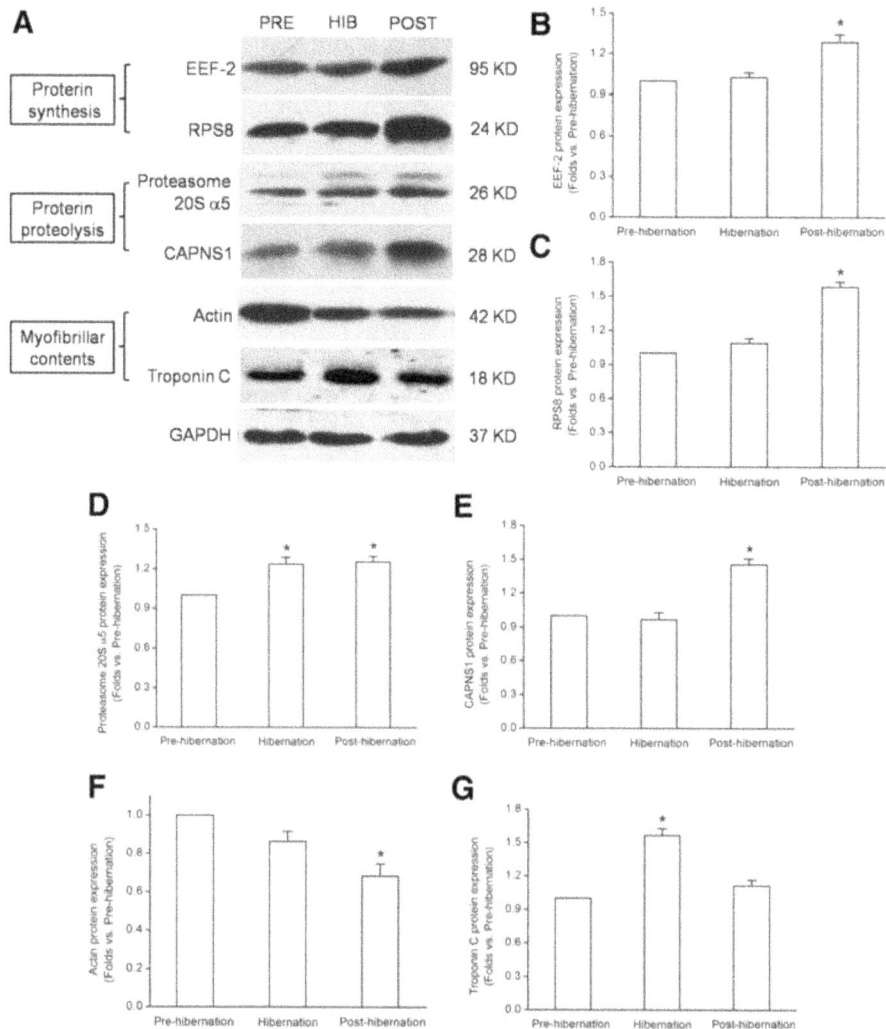

Fig. 9 Western blot analysis of 6 differentially expressed proteins. **a** Representative bands of EEF-2, RPS8, proteasome 20S α5, CAPNS1, actin and troponin C in the soleus of pre-hibernation (PRE), hibernation (HIB) and post-hibernation (POST) squirrels. **b-g** Summarized data of protein quantity normalized by GAPDH in different groups. Values are mean ± SEM. *$P < 0.05$ versus pre-hibernation group

(ribosome modulation factor), HPF (hibernation promoting factor) and YfiA (protein which inactivates ribosomes as 70S monomers) turn off protein synthesis via binding to the ribosome [45]. However, since the protein synthesis is a process of energy dissipation and the hibernating animals live in low temperature, fasting, or eating less state, we consider that the theory that the hibernating animals during prolonged period of immobilization and starvation by promoting protein synthesis to overcome muscular atrophy may not occupy a dominant position. Indeed, the rate of protein synthesis in vivo in the brain of torpid ground squirrels was just 0.04 % of that in active squirrels [46]. Together these findings suggest that protein synthesis was not inhibited during prolonged hibernation, which might play an important role in preventing atrophy in SOL of ground squirrels during hibernation.

The rates of protein degradation decreased significantly in hibernation and post-hibernation group as compared with the pre-hibernation group (Fig. 10). Different protein degradation pathways may be involved in sarcomeric protein loss in muscle atrophy. The ubiquitin-proteasome system and the autophagy-lysosome pathway are the major protein degradation systems involved in this process [47]. Moreover, calpains, a family of Ca^{2+}-dependent proteases, play an initiating role in the protein degradation process and cause rapid and complete loss of Z-disk while activated by calcium.

In this study, we investigated that 10 and 14 proteins with protein proteolysis function were found significantly up-regulated in the 60-d hibernation group and post-hibernation group, respectively, relative to the pre-hibernation group (Fig. 2). Two-thirds of the dysregulated

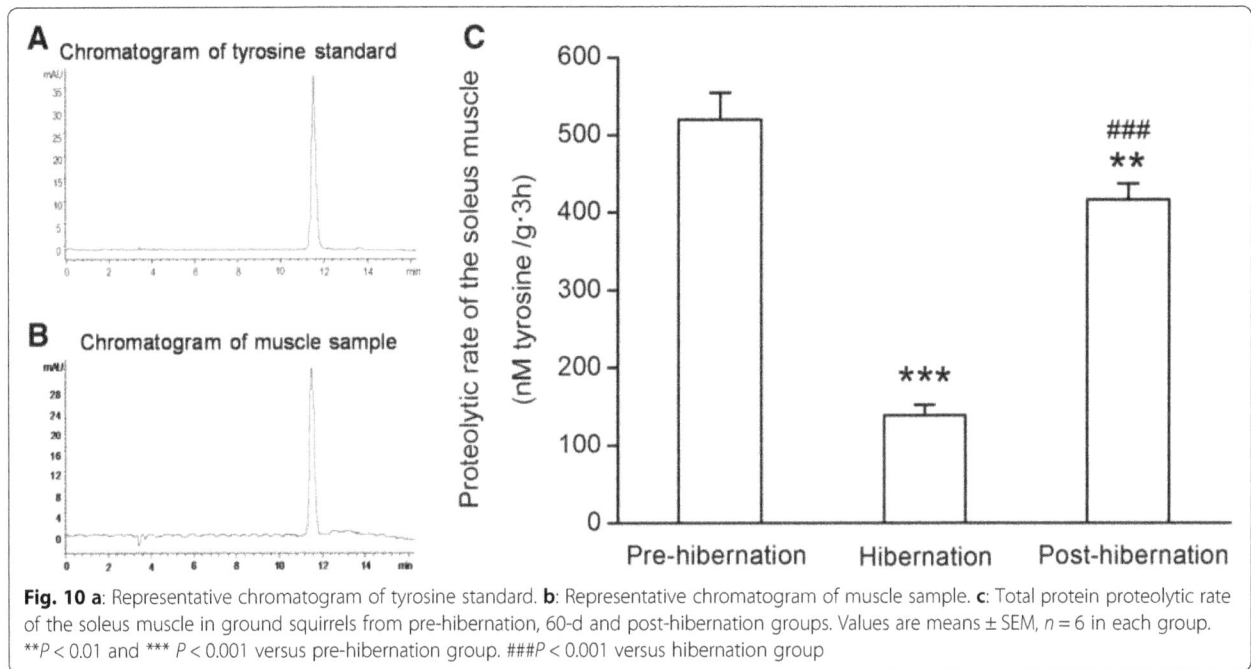

Fig. 10 a: Representative chromatogram of tyrosine standard. **b**: Representative chromatogram of muscle sample. **c**: Total protein proteolytic rate of the soleus muscle in ground squirrels from pre-hibernation, 60-d and post-hibernation groups. Values are means ± SEM, $n = 6$ in each group. **P < 0.01 and *** $P < 0.001$ versus pre-hibernation group. ###$P < 0.001$ versus hibernation group

proteins were involved in ubiquitin-proteasome pathway and calpains pathway. Besides, proteasome subunit alpha type-5, proteasome subunit alpha type-6 and 26S protease regulatory subunit 10B were found continuously upregulated in the 60-d and post-hibernation groups relative to the pre-hibernation ground squirrels (Table 2). These results indicated that ubiquitin-proteasome proteolysis was enhanced during prolonged hibernation. Actin, myosin heavy chains (in atrophying skeletal muscle), myosin light chains, members of the troponin family and telethonin were confirmed to be the ubiquitin-proteasome system substrates [48]. Telethonin, a substrate of the ubiquitin-proteasome system, was significantly down-regulated in the 60-d hibernation group compared to the pre-hibernation group. It is noteworthy that ubiquitin carboxyl-terminal hydrolase isozyme L3 was 1.591-fold increase in post-hibernation ground squirrels compared to pre-hibernation group. Similar with non-hibernators, the ubiquitin carboxy-terminal hydrolase L1 (UCHL1), functioning as an ubiquitin ligase and a mono-ubiquitin stabilizer, were also up-regulated in mice during hindlimb unloading [14], which suggest that ubiquitin carboxyl-terminal hydrolase might be activated in disuse. However, torpor may limit proteolysis in accordance with lower metabolic demands in livers of hibernating golden-mantled ground squirrels (*Spermophilus lateralis*) [49]. Importantly, calpastatin, an endogenous inhibitor of calpains, was found significantly up-regulated in the post-hibernation ground squirrels compared to the pre-hibernation group in present study, which is in consistent with our previous experimental report, in which the protein expression validation of calpastatin had been

detected by western blot analyses, demonstrating the same result that inhibition of calpain activity and consequently calpain-mediated protein degradation by highly elevated calpastatin protein expression levels may be an important mechanism for preventing muscle protein loss during hibernation [15]. Nevertheless, calpain small subunit 1 also showed a significant increase in the post-hibernation ground squirrels compared to the pre-hibernation group from proteomic analysis. Calpains are heterodimers containing an identical 28-kDa regulatory small subunit and a distinct 80-kDa catalytic large subunit. Effective inhibition of the calpains by calpastatin requires that calpastatin binds specifically to the domain II or IV of large subunit of calpains in a Ca^{2+} dependent manner [50]. Hence, it appears reasonable to assume that calpains are inhibited at the level of enzymatic activity rather than the protein expression, and the increase of calpains expression make it possible that hibernating ground squirrels can get energy via the proteolysis of myofibrillar proteins to get through the energy crisis, while the degree of protein proteolysis can be also regulated by the enzymatic activity when squirrels do not need the muscle protein supply energy, thereby minimizing the muscular atrophy of disuse.

Collectively, these results suggested that the ubiquitin–proteasome catabolic pathway might be strengthened and be responsible for limited atrophy in the late hibernating period of Daurian ground squirrels. Moreover, calpains pathway might be regulated via the enzyme activity in hibernating ground squirrels, which subsequently regulate the ubiquitin–proteasome catabolic pathway and even the level of protein proteolysis of the whole body. Although recent study emphasizes autophagy-lysosome as another important proteolytic pathway in triggering the early stages of atrophy [47], no protein related to autophagy-lysosome pathway was detected proteomics differentially expressed in our study. Taken together, our study suggested that the limited SOL atrophy happened in the late hibernating period might be the result of activation of ubiquitin–proteasome catabolic pathway, and the effect of protecting atrophy in SOL during hibernation might be due to the inhibition of calpains and autophagy-lysosome proteolytic pathways.

Conclusions

In this study, we present a proteomic analysis in soleus muscle of Daurian ground squirrels for the first time. These findings not only provide novel insights into the myofibrillar remodeling, which was marked by dysregulated contractile proteins (myosins and actin) and regulatory proteins (troponin and tropomodulin) and maintain of most sarcomeric structural proteins, contributes to a protective effect that prevents muscle atrophy in spite of prolonged disuse during hibernation, but also provide the first experimental evidence that the total proteolysis rates of soleus in hibernating ground squirrels is decreased. Moreover, the strengthened ubiquitin-proteasome pathway and calpains pathway related to the protein degradation, associated with higher level of calpastatin, contributes to maintain of most myofibrillar proteins in hibernation. Although the number of differentially expressed proteins associated with protein degradation is much more than those associated with protein synthesis in hibernation, the increased part may be partially offset by the strengthened ubiquitin–proteasome catabolic pathway,which together with partial inhibition of calpains might play a critical role in maintain of myofibrillar proteins and provide a foundation for elucidating the mechanisms of prevention of the disuse atrophy in skeletal muscle in non-hibernation animals.

Limitations of the study

The iTRAQ-based proteomic analysis should have biological triplicates, however, we have validated the expression levels of some proteins using western blot technology to make up for this deficiency. In addition, iTRAQ-based quantitative proteomic analysis does not involve the differentially modulation of post-translational modification of proteins which are critical in adaptation to hibernating state. Besides, the proteins and pathways identified by computer-based statistical algorithms also need to be verified experimentally in future.

Acknowledgements

This study was supported by funds from the National Nature Science Foundation of China (31270455 and 31200862), the Postdoctoral Research Fund of Shaanxi Province,the Shaanxi Province Natural Science Basic Research Program (2016JQ3014),the Postdoctoral Science Foundation of China (2015 M580869).

Funding

The National Nature Science Foundation of China (31270455)–Gao YF. The National Nature Science Foundation of China (31200862)–Chang H. The Postdoctoral Research Fund of Shaanxi Province–Chang H. The Shaanxi Province Natural Science Basic Research Program (2016JQ3014) –Chang H. The Postdoctoral Science Foundation of China (2015 M580869) –Chang H. Gao YF designed the study wrote the manuscript. Chang H performed the experiments, analyzed the data and wrote the manuscript.

Authors' contributions

Conceived and designed the experiments: SFJ, YFG, HC. Performed the experiments: HC, SFJ, KD, HPW, SHX. Analyzed the data: HC and SFJ. Wrote the paper: HC, YFG. All authors read and approved the final manuscript.

Competing interests

The authors declare that they have no competing interests.

References

1. Wang Y, Pessin JE. Mechanisms for fiber-type specificity of skeletal muscle atrophy. Curr Opin Clin Nutr Metab Care. 2013;16(3):243–50.
2. Narici MV, de Boer MD. Disuse of the musculo-skeletal system in space and on earth. Eur J Appl Physiol. 2011;111(3):403–20.
3. Adams GR, Caiozzo VJ, Baldwin KM. Skeletal muscle unweighting: spaceflight and ground-based models. J Appl Physiol (1985). 2003;95(6): 2185–201.
4. Baldwin KM, White TP, Arnaud SB, et al. Musculoskeletal adaptations to weightlessness and development of effective countermeasures. Med Sci Sports Exerc. 1996;28(10):1247–53.
5. Hvid LG, Ortenblad N, Aagaard P, Kjaer M, Suetta C. Effects of ageing on single muscle fibre contractile function following short-term immobilisation. J Physiol. 2011;589(Pt 19):4745–57.
6. Hvid LG, Suetta C, Aagaard P, Kjaer M, Frandsen U, Ortenblad N. Four days of muscle disuse impairs single fiber contractile function in young and old healthy men. Exp Gerontol. 2013;48(2):154–61.
7. Lawler JM, Kunst M, Hord JM, et al. EUK-134 ameliorates nNOSmu translocation and skeletal muscle fiber atrophy during short-term mechanical unloading. Am J Physiol Regul Integr Comp Physiol. 2014; 306(7):R470–82.
8. Seo Y, Lee K, Park K, Bae K, Choi I. A proteomic assessment of muscle contractile alterations during unloading and reloading. J Biochem. 2006; 139(1):71–80.
9. Steffen JM, Musacchia XJ. Disuse atrophy, plasma corticosterone, and muscle glucocorticoid receptor levels. Aviat Space Environ Med. 1987; 58(10):996–1000.
10. Thomason DB, Herrick RE, Surdyka D, Baldwin KM. Time course of soleus muscle myosin expression during hindlimb suspension and recovery. J Appl Physiol (1985). 1987;63(1):130–7.
11. Kachaeva EV, Shenkman BS. Various jobs of proteolytic enzymes in skeletal muscle during unloading: facts and speculations. J Biomed Biotechnol. 2012;2012:493618.
12. Tinker DB, Harlow HJ, Beck TD. Protein use and muscle-fiber changes in free-ranging, hibernating black bears. Physiol Zool. 1998;71(4):414–24.
13. Shanfeng Jiang YG, Yangmei Zhang KL, Goswami HWN. The research on the formation mechanism_x000A_of extraordinary oxidative capacity of_ x000A_skeletal muscle in hibernating ground_x000A_squirrels (Spermophilus dauricus). Zool Stud. 2015;54:46.
14. Wang F, Zhang P, Liu H, Fan M, Chen X. Proteomic analysis of mouse soleus muscles affected by hindlimb unloading and reloading. Muscle Nerve. 2015;52(5):803–11.

15. Yang CX, He Y, Gao YF, Wang HP, Goswami N. Changes in calpains and calpastatin in the soleus muscle of Daurian ground squirrels during hibernation. Comp Biochem Physiol A Mol Integr Physiol. 2014;176:26–31.

16. Gao YF, Wang J, Wang HP, et al. Skeletal muscle is protected from disuse in hibernating dauria ground squirrels. Comp Biochem Physiol A Mol Integr Physiol. 2012;161(3):296–300.

17. Lohuis TD, Harlow HJ, Beck TD. Hibernating black bears (Ursus americanus) experience skeletal muscle protein balance during winter anorexia. Comp Biochem Physiol B Biochem Mol Biol. 2007;147(1):20–8.

18. Fedorov VB, Goropashnaya AV, Toien O, et al. Elevated expression of protein biosynthesis genes in liver and muscle of hibernating black bears (Ursus americanus). Physiol Genomics. 2009;37(2):108–18.

19. Andres-Mateos E, Brinkmeier H, Burks TN, et al. Activation of serum/glucocorticoid-induced kinase 1 (SGK1) is important to maintain skeletal muscle homeostasis and prevent atrophy. EMBO Mol Med. 2013;5(1):80–91.

20. Dang K, Li YZ, Gong LC, et al. Stable atrogin-1 (Fbxo32) and MuRF1 (Trim63) gene expression is involved in the protective mechanism in soleus muscle of hibernating Daurian ground squirrels (Spermophilus dauricus). Biol Open. 2016;5(1):62–71.

21. Dang K, Feng B, Gao YF, et al. Muscle protection during hibernation: Role of atrogin-1 and MuRF1, and fiber type transition in Daurian ground squirrels. Can J Zool. 2016;94(9):619–29.

22. Li T, Brouwer M. Gene expression profile of hepatopancreas from grass shrimp Palaemonetes pugio exposed to cyclic hypoxia. Comp Biochem Physiol Part D Genomics Proteomics. 2013;8(1):1–10.

23. Cui J, Wang H, Liu S, Qiu X, Jiang Z, Wang X. Transcriptome analysis of the gill of Takifugu rubripes using Illumina sequencing for discovery of SNPs. Comp Biochem Physiol Part D Genomics Proteomics. 2014;10:44–51.

24. Chang H, Sheng JJ, Zhang L, et al. ROS-induced nuclear translocation of calpain-2 facilitates cardiomyocyte apoptosis in tail-suspended rats. J Cell Biochem. 2015;116(10):2258–69.

25. Tiao G, Fagan J, Roegner V, et al. Energy-ubiquitin-dependent muscle proteolysis during sepsis in rats is regulated by glucocorticoids. J Clin Invest. 1996;97(2):339–48.

26. Long M, Zhao J, Li T, et al. Transcriptomic and proteomic analyses of splenic immune mechanisms of rainbow trout (Oncorhynchus mykiss) infected by Aeromonas salmonicida subsp. salmonicida. J Proteomics. 2015;122:41–54.

27. Rahimov F, Kunkel LM. The cell biology of disease: cellular and molecular mechanisms underlying muscular dystrophy. J Cell Biol. 2013;201(4):499–510.

28. Lee K, Park JY, Yoo W, et al. Overcoming muscle atrophy in a hibernating mammal despite prolonged disuse in dormancy: proteomic and molecular assessment. J Cell Biochem. 2008;104(2):642–56.

29. Miller MS, Toth MJ. Myofilament protein alterations promote physical disability in aging and disease. Exerc Sport Sci Rev. 2013;41(2):93–9.

30. Akhter S, Jin JP. Distinct conformational and functional effects of two adjacent pathogenic mutations in cardiac troponin I at the interface with troponin T. FEBS Open Bio. 2015;5:64–75.

31. Salanova M, Gelfi C, Moriggi M, et al. Disuse deterioration of human skeletal muscle challenged by resistive exercise superimposed with vibration: evidence from structural and proteomic analysis. FASEB J. 2014;28(11):4748–63.

32. Rao JN, Madasu Y, Dominguez R. Mechanism of actin filament pointed-end capping by tropomodulin. Science. 2014;345(6195):463–7.

33. Gokhin DS, Tierney MT, Sui Z, Sacco A, Fowler VM. Calpain-mediated proteolysis of tropomodulin isoforms leads to thin filament elongation in dystrophic skeletal muscle. Mol Biol Cell. 2014;25(6):852–65.

34. Sussman MA, Baque S, Uhm CS, et al. Altered expression of tropomodulin in cardiomyocytes disrupts the sarcomeric structure of myofibrils. Circ Res. 1998;82(1):94–105.

35. Udaka J, Ohmori S, Terui T, et al. Disuse-induced preferential loss of the giant protein titin depresses muscle performance via abnormal sarcomeric organization. J Gen Physiol. 2008;131(1):33–41.

36. Salmov NN, Vikhlyantsev IM, Ulanova AD, et al. Seasonal changes in isoform composition of giant proteins of thick and thin filaments and titin (connectin) phosphorylation level in striated muscles of bears (Ursidae, Mammalia). Biochemistry (Mosc). 2015;80(3):343–55.

37. Garton FC, Seto JT, Quinlan KG, Yang N, Houweling PJ, North KN. alpha-Actinin-3 deficiency alters muscle adaptation in response to denervation and immobilization. Hum Mol Genet. 2014;23(7):1879–93.

38. Thijssen VL, Borgers M, Lenders MH, et al. Temporal and spatial variations in structural protein expression during the progression from stunned to hibernating myocardium. Circulation. 2004;110(21):3313–21.

39. Mitch WE, Goldberg AL. Mechanisms of muscle wasting. The role of the ubiquitin-proteasome pathway. N Engl J Med. 1996;335(25):1897–905.

40. Fitts RH, Riley DR, Widrick JJ. Physiology of a microgravity environment invited review: microgravity and skeletal muscle. J Appl Physiol (1985). 2000;89(2):823–39.

41. Hindle AG, Otis JP, Epperson LE, et al. Prioritization of skeletal muscle growth for emergence from hibernation. J Exp Biol. 2015;218(Pt 2):276–84.

42. Gordon BS, Kelleher AR, Kimball SR. Regulation of muscle protein synthesis and the effects of catabolic states. Int J Biochem Cell Biol. 2013;45(10):2147–57.

43. Doig J, Griffiths LA, Peberdy D, et al. In vivo characterization of the role of tissue-specific translation elongation factor 1A2 in protein synthesis reveals insights into muscle atrophy. FEBS J. 2013;280(24):6528–40.

44. Ivakine EA, Cohn RD. Maintaining skeletal muscle mass: lessons learned from hibernation. Exp Physiol. 2014;99(4):632–7.

45. Polikanov YS, Blaha GM, Steitz TA. How hibernation factors RMF, HPF, and YfiA turn off protein synthesis. Science. 2012;336(6083):915–8.

46. Frerichs KU, Smith CB, Brenner M, et al. Suppression of protein synthesis in brain during hibernation involves inhibition of protein initiation and elongation. Proc Natl Acad Sci U S A. 1998;95(24):14511–6.

47. Sandri M. Protein breakdown in muscle wasting: role of autophagy-lysosome and ubiquitin-proteasome. Int J Biochem Cell Biol. 2013;45(10):2121–9.

48. Polge C, Attaix D, Taillandier D. Role of E2-Ub-conjugating enzymes during skeletal muscle atrophy. Front Physiol. 2015;6:59.

49. Velickovska V, van Breukelen F. Ubiquitylation of proteins in livers of hibernating golden-mantled ground squirrels, Spermophilus lateralis. Cryobiology. 2007;55(3):230–5.

50. Goll DE, Thompson VF, Li H, Wei W, Cong J. The calpain system. Physiol Rev. 2003;83(3):731–801.

Proteomic based approach for characterizing 4-hydroxy-2-nonenal induced oxidation of buffalo (*Bubalus bubalis*) and goat (*Capra hircus*) meat myoglobins

Naveena B. Maheswarappa[1*], K. Usha Rani[1], Y. Praveen Kumar[1], Vinayak V. Kulkarni[1] and Srikanth Rapole[2]

Abstract

Background: Myoglobin (Mb) is a sarcoplasmic heme protein primarily responsible for meat color and its chemistry is species specific. 4-hydroxy-2-nonenal (HNE) is a cytotoxic lipid derived aldehyde detected in meat and was reported to covalently adduct with nucleophilic histidine residues of Mb and predispose it to greater oxidation. However, no literature is available on characterization of lipid oxidation induced oxidation of Indian water buffalo (*Bubalus bubalis*) and goat (*Capra hircus*) myoglobins.

Methods: Present study characterize the Mb extracted from water buffalo and goat cardiac muscles using two-dimensional gel electrophoresis (2DE), OFFGEL electrophoresis and mass spectrometry (MS). Purified buffalo and goat bright red oxymyoglobin were reacted with HNE in-vitro at physiological pH (7.4) and temperature (37 °C) conditions and the formation of oxidised brown metmyoglobin was measured. The Mb-HNE adducts were detected using MALDI-TOF MS, whereas specific sites of adduction was determined using ESI-QTOF MS/MS.

Results: Purified buffalo and goat Mb samples revealed a molecular mass of 17,043.6 and 16,899.9 Daltons, respectively. The 2DE analysis exhibited 65 (sarcoplasmic protein extract) and 6 (pure Mb) differentially expressed ($P < 0.05$) protein spots between buffalo and goat samples. OFFGEL electrophoresis revealed an isoelectric point of 6.77 and 7.35 respectively, for buffalo and goat Mb's. In-vitro incubation of HNE with bright red buffalo and goat oxymyoglobin's at pH 7.4 and 37 °C resulted in pronounced ($P < 0.05$) oxidation and formation of brown metmyoglobin. MALDI-TOF MS analysis of Mb-HNE reaction mix revealed covalent binding (via Michael addition) of 3 and 5 molecules of HNE with buffalo and goat Oxy-Mb's, respectively. ESI-QTOF MS/MS identified seven and nine histidine (HIS) residues of Mb that were readily adducted by HNE in buffalo and goat, respectively.

Conclusion: The study demonstrated better redox stability of buffalo Mb than goat Mb. Our findings confirm the hypothesis that relative effect of HNE was greater for Mb's with 12 ± 1 HIS residues than Mb's with 9 HIS residues and helps meat processors in developing species-specific processing strategies to reduce the color variability.

Keywords: Myoglobin-HNE adduction, Redox instability, OFFGEL electrophoresis, 2-Dimensional gel electrophoresis, Mass spectrometry

* Correspondence: naveenlpt@rediffmail.com
[1]National Research Centre on Meat, Chengicherla, Hyderabad, Telangana 500092, India
Full list of author information is available at the end of the article

Background

India is the largest producer and exporter of water buffalo (*Bubalus bubalis*) meat in the world and accounts for 23.5% of global bovine meat exports [1]. Indian water buffalo meat is considered lean with higher protein and ash content relative to cattle meat (beef) [2]. Owing to its healthier properties, buffalo meat is emerging as an important red meat source in several Middle-east, South-east and European countries. India is the second largest producer of goat (*Capra hircus*) meat next to China. Goat meat is a popular red meat in many regions of the world and the consumer preference of goat meat is almost universal depending on cultural traditions and social and economic conditions [3]. Buffalo meat is darker compared to beef and the darkness is attributed to higher myoglobin (Mb) content (5.0 mg/g) [4]. Dosi et al. (2006) have studied the primary structure of buffalo Mb using a combined approach of Edman degradation and MALD-TOF mass spectrometry and found a difference of three amino acids out of 153 compared to beef Mb [5]. These authors have also studied the stability, autoxidation and percent metmyoglobin formation in beef and buffalo Mb's and reported identical results between them. Another important livestock species, goat was reported to share 98.7% sequence similarity with sheep than with buffalo which has got 95.4% sequence similarity. Goat meat was reported to be darker, more red and have higher sarcoplasmic protein content than sheep meat [6]. Primary structure of goat Mb was determined by Suman et al. (2009) who reported that goat Mb shared 98.7% similarity with sheep Mb and the distal (64) and proximal (93) histidines responsible for coordinating the heme group and reversible binding of oxygen are conserved in goat Mb, similar to other meat-producing livestock [7].

Molecular properties and overall structure of the Mb protein from different species have been reported to be very similar although there are differences in the amino acid composition and the exact length of the polypeptide chain. Primary structure of Mb influences meat color stability via mechanisms such as autoxidation, heme retention, structural stability, thermostability, and oxygen affinity [8]. Oxidation of ferrous oxymyolobin (Oxy-Mb) to ferric metmyoglobin (Met-Mb) will result in conversion of desirable bright red color of the meat into undesirable brown color [9]. Several researchers have demonstrated that myoglobin oxidation and lipid oxidation are interrelated and the acceleration of one will exacerbate the other [9–12]. These researchers have demonstrated acceleration of heme protein oxidation in the presence of unsaturated aldehydes generated by free radical-induced lipid peroxidation of polyunsaturated fatty acids. 4-Hydroxy-2-nonenal (HNE) is an α, β-unsaturated aldehyde derived from the oxidation of ω-6 polyunsaturated fatty acids and has been reported in

meat [13]. It was reported to involve in the onset and progression of many pathological conditions such as cardiovascular and neurodegenerative diseases due to its ability to react with the nucleophilic sites of proteins and peptides to form covalently modified biomolecules [14]. Because of high reactivity of HNE with proteins, researchers have concluded the possibility of using HNE-adducted proteins as markers for oxidative damage [15]. HNE was also reported to induce redox instability of Mb's from horse, cattle, pig, sheep, deer, chicken, turkey, emu and ostrich meats under different conditions and in all cases Oxy-Mb oxidation was promoted [9, 10, 12, 16–18]. These studies suggest that, HNE accelerates Oxy-Mb oxidation in-vitro by covalent modification at histidine residues. Using LC-MS/MS analysis Alderton et al. (2003) identified six histidine residues of beef Mb that were readily adducted by HNE, including the proximal (HIS 93) and distal (HIS 64) histidines associated with the heme group at pH 7.4 and 37 °C [10]. However, Suman et al. (2007) demonstrated the adduction of HNE with three and seven histidine residues in porcine and beef Mb's under similar conditions [12]. These authors have hypothesized that, the lesser number of nucleophilic HIS residues and their specific locations in porcine Mb are responsible for the greater redox stability of porcine Mb in the presence of lipid oxidation products relative to beef. They further concluded that, preferential HNE adduction at proximal HIS 93 was observed only in bovine Oxy-Mb and hence lipid oxidation induced Oxy-Mb oxidation was potentially more critical in beef than pork. Yin et al. (2011) studied the HNE induced oxidation of Oxy-Mb's from seven different meat animals and concluded that, the relative effect of HNE was greater for Mb's that contained 12 ± 1 HIS residues than for those that contained nine HIS residues [18]. All these researchers suggested a species specific effect of HNE on Mb. In the present study we propose to characterize the Mb-HNE adductions in two ruminant species viz, water buffalo and goat with 13 and 12 HIS residues in their Mb, respectively.

Almost all the studies related to meat color have used liquid chromatography-mass spectrometry (LC-MS) based tools to characterize the primary structure of Mb, to understand the lipid oxidation induced oxidation of Oxy-Mb and to determine the redox instability of mutant sperm whale Mb [5, 9, 10, 19]. Investigations have documented the contribution of sarcoplasmic proteome on muscle-specificity in beef color [20, 21]. However, to our knowledge no studies have been reported on use of two-dimensional gel electrophoresis (2DE) coupled with mass spectrometry to characterize purified Mb's especially from water buffalo and goat meats. Further, to our knowledge HNE induced oxidation of buffalo and goat Mb's and identification of specific amino acids with which it reacts was not previously reported. Therefore, our objectives were 1) to characterize the Mb using 2DE, OFFGEL electrophoresis and mass spectrometry from two important emerging meat animals, buffalo and goat and 2) to investigate the potential binding of HNE to buffalo and goat meat Mb's and determine how it affects redox stability under different temperature and pH conditions. The present study is the first to characterize the livestock Mb's using 2DE and OFFGEL electrophoresis.

Results and discussion
Purification of buffalo and goat Mb's
Isolation and purification of beef, pork, sheep, turkey and chicken myoglobins (Mbs) either from skeletal, heart or smooth muscles using ammonium sulfate precipitation and gel filtration has been reported by different authors [7, 12, 16, 18]. The extraction and purification of Mb from buffalo and goat is minimally investigated and to our knowledge only two papers are available in the literature for buffalo and goat Mb extraction and characterization [5, 7]. Dosi et al. (2006) have extracted buffalo Mb in Milli Q water and further purified through dialysis, Sephacryl S-100 column chromatography followed by Diethylaminoethyl (DEAE) anion exchange chromatography [5]. Goat Mb was extracted and purified by Suman et al. (2009) using the modified procedure of Faustman and Phillips (2001) [7, 22]. These authors have used 50% ammonium sulfate saturation in place of 70% as originally suggested by Faustman and Phillips (2001) [22]. In the present study, we have successfully extracted and purified both buffalo and goat Mb's as per the procedure suggested by Faustman and Phillips (2001) with 70% ammonium sulfate precipitation and gel filtration chromatography using Sephacryl S-200 HR [22]. Purification of ammonium sulfate precipitated and dialyzed Mb on Sephacryl S-200 HR gel-filtration column revealed two major peaks at A_{540} for buffalo (Fig. 1a) and goat (Fig. 1b). This elution pattern suggests clear separation of hemoglobin and Mb on gel filtration column. Hemoglobin (~64 kDa) elutes first followed by Mb (~17 kDa) and these findings confirm with our earlier reports for chicken and turkey Mb's [16].

SDS-PAGE and two-dimensional gel electrophoresis (2DE) of buffalo and goat Mb's
The SDS-PAGE of pooled fractions from second peak which is supposed to be Mb, consistently revealed the presence of single band at approximately 17 kDa level in both buffalo and goat (Fig. 2) samples. Dosi et al. (2006) also reported a single protein band for purified buffalo Mb on SDS-PAGE without any detectable contaminating protein bands [5]. Our findings confirm that buffalo and goat Mb's can be purified using different ammonium sulphate precipitation, dialysis, filtration and gel filtration chromatography steps as suggested by Faustman and Phillips (2001) [22]. To further check the purity of

Mb, 2DE of sarcplasmic extract and the purified Mb's from buffalo and goat was carried out. Wu et al. (2016) used 2DE and tandem MS to differentiate sarcoplasmic proteome of *Longissimus lumborum* (LL) and *Psoas major* (PM) muscles of Chinese *Luxi* yellow cattle and identified the proteins mainly involved in glycolytic metabolism which contributed to better meat color stability in LL compared to PM [23]. In the present study, sarcoplasmic proteome extract and gel-filtered purified Mb from buffalo and goat were characterized using 2DE. The 2DE was done to separate differentially expressed sarcoplasmic proteins mainly consist of myoglobin, hemoglobin, cytochrome and wide variety of endogenous enzymes that are associated with meat color. The analysis of 2DE gels revealed separation of 508 and 563 spots respectively, in buffalo and goat sarcoplasmic extracts (Fig. 3a and c). The class analysis table by analysis of variance (ANOVA) of buffalo and goat sample gels indicated 65 differential spots ($p < 0.05$) which had protein spot expression of 1.5 fold or more between them. Variation in spot intensity/abundance was consistently observed across three gels each for buffalo and goat samples. Joseph et al. (2012) observed a total of 180 protein spots for sarcoplasmic protein extract from beef *Longissimus lumborum* and indicated 17 differentially abundant protein spots in comparison to *Psoas major* muscle [21]. Purified Mb's revealed 19 and 20 spots, respectively for buffalo (Fig. 3b) and goat (Fig. 3d) samples with six spots being differentially expressed ($p < 0.05$) between them. The over-abundance of spots in goat sarcoplasmic proteins might be due to post-translational modification of some proteins at a greater degree in goat than in buffalo which may compromise color stability. Sarcoplasmic proteome governs different biochemical processes influencing meat color stability and their interactions with myoglobin are critical to meat color. Present study established the differential abundance of sarcoplasmic proteome between buffalo meat and goat meats. A beef color stability variation due to differences in sarcoplasmic proteome was reported by Joseph et al. (2012) [21]. Differential analysis of sarcoplasmic proteome using image analyses of the Coomassie-stained 2DE gels was reported to contain twelve differentially abundant protein spots in color-stable and color-labile beef longissimus muscles [20]. Our findings suggest significant variation in sarcoplasmic proteome between buffalo and goat samples. Eventhough, purification of buffalo and goat Mb's using gel-filtration resulted in single band at 17 kDa in SDS-PAGE anlaysis, the 2DE revealed few other contaminating proteins with very little intensity/abundance. However, in both buffalo and goat 2DE gels, single large Mb fraction was observed at same location. This spot was further analysed using peptide mass fingerprinting (PMF) for confirmation of Mb.

Proteomic based approach for characterizing 4-hydroxy-2-nonenal induced oxidation of buffalo (Bubalus bubalis)...

85

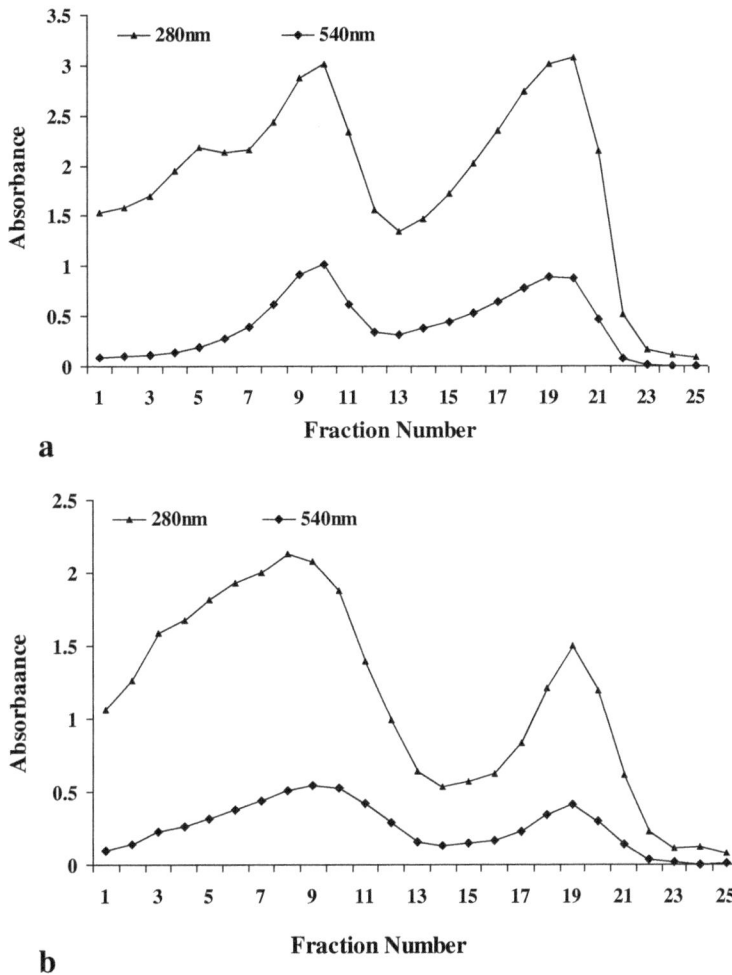

Fig. 1 Elution profile of (**a**) buffalo and **b** goat cardiac myoglobin from Sephacryl S-200 HR gel filtration chromatography. The column was equilibrated and eluted with 5 mM Tris–HCl + 1 mM EDTA buffer, pH 8.0 at a flow rate of 0.1 mL/min

Determination of molecular mass and peptide mass fingerprinting using MALDI-TOF MS

Mass spectrometry analysis was performed only for the purified Mb's to determine molecular mass and to confirm their identity through PMF. The MALDI-TOF MS analysis of intact buffalo Mb revealed the mass of 17043.6 Daltons (Fig. 4a) which is 97.6 Daltons more than beef Mb [24]. The MS analysis of goat Mb revealed a mass of 16899 Daltons (Fig. 4b) which is 24 Daltons less than sheep Mb [25]. The observed molecular mass for buffalo Mb in the present study is 9.6 Daltons higher than the report of Dosi et al. (2006) for Italian water buffalo Mb extracted from skeletal muscles [5]. The molecular mass of goat Mb observed in the present study is 75 Daltons higher than the report of Suman et al. (2009) [7]. In general it is agreed that the mass of Mb for all the livestock and poultry species is around 17000 Daltons.

Purified Mb protein separated by 2DE gel was identified by MALDI-TOF MS. The analysis of generated peptides along with their m/z values and intensities revealed an abundant peptide with a m/z value of 748 for buffalo Mb compared to goat Mb in which a peptide with m/z of 1592 was most abundant (data not shown). The PMF results with details of molecular weight search score, number of matched peptides, nominal and observed mass along with % sequence coverage has been given in Table 1. Clear distinction can be made from the result where there is a reducing ion score for goat samples relative to buffalo samples. Peptide masses detected in the present study were matched for buffalo and goat Mb proteins in a database which showed the sequence coverage of 39.61% for buffalo and 41.55% for goat. Four peptides matched for buffalo Mb, whereas for goat Mb six peptides were matched. Furthermore, MALDI-TOF analysis consistently gave information on the peptide-mass fingerprint from position 2–119 for buffalo and 2–153 for goat samples. These results confirmed the identity of purified protein spot on 2DE gel as buffalo and

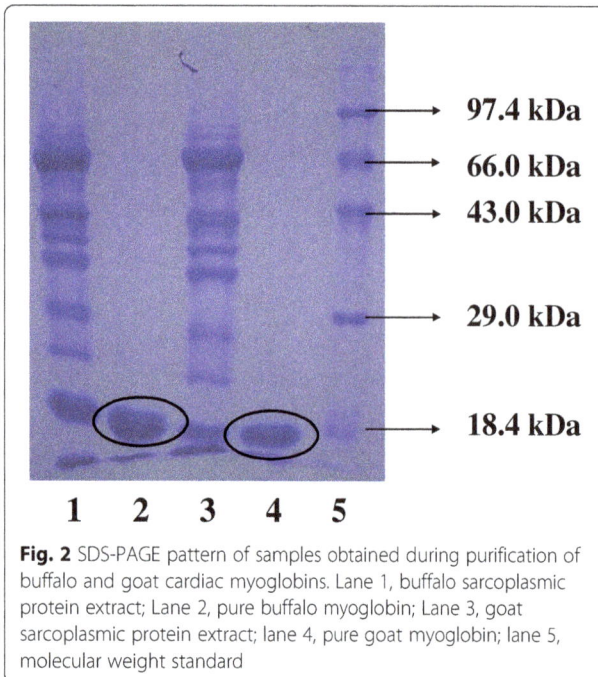

Fig. 2 SDS-PAGE pattern of samples obtained during purification of buffalo and goat cardiac myoglobins. Lane 1, buffalo sarcoplasmic protein extract; Lane 2, pure buffalo myoglobin; Lane 3, goat sarcoplasmic protein extract; lane 4, pure goat myoglobin; lane 5, molecular weight standard

multi-compartment chamber on top of an IPG strip which provides the desired pH gradient and solubilized proteins migrate through the strip until they reach their p*I* at a given compartment and then return into solution [26]. Using the Agilent 3100 OFFGEL fractionator high resolution mode (24 fractions per 24 cm, 3–10 IPG strip) with the Agilent's proprietary starter kit, buffalo and goat Mb's were fractionated separately. The p*I* was calculated by dividing the seven pH units of 3–10 gradient by the number of compartments, which is 24 in the present experiment. Each compartment corresponds to a pH resolution of 0.29. After completion of OFFGEL electrophoresis, the fractionated samples from each well, in solution form were taken out and subjected to SDS-PAGE. Out of 24 collected fractions each for buffalo and goat Mb's, single clear band corresponding to 17 kDa molecular mass was obtained (figure not shown) at 14 and 16th fractions, respectively for buffalo and goat. Based on the fraction number corresponding to 17 kDa band, the p*I* of buffalo and goat Mb's was calculated to be 6.77 and 7.35, respectively. Post-translational modification of proteins via phosphorylation leads to an acidic shift in the isoelectric pH which may result in variation [27]. The p*I* values investigated in the present study are almost similar to theoretical p*I* values of 6.71 and 7.38 for buffalo and goat Mb's respectively (web.expasy.org).

goat Mb's. The observed differences in peptides and their masses between buffalo and goat samples is because of the fact that buffalo Mb has 95.4% sequence similarity with goat Mb (expacy.com).

OFFGEL electrophoresis
For the first time, we have attempted to use the OFFGEL electrophoresis to determine the isoelectric point (p*I*) of buffalo and goat Mb's. OFFGEL electrophoresis uses

HNE induced oxidation of buffalo and goat Oxy-Mb's
HNE is a well-documented secondary product of ω-6 polyunsaturated fatty acids oxidation. Various researchers have utilized HNE as a model oxidation product to study its interaction with Oxy-Mb from different livestock and poultry species as mentioned earlier [28]. To understand

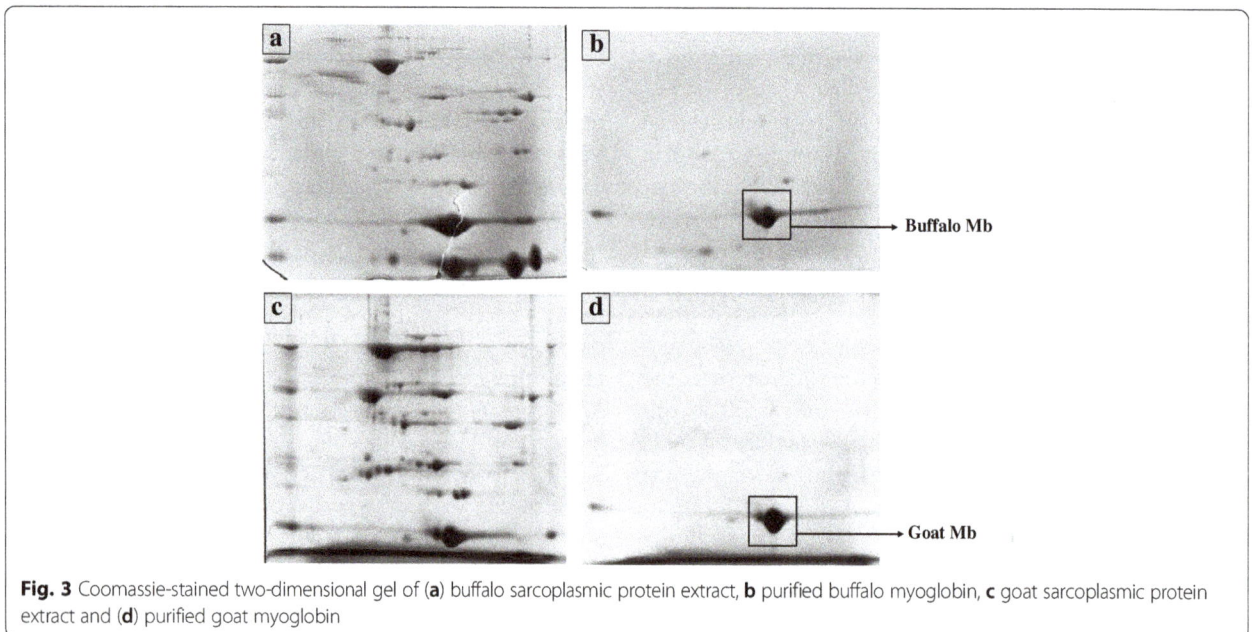

Fig. 3 Coomassie-stained two-dimensional gel of (**a**) buffalo sarcoplasmic protein extract, **b** purified buffalo myoglobin, **c** goat sarcoplasmic protein extract and (**d**) purified goat myoglobin

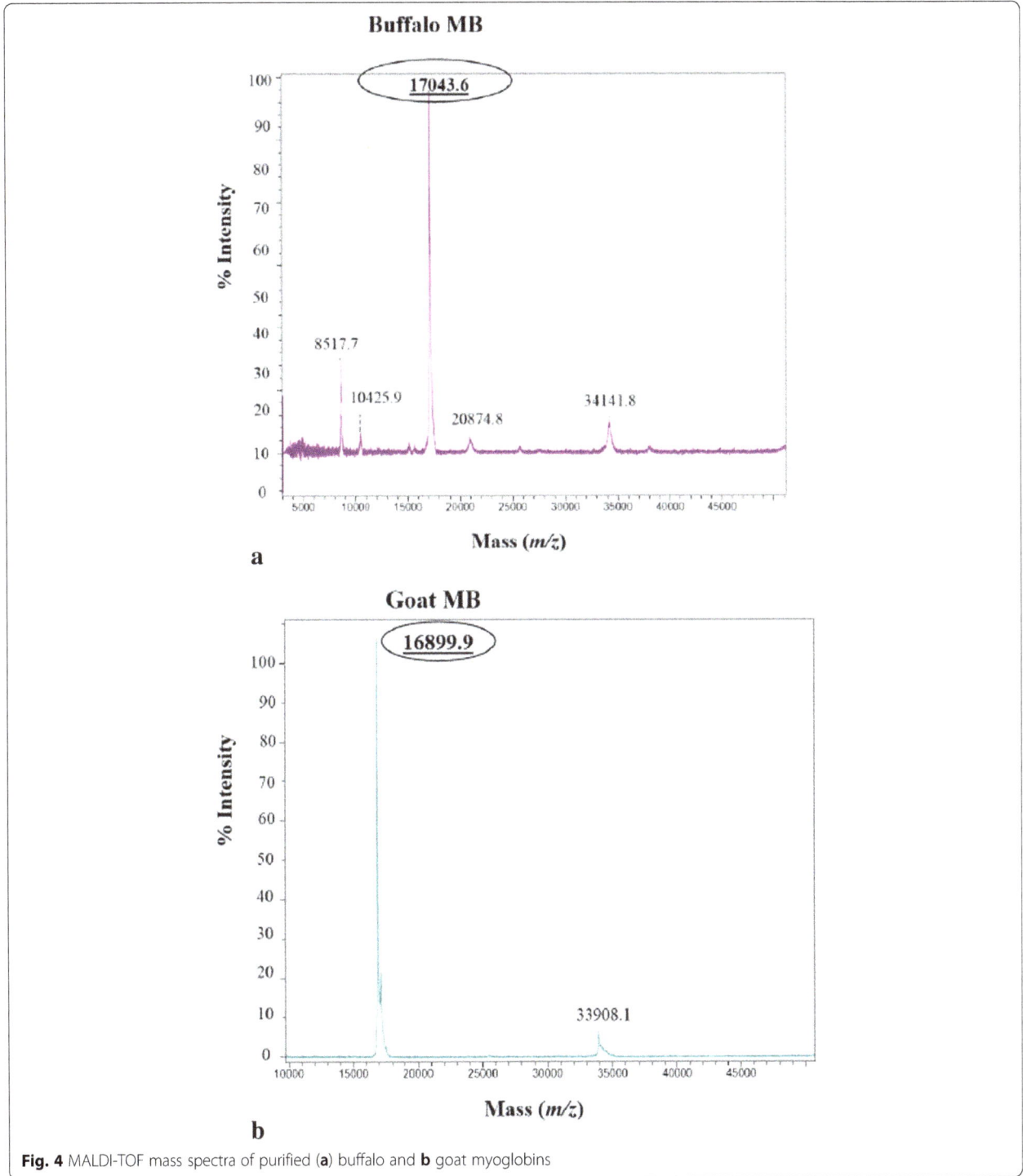

Fig. 4 MALDI-TOF mass spectra of purified (**a**) buffalo and **b** goat myoglobins

the proxidative effect of α,β-unsaturated aldehyde, HNE was incubated with buffalo and goat Oxy-Mb's for 3.5 h at 25 °C, pH 5.6 and 37 °C, pH 7.4 to represent the pH values of post-mortem muscle and physiological conditions, respectively. Pure Mb without HNE (control) was also incubated along with HNE reacted Mb (treatment) for comparison under similar conditions. The %Met-Mb formed during incubation of buffalo and goat Oxy-Mb's with

HNE (treatment) and without HNE (control) at 25 °C, pH 5.6 and 37 °C, pH 7.4 are presented in Figs. 5 and 6, respectively. When HNE was reacted with Oxy-Mb solutions, the %Met-Mb formation was greater ($p < 0.05$) than controls at all the temperatures and pH conditions for both buffalo and goat. As expected, the %Met-Mb formation is higher ($p < 0.05$) at 37 °C compared to 25 °C. Oxy-Mb oxidation of both buffalo and goats were higher

Table 1 Results of peptide mass fingerprinting (PMF) for identification of water buffalo (*Bubalus bubalis*) and goat meat (*Capra hircus*) myoglobins

Species	Accession No.	Score	No. matched peptides	Nominal mass	Observed mass	% Sequence coverage highlighting the matched peptides
Buffalo (*Bubalusbubalis*)	gil116248552	289	4	17155.0	17043.6	Sequence coverage = 39.61% M**GLSDGEWQLVLNAWGKVETDVAGHGQEVL IR**LFTGHPET LEKFDKFKHL KTEAEMKASE DLKK**HGNTVLTALGGILK**KK GHHEAEVKHL AESHANKHKI PVK**YLEFISDAIIHVLHDK**H PSDFGADAQA AMSKALELFR NEMAAQYKVL GFHG
Goat (*Capra hircus*)	gil118595750	79	6	17043.0	16899.9	Sequence coverage = 41.55% M**GLSDGEWTLVLNAWGKVEADVAGHGQEVLIR**LFTGHPET LEKFDKFKHL KTGAEMKASE DLKKH**GNTVLTALGGILK**KK GHHEAEVKHL AESHANKHKI PVKYLEFISD AIIHVLHAKH PSDFGADAQG AMSK**ALELFRNDMAAQYKVL GFQG**

$(p < 0.05)$ at pH 5.6 relative to pH 7.4, a result attributed to greater autoxidation. Similar results were observed during the incubation of HNE with equine, bovine and porcine Oxy-Mb's [9, 10, 12]. In general, the oxidising effect of HNE is greater at 7.4 compared to 5.6 potentially because of decreased nucleophilicity of candidate histidine residues at this pH [29]. Oxymyoglobin-HNE adduct formation was greater at pH 7.4 than at pH 5.6, because at acidic pH several ionizable imidazole groups (in HIS) are protonated and therefore, less reactive with HNE. The pKa value for the imidazole group of histidine is approximately 6.0, and therefore HIS residues would act as suitable nucleophilic targets for aldehyde adduction at pH 7.4. However, protonation of the imidazole ring at pH 5.6 would render HIS residues a less favorable candidate for aldehyde adduction via Michael addition. This will contribute to the greater number of HNE adduction at pH 7.4 than at typical meat pH (5.6) [12]. Further, Oxy-Mb oxidation at pH 5.6 will progress significantly rapid enough to mask any observable redox destabilizing effect of HNE. Considering these reasons and the earlier studies we used pH 7.4 condition for further characterization of Mb-HNE adduction.

MALDI-TOF MS analysis of Mb-HNE reaction mix (intact protein) at the end of incubation period (after 3.5 h) at pH 7.4 and 37 °C revealed covalent binding of upto three and five molecules of HNE to buffalo (Fig. 7a) and goat (Fig. 7b) Oxy-Mb's, respectively. These results indicate that HNE adducts were formed via Michael addition as the adduct peaks corresponded to the mass of myoglobin plus 156 Da, the molecular mass of HNE. These results suggest that buffalo myoglobin was less susceptible to nucleophilic attack and subsequent adduction with HNE, compared to goat Mb. This explains the reason for higher %Met-Mb formation observed for goat Oxy-Mb than buffalo Oxy-Mb when incubated with HNE. Research to date has demonstrated that a multitude of factors (endogenous and exogenous) contribute to meat color stability and biochemistry. Among several factors, adduction of lipid derived aldehydes with Oxy-Mb was reported to cause increased %Met-Mb formation and redox instability in different livestock and poultry species [8]. Michael adduction of HNE with apomyoglobin (model heme protein lacking cysteine residue) was confirmed by Bolgar and Gaskell (1996) using ESI-MS and indicated

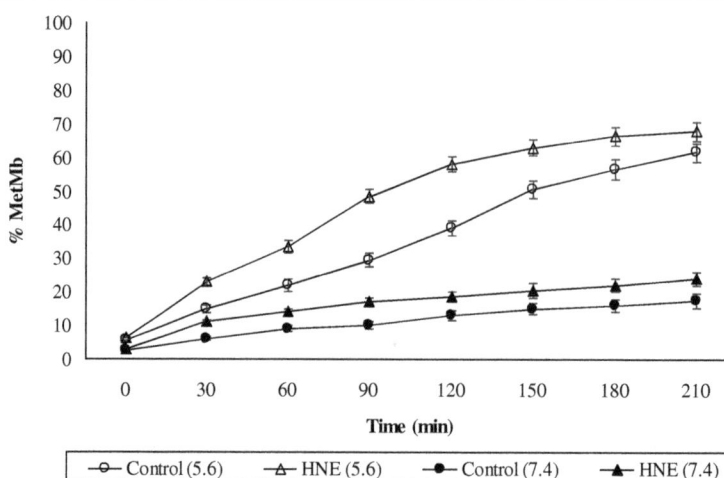

Fig. 5 Oxidation of buffalo meat oxymyoglobin reacted with HNE (treatment) and without HNE (control) at 25 °C, pH 5.6 and 37 °C, pH 7.4. Standard error bars are indicated ($n = 3$)

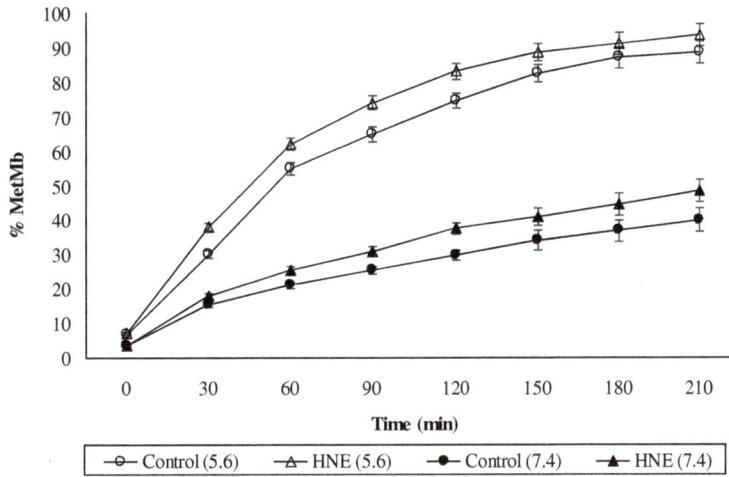

Fig. 6 Oxidation of goat meat oxymyoglobin reacted with HNE (treatment) and without HNE (control) at 25 °C, pH 5.6 and 37 °C, pH 7.4. Standard error bars are indicated ($n = 3$)

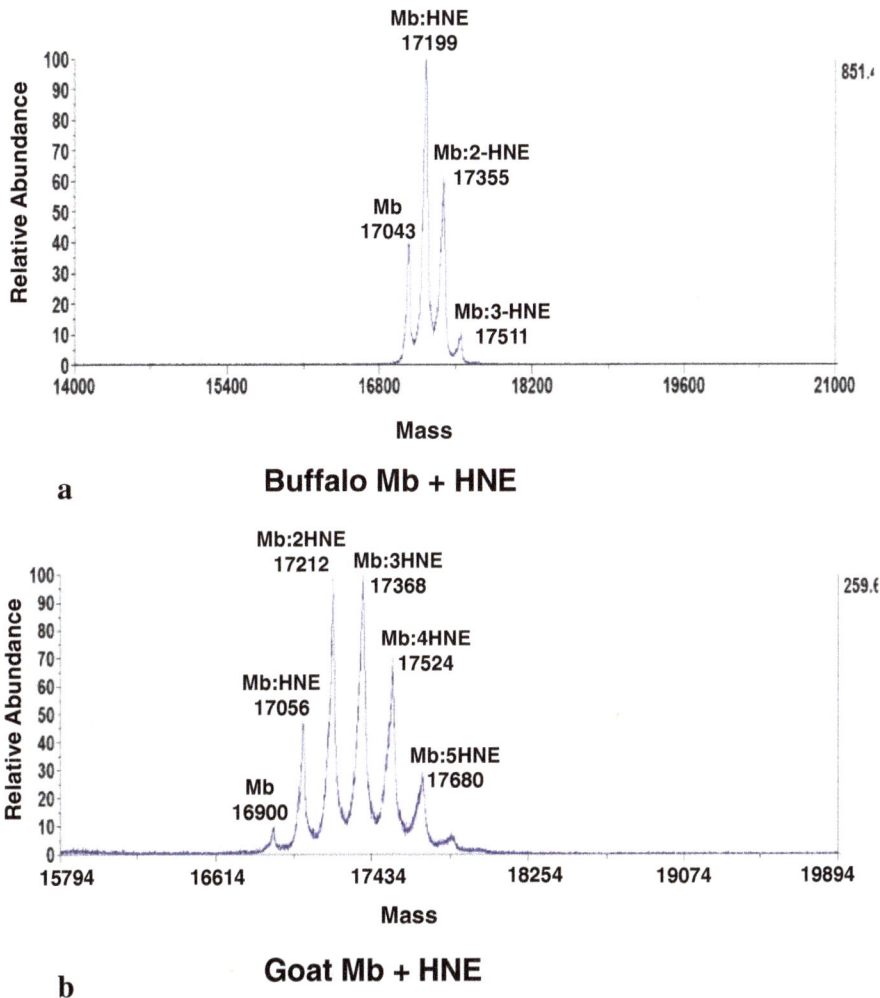

Fig. 7 MALDI-TOF MS spectrum of (**a**) buffalo meat oxymyoglobin (0.15 mM) reacted with HNE (1.0 mM) and **b** goat meat oxymyoglobin (0.075 mM) reacted with HNE (0.5 mM) at pH 7.4 and 37 °C for 3.5 h

three to 10 HNE adducts per protein molecule [30]. Faustman et al. (1999) and Alderton et al. (2003) have reported tri-adducts of HNE with equine and beef Oxy-Mb's incubated at pH 7.4 and 37 °C after 3 hours of incubation [9, 10]. Species specific effect of Mb-HNE adducts were observed by previous researchers as indicated by mono-adducts in porcine Mb at pH 5.6 [12], di-adducts for chicken and turkey at pH 7.4 [16], mono-adducts for emu and ostrich at pH 7.4 [17] during incubation for different time periods. Our present findings synergistically add to the existing knowledge that lipid oxidation induced oxidation of myoglobin from red meat species (equine meat, beef and sheep meat) is more pronounced relative to white meats (poultry species).

Adduction of Oxy-Mb with 4-hydroxy-2-nonenal would be expected to alter the protein's tertiary structure and predispose it to greater oxidation [14]. The specific sites of HNE adduction in buffalo and goat Mb's incubated with and without (control) HNE were further investigated using ESI-QTOF MS/MS specifically at histidine (HIS) site as the earlier researchers have confirmed that HIS was the sole amino acid adducted by HNE [10, 12]. The spectra corresponding to unadducted and adducted Mb peptides were identified and the b- and y-series ions for seven buffalo Mb peptides with HNE adductions are summarized in Table 2. In the present study, we identified seven nucleophilic histidine residues for buffalo Mb confirmed to be adducted with HNE. The HIS 24, 36, 81, 88, 93, 119 and 152 each had a mass addition of 158 Da which is 2-Da higher than the mass of HNE. Previous study by Alderton et al. (2003) revealed a mass increment of 156 Da due to HNE adduction via Michael addition [10]. The difference of 2-Da (156 vs. 158) in the present study could be due to protonation of HNE-adducted HIS which might have resulted in a 2-Da addition to the adduct's molecular mass and it is in agreement with Suman et al. (2007) [12]. For goat Mb we identified nine nucleophilic histidine residues confirmed to be adducted with HNE. The HIS 24, 36, 64, 81, 88, 97, 113, 116 and 119 each had a mass addition of 158 Da. The spectra corresponding to unadducted and adducted Mb peptides were identified and the b- and y-series ions for goat Mb peptides with HNE adductions are summarized in Table 3. Compared to buffalo Mb, goat Mb had two additional HNE adductions. Our results for buffalo Mb-HNE adductions were similar to the findings of Suman et al. (2007) who has also observed seven beef Mb-HNE adductions under similar pH and temperature conditions [12]. This is because of the fact that both buffalo and cattle shares 98% sequence similarity with 13 total histidine residues in each located at the same position. Similar to the findings of Suman et al. (2007), HIS 88 and HIS 93 originating from the same peptide (HLAESHANK, position 88–96) were found adducted to HNE [12]. Simultaneous adduction at histidine 88 and histidine 93 was not detected because of the fact that alkylation of one histidine prevents the other nearby histidine residue from being alkylated due to steric hindrance [19].

An interesting finding in our study is that, goat Mb with 95.4% and 97.4% sequence identity with buffalo and beef Mb, respectively with 12 total histidine residues exhibited 9 HNE adductions. Similar to buffalo Mb, even for goat Mb HIS 113 and HIS 116 originating from same peptide (YLEFISDAIIHVLHAK, position 103–118) were found adducted to HNE. The difference in HNE adductions between buffalo and goat Mb's is presented in Table 4. These results indicate a difference in HNE adduction between buffalo and goat Mb's both in terms of number and location of adductions. We observed the adduction of proximal HIS 93 in buffalo, whereas for goat the adduction of distal HIS 64 was found. This is of great significance as HIS 93, the proximal histidine, is bound to the heme moiety of Mb, whereas distal HIS 64, coordinates with oxygen or other molecules associated with the sixth ligand during the interconversion of Mb redox forms (i.e., Oxy-Mb and Met-Mb). Because HIS 93 and 64 lie in close proximity to the heme group, their modification by HNE could be expected to alter the protein structure around the heme cleft and subsequently impact redox stability. In their study comparing HNE induced oxidation of beef and pork Oxy-Mb's Suman et al. (2007) observed that fewer HIS residues were adducted by HNE in porcine Mb when compared to bovine Mb, which suggested an apparently lower susceptibility of porcine Mb to the redox destabilizing effect of HNE [12]. They concluded that preferential HNE adduction at proximal residue (HIS 93), exclusively observed in bovine Oxy-Mb might result in more pronounced Oxy-Mb oxidation in beef relative to pork. Our findings indicate that goat Oxy-Mb with HNE adduction at distal histidine (HIS 64) results in more pronounced Oxy-Mb oxidation resulting in higher % Met-Mb than beef Mb with HNE adduction at proximal histidine (HIS 93). Yin et al. (2011) studied the HNE induced oxidation of Oxy-Mb's from seven different meat animals and concluded that, the relative effect of HNE was greater for Mb's that contained 12 ± 1 histidine residues than for those that contained nine histidine residues [31]. Table 4 clearly indicate greater number of adductions and alkylation of more HIS residues in red meat producing livestocks (12 or 13 HIS) except pork (nine histidine) relative to poultry birds (eight or nine histidine). Present study confirm these findings, wherein livestock Mb's with higher number of histidine residues results in greater Mb-HNE adducts and more covalently modified histidine residues leading to increased susceptibility for lipid oxidation induced oxidation.

Conclusions

In conclusion, results obtained from this study revealed that buffalo and goat Mb's were similar in behaviour during isolation and purification and exhibited a molecular mass of

Table 2 MS/MS spectral features of unadducted and HNE-adducted water buffalo (*Bubalus bubalis*) myoglobin peptides

Peptide position[a]	Peptide sequence[b]	Modification and mass shift	Precursor m/z	b and y ions identified[c]
17–31	VETDVAGHGQEVLIR	Unadducted	1604.83	b ions: 100.08 (b1), 229.12 (b2), 300.16 (b3), 415.18 (b4), 514.25 (b5), 585.29 (b6), 642.31 (b7), 779.37 (b8), 836.39 (b9), 964.45 (b10), 1093.49 (b11), 1192.56 (b12), 1305.64 (b13), 1418.73 (b14) y ions: 173.10 (y1), 286.19 (y2), 399.27 (y3), 498.34 (y4), 627.38 (y5), 755.44 (y6), 812.46 (y7), 949.52 (y8), 1006.54 (y9), 1077.58 (y10), 1176.65 (y11), 1291.68 (y12), 1362.71 (y13), 1491.76 (y14)
17–31	VETDVAG**H**GQEVLIR	HNE + 158.1	1604.84	b ions: 100.08 (b1), 229.12 (b2), 300.16 (b3), 415.18 (b4), 514.25 (b5), 585.29 (b6), 642.31 (b7), **937.50 (b8)**, **994.52 (b9)**, **1122.58 (b10)**, **1251.62 (b11)**, **1350.69 (b12)**, **1463.77 (b13)**, **1576.86 (b14)** y ions: 173.10 (y1), 286.19 (y2), 399.27 (y3), 498.34 (y4), 627.38 (y5), 755.44 (y6), 812.46 (y7), **1107.65 (y8)**, **1164.67 (y9)**, **1235.71 (y10)**, **1334.78 (y11)**, **1449.81 (y12)**, **1520.84 (y13)**, **1649.89 (y14)**
32–42	LFTGHPETLEK	Unadducted	1271.66	b ions: 114.09 (b1), 261.16 (b2), 362.21 (b3), 419.23 (b4), 556.29 (b5), 653.34 (b6), 782.38 (b7), 883.43 (b8), 996.52 (b9), 1125.56 (b10) y ions: 145.10 (y1), 274.14 (y2), 387.22 (y3), 488.27 (y4), 617.31 (y5), 714.37 (y6), 851.43 (y7), 908.45 (y8), 1009.50 (y9), 1156.56 (y10)
32–42	LFT**GH**PETLEK	HNE + 158.1	1253.65	b ions: 114.09 (b1), 261.16 (b2), 362.21 (b3), 419.23 (b4), **714.42 (b5)**, **811.47 (b6)**, **940.51 (b7)**, **1041.56 (b8)**, **1154.65 (b9)**, **1283.69 (b10)** y ions: 145.10 (y1), 274.14 (y2), 387.22 (y3), 488.27 (y4), 617.31 (y5), 714.37 (y6), **1009.56 (y7)**, **1066.58 (y8)**, **1167.63 (y9)**, **1314.69 (y10)**
80–87	GHHEAEVK	Unadducted	1893.90	b ions: 58.03 (b1), 195.09 (b2), 332.15 (b3), 461.19 (b4), 532.23 (b5), 661.27 (b6), 760.34 (b7) y ions: 145.10 (y1), 244.17 (y2), 373.21 (y3), 444.25 (y4), 573.29 (y5), 710.35 (y6), 847.41 (y7)
80–87	**G**HHEAEVK	HNE + 158.1	1893.93	b ions: 58.03 (b1), **353.22 (b2)**, **490.28 (b3)**, **619.32 (b4)**, **690.36 (b5)**, **819.40 (b6)**, **918.47 (b7)** y ions: 145.10 (y1), 244.17 (y2), 373.21 (y3), 444.5 (y4), 573.29 (y5),710.35 (y6), **1005.54 (y7)**
88–96	HLAESHANK	Unadducted	1253.65	b ions: 138.07 (b1), 251.15 (b2), 322.19 (b3), 451.23 (b4), 538.26 (b5), 675.32 (b6), 746.36 (b7), 860.40 (b8) y ions: 145.10 (y1), 259.14 (y2), 330.18 (y3), 467.24 (y4), 554.27 (y5), 683.70 (y6), 754.78 (y7), 867.94 (y8)
88–96	**H**LAESHANK	HNE + 158.1	1253.65	b ions: **296.20 (b1)**, **409.28 (b2)**, **480.32 (b3)**, **609.74 (b4)**, **696.82 (b5)**, **833.96 (b6)**, **905.04 (b7)**, **1019.15 (b8)** y ions: 145.10 (y1), 259.14 (y2), 330.18 (y3), 467.24 (y4), 554.27 (y5), 683.70 (y6), 754.78 (y7), 867.94 (y8)
88–96	HLAESHANK	Unadducted	1271.66	b ions: 138.07 (b1), 251.15 (b2), 322.19 (b3), 451.23 (b4), 538.26 (b5), 675.32 (b6), 746.36 (b7), 860.40 (b8) y ions: 145.10 (y1), 259.14 (y2), 330.18 (y3), 467.24 (y4), 554.27 (y5), 683.70 (y6), 754.78 (y7), 867.94 (y8)
88–96	HLAE**SH**ANK	HNE + 158.1	1271.66	b ions: 138.07 (b1), 251.15 (b2), 322.19 (b3), 451.23 (b4), 538.26 (b5), **833.96 (b6)**, **905.04 (b7)**, **1019.15 (b8)** y ions: 145.10 (y1), 259.14 (y2), 330.18 (y3), **625.75 (y4)**, **712.82 (y5)**, **841.94 (y6)**, **913.02 (y7)**, **1026.18 (y8)**
119–133	HPSDFGADAQAAMSK	Unadducted	1514.67	b ions: 138.07 (b1), 235.12 (b2), 322.15 (b3), 437.18 (b4), 584.25 (b5), 641.27 (b6), 712.31 (b7), 827.33 (b8), 898.37 (b9), 1026.43 (b10), 1097.47 (b11), 1168.50 (b12) y ions: 363.17 (y1), 434.21 (y2), 505.24 (y3), 633.30 (y4), 704.34 (y5), 819.37 (y6), 890.40 (y7), 947.43 (y8), 1094.49 (y9), 1209.52 (y10), 1296.55 (y11), 1393.61 (y12)
119–133	**H**PSDFGADAQAAMSK	HNE + 158.1	1532.68	b ions: **296.20 (b1)**, **393.25 (b2)**, **480.28 (b3)**, **595.67 (b4)**, **742.85 (b5)**, **799.90 (b6)**, **870.98 (b7)**, **986.07 (b8)**, **1057.15 (b9)**, **1185.28 (b10)**, **1256.36 (b11)**, **1327.44 (b12)** y ions: 363.17 (y1), 434.21 (y2), 505.24 (y3), 633.75 (y4), 704.83 (y5), 819.91 (y6), 890.99 (y7), 948.05 (y8), 1095.22 (y9), 1210.31 (y10), 1297.39 (y11), 1394.51 (y12)
148–153	VLGFHG	Unadducted	629.34	b ions: 100.08 (b1), 213.16 (b2), 270.18 (b3), 417.25 (b4), 554.31 (b5) y ions: 74.02 (y1), 211.08 (y2), 358.15 (y3), 415.17 (y4), 528.26 (y5)
148–153	VLG**FH**G	HNE + 158.1	629.34	b ions: 100.08 (b1), 213.16 (b2), 270.18 (b3), 417.25 (b4), **712.91 (b5)** y ions: 74.02 (y1), **369.21 (y2)**, **516.28 (y3)**, **573.30 (y4)**, **686.83 (y5)**

a) Amino acid position in the water buffalo Mb; b) Amino acid sequence in the water buffalo Mb; c) Observed signals assigned as *b* or *y* ions. Ions showing shift in the mass by 156 Da indicating HNE adduction with respect to corresponding ions in unmodified peptides are highlighted in bold fonts

Table 3 MS/MS spectral features of unadducted and HNE-adducted goat (*Capra hircus*) meat myoglobin peptides

Peptide position[a]	Peptide sequence[b]	Modification and mass shift	Precursor m/z	b and y ions identified[c]
17–31	VEADVAGHGQEVLIR	Unadducted	1592.77	b ions: 100.08 (b1), 229.12 (b2), 300.16 (b3), 415.16 (b4), 514.25 (b5), 585.29 (b6), 642.31 (b7), 779.37 (b8), 836.39 (b9), 964.45 (b10), 1093.49 (b11), 1192.56 (b12), 1305.64 (b13), 1418.73 (b14) y ions: 174.11 (y1), 287.20 (y2), 400.28 (y3), 499.35 (y4), 628.39 (y5), 756.45 (y6), 813.47 (y7), 950.53 (y8), 1107.55 (y9), 1078.59 (y10), 1177.66 (y11), 1292.68 (y12), 1363.72 (y13), 1492.76 (y14)
17–31	VEADVAG**H**GQEVLIR	HNE + 158.1	1751.01	b ions: 100.08 (b1), 229.12 (b2), 300.16 (b3), 415.16 (b4), 514.25 (b5), 585.29 (b6), 642.31 (b7), **937.50 (b8)**, **994.52 (b9)**, **1122.58 (b10)**, **1251.62 (b11)**, **1350.69 (b12)**, **1463.77 (b13)**, **1576.86 (b14)** y ions: 174.11 (y1), 287.20 (y2), 400.28 (y3), 499.35 (y4), 628.39 (y5), 756.45 (y6), 813.47 (y7), **1108.66 (y8)**, **1165.68 (y9)**, **1236.72 (y10)**, **1335.79 (y11)**, **1450.81 (y12)**, **1521.85 (y13)**, **1650.89 (y14)**
32–42	LFTGHPETLEK	Unadducted	1271.43	b ions: 114.09 (b1), 261.16 (b2), 362.21 (b3), 419.23 (b4), 556.29 (b5), 653.34 (b6), 782.38 (b7), 883.43 (b8), 996.52 (b9), 1125.56 (b10) y ions: 146.11 (y1), 275.15 (y2), 388.23 (y3), 489.28 (y4), 618.32 (y5), 715.38 (y6), 852.43 (y7), 909.46 (y8), 1010.50 (y9), 1157.57 (y10)
32–42	LFTG**H**PETLEK	HNE + 158.1	1429.67	b ions: 114.09 (b1), 261.16 (b2), 362.21 (b3), 419.23 (b4), **714.42 (b5)**, **811.47 (b6)**, **940.51 (b7)**, **1041.56 (b8)**, **1154.65 (b9)**, **1283.69 (b10)** y ions: 146.11 (y1), 275.15 (y2), 388.23 (y3), 489.28 (y4), 618.32 (y5), 715.38 (y6), **1010.56 (y7)**, **1067.59 (y8)**, **1168.63 (y9)**, **1315.70 (y10)**
63–77	KHGNTVLTALGGILK	Unadducted	1521.82	b ions: 129.10 (b1), 266.16 (b2), 323.18 (b3), 437.23 (b4), 538.27 (b5), 637.34 (b6), 750.43 (b7), 851.47 (b8), 922.51 (b9), 1035.60 (b10), 1092.62 (b11), 1149.64 (b12), 1262.72 (b13), 1375.81 (b14) y ions: 146.11 (y1), 259.19 (y2), 372.27 (y3), 429.30 (y4), 486.32 (y5), 599.40 (y6), 670.44 (y7), 771.49 (y8), 884.57 (y9), 983.64 (y10), 1084.69 (y11), 1198.73 (y12), 1255.75 (y13), 1392.81 (y14)
63–77	K**H**GNTVLTALGGILK	HNE + 158.1	1680.06	b ions: 129.10 (b1), **424.29 (b2)**, **481.31 (b3)**, **595.36 (b4)**, **696.40 (b5)**, **795.47 (b6)**, **908.56 (b7)**, **1009.60 (b8)**, **1080.64 (b9)**, **1193.73 (b10)**, **1250.75 (b11)**, **1307.77 (b12)**, **1420.85 (b13)**, **1533.94 (b14)** y ions: 146.11 (y1), 259.11 (y2), 372.27 (y3), 429.30 (y4), 486.32 (y5), 599.40 (y6), 670.44 (y7), 771.49 (y8), 884.57 (y9), 983.64 (y10), 1084.69 (y11), 1198.73 (y12), 1255.75 (y13), **1550.94 (y14)**
80–87	GHHEAEVK	Unadducted	905.96	b ions: 58.03 (b1), 195.09 (b2), 332.15 (b3), 461.19 (b4), 532.23 (b5), 661.27 (b6), 760.34 (b7) y ions: 146.11 (y1), 245.17 (y2), 374.22 (y3), 445.25 (y4), 574.30 (y5), 711.36 (y6), 848.41 (y7)
80–87	G**H**HEAEVK	HNE + 158.1	1064.20	b ions: 58.03 (b1), **353.22 (b2)**, **490.28 (b3)**, **619.32 (b4)**, **690.36 (b5)**, **819.40 (b6)**, **918.47 (b7)** y ions: 146.11 (y1), 245.17 (y2), 374.22 (y3), 445.25 (y4), 574.30 (y5), 711.36 (y6), **1006.54 (y7)**
88–96	HLAESHANK	Unadducted	1271.40	b ions: 138.07 (b1), 251.15 (b2), 322.19 (b3), 451.23 (b4), 538.26 (b5), 675.32 (b6), 746.36 (b7), 860.40 (b8), 988.50 (b9), 1125.56 (b10) y ions: 146.11 (y1), 283.16 (y2), 411.26 (y3), 525.30 (y4), 596.34 (y5), 733.40 (y6), 820.43 (y7), 949.47 (y8), 1020.51 (y9), 1133.59 (y10)
88–96	**H**LAESHANK	HNE + 158.1	1164.32	b ions: **296.20 (b1)**, **409.28 (b2)**, **480.38 (b3)**, **609.36 (b4)**, **696.39 (b5)**, **833.45 (b6)**, **904.49 (b7)**, **1018.53 (b8)** y ions: 146.11 (y1), 260.15 (y2), 331.19 (y3), 468.24 (y4), 555.28 (y5), 684.32 (y6), 755.36 (y7), 868.44 (y8)
88–98	HLAESHANKHK	Unadducted	1271.40	b ions: 138.07 (b1), 251.15 (b2), 322.19 (b3), 451.23 (b4), 538.26 (b5), 675.32 (b6), 746.36 (b7), 860.40 (b8), 988.50 (b9), 1125.56 (b10) y ions: 146.11 (y1), 283.16 (y2), 411.26 (y3), 525.30 (y4), 596.34 (y5), 733.40 (y6), 820.43 (y7), 949.47 (y8), 1020.51 (y9), 1133.59 (y10)
88–98	HLAESHANK**H**K	HNE + 158.1	1429.64	b ions:138.07 (b1), 251.15 (b2), 322.19 (b3), 451.23 (b4), 538.26 (b5), 675.32 (b6), 746.36 (b7), 860.40 (b8), 988.50 (b9), **1283.69 (b10)** y ions: 146.11 (y1), **441.30 (y2)**, **569.39 (y3)**, **683.43 (y4)**, **754.47 (y5)**, **891.53 (y6)**, **978.56 (y7)**, **1107.60 (y8)**, **1178.64 (y9)**, **1291.72 (y10)**

Table 3 MS/MS spectral features of unadducted and HNE-adducted goat (*Capra hircus*) meat myoglobin peptides (*Continued*)

Position[a]	Sequence[b]	Adduct	Mass	Ions[c]
103–118	YLEFISDAIIHVLHAK	Unadducted	1869.19	b ions: 164.07 (b1), 277.16 (b2), 406.20 (b3), 553.27 (b4), 666.35 (b5), 753.38 (b6), 868.41 (b7), 939.45 (b8), 1052.53 (b9), 1165.61 (b10), 1302.67 (b11), 1401.74 (b12), 1514.83 (b13), 1651.88 (b14), 1722.92 (b15) y ions: 146.11 (y1), 217.14 (y2), 354.20 (y3), 467.29 (y4), 566.35 (y5), 703.41 (y6), 816.50 (y7), 929.58 (y8), 1000.62 (y9), 1115.65 (y10), 1202.68 (y11), 1315.76 (y12), 1462.83 (y13), 1591.95 (y14), 1704.96 (y15)
103–118	YLEFISDAII**H**VLHAK	HNE + 158.1	2027.43	b ions: 164.07 (b1), 277.16 (b2), 406.20 (b3), 553.27 (b4), 666.35 (b5), 753.38 (b6), 868.41 (b7), 939.45 (b8), 1052.53 (b9), 1165.61 (b10), 1460.80 (b11), **1559.87 (b12)**, **1672.96 (b13)**, 1810.02 (b14), **1881.05 (b15)** y ions: 146.11 (y1), 217.14 (y2), 354.20 (y3), 467.29 (y4), 566.35 (y5), **861.54 (y6)**, **974.63 (y7)**, **1087.71 (y8)**, **1158.75 (y9)**, **1273.78 (y10)**, **1360.81 (y11)**, 1473.89 (y12), **1620.96 (y13)**, 1750.00 (y14), **1863.09 (y15)**
103–118	YLEFISDAIIHVLHAK	Unadducted	1869.19	b ions: 164.07 (b1), 277.16 (b2), 406.20 (b3), 553.27 (b4), 666.35 (b5), 753.38 (b6), 868.41 (b7), 939.45 (b8), 1052.53 (b9), 1165.61 (b10), 1302.67 (b11), 1401.74 (b12), 1514.83 (b13), 1651.88 (b14), 1722.92 (b15) y ions: 146.11 (y1), 217.14 (y2), 354.20 (y3), 467.29 (y4), 566.35 (y5), 703.41 (y6), 816.50 (y7), 929.58 (y8), 1000.62 (y9), 1115.65 (y10), 1202.68 (y11), 1315.76 (y12), 1462.83 (y13), 1591.95 (y14), 1704.96 (y15)
103–118	YLEFISDAIIHVL**H**AK	HNE + 158.1	2027.43	b ions: 164.07 (b1), 277.16 (b2), 406.20 (b3), 553.27 (b4), 666.35 (b5), 753.38 (b6), 868.41 (b7), 939.45 (b8), 1052.53 (b9), 1165.61 (b10), 1302.67 (b11), 1401.74 (b12), 1514.83 (b13), **1810.02 (b14)**, **1881.05 (b15)** y ions: 146.11 (y1), 217.14 (y2), **512.33 (y3)**, **625.42 (y4)**, **724.48 (y5)**, **861.54 (y6)**, **974.63 (y7)**, **1087.71 (y8)**, **1158.75 (y9)**, **1273.78 (y10)**, **1360.81 (y11)**, 1473.89 (y12), **1620.96 (y13)**, 1750.00 (y14), **1863.09 (y15)**
119–133	HPSDFGADAQGAMSK	Unadducted	1518.62	b ions: 138.07 (b1), 235.12 (b2), 322.15 (b3), 437.18 (b4), 584.25 (b5), 641.27 (b6), 712.31 (b7), 827.33 (b8), 898.37 (b9), 1026.43 (b10), 1083.45 (b11), 1154.49 (b12), 1285.53 (b13), 1372.56 (b14) y ions: 146.11 (y1), 233.14 (y2), 364.18 (y3), 435.22 (y4), 492.24 (y5), 620.30 (y6), 691.33 (y7), 806.36 (y8), 877.40 (y9), 934.42 (y10), 1081.49 (y11), 1196.51 (y12), 1283.55 (y13), 1380.60 (y14)
119–133	**H**PSDFGADAQGAMSK	HNE + 158.1	1676.86	b ions: **296.20 (b1)**, **393.25 (b2)**, **480.28 (b3)**, **595.67 (b4)**, **742.85 (b5)**, **799.40 (b6)**, **870.44 (b7)**, **985.46 (b8)**, **1056.50 (b9)**, **1184.56 (b10)**, **1241.58 (b11)**, **1312.62 (b12)**, **1443.66 (b13)**, **1530.69 (b14)** y ions: 146.11 (y1), 233.14 (y2), 364.18 (y3), 435.22 (y4), 492.24 (y5), 620.30 (y6), 691.33 (y7), 806.36 (y8), 877.40 (y9), 934.42 (y10), 1081.49 (y11), 1196.51 (y12), 1283.55 (y13), 1380.60 (y14)

a) Amino acid position in the goat Mb; b) Amino acid sequence in the goat Mb; c) Observed signals assigned as *b* or *y* ions. Ions showing shift in the mass by 156 Da indicating HNE adduction with respect to corresponding ions in unmodified peptides are highlighted in bold fonts

Table 4 Comparison of Mb-HNE adductions and number of covalently modified histidine (HIS) residues in different meat animals and poultry

Meat animals and poultry	Total No. HIS residues	Mb-HNE incubation conditions	Incubation time	No. Mb-HNE adductions	Modified HIS residues	Position of adducted HIS residues	Reference
Meat animals							
Water buffalo (bovine)	13	pH 7.4, 37 °C	210 min	3	7	24,36,81,88,93,119,152	Present study
Goat (caprine)	12	pH 7.4, 37 °C	210 min	5	9	24,36,64,81,88,97,113, 116,119	Present study
Horse (equine)	11	pH 7.4, 37 °C	160 min	3	——————	——————	[9]
Cattle (bovine)	13	pH 7.4, 37 °C	160 min	3	7	24,36,81,88,93,119,152	[12]
				3	6	24,64,93,116,119,152	[10]
Pig (porcine)	9	pH 7.4, 37 °C	240 min	1	3	24,36,119	[12]
Sheep (ovine)	12	pH 7.4, 37 °C	240 min	3	3	25,65,120	[18]
Poultry							
Chicken	9	pH 5.8, 25 °C	240 min	2	2	64,93	[11]
Turkey	9	pH 5.8, 25 °C	240 min	2	——————	——————	[11]
Ostrich	8	pH 7.4, 37 °C	360 min	1	1	36	[17]
Emu	9	pH 7.4, 37 °C	360 min	1	2	34,36	[17]

17,043.6 Daltons and 16,899 Daltons, respectively. The study has demonstrated the species-specific variation in 2DE properties of buffalo and goat Mb's. Present study demonstrated the suitability of OFFGEL electrophoresis for determining the p*I* of Mb proteins. The study also reinforces the potential interaction between water buffalo and goat Mb's with a lipid derived aldehyde (HNE) via covalent modification of seven and nine histidine residues, respectively. The results suggest lower susceptibility of water buffalo Mb to the redox destabilizing effect of HNE compared to goat Mb. Our findings provide explanation for the previously noted observation that more number of Mb-HIS adducts and increased alkylation of HIS residues will exacerbate the Oxy-Mb oxidation.

Methods
Materials
Sephacryl S-200 HR and sodium hydrosulfite were obtained from Sigma-Aldrich chemicals Co., Sweden. The disposable PD-10 columns, IPG strips, DTT, Iodoacetamide, Acrylamide, Bis-acrylamide, Glycine, Methanol, Glacial Acetic acid, β-mercaptoenthanol and Coomassie blue, were obtained from GE Health Care. HNE was obtained from Cayman Chemical Company. Dialysis tubing was sourced from Spectrum laboratories, Inc. (Rancho Domingues, CA, USA). OFFGEL starter kit was procured from Agilent Technologies. All chemicals were of reagent grade or greater purity.

Isolation and purification of buffalo and goat myoglobins
Buffalo and goat hearts procured from municipal abattoir, Chengicherla, Hyderabad, India were trimmed off fat and connective tissue, vacuum packaged and stored at −80 °C till extraction. Myoglobin (Mb) was isolated

from cardiac muscles [22]. Frozen cardiac muscle samples were chilled overnight in refrigerator, cut into small pieces and homogenized with buffer containing 10 mM Tris–HCl with 1 mM EDTA (pH 8.0) and centrifuged at 5000 g for 10 min at 4 °C. The supernatant was brought to 70% ammonium sulfate saturation and centrifuged at 18000 g for 20 min 4 °C. The resulting supernatant was brought to 100% saturation with ammonium sulfate and centrifuged at 20000 g for 1 h 4 °C. The precipitate was resuspended in homogenization buffer and dialyzed against ten volumes of dialysis buffer (5 mM Tris–HCl with 1 mM EDTA, pH 8.0) for 24 h, with buffer changes at every 8 h interval. The dialysate was filtered through 0.45 μm syringe filter followed by and 0.22 μm filter. The 3 mL filtrate with a protein concentration of approximately 10 mg/mL was loaded on to Sephacryl S-200 HR gel filtration column (Econo column, 1.5 × 100 cm, BIORAD) equilibrated with elution buffer (5 mM Tris–HCl with 1 mM EDTA, pH 8.0) at a flow rate of 0.1 mL/min. Purified Mb fractions (1 mL) were collected and absorbance was measured at 280 and 540 nm using UV–VIS spectrophotometer (Model: UV-1700, PharmaSpec, SHIMADZU, Japan). To confirm the purity of Mb, fractions were subjected to sodium dodecyl sulphate-polyacrylamide gel electrophoresis (SDS-PAGE) and 2-dimensional gel electrophoresis (2DE).

Isolation of sarcoplasmic proteome
The sarcoplasmic proteome from buffalo and goat samples were extracted as described by Sentandreu et al. (2010) [32]. Frozen samples (1 g) were cut and homogenized in 10 mL cold extraction buffer (50 mM Tris, and pH 8.0). The homogenate was centrifuged at 10,000 g for 20 min at 4 °C. The supernatant constituting the

sarcoplasmic extract in which all soluble proteins were contained was filtered and used for 2DE.

Sodium dodecyl sulfate polyacrylamide gel electrophoresis
SDS-PAGE analyses of buffalo and goat Mb fractions from different steps of purification were performed using the method of Laemmli (1970) under reducing conditions using a 4% stacking gel and 12% separating gel in midi-electrophoresis apparatus (Model: SE-600 Ruby; GE Healthcare, Uppsala, Sweeden) [33]. The current for each gel was maintained at 10 mA. After separation, the proteins were stained with 0.1% (w/v) coomassie brilliant blue R-250 in 50% (v/v) methanol and 10% (v/v) acetic acid, and destained with 40% methanol (v/v) and 10% (v/v) acetic acid. Destained images were scanned using Image Scanner-III, LabScan 6.0 (GE Healthcare, Uppsala, Sweeden) and IQTL calibration converter was used to obtain image.

Two-dimensional gel electrophoresis and image analysis
Sarcoplasmic extract and gel-filtered (pure) Mb samples were passed through PD-10 desalting columns equilibrated with double distilled water to remove the salts. Later, protein concentration was determined using 2-D Quant Kit (GE Healthcare, USA) and 2DE was performed as described by Lametsch and Benedixen (2001) with few modifications [34]. The immobilized pH gradient (IPG) strips (Immobiline®drystrip, 11 cm, pH range 3–10) were passively rehydrated in 200 µL of myoglobin samples containing 750–800 µg proteins for 12 h. Rehydrated IPG strips were subjected to Iso-electric focusing (IEF) in a Ettan IPGPhor-3 (GE health care, Uppsala, Sweeden) gel apparatus at 18 °C for a total of 10800 Vh. Focused IPG strips were equilibrated in 6 M urea, 20% glycerol, 2% SDS, 0.375 M Tris (pH 8.8), for 15 min in 2% dithiothreitol (DTT) and 5% iodoacetamide respectively. After equilibration, proteins were separated in the second dimension, with the SE 600 Ruby apparatus at 100 V with 60 mA/gel until tracking dye reached lower end of gel. The gel was removed and stained with colloidal Coomassie blue for 3 h followed by overnight destaining. Myoglobin samples from buffalo and goat were run under the same conditions and the three gels were produced for each.

Stained gels were scanned on an Image Scanner III using labscan 6.0 software. Spot detection and quantification were performed with Image Master Platinum7.0 software (GE Healthcare, Uppsala, Sweeden). The images were grouped into different groups according to required study and the intensity and area of individual spots were analyzed for comparative image analysis. For each spot in buffalo or goat Mb samples, spot quantity values in triplicate gels were averaged for statistical

analysis. A spot was considered to be significant in differential abundance when it was associated with $P < 0.05$.

Determination of molecular mass using MALDI-TOF MS and protein identification through peptide mass fingerprinting (PMF)
Purified Mb was mixed 1:1 with 50% sinapinic acid and the mixture was allowed to crystallize on the MALDI target plate for 10 min. Protein molecular ions were analyzed in linear, positive ion mode using MALDI-TOF ULTRAFLEX III instrument (MALDI-TOF MS Bruker Daltonics, ULTRAFLEX III), using an acceleration voltage of 2.2 kV. The instrument was calibrated using lysozyme and beta-lactoglobulin as protein standards. The resulting spectra (from 1000 laser shots) were averaged, noise-smoothed, baseline-corrected, and analysed [12].

In-gel digestion of the myoglobin protein separated by 2DE was performed as described by Shevchenko et al. (2006) with slight modifications [35]. The PMF measured mass-to-charge ratio (m/z) values of peptides resulting from a protein digest form the basis for a characteristic fingerprint of the intact protein. Desalted peptide extracts (0.3 µL) were spotted onto MALDI target plate with 0.3 µL of 5 mg/mL α-cyano-4-hydroxycinnamic acid (Aldrich, St. Louis, MO, USA) in 50% CH$_3$CN/50% 0.1% trifluoroacetic acid. Crystallized samples were washed with cold 0.1% trifluoroacetic acid and were analyzed in linear, positive ion mode on the MALDI-TOF ULTRAFLEX III instrument (MALDI-TOF MS Bruker Daltonics ULTRAFLEX III). The resulting spectra (from 600 laser shots) were averaged; noise-smoothed, baseline-corrected and further analysis was done with FLEX ANALYSIS SOFTWARE for obtaining the peptide mass fingerprint [36]. The masses obtained in the peptide mass fingerprint were submitted for MASCOT 2.2 search in NCBI database for identification of the protein.

Determination of isoelectric point using OFFGEL electrophoresis
Buffalo and goat Mb protein samples were fractionated by isoelectric focusing using an Agilent 3100 OFFGEL Fractionator (Agilent Technologies, Santa Clara, CA, USA), following manufacturer's instructions. Parts were assembled, Immobilised pH gradient (IPG) strips were rehydrated, and protein samples were diluted into the 1.25× OFFGEL stock solution as instructed in the Agilent Quick Start Guide. Mb proteins were focused based on their pI using two different 24 cm long IPG strips with a linear pH gradient ranging from 3.0 to 10.0 placed in the same tray following the Agilent method. A total of 0.5 mg of Mb protein was loaded onto each IPG strip. At the end of the run, samples from each well were directly collected and subjected to SDS-PAGE analysis.

Reaction of buffalo and goat OxyMb's with HNE

Buffalo (0.15 mM) and goat (0.075 mM) Oxy-Mb's were each combined with 1.0 and 0.5 mM of 4-Hydroxy-2-nonenal, respectively, at 37 °C, pH 7.4 and 25 °C, pH 5.6, the typical pH in live animals and post-mortem muscles. These Mb concentrations were selected according to our earlier studies as they reflect the average Mb levels found in buffalo (5.0 mg Mb/g tissue) and goat (2.5 mg Mb/g tissue) meats. The molar ratio of Mb: aldehydes was maintained at 1:7 for all reactions [9]. Controls consisted of Oxy-Mb and a volume of ethanol equivalent to that used to deliver the HNE to the treatment mixture. During incubation, the control (Oxy-Mb + ethanol) and treatment (Oxy-Mb + HNE) solutions were scanned spectrophotometrically from 650 to 450 nm using a UV–VIS spectrophotometer (Model: UV-1700, PharmaSpec, SHIMADZU, Japan). The blank contained only buffer (10 mM Tris–HCl with 1 mM EDTA). Metmyoglobin formation was calculated using absorbance values at 503, 557, and 582 nm [37]. At the end of incubation, Mb-HNE mix was directly submitted for MALDI-TOF analysis, whereas for ESI-QTOF MS/MS, the mix was subjected to SDS-PAGE and the gel bands were excised, digested and analysed.

MALDI-TOF MS and ESI-QTOF MS/MS

Native and HNE-treated buffalo and goat myoglobins (intact proteins) were analyzed by MALDI-TOF MS to detect changes in the total mass resulting from potential HNE adduction. One mL of the reaction solution was passed over a PD-10 desalting column calibrated with distilled water to remove unreacted 4-hydroxynonenal. Buffalo and goat Mb-HNE adducts were detected using MALDI-TOF MS as explained earlier.

SDS–PAGE of native and HNE-treated buffalo and goat Mb's (after 3 h at 37 °C and pH 7.4) was performed under reducing conditions using a midi-gel electrophoresis unit. Each band of native and HNE-treated buffalo and goat Mb's on Coomassie-stained gels were excised, destained and digested [35] with sequencing-grade trypsin at 37 °C for 18 h . After digestion, the sample was vacuum dried and was dissolved in 50 μl of 0.1% Formic acid. One μl of each of the samples were separated on the Nano-Acquity BEH C18 column (100 μm i.d. × 100 mm) connected to WATERS nanoUPLC system for 150 min with 50% gradient of water 0.1%formic acid (buffer A) and ACN, 0.1% formic acid at a flow rate of 400 nL/min. The nano LC separated peptides were analysed for MS and MS/MS fragmentation on SYNAPT G2 WATERS nLC coupled QTOF mass spectrometer with ESI source.

The MS/MS spectra were acquired in data-dependent scanning mode with one full MS scan with a m/z of 50–1000 followed by four MS/MS scans on the first four most intense ions with dynamic exclusion of previously

selected precursors for a period of 1 min. MS/MS spectra were matched to Mb database sequences using the protein database search software MASCOT search engine (http://www.matrixscience.com) with a parent ion tolerance of 2.5 amu and fragment ion tolerance of 1 Da. Protein identification was done using the WATERS Protein Lynx Global Server (PLGS) v4.1 against the NCBI or UNIPROT database of *Bubbalis bubbalis* and *Capra hircus* Myoglobin. The raw data was obtained for the samples and was analysed in MASSLYNX (v4.1) using the BIOLYNX Tool and were checked for HNE modifications in the HISTIDINE sites of the respective myoglobin. The option Chordata (vertebrates and relatives) was selected as taxonomy restriction parameter. Electrophile adduction in several proteins has been identified successfully utilizing SALSA (Scoring Algorithm for Spectral Analysis).

Statistical analysis

Statistical analysis was performed with the analysis of variance (ANOVA) using SPSS (SPSS version 13.0 for windows; SPSS, Chicago, IL, USA). Least square means for F-tests were calculated by using Duncan's multiple range tests and were considered significant at $p < 0.05$. The experimental design for evaluation of Mb oxidation study was a completely randomized design with three replicates ($n = 3$). The data for buffalo and goat were analyzed separately using two-way analysis of variance (ANOVA) option in SPSS and the differences among means were detected using the least significance difference (LSD) at a 5% level.

Abbreviations

2DE: Two-dimensional gel electrophoresis; ESI-QTOF MS/MS: Electrospray ionization quadrupole time-of-flight tandem mass spectrometry; HNE: 4-Hydroxy-2-nonenal; IPG: Immobilized pH gradient; *m/z*: Mass-to-charge ratio; MALDI-TOF-MS: Matrix-assisted laser desorption ionization-time of flight mass spectrometry; Mb: Myoglobin; MS: Mass spectrometry; OxyMb: Oxymyoglobin; PMF: Peptide mass fingerprinting; SDS-PAGE: Sodium dodecyl sulfate polyacrylamide gel electrophoresis

Acknowledgements

The financial assistance from Department of Science and Technology (DST), Government of India (DST-FAST Track Project No. SR/FT/LS-149/2009) is acknowledged.

Funding

Department of Science and Technology (DST), Government of India provided funding to carry out this research project including design of the study and collection, analysis, and interpretation of data and in writing the manuscript through its grant DST-FAST Track Project No. SR/FT/LS-149/2009.

Authors' contributions

NM is responsible for the conception of the study data analysis and overall execution of the work; UK is responsible for myoglobin extraction, purification, SDS-PAGE characterization; PY performed the OFFGEL electrophoresis, Mb-HNE reaction and estimation of different Mb forms; SR is responsible for trypsin digestion, MALDI-TOF MS, PMF, Q-TOF MS/MS and all other MS related analysis;

VK provided all the facilities and extended support for the study; NM and VK reviewed the data, compilation, manuscript writing, critical reviewing and submission; All authors wrote or contributed to the writing of the manuscript and approved the final version.

Competing interests
The authors have no known competing interests either financial or personal between themselves and others that might bias the work.

Author details
[1]National Research Centre on Meat, Chengicherla, Hyderabad, Telangana 500092, India. [2]Proteomics Lab, National Centre for Cell Science, Pune 411007, India.

References
1. FAOSTAT. http://faostat.fao.org/site/291/default.aspx. 2013.
2. Naveena BM, Kiran M. Anim Front. 2014;4:18–24.
3. Webb EC, Casey NH, Simela L. Small Rum Res. 2005;60:153–66.
4. Naveena BM, Sen AR, Muthukumar M, Babji Y, Kondaiah N. Effects of salt and ammonium hydroxide on the quality of ground buffalo meat. Meat Sci. 2011;87:315–20.
5. Dosi R, Di-Moro A, Chambery A, Colonna G, Costantini S, Geraci G, Parente A. Characterization and kinetics studies of water buffalo (Bubalus bubalis) myoglobin. Comp Biochem Phys Part B. 2006;145:230–8.
6. Babiker SA, El Khider IA, Shafie SA. Chemical composition and quality attributes of goat meat and lamb. Meat Sci. 1990;28:273–7.
7. Suman SP, Joseph P, Li S, Steinke L, Fontaine M. Primary structure of goat myoglobin. Meat Sci. 2009;82:456–60.
8. Suman SP, Joseph P. Myoglobin chemistry and meat color. The Ann Rev Food Sci Tech. 2013;4:79–99.
9. Faustman C, Liebler DC, McClure TD, Sun Q. alpha,beta-unsaturated aldehydes accelerate oxymyoglobin oxidation. J Agric Food Chem. 1999;47:3140–4.
10. Alderton AL, Faustman C, Liebler DC, Hill DW. Induction of redox instability of bovine myoglobin by adduction with 4-hydroxy-2-nonenal. Biochem. 2003;42:4398–405.
11. Naveena BM, Faustman C, Tatiyaborwornatham N, Yin S, Ramanathan R, Mancini RA. Food Chem. 2010;122:836–40.
12. Suman SP, Faustman C, Stamer SL, Liebler DC. Proteomics of lipid oxidation-induced oxidation of porcine and bovine oxymyoglobins. Proteomics. 2007;7:628–40.
13. Sakai T, Kuwazuru S, Yamauchi K, Uchida K. A lipid peroxidation-derived aldehyde, 4-hydroxy-2-nonenal and omega 6 fatty acids contents in meats. Biosci Biotech Biochem. 1995;59:1379–80.
14. Carini M, Aldini G, Facino RM. Mass spectrometry for detection of 4-hydroxy-trans-2-nonenal (HNE) adducts with peptides and proteins. Mass Spectrometry Rev. 2004;23:281–305.
15. Petersen DR, Doorn JA. Reactions of 4-hydroxynonenal with proteins and cellular targets. Free Rad Biol Med. 2004;37:937–45.
16. Maheswarappa NB, Faustman C, Tatiyaborworntham N, Yin S, Ramanathan R, Mancini RA. Mass spectrometric characterization and redox instability of turkey and chicken myoglobins as induced by unsaturated aldehydes. J Agric Food Chem. 2009;57:8668–76.
17. Nair M, Suman SP, Li S, Joseph P, Beach CM. Lipid oxidation-induced oxidation in emu and ostrich myoglobins. Meat Sci. 2014;96:984–93.
18. Yin S, Faustman C, Tatiyaborwornatham N, Ramanathan R, Sun Q. The effects of HNE on ovine oxymyoglobin redox stability in a microsome model. Meat Sci. 2013;95:224–8.
19. Tatiyaborwornatham N, Faustman C, Yin S, Ramanathan R, Mancini RA, Suman SP, Beach CM, et al. Redox instability and hemin loss of mutant sperm whale myoglobins induced by 4-hydroxynonenal in vitro. J Agric Food Chem. 2012;60:8473–83.
20. Canto ACVCS, Suman SP, Nair MN, Li S, Rentfrow G, Beach CM, Silva TJP, et al. Differential abundance of sarcoplasmic proteome explains animal effect on beef Longissimus lumborum color stability. Meat Sci. 2015;102:90–8.
21. Joseph P, Suman SP, Rentfrow G, Li S, Beach CM. Proteomics of muscle-specific beef color stability. J Agric Food Chem. 2012;60:3196–203.
22. Faustman C, Phillips AL. Ch. F3 Unit F3.3. In: Current protocol in food analytical chemistry. 2001.
23. Wu W, Yu QQ, Fu Y, Tian XJ, Jia F, Li XM, Dai RT. J Proteomics. 2016;doi: 10.1016/j.jprot.2015.10.027.
24. Han K, Dautrevaux M, Chaila X, Biserte G. The covalent structure of beef heart myoglobin. Eur J Biochem. 1970;16:465–71.
25. Han K, Tetaert D, Moschetto Y, Dautrevaux M, Kopeyan C. The covalent structure of sheep-heart myoglobin. Eur J Biochem. 1972;27:585–92.
26. Ros A, Faupel M, Mees M, Van Oostrum J, Ferrigno R, Reymond F, Michel P, et al. Protein purification by Off-Gel electrophoresis. Proteomics. 2002;2:151–6.
27. Zhu K, Zhao J, Lubman DM, Miller FR, Barder TJ. Protein pI shifts due to posttranslational modifications in the separation and characterization of proteins. Anal Chem. 2005;77:2745–55.
28. Faustman C, Sun Q, Mancini RA, Suman SP. Myoglobin and lipid oxidation interactions: mechanistic bases and control. Meat Sci. 2010;86:86–94.
29. Lynch MP, Faustman C. Effect of aldehyde lipid oxidation products on myoglobin. J Agric Food Chem. 2000;48:600–4.
30. Bolgar MS, Yang CY, Gaskell SJ. First direct evidence for lipid/protein conjugation in oxidized human low density lipoprotein. J Biol Chem. 1996;271:27999–8001.
31. Yin S, Faustman C, Tatiyaborwornatham N, Ramanathan R, Maheswarappa NB, Mancini RA, Joseph P, et al. Species-specific myoglobin oxidation. J Agric Food Chem. 2011;59:12198–203.
32. Sentandreu MA, Fraser PD, Halket J, Patel R, Bramley PM. A proteomic-based approach for detection of chicken in meat mixes. J Proteome Res. 2010;9:3374–83.
33. Laemmli UK. Cleavage of structural proteins during the assembly of the head of bacteriophage T4. Nature. 1970;227:680–5.
34. Lametsch R, Benedixen E. Proteome analysis applied to meat science: characterizing postmortem changes in porcine muscle. J Agric Food Chem. 2001;49:4531–7.
35. Shevchenko A, Tomas H, Havlis J, Olsen JV, Mann M. In-gel digestion for mass spectrometric characterization of proteins and proteomes. Nature Prot. 2006;1:2856–60.
36. Padliya ND, Wood TD. Improved peptide mass fingerprinting matches via optimized sample preparation in MALDI mass spectrometry. Anal Chim Acta. 2008;627:162–8.
37. Tang J, Faustman C, Hoagland TA. J Food Sci. 2004;69:C717–20.

9

Proteome changes in the small intestinal mucosa of growing pigs with dietary supplementation of non-starch polysaccharide enzymes

Jize Zhang[1,2], Yang Gao[3], Qingping Lu[2], Renna Sa[2] and Hongfu Zhang[2*]

Abstract

Background: Non-starch polysaccharide enzymes (NSPEs) have long been used in monogastric animal feed production to degrade non-starch polysaccharides (NSPs) to oligosaccharides in order to promote growth performance and gastrointestinal (GI) tract health. However, the precise molecular mechanism of NSPEs in the improvement of the mammalian small intestine remains unknown.

Methods: In this study, isobaric tags were applied to investigate alterations of the small intestinal mucosa proteome of growing pigs after 50 days of supplementation with 0.6% NSPEs (mixture of xylanase, β-glucanase and cellulose) in the diet. Bioinformatics analysis including gene ontology annotation was performed to determine the differentially expressed proteins. A protein fold-change of ≥ 1.2 and a P-value of < 0.05 were selected as thresholds.

Results: Dietary supplementation of NSPEs improved the growth performance of growing pigs. Most importantly, a total of 90 proteins were found to be differentially abundant in the small intestinal mucosa between a control group and the NSPE group. Up-regulated proteins were related to nutrient metabolism (energy, lipids, protein and mineral), immunity, redox homeostasis, detoxification and the cell cytoskeleton. Down-regulated proteins were primarily related to transcriptional and translational regulation. Our results indicate that the effect of NSPEs on the increase of nutrient availability in the intestinal lumen facilitates the efficiency of nutrient absorption and utilization, and the supplementation of NSPEs in growing pigs also modulates redox homeostasis and enhances immune response during simulating energy metabolism due to a higher uptake of nutrients in the small intestine.

Conclusions: These findings have important implications for understanding the mechanisms of NSPEs on the small intestine of pigs, which provides new information for the better utilization of this feed additive in the future.

Keywords: Non-starch polysaccharide enzymes, Small intestinal mucosa, Proteomics, Growing pigs

Background

Many cereals such as soybean and wheat contain up to 15% non-starch polysaccharides (NSPs) in their outer or inner cell walls [1]. Monogastric animals lack enzymes to degrade the cell wall and NSP in these feeds. Thus, these anti-nutritive factors may interfere with digestion, nutrient absorption, and intestinal tract health by encapsuling starch and protein, as well as increase the viscosity of the chymus, which may elevate the proliferation of pathological bacteria in the small intestine and reduce the feed conversion ratio of monogastric livestock species [2–4].

The supplementation of exogenous enzymes such as xylanases and β-glucanases in pig diets may facilitate the hydrolysis of the main NSPs and increase the utilization of available raw materials [5, 6]. Adding exogenous enzymes to cereal diets improves both nutrient digestibility and growth performance in pigs [7, 8]. However, the exact molecular mechanisms of NSPEs, particularly in the gastrointestinal (GI) tract, are unknown [9]. There are several

* Correspondence: zhanghf6565@vip.sina.com
[2]State Key Laboratory of Animal Nutrition, Institute of Animal Sciences, Chinese Academy of Agricultural Sciences, Beijing 100193, People's Republic of China
Full list of author information is available at the end of the article

indications that exogenous enzymes may function in the GI tract of animals to aid digestion. The supplementation of NSPEs in the diets could increase the activities of certain types of digestive enzymes in vivo including protease, trypsin, and α-amylase [2, 4, 10]. These enzymes reduce the degradation of NSPs within the small intestine, thereby decreasing the viscosity of the digesta, which leads to a reduced bacterial load in the gut, especially potential pathogens [11]. Furthermore, the degradation of NSPs due to the supplementation of NSPEs promotes the higher availability of digestible nutrients such as energy substrates [12]. Additionally, the intestinal morphological structure and some physiological functions in animals benefit from the improvement of the changing intestinal environment due to the supplementation of NSPEs. Some research demonstrated that intestinal morphologies, including the villus height, the ratio of villus height to crypt depth, and the number of crypts and goblet cells, were changed due to the addition of xylanases alone or multiple enzymes [13, 14]. In addition to the effects of NSPEs observed on the GI tract, alterations of blood parameters related to the nutrient metabolism were also noted [15].

Previous studies reported that diet composition affected gene expression in animals [9, 16]. It is assumed that the improvement of the intestinal environment due to the supplementation of NSPEs in the diet may influence the gene expression and subsequent protein expression of epithelial-cell nutrient transporters in the GI tract mucosa, which has not been studied before. However, RNA editing and numerous options for posttranslational modifications should be taken into account [17, 18]. Hence, elucidation protein expression is important [19].

It is impractical to simultaneously measure all protein expression in the GI mucosa by classical method, such as western blotting. More research has yielded high throughput mass spectrometric proteomic technologies that can simultaneously detect hundreds of proteins [20, 21]. A proteomic analysis of the rat small intestinal proteome showed the presence of previously unrecognized proteins involved in various functions including the absorption and transport of nutrients and the maintenance of cell structure, as well as intestinal molecular chaperones [22]. There remains a great need to pursue proteomic technology to elucidate the beneficial effects of NSPEs in the GI tract mucosa. Therefore, we utilized a label-based iTRAQ (isobaric tags for relative and absolute quantitation) method, followed by LC-MS/MS, to quantitate proteins that are differentially induced in the small intestinal mucosa of growing pigs supplemented with NSPEs in the diet.

Methods
Enzyme preparation
The NSP enzyme mixture preparation supplemented in the diet was provided by the State Key Laboratory of Animal Nutrition, Institute of Animal Sciences, Chinese Academy of Agricultural Sciences (Beijing, China); the mixture contained 7×10^5 U/g xylanase activity (EC 3.2.1.8), 1×10^5 U/g β-glucanase activity (EC 3.2.1.6), and 9000 U/g cellulase activity (EC 3.2.1.4). The activities of the enzymes used in the present study was measured according the methods mentioned in previous research [23].

Animals and treatments
Forty-eight crossbred (Duroc × Landrace × Large White) growing pigs had similar initial body weights (39.18 ± 0.98 kg); the pigs were obtained from a commercial farm in Beijing (Shunliang pig farm, Beijing). The pigs were randomly divided into two groups according to their littermates, sex and mean initial body weights with four replicates in each group and six pigs in each replicate (half females and half males). The following two groups were a control group (CTRL, basal diet) and a treatment group (NSPE, basal diet + 0.6% NSP enzymes). The amount of NSPEs supplementation in the present study was based on the previous results from our group [24]. Both diets were formulated to meet NRC (2012) recommendations (Table 1). All pigs were kept in eight adjacent pens covered in a fermentation bed facility. Feed and water were provided *ad libitum* during the 50 day experimental period. The individual pig weight and feed intake were recorded at the initiation and the termination of the experiment for the measurement of the average daily gain (ADG), average daily feed intake (ADFI) and feed conversion ratio (FCR). All procedures involving animals were evaluated and approved by the Animal Ethics Committee of the Institute of Animal Sciences, Chinese Academy of Agricultural Sciences.

Sample collection
At the end of the experiment (Day 50), all pigs were weighted after 12 h of fasting. One pig per replicate, a total of eight pigs ($n = 8$), were sacrificed by CO_2 asphyxiation and then exsanguinated. Blood samples were obtained from the cervical vein by syringe before sacrifice. The whole blood was centrifuged at 2000 g for 30 min at 4 °C, followed by centrifugation at 400 g for 10 min at 4 °C. Then, the resulting supernatant was collected as sera samples, which were stored at –20 °C for further analysis. A 20-cm tissue section was rapidly excised at 50% of the length of the small intestine, rinsed with cold phosphate buffer saline, and blotted dry on paper. Mucosa from this small intestine section was sequentially obtained by careful scraping of the mucosal layer using a glass microscope slide as previously described [25]. Then, the collected mucosal samples were snap-frozen in liquid nitrogen and stored at –80 °C for proteomic analysis.

Table 1 Composition of the basal diet and calculated proximate composition of the diet

Ingredients	Proportion (%)[a]
Corn	70.70
Soybean meal	19.82
Soybean oil	2.10
Wheat bran	5.00
Limestone	0.51
Calcium hydrophosphate	0.56
L-Lysine	0.01
Sodium chloride	0.30
Premix[b]	1.00
Total	100
Nutrient	
ME	13.65 (MJ/kg)
Ether extract (EE)	4.82
Crude protein (CP)	15.50
Calcium	0.50
Total phosphorus	0.45
Available phosphorus	0.24
Total lysine	0.75
Total methionine	0.25

[a]All data is expressed in g/kg dry weight except for metabolizable energy (ME) in MJ/kg. The amounts of nutrient were estimated based on the NRC 11th ed. swine feedstuff composition table
[b]Providing the following (g/kg fresh weight), Vitamin A, 8250 IU; Vitamin D$_3$: 825 IU; Vitamin E: 40 IU; Vitamin K$_3$, 4.0 mg; Vitamin B$_1$, 1.0 mg; Vitamin B$_2$, 5.0 mg; Vitamin B$_6$, 2.0 mg; Vitamin B$_{12}$, 25 μg; choline chloride, 600 mg; nicotinic acid, 35 mg; folic acid, 2.0 mg; biotin, 4.0 mg; Cu, 50.0 mg; Fe, 80.0 mg; Zn, 100.0 mg; Mn, 25.0 mg; Se, 0.15 mg; I, 0.5 mg

Serum biochemical analyses

Important serum biochemical parameters, including alanine aminotransferase (ALT), aspartate aminotransferase (AST), total protein (TP), alkaline phosphatase (ALP), glucose (GLU), and creatine kinase (CK), were analyzed using an automatic biochemical analyzer (Hitachi 7020, Tokyo, Japan). Serum levels of total superoxide dismutase (T-SOD) and immunoglobulin G (IgG) were measured using a corresponding kit (Nanjing Jiancheng Bioengineering Institute, Nanjing, China) according to the manufacturer's instructions.

Protein extraction and sample preparation

Small intestinal mucosa samples (500 μg) were ground in liquid nitrogen using a Dounce glass grinder. Grinded powder was precipitated with 10% trichloroacetic acid (TCA) (w/v) and 90% ice-cold acetone at –20 °C for 2 h. The precipitate was obtained by centrifugation at 20,000 g for 30 min at 4 °C and subsequently washed with ice-cold acetone. Then, the precipitate was lysed in lysis buffer [8 M urea, 30 mM 4-(2-hydroxyethyl)-1-piperazineethane-sulfonic acid (HEPES), 1 mM phenylmethanesulfonyl

fluoride (PMSF), 2 mM ethylene diamine tetraacetic acid (EDTA), and 10 mM dithiothreitol (DTT)]. The crude tissue extracts were centrifuged to remove the remaining debris. The tissue lysates were reduced for 1 h at 56 °C in a water bath using 10 mM DTT and then alkylated with 55 mM iodoacetamide for 1 h in the dark. Afterwards, the lysates were precipitated by adding four volumes of pre-chilled acetone. The pellets were then washed three times with pre-chilled pure acetone and resuspended in the buffer (50% TEAB and 0.1% SDS). The centrifugation was repeated to remove the undissolved pellets. Subsequently, protein quantitation was determined using a Bio-Rad Bradford Protein Assay Kit (Hercules, CA, USA). Each sample was digested with modified sequence grade trypsin (Promega Corporation, Madison, WI) at a 1: 30 ratio (3.3 μg trypsin : 100 μg target) overnight at 37 °C. Each isobaric tag (113, 114, 115, 116, 117, 118, 119, and 121) was solubilized in 70 μL isopropanol and then added to each respective sample (4 samples per group). Incubation continued for 2 h at room temperature.

Strong cation exchange chromatography

The strong cation exchange fractionation was performed according to a previous report [26] with slight modification. Briefly, 800 μg of labeled sample was loaded onto a strong cation exchange column (Phenomenex Luna SCX 100A) installed in an Agilent 1100 (Santa Clara, CA) system and equilibrated with buffer A (25% acetonitrile and 10 mM KH$_2$PO$_4$, pH 3.0). The peptides were separated by a linear gradient of buffer B (25% acetonitrile, 2 M KCl and 10 mM KH$_2$PO$_4$, pH 3.0) according to this procedure (increasing to 5% after 41 min, 50% after 66 min and 100% after 71 min with a flow rate of 1 ml/min). Elution was monitored by setting the absorbance at 214 nm. A total of 10 fractions were obtained, then desalted with a Strata X C18 column (Phenomenex) and dried under a vacuum. The pellets were resuspended by adding 0.1% formic acid before the LC-MS/MS run.

Mass spectrometry

LC-MS/MS was conducted according to a previous report [27], and the detailed process and parameters are shown in Additional file 1.

Data processing and protein quantification

All the detailed parameters are shown in the Supporting Information (Additional file 1). MS/MS data for iTRAQ protein identification and quantitation were analyzed using Proteome Discover 1.3 (Thermo Fisher Scientific, Bremen, Germany) and in-house MASCOT software (Matrix Science, London, UK; Version 2.3.0) against the database Uniprot_pig (Apr. 11th, 2014). Median ratio normalization was performed in intra-sample channels to normalize each channel across all proteins. Protein quantitative ratios for each iTRAQ labeled sample were obtained, using a sample in the control group (sample tagged with 113) as the denominator. Quantitative ratios were then log transformed to base two and presented as the fold change relative to the denominator in the control group for final quantitative testing. Differentially expressed proteins were identified using Student's t-test corrected for multiple testing using the Benjamini and Hochberg correction [21, 25, 28, 29]. Based on above the relative quantification, statistical analysis, and a number of previous reports regarding to iTRAQ experiments [29–31], we set a 1.2-fold change or greater as the threshold for differentially expressed proteins.

Bioinformatics analysis and validation of protein expression

The databases and software for bioinformatics analysis are shown in Additional file 1. Real-time qPCR was used to verify six small intestinal mucosal proteins of differential abundance at the mRNA level. All detailed procedures are described in the Supporting Information (Additional file 1). The primer sequences used in this study are shown in Additional file 2: Table S1.

Statistical analysis

The data for growth parameters, serum parameters, and gene expression were analyzed by one-way ANOVA using block as a covariate (SAS Version 9.2, SAS institute Inc., Cary, NC) according the previous studies [21, 31], and a t-test was used for independent samples in MS data analysis. A group difference was assumed statistically significant when $P < 0.05$.

Results

Growth performance of growing pigs

During the entire experimental period (50 days), NSPE pigs had 15.5% greater ADG ($P < 0.05$) compared with the control group; however, the ADFI between the two groups was not significantly different ($P > 0.05$). It is notable that pig fed NSPEs had an 8.7% greater FCR compared with the control group ($P < 0.05$; Table 2).

Serum parameters of growing pigs

In NSPE pigs, serum concentration of CK was significantly lower ($P < 0.05$) than the control group (Table 3). Furthermore, the serum concentrations of T-SOD, IgG, and glucose were significantly elevated compared with the control group ($P < 0.05$) (Table 3). Serum levels of TP, ALT and AST were similar between the two groups (Table 3).

Identification and comparison of proteins of differential abundance

Using iTRAQ analysis, a total of 2634 proteins were identified within the FDR (false discovery rate) of 1%

Table 2 Effects of NSP enzymes on growth performance of growing pigs

	Groups		
	Control	Treatment	P value
Initial weight (kg)	38.80 ± 0.99	39.55 ± 0.63	0.1245
Final weight (kg)	74.04 ± 1.77[b]	78.42 ± 1.06[a]	0.0318
ADG (kg/d)[c]	0.71 ± 0.05[b]	0.82 ± 0.05[a]	0.0437
ADFI (kg/d)[d]	1.97 ± 0.09	2.07 ± 0.06	0.0423
FCR (kg feed/kg weight gain)[e]	2.77 ± 0.02[a]	2.53 ± 0.03[b]	0.0352

[a, b] Values within a column having different superscript letters indicate a significant difference at $P < 0.05$. Numbers are mean ± S.D. ($n = 24$ for ADG; $n = 4$ for ADFI and FCR)
[c] ADG = average daily gain
[d] ADFI = average daily feed intake
[e] FCR = feed conversion ratio

(Additional file 3: Table S2). Following statistical analysis, 104 proteins were found to be differentially expressed in the small intestinal mucosa between CTRL and NSPE pigs, with 43 up-regulated and 61 down-regulated (Additional file 4: Table S3).

A total of 90 proteins of differential abundance were grouped into eight classes based on putative functions: transcriptional and translational regulation (44.4%), miscellaneous (16.7%), redox homeostasis and detoxification (10.0%), immune response and inflammation (8.9%), energy metabolism (7.8%), protein metabolism and modification (5.6%), lipid metabolism (3.3%), and cell cytoskeleton (3.3%) (Fig. 1). Those related to transcriptional and translational regulation, redox homeostasis, and immune response were predominant, accounting for approximately 63% of the differentially expressed proteins. A comparison

Table 3 Effect of NSPEs on serum biochemical parameters of growing pigs

	Groups		
	Control	Treatment	P value
ALT (IU/L)[c]	49.01 ± 7.96	49.00 ± 9.30	0.4768
AST (IU/L)[d]	79.60 ± 10.70	63.80 ± 16.05	0.2240
TP (mmol/L)[e]	67.31 ± 5.44	69.50 ± 2.44	0.5331
ALP (U/L)[f]	131.83 ± 36.14	126.40 ± 22.06	0.2565
GLU (mmol/L)[g]	6.37 ± 2.24[b]	9.73 ± 2.34[a]	0.0479
T-SOD (U/mL)[h]	61.55 ± 2.67[b]	67.44 ± 3.64[a]	0.0002
CK (U/L)[i]	3117 ± 274[a]	2188 ± 218[b]	0.0089
IgG (g/L)[j]	3.19 ± 0.16[b]	3.43 ± 0.20[a]	0.0392

[a, b] Values within a column not sharing a common superscript letter indicate significant difference at $P < 0.05$. Numbers are means ± S.D. ($n = 4$)
[c] ALT = alanine aminotransferase
[d] AST = aspartate aminotransferase
[e] TP = total protein
[f] ALP = alkaline phosphatase
[g] GLU = glucose
[h] T-SOD = total superoxide dismutase
[i] CK = creatine kinase
[j] IgG = immunoglobulin G

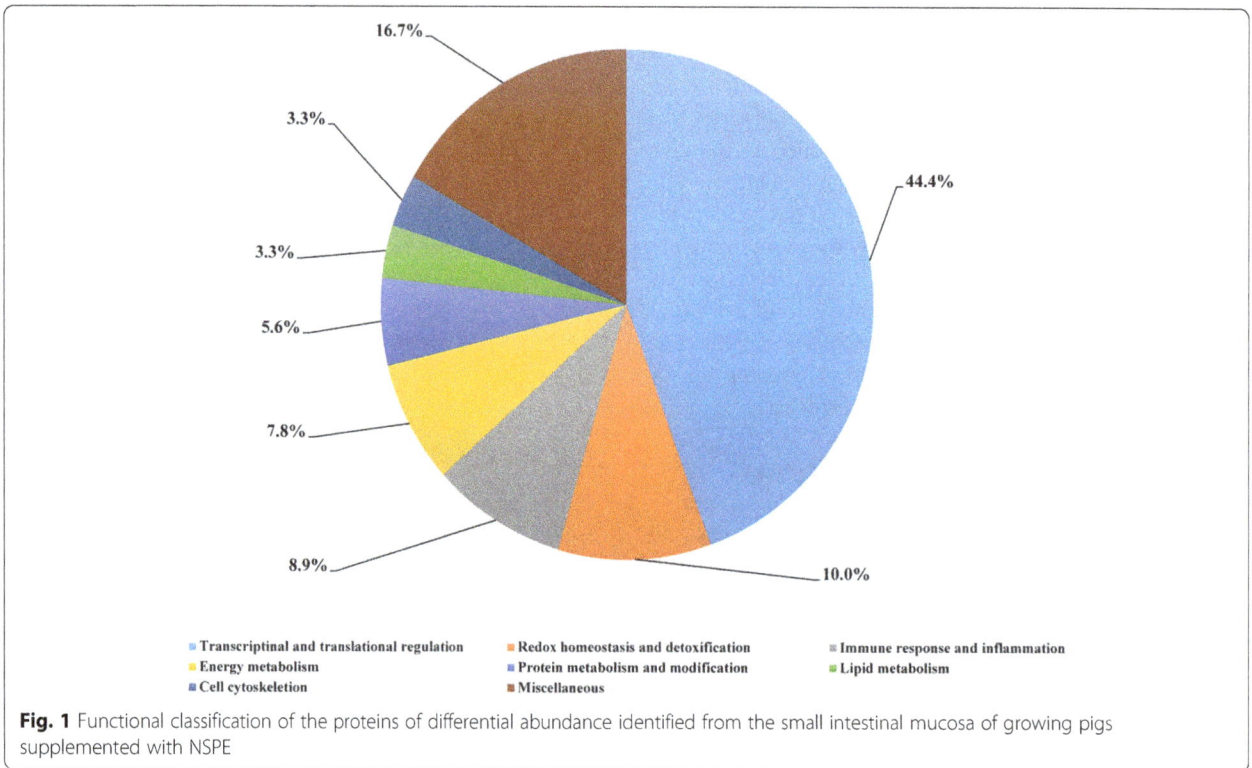

Fig. 1 Functional classification of the proteins of differential abundance identified from the small intestinal mucosa of growing pigs supplemented with NSPE

of proteins of differential abundance with functional groupings between the two groups indicated that a smaller number of protein species were up-regulated in NSPE pigs (36 versus 54) (Table 4).

GO annotations of proteins of differential abundance
In the cellular component group, the differentially expressed proteins were concentrated in the intracellular part and membrane-bounded organelles (Fig. 2). In the molecular functional group, the differentially expressed proteins that are binding proteins (protein, nucleotide, or nucleic acid binding) and metabolic enzymes (hydrolase, oxidoreductase, or transferase activity) were ranked at the top of the category (Fig. 2). In the biological process category, the proteins that participate in cellular process (organelle organization process), metabolic process (nitrogen compound metabolic and biosynthetic process), and biological regulation (transcriptional and translational regulation, redox homeostasis, and immune response) had the highest ratios among the differentially expressed proteins.

Validation of proteins of differential abundance
Six differentially expressed proteins superoxide dismutase (SOD1) involved in redox homeostasis; calmodulin (CALM1) involved in calcium ion binding; MHC class I antigen (SLA-1) involved in immune response; acyl-coenzyme A oxidase (ACOX1) involved in energy metabolism; 40S ribosomal protein S6 (RPS6) involved in transcriptional and translational regulation; and

apolipoprotein C-III (APOC3) involved in lipid absorption, were selected for the validation of proteomic data at the mRNA level using qPCR (Fig. 3). Most protein levels were consistent with their mRNA expression levels, except for RPS6.

Discussion
The benefit of NSPEs supplementation is well recognized in monogastric animal production; NSPEs supplementation promotes growth performance and GI tract health, including the efficiency of nutrient utilization [2, 3, 8]. A number of studies have proven that the addtion of NSPEs to the diet reduces digesta viscosity by the partial or complete hydrolysis of soluble NSPs, which triggers the changes in microbial composition, especially the reduction of the amount of pathological bacteria within the small intestine [11, 32]. Moreover, the supplementation of NSPEs could increase the nutrient availability in the intestinal lumen (for example, energy substrates and proteins) [12, 33]. All above effects of NSPE supplementation are due to the improvements of the intestinal environment. However, it is still largely unknown how the small intestinal mucosa of the hosts responds to alterations in the luminal environment triggered by the addition of NSPEs. The present study marks the first time that the well-established quantitative iTRAQ label-based technology was applied for the proteomic analysis of the small intestinal mucosa of growing pigs with dietary supplementation of NSPEs. Various functional groupings of differentially expressed mucosal proteins

Table 4 List of differentially expressed proteins in small intestinal mucosal samples from treatment group and control group

Accession[a]	Description[b]	Gene symbol	Score[c]	Pep. No[d]	Log$_2$ fold change	P-value[e]	Biological process GO term
Transcriptional and translational regulation							
F1S419	Uncharacterized protein OS = Sus scrofa GN = SF3B3 PE = 4 SV = 2 - [F1S419_PIG]	None	85.61	3	−0.37	0.0007	RNA binding
K9J4V0	U5 small nuclear ribonucleoprotein 200 kDa helicase OS = Sus scrofa GN = SNRNP200 PE = 2 SV = 1 - [K9J4V0_PIG]	SNRNP200	248.18	9	−0.32	0.0012	Nucleic acid binding
F2Z5Q6	40S ribosomal protein S6 (Fragment) OS = Sus scrofa GN = RPS6 PE = 3 SV = 2 - [F2Z5Q6_PIG]	RPS6	140.42	4	−0.81	0.0013	Structural constituent of ribosome
F1SD96	Uncharacterized protein (Fragment) OS = Sus scrofa GN = RAD23A PE = 4 SV = 1 - [F1SD96_PIG]	RAD23A	85.25	3	1.06	0.0026	Nucleotide excision repair
F1S8K5	Uncharacterized protein OS = Sus scrofa GN = SUPT16H PE = 4 SV = 1 - [F1S8K5_PIG]	SUPT16H	35.43	2	−0.40	0.0028	RNA binding
F1RZH4	Uncharacterized protein OS = Sus scrofa PE = 4 SV = 1 - [F1RZH4_PIG]	ADAM10	32.67	1	−0.82	0.0048	Structural constituent of ribosome
F1SD98	Uncharacterized protein OS = Sus scrofa GN = TRMT1 PE = 4 SV = 2 - [F1SD98_PIG]	TRMT1	27.72	1	−0.30	0.0065	Poly(A) RNA binding
I3LHZ6	Uncharacterized protein OS = Sus scrofa GN = DHX9 PE = 4 SV = 1 - [I3LHZ6_PIG]	DHX9	994.71	27	−0.30	0.0075	ATP-dependent RNA helicase activity
F1SDV7	Uncharacterized protein (Fragment) OS = Sus scrofa GN = TOP1 PE = 4 SV = 1 - [F1SDV7_PIG]	TOP1	99.93	4	−0.44	0.0075	DNA binding
P62802	Histone H4 OS = Sus scrofa PE = 1 SV = 2 - [H4_PIG]	None	358.11	7	−0.67	0.0103	DNA binding
F1S1V1	Uncharacterized protein OS = Sus scrofa GN = SSB PE = 4 SV = 2 - [F1S1V1_PIG]	SSB	196.5	6	−0.73	0.0111	Nucleotide binding
F1RS45	DNA topoisomerase 2 OS = Sus scrofa PE = 3 SV = 2 - [F1RS45_PIG]	TOP2B	116.62	6	−0.27	0.0117	DNA binding
F1S1X3	Uncharacterized protein OS = Sus scrofa GN = NARS PE = 3 SV = 2 - [F1S1X3_PIG]	NARS	262.65	7	−0.30	0.0119	Nucleotide binding
F2Z576	Histone H3 OS = Sus scrofa GN = LOC100525821 PE = 2 SV = 1 - [F2Z576_PIG]	HIST1H3E	159.44	6	−0.77	0.0120	DNA binding
Q29194	Ribosomal protein S2 (Fragment) OS = Sus scrofa PE = 2 SV = 1 - [Q29194_PIG]	None	46.59	1	−0.45	0.0138	Structural constituent of ribosome
I3LFV4	Uncharacterized protein OS = Sus scrofa GN = YBX1 PE = 4 SV = 1 - [I3LFV4_PIG]	YBX1	157.89	4	0.41	0.0148	DNA repair
I3LIN8	Histone H2A OS = Sus scrofa GN = H2AFY PE = 3 SV = 1 - [I3LIN8_PIG]	H2AFY	224.89	6	−0.52	0.0149	Chromatin DNA binding
B0FWK5	Ribosomal protein L5 OS = Sus scrofa GN = RPL5 PE = 2 SV = 1 - [B0FWK5_PIG]	RPL5	178.57	8	−0.34	0.0165	Structural constituent of ribosome
I3LCI4	Uncharacterized protein OS = Sus scrofa GN = ZFR PE = 4 SV = 1 - [I3LCI4_PIG]	ZFR	41.93	2	−0.28	0.0167	Poly(A) RNA binding
F1S8A5	Uncharacterized protein OS = Sus scrofa GN = MRPS26 PE = 4 SV = 1 - [F1S8A5_PIG]	MRPS26	38.63	1	−0.36	0.0181	Poly(A) RNA binding
A5GFY4	Negative elongation factor D OS = Sus scrofa GN = NELFCD PE = 3 SV = 1 - [NELFD_PIG]	NELFCD	43.67	1	−0.32	0.0189	Negative regulation of transcription
F1S5A8	Uncharacterized protein OS = Sus scrofa GN = DHX15 PE = 4 SV = 1 - [F1S5A8_PIG]	DHX15	259.43	8	−0.26	0.0198	ATP-dependent RNA helicase activity
F1RRG9	Uncharacterized protein OS = Sus scrofa GN = SMARCA5 PE = 4 SV = 1 - [F1RRG9_PIG]	SMARCA5	99.44	3	−0.39	0.0201	DNA binding
F1RGP1	Uncharacterized protein OS = Sus scrofa GN = MYBBP1A PE = 4 SV = 1 - [F1RGP1_PIG]	MYBBP1A	445.89	12	−0.50	0.0208	Poly(A) RNA binding
F2Z5Q8	Uncharacterized protein OS = Sus scrofa GN = LOC100519675 PE = 4 SV = 1 - [F2Z5Q8_PIG]	RPL35A	57.33	2	−0.45	0.0209	Structural constituent of ribosome

Table 4 List of differentially expressed proteins in small intestinal mucosal samples from treatment group and control group *(Continued)*

I3L7T6	Histone H2A OS = Sus scrofa GN = H2AFX PE = 3 SV = 1 - [I3L7T6_PIG]	H2AFX	357.7	7	−0.56	0.0231	DNA binding
F1SMZ9	Uncharacterized protein (Fragment) OS = Sus scrofa GN = SF3B1 PE = 4 SV = 2 - [F1SMZ9_PIG]	SF3B1	267.33	9	−0.26	0.0245	mRNA binding
F2Z5K9	Histone H3 OS = Sus scrofa GN = LOC100622412 PE = 3 SV = 1 - [F2Z5K9_PIG]	LOC100622412	178.75	6	−0.76	0.0270	DNA binding
P53027	60S ribosomal protein L10a (Fragment) OS = Sus scrofa GN = RPL10A PE = 2 SV = 3 - [RL10A_PIG]	RPL10A	154.25	5	−0.34	0.0272	RNA binding
K9IVG8	DEAD (Asp-Glu-Ala-Asp) box helicase 21 OS = Sus scrofa GN = DDX21 PE = 2 SV = 1 - [K9IVG8_PIG]	DDX21	44.62	1	−0.38	0.0292	RNA binding
F2Z554	Uncharacterized protein OS = Sus scrofa GN = RPL30 PE = 3 SV = 1 - [F2Z554_PIG]	RPL30	105.87	4	−0.26	0.0323	RNA binding
Q29195	60S ribosomal protein L10 OS = Sus scrofa GN = RPL10 PE = 2 SV = 3 - [RL10_PIG]	RPL10	105.8	4	−0.39	0.0350	Structural constituent of ribosome
P67985	60S ribosomal protein L22 OS = Sus scrofa GN = RPL22 PE = 2 SV = 2 - [RL22_PIG]	RPL22	113.83	3	−0.48	0.0355	Structural constituent of ribosome
I7GF95	Guanine nucleotide binding protein-like 1 OS = Sus scrofa GN = GNL1 PE = 4 SV = 1 - [I7GF95_PIG]	GNL1	58.56	1	−0.35	0.0371	Ribosome biogenesis
F1S8L9	Uncharacterized protein OS = Sus scrofa GN = HNRNPU PE = 4 SV = 2 - [F1S8L9_PIG]	HNRNPU	883.61	23	−0.34	0.0377	Poly(A) RNA binding
Q53DY5	Histone H1.3-like protein OS = Sus scrofa GN = LOC595122 PE = 2 SV = 1 - [Q53DY5_PIG]	HIST1H1D	251.92	7	1.29	0.0384	Chromatin DNA binding
F1S2G3	Uncharacterized protein (Fragment) OS = Sus scrofa GN = TBCA PE = 4 SV = 1 - [F1S2G3_PIG]	TBCA	78.18	2	0.31	0.0389	Poly(A) RNA binding
F2Z5P1	Histone H2A (Fragment) OS = Sus scrofa GN = H2AFV PE = 3 SV = 1 - [F2Z5P1_PIG]	LOC100512448	256.74	5	−0.43	0.0427	DNA binding
F2Z553	Uncharacterized protein OS = Sus scrofa GN = EIF1 PE = 4 SV = 1 - [F2Z553_PIG]	EIF1	103.66	2	0.82	0.0437	Translation initiation factor activity
F2Z5L5	Histone H2A OS = Sus scrofa GN = HIST2H2AC PE = 3 SV = 1 - [F2Z5L5_PIG]	HIST2H2AC	322.1	5	−0.62	0.0448	DNA binding
Redox homeostasis and detoxification							
F1SKJ2	Uncharacterized protein OS = Sus scrofa GN = TXN2 PE = 4 SV = 1 - [F1SKJ2_PIG]	TXN2	29.86	1	0.39	0.0043	Cell redox homeostasis
F1SGS9	Catalase OS = Sus scrofa GN = CAT PE = 3 SV = 1 - [F1SGS9_PIG]	CAT	923.56	23	0.58	0.0151	Protect cells from the toxic effects of hydrogen peroxide
I3LDJ8	Uncharacterized protein OS = Sus scrofa PE = 3 SV = 1 - [I3LDJ8_PIG]	None	303.51	10	0.77	0.0202	Oxidoreductase activity
P12309	Glutaredoxin-1 OS = Sus scrofa GN = GLRX PE = 1 SV = 2 - [GLRX1_PIG]	GLRX	277.83	6	0.64	0.0208	Cell redox homeostasis
F1SCF9	Uncharacterized protein (Fragment) OS = Sus scrofa GN = TECR PE = 4 SV = 2 - [F1SCF9_PIG]	TECR	38.34	1	−0.37	0.0242	Oxidoreductase activity
A5J2A8	Thioredoxin (Fragment) OS = Sus scrofa GN = TRX PE = 4 SV = 1 - [A5J2A8_PIG]	TRX	128.36	3	0.34	0.0303	Cell redox homeostasis
F1SMY1	Uncharacterized protein OS = Sus scrofa GN = TMX3 PE = 4 SV = 2 - [F1SMY1_PIG]	TMX3	39.1	2	0.30	0.0345	Cell redox homeostasis
P16549	Dimethylaniline monooxygenase [N-oxide-forming] 1 OS = Sus scrofa GN = FMO1 PE = 1 SV = 3 - [FMO1_PIG]	FMO1	39.55	2	1.64	0.0084	Oxidative metabolism of a variety of xenobiotics
P04178	Superoxide dismutase [Cu-Zn] OS = Sus scrofa GN = SOD1 PE = 1 SV = 2 - [SODC_PIG]	SOD1	459.04	9	0.35	0.0424	Superoxide dismutase activity
Immune response and inflammation							
A3FJ41	MHC class I antigen (Fragment) OS = Sus scrofa GN = SLA-1 PE = 4 SV = 1 - [A3FJ41_PIG]	SLA-1	120.03	5	0.35	0.0050	Immune response

Table 4 List of differentially expressed proteins in small intestinal mucosal samples from treatment group and control group (Continued)

F1RGC8	Uncharacterized protein OS = Sus scrofa GN = NLRP6 PE = 4 SV = 3 - [F1RGC8_PIG]	NLRP6	119.58	4	−0.32	0.0061	Activation of NF-κB
F1RFM7	Uncharacterized protein OS = Sus scrofa GN = AIMP2 PE = 4 SV = 1 - [F1RFM7_PIG]	AIMP2	232.75	6	−0.29	0.0076	Metabolism of xenobiotics
A2SZV5	Tax1 binding protein 3 (Fragment) OS = Sus scrofa PE = 4 SV = 1 - [A2SZV5_PIG]	None	55.14	1	0.29	0.0133	Negative regulation of NF-κB
B8XX91	DNA-dependent activator of IFN-regulatory factor OS = Sus scrofa GN = DAI PE = 2 SV = 1 - [B8XX91_PIG]	DAI	100.5	4	0.70	0.0137	Innate immune responses
Q8WNQ7	N-acetylgalactosamine-6-sulfatase OS = Sus scrofa GN = GALNS PE = 2 SV = 1 - [GALNS_PIG]	GALNS	52.52	1	0.60	0.0311	Degradation of the glycosaminoglycans keratan sulfate
B8XTR8	Granzyme H OS = Sus scrofa GN = gzmH PE = 2 SV = 1 - [B8XTR8_PIG]	gzmH	168.84	6	−0.67	0.0272	Serine-type endopeptidase activity
A5GFQ5	Protein canopy homolog 3 OS = Sus scrofa GN = CNPY3 PE = 3 SV = 1 - [CNPY3_PIG]	CNPY3	40.13	2	−0.63	0.0376	Receptor binding for proper TLR folding
Energy metabolism							
Q1ACV5	Transporter associated with antigen processing 1 OS = Sus scrofa PE = 2 SV = 1 - [Q1ACV5_PIG]	None	298.67	7	−0.32	0.0030	Triggers ATP hydrolysis
F1RIG0	Uncharacterized protein (Fragment) OS = Sus scrofa PE = 4 SV = 1 - [F1RIG0_PIG]	None	47.28	2	−0.27	0.0169	ATP binding
Q7SIB7	Phosphoglycerate kinase 1 OS = Sus scrofa GN = PGK1 PE = 1 SV = 3 - [PGK1_PIG]	PGK1	850.39	23	0.30	0.0160	Conversion of 1,3-diphosphoglycerate to 3-phosphoglycerate
H9BYW2	Acyl-coenzyme A oxidase OS = Sus scrofa GN = ACOX1 PE = 2 SV = 1 - [H9BYW2_PIG]	ACOX1	370.35	10	0.91	0.0200	Fatty acid beta-oxidation
I3LEN7	Uncharacterized protein OS = Sus scrofa GN = ALDH1L1 PE = 3 SV = 1 - [I3LEN7_PIG]	ALDH1L1	49.04	2	0.40	0.0245	Formate oxidation
F1S0Y8	Uncharacterized protein OS = Sus scrofa GN = ADH4 PE = 3 SV = 2 - [F1S0Y8_PIG]	ADH4	40.7	2	0.67	0.0309	Oxidation of long-chain aliphatic alcohols
A7UIU7	ATP citrate lyase OS = Sus scrofa GN = ACL PE = 2 SV = 1 - [A7UIU7_PIG]	ACL	468.98	14	−0.38	0.0374	ATP binding
Protein metabolism and modification							
F1RIF3	Uncharacterized protein OS = Sus scrofa GN = FAH PE = 4 SV = 1 - [F1RIF3_PIG]	FAH	38.37	2	0.39	0.0010	Catabolism of the amino acid phenylalanine
Q9GK25	Peptidyl-prolyl cis-trans isomerase (Fragment) OS = Sus scrofa PE = 2 SV = 1 - [Q9GK25_PIG]	None	266.1	7	1.43	0.0025	Accelerate the folding of proteins
I3L739	Uncharacterized protein OS = Sus scrofa GN = JMJD6 PE = 4 SV = 1 - [I3L739_PIG]	JMJD6	39.99	1	−0.29	0.0193	Protein hydroxylases
I3LK37	Uncharacterized protein (Fragment) OS = Sus scrofa PE = 3 SV = 1 - [I3LK37_PIG]	GALNT7	33.39	2	−0.30	0.0248	Protein glycosylation
F1RNR6	4-hydroxyphenylpyruvate dioxygenase OS = Sus scrofa GN = HPD PE = 3 SV = 2 - [F1RNR6_PIG]	HPD	31	1	0.35	0.0391	Aromatic amino acid family metabolic process
Lipid metabolism							
I3LM15	Uncharacterized protein OS = Sus scrofa GN = AGPS PE = 4 SV = 1 - [I3LM15_PIG]	AGPS	48.77	1	−0.36	0.0019	Lipid biosynthetic process
Q9GJX2	Diazepam binding inhibitor (Fragment) OS = Sus scrofa GN = DBI PE = 2 SV = 1 - [Q9GJX2_PIG]	DBI	80.13	3	0.91	0.0057	Long-chain fatty acyl-CoA binding, triglyceride metabolic process
P27917	Apolipoprotein C-III OS = Sus scrofa GN = APOC3 PE = 1 SV = 2 - [APOC3_PIG]	APOC3	226.39	7	0.78	0.0241	High-density lipoprotein particle receptor binding
Cell cytoskeleton							
P10668	Cofilin-1 OS = Sus scrofa GN = CFL1 PE = 1 SV = 3 - [COF1_PIG]	CFL1	704.02	15	0.31	0.0059	Cytoskeleton organization

Table 4 List of differentially expressed proteins in small intestinal mucosal samples from treatment group and control group
(Continued)

Q5G6W0	Cofilin-2 (Fragment) OS = Sus scrofa PE = 2 SV = 1 - [Q5G6W0_PIG]	CFL1	48.67	2	0.43	0.0073	Cytoskeleton organization
B5APV0	Actin-related protein 2/3 complex subunit 5 OS = Sus scrofa GN = ARPC5 PE = 2 SV = 1 - [B5APV0_PIG]	ARPC5	170.99	6	0.30	0.0167	Structural constituent of cytoskeleton
Miscellaneous							
Q9TSA7	Calmodulin (Fragments) OS = Sus scrofa PE = 4 SV = 1 - [Q9TSA7_PIG]	None	108.72	4	1.11	0.0008	Calcium ion binding
K7GKQ1	Uncharacterized protein OS = Sus scrofa GN = RAB9A PE = 3 SV = 1 - [K7GKQ1_PIG]	RAB9A	26.6	1	−0.40	0.0071	Cytoskeletal signaling
F1RKI3	Uncharacterized protein OS = Sus scrofa GN = HINT1 PE = 4 SV = 1 - [F1RKI3_PIG]	HINT1	80.55	3	0.32	0.0073	Tumor suppressing
I3LSY0	Uncharacterized protein OS = Sus scrofa GN = ACSM4 PE = 4 SV = 1 - [I3LSY0_PIG]	ACSM4	21.13	1	0.86	0.0179	Catalytic activity
D0G6R8	Phosphatidate cytidylyltransferase OS = Sus scrofa GN = CDS2 PE = 2 SV = 1 - [D0G6R8_PIG]	CDS2	33.01	1	−0.39	0.0192	Synthesis of phosphatidylglycerol
Q95332	Betaine–homocysteine S-methyltransferase 1 (Fragment) OS = Sus scrofa GN = BHMT PE = 1 SV = 3 - [BHMT1_PIG]	BHMT	110.41	4	1.10	0.0193	Regulation of homocysteine metabolism
F1RS34	Uncharacterized protein OS = Sus scrofa GN = GAPVD1 PE = 4 SV = 2 - [F1RS34_PIG]	GAPVD1	22.69	1	−0.40	0.0207	Signal transduction
F1ST01	Uncharacterized protein OS = Sus scrofa GN = SELENBP1 PE = 4 SV = 1 - [F1ST01_PIG]	SELENBP1	936.42	22	0.33	0.0209	Selenium binding
Q9TV62	Myosin-4 OS = Sus scrofa GN = MYH4 PE = 2 SV = 1 - [MYH4_PIG]	MYH4	192.94	7	−0.83	0.0336	Motor activity
F1RN91	Uncharacterized protein (Fragment) OS = Sus scrofa PE = 4 SV = 2 - [F1RN91_PIG]	MYO18A	35.04	2	0.28	0.0355	Cell migration
F1RPC8	Uncharacterized protein OS = Sus scrofa GN = CRYM PE = 4 SV = 2 - [F1RPC8_PIG]	CRYM	59.33	2	0.49	0.0392	Thyroid hormone binding
F2Z5W6	Uncharacterized protein OS = Sus scrofa GN = LAMTOR1 PE = 4 SV = 1 - [F2Z5W6_PIG]	LAMTOR1	26.54	1	−0.37	0.0410	Guanyl-nucleotide exchange factor activity
Q29069	Myosin light chain OS = Sus scrofa PE = 2 SV = 2 - [Q29069_PIG]	None	58.61	3	−0.38	0.0458	Calcium ion binding
O19175	Casein kinase I isoform alpha (Fragment) OS = Sus scrofa GN = CSNK1A1 PE = 2 SV = 1 - [KC1A_PIG]	CSNK1A1	51.13	1	−0.44	0.0473	Protein kinase activity
N0E654	Casein kinase II b subunit splicing isoform 476 (Fragment) OS = Sus scrofa GN = Csnk2b PE = 2 SV = 1 - [N0E654_PIG]	Csnk2b	63.97	2	−0.27	0.0039	Cell proliferation and cell differentiation

[a]Uniprot_ Sus scrofa_9823 database accession number
[b]The name of the protein exclusive of the identifier that appears in the database
[c]The sum of the scores of the individual peptides
[d]The number of distinct peptide sequences in the protein group
[e]Differential protein expression in the treatment group was presented as a $\log_2$ fold change relative to the control group

related to nutrient metabolism, transcriptional and translational regulation, immune, and redox homeostasis were identified in response to NSPEs.

In former research, the utilization of β-glucanase and xylanase in the diet demonstrated that enzymes tended to increase the absorptive area and reduce cell proliferation and intraepithelial lymphocytes in the gut of pigs [34]. Both cereal grains and enzymes would affect components of gut health, including intestine morphology, bacteria populations, and microbial metabolites in the gut content [35]. It has been demonstrated that enhanced cell proliferation in the intestinal mucosa is associated with bowel

diseases, cellular repair, and apoptosis [36, 37]. As shown in the present study, 89% of proteins related to transcriptional and translational regulation were down-regulated in NSPE pigs. We speculate that supplementation with NSPEs in the diet of growing pigs can reduce the possibility of intestinal infection. This is consistent with the former research result that NSPEs reduce the amount of pathological bacteria within the small intestine by lowering the viscosity of intestinal digesta [11].

The abundance of proteins CFL1 (cofilin-1), CFL2 (cofilin-2) and ARPC5 (actin-related protein 2/3 complex subunit 5), which are classified as cell cytoskeleton

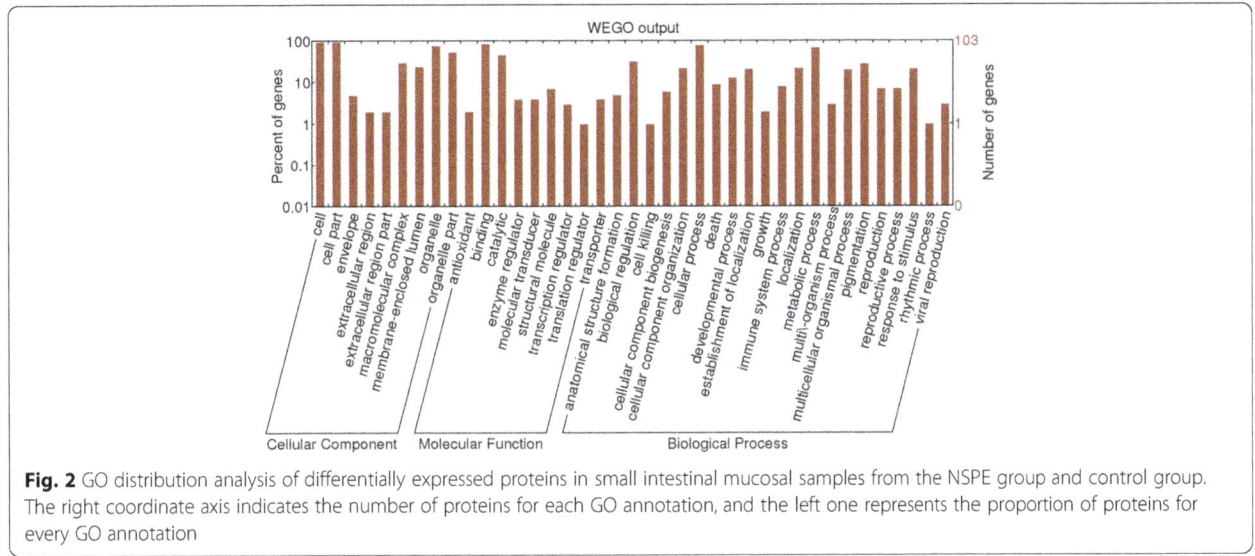

Fig. 2 GO distribution analysis of differentially expressed proteins in small intestinal mucosal samples from the NSPE group and control group. The right coordinate axis indicates the number of proteins for each GO annotation, and the left one represents the proportion of proteins for every GO annotation

proteins relevant to cell structure and mobility, was increased. CFL1 and CFL2 are widely distributed intracellular actin-modulating proteins [38]. These two proteins can cause actin cytoskeleton rearrangement and membrane remodeling to the formation of phagosomes, which are recognized by Fc gamma receptors and beneficial for the host-defense in animals [39]. ARPC5 has a similar function as cofilin in the actin cytoskeleton, which is required for phagocytosis in mammals [40]. The up-regulation of these proteins might reflect the improved integrity of the intestinal mucosa.

As an important immune organ, the small intestine participates in the inflammatory response and the prevention of bacterial infection. SLA-1 (MHC class I antigen), GALNS (N-acetylgalactosamine-6-sulfatase), and DAI

(DNA-dependent activator of IFN-regulatory factor) are considered to be involved in the immune response. SLA-1 alerts the immune system to virus-infected cells by presenting peptide fragments derived from intracellular proteins [41]. GALNS is located in lysosomes that digest different types of molecules and engulf viruses or bacteria within cells [42, 43]. DAI selectively enhances the DNA-mediated induction of type I IFN and other genes involved in innate immunity [44, 45]. The abundance of these proteins was up-regulated in NSPE pigs, suggesting that the supplementation of NSPEs may improve potential immunity and reduce the chance of bacterial infection in the small intestine. This is consistent with the elevated serum level of IgG in the NSPE group. However, challenges with exogenous pathogens are still required to verify the effect

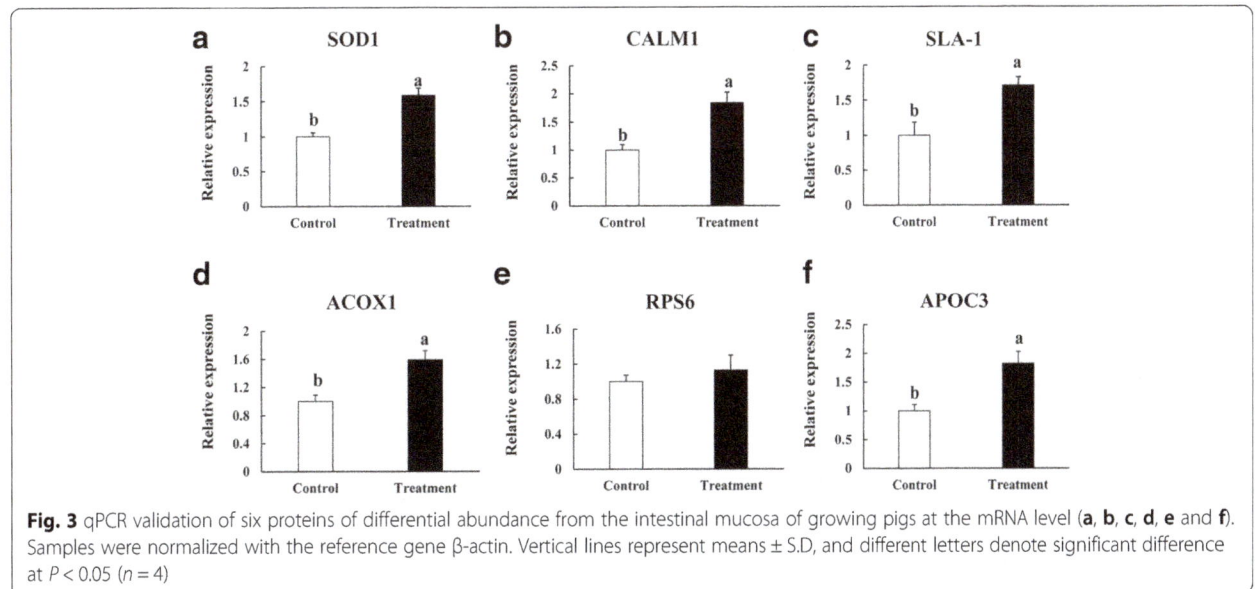

Fig. 3 qPCR validation of six proteins of differential abundance from the intestinal mucosa of growing pigs at the mRNA level (**a**, **b**, **c**, **d**, **e** and **f**). Samples were normalized with the reference gene β-actin. Vertical lines represent means ± S.D, and different letters denote significant difference at $P < 0.05$ ($n = 4$)

of NSPEs supplementation on immunity. In contrast, proteins involved in an inflammatory response, including NLRP6 (NLR family, pyrin domain containing 6) and CNPY3 (protein canopy homolog 3), are down-regulated, which indicates that inflammation is attenuated in the small intestinal mucosa due to the supplementation of NSPEs [46]. It has been suggested that one of the performance improvement attributes of NSPEs is due to the reduced local inflammation by controlling pathogens within the small intestine [32].

In addition to affecting the immune response, the up-regulated proteins catalase (CAT), glutaredoxin (GRXS), thioredoxin (TRX), superoxide dismutase (SOD), dimethylaniline monooxygenase [N-oxide-forming] 1 (FMO1) and 4-hydroxyphenylpyruvate dioxygenase (HPPD) are classified as redox homeostasis and detoxification proteins based on their primary functions. The up-regulation of CAT, GRXS, TRX and SOD may suggest that NSPE pigs had more potential to keep redox homeostasis in vivo [47–52]. This is consistent with the increased serum level of T-SOD in the NSPE group of this study. The reason for the up-regulation of these oxidoreductases and immune factors in the present study may be the increased abundance of reactive oxygen species (ROS) and inflammatory factors during stimulating energy metabolism due to a higher uptake of nutrients with NSPEs supplementation. However, further study is required to prove the effect of NSPEs on redox homeostasis. As one of the detoxification enzymes, FMO1 is regulated by xenobiotics, as the enzyme activity markedly increases in response to the invading harmful chemicals [53]. The up-regulation of this protein suggests that the supplementation of NSPEs is helpful to eliminate xenobiotics in the small intestine, which also could be related to the improvement of the intestinal lumen due to NSPEs.

Furthermore, the up-regulated abundance of proteins was observed in the NSPE group, including multiple nutrient metabolism processes such as energy, lipid, amino acid and mineral. These proteins included phosphoglycerate kinase 1 (PGK1), diazepam binding inhibitor (DBI), and acyl-coenzyme A oxidase (ACOX1). PGK1 plays a vital role in glycolysis or gluconeogenesis [54]. The up-regulation of ACOX1 indicates the elevation of glucose synthesis in the small intestine, which is consistent with the increased serum glucose level in the NSPE group. Likewise, higher abundance of DBI and ACOX1 was observed in this study, suggesting the stimulation of lipids β-oxidation for nutrient absorption to meet the energy requirement in the small intestine of NSPE pigs [55, 56]. Apolipoprotein C-III (APOC3) is an important modulator that is secreted from the intestine on the chylomicron upon lipid absorption [57]. The up-regulation of APOC3 implies the enhanced absorption of dietary lipids in the NSPE group.

Two differentially expressed proteins related to the permeability of the tight junction (TJ), including casein kinase II beta subunit splicing isoform 476 (Csnk2b) and myosin-4 (MYH4), were identified in the present study. The tight junctions (TJs) in the small intestine are not only a physical and biological barrier but also a passive diffusion system that depends on the permeability of the TJs [58]. Paracellular transport is one of the passive diffusion systems providing an absorption way for small molecular compounds [59], which are regulated by the permeability of the TJs and are thought to be important for mineral absorption [60]. Additionally, the transepithelial transport of oligosaccharides, but not polysaccharides, also occurs *via* the paracellular pathway [61]. Previous research has demonstrated that NSPEs are capable of hydrolyzing polysaccharides from the food to oligosaccharides in the gut [62]. Thus, the down-regulation of these two proteins in this study, in addition to former studies, indicates an increased permeability of the TJs in the NSPE group, which is beneficial to small molecular compounds absorption in the small intestine.

Calmodulin regulates cellular calcium concentration as a primary calcium-binding protein [63]. Calcium absorption is reduced if the bioavailability of dietary calcium is lowered by calcium-binding agents like cellulose because nearly all dietary calcium intake occurs in the upper intestine [64]. The up-regulation of this protein observed in this study suggests that calcium absorption in the small intestine is facilitated in the NSPE group by the degradation of calcium-binding agents in the diet, which could be conductive to bone health.

It has been demonstrated that one of the important roles of NSPEs within the small intestine is the elimination of the nutrient-encapsulating effect of cell wall polysaccharides, which increases the availability of starches, amino acids, and minerals. These results are consistent with our results from the present study that the levels of proteins related to nutrient absorption and utilization (energy, lipid, amino acid and mineral) are up-regulated. A fully understanding of the mechanisms of NSPEs supplementation will require the determination of protein modifications and protein regulation such as phosphorylation or glycosylation [65]. However, this part was not involved in the present study due to the technical limitation. Thus, further study is required to prove the effect of NSPEs on regulatory proteins using specific method, for example, the phosphoproteome.

Conclusions

The results of this study provide the first evidence that the small intestinal mucosa proteome is altered in growing pigs supplemented with NSPEs. Growing pigs most likely responded to the increased reactive oxygen species (ROS) and inflammatory factors during stimulating energy metabolism due to NSPEs supplementation by changing the abundance of certain mucosal proteins that modulate redox homeostasis and enhance immune response. Most important of all, the effect of NSPEs on the increase of nutrient availability in the intestinal lumen provided additional benefits to facilitate protein expressions related to the efficiency of nutrient absorption and utilization, such as energy metabolism, amino acid metabolism, mineral metabolism, lipid absorption, and cell structure and mobility. These novel findings show the mechanisms whereby dietary supplementation with NSPEs promotes growth performance and improves the GI health of growing pigs, which also has important implications for the better utilization of this feed additive.

Additional files

Additional file 1: The detailed description of the experiment methods, including mass spectrometric analysis procedures and parameters, bioinformatics analysis softwares, websites and real-time qPCR procedures. (DOCX 15 kb)

Additional file 2: Table S1. qPCR primers used for verification of the differentially expressed genes of the small intestinal mucosa in growing pigs. (DOCX 14 kb)

Additional file 3: Table S2. List of all proteins (n = 2634) identified in the study. (XLSX 199 kb)

Additional file 4: Table S3. List of all differentially expressed proteins (n = 104) identified in the study. (XLSX 31 kb)

Abbreviations

ACOX1: Acyl-coenzyme A oxidase; ADFI: Average daily feed intake; ADG: Average daily gain; ALP: Alkaline phosphatase; ALT: Alanine aminotransferase; APOC3: Apolipoprotein C-III; ARPC5: Actin-related protein 2/3 complex subunit 5; AST: Aspartate aminotransferase; CALM1: Calmodulin; CAT: Catalase; CFL: Cofilin; CK: Creatine kinase; CNPY3: Protein canopy homolog 3; Csnk2b: Casein kinase II beta subunit splicing isoform 476 MYH4: myosin-4; DAI: DNA-dependent activator of IFN-regulatory factor; DTT: Dithiothreitol; FCR: Feed conversion ratio; FMO1: Dimethylaniline monooxygenase [N-oxide-forming] 1; GALNS: N-acetylgalactosamine-6-sulfatase; GI: Gastrointestinal; GRXS: Glutaredoxin; HEPES: 4-(2-hydroxyethyl)-1-piperazineethanesulfonic acid; HPPD: 4-hydroxyphenylpyruvate dioxygenase; IgG: Immunoglobulin G; iTRAQ: Isobaric tags for relative and absolute quantitation; NSPEs: Non-starch polysaccharide enzymes; PGK1: Phosphoglycerate kinase 1; PMSF: Phenylmethanesulfonyl fluoride; RPS6: 40S ribosomal protein S6; SLA-1: MHC class I antigen; SOD1: Superoxide dismutase; TP: Total protein; TRX: Thioredoxin; T-SOD: Total superoxide dismutase

Acknowledgements
This research was supported by the Chinese National Science and Technology Pillar Program (No: 2012BAD39B0), the Special Fund for Innovation Team of the Chinese Academy of Agricultural Sciences (No: ASTTP-IAS07), and the Chinese National Key Basic Research and Development Program (No: 2014CB138804).

Funding
This research was supported by the Chinese National Science and Technology Pillar Program (No: 2012BAD39B0), the Special Fund for Innovation Team of the Chinese Academy of Agricultural Sciences (No: ASTTP-IAS07), and the Chinese National Key Basic Research and Development Program (No: 2014CB138804). The funders had no role in study design, data collection and analysis, decision to publish, or preparation of the manuscript.

Authors' contributions
JZ and HZ designed the study. JZ and YG performed the experiments and analyzed the data. JZ, QL and RS contributed reagents/materials/analysis tools. JZ prepared the manuscript and all of the authors contributed to, read and approved the final manuscript.

Competing interest
The authors declare that there is no competing interest.

Author details
[1]Institute of Grassland Research, Chinese Academy of Agricultural Sciences, Hohhot 010010, People's Republic of China. [2]State Key Laboratory of Animal Nutrition, Institute of Animal Sciences, Chinese Academy of Agricultural Sciences, Beijing 100193, People's Republic of China. [3]College of Animal Science and Technology, Jilin Agricultural University, Changchun 130118, People's Republic of China.

References
1. Sterk A, Verdonk JMAJ, Mul AJ, Soenen B, Bezençon ML, Frehner M, et al. Effect of xylanase supplementation to a cereal-based diet on the apparent faecal digestibility in weanling piglets. Livest Sci. 2007;108(1–3):269–71.
2. Wang MQ, Xu ZR, Sun JY, Kim BG. Effects of enzyme supplementation on growth, intestinal content viscosity, and digestive enzyme activities in growing pigs fed rough rice-based diet. Asian-Aust J Anim Sci. 2008;21(2):270–76.
3. Lindberg JE. Fiber effects in nutrition and gut health in pigs. J Anim Sci Biotechnol. 2014;5(1):15.
4. Li WF, Feng J, Xu ZR, Yang CM. Effects of non-starch polysaccharides enzymes on pancreatic and small intestinal digestive enzyme activities in piglet fed diets containing high amounts of barley. World J Gastroenterol. 2004;10(6):856–9.
5. Willamil J, Badiola I, Devillard E, Geraert PA, Torrallardona D. Wheat-barley-rye-or corn-fed growing pigs respond differently to dietary supplementation with a carbohydrase complex. J Anim Sci. 2012;90(3):824–32.
6. Susenbeth A, Naatjes M, Blank B, Kühl R, Ader P, Dickhoefer U. Effect of xylanase and glucanase supplementation to a cereal-based, threonine-limited diet on the nitrogen balance of growing pigs. Arch Anim Nutr. 2011;65(2):123–33.
7. Prandini A, Sigolo S, Morlacchini M, Giuberti G, Moschini M, Rzepus M, et al. Addition of non-starch polysaccharides degrading enzymes to two hulless barley varieties fed in diets for weaned pigs. J Anim Sci. 2014;92(5):2080–6.
8. Kim JS, Ingale SL, Hosseindoust AR, Lee SH, Lee JH, Chae BJ. Effects of mannan level and β-mannanase supplementation on growth performance, apparent total tract digestibility and blood metabolites of growing pigs. Animal. 2016. doi:10.1017/S1751731116001385.
9. Guo S, Liu D, Zhao X, Li C, Guo Y. Xylanase supplementation of a wheat-based diet improved nutrient digestion and mRNA expression of intestinal nutrient transporters in broiler chickens infected with Clostridium perfringens. Poult Sci. 2014;93(1):94–103.
10. Lin PH, Shih BL, Hsu JC. Effects of different sources of dietary non-starch polysaccharides on the growth performance, development of digestive tract and activities of pancreatic enzymes in goslings. Br Poult Sci. 2010;51(2):270–7.
11. Kiarie EG, Slominski BA, Nyachoti CM. Effect of products derived from hydrolysis of wheat and flaxseed non starch polysaccharides by carbohydrate enzymes on net absorption in enterotoxigenic Escherichia coli (K88) challenged piglet jejunal segments. Anim Sci J. 2010;81(1):63–71.
12. Aulrich K, Flachowsky G. Studies on the mode of action of non-starch-polysaccharides (NSP)-degrading enzymes in vitro. 2. Communication: effects on nutrient release and hydration properties. Arch Tierernahr. 2001; 54(1):19–32.
13. Silva SS, Smithard RR. Effect of enzyme supplementation of a rye-based diet on xylanase activity in the small intestine of broilers, on intestinal crypt cell proliferation and on nutrient digestibility and growth performance of the birds. Br Poult Sci. 2002;43(2):274–82.
14. Balamurugan R, Chandrasekaran D, Kirubakaran A. Effects of multi-enzyme supplementation on gut morphology and histomorphology in broilers. Indian J Anim Sci. 2011;4(1):15–8.
15. Ao X, Meng QW, Yan L, Kim YH, Hong SM, Cho JH, et al. Effects of non-

starch polysaccharide-degrading enzymes on nutrient digestibility, growth performance and blood profiles of growing pigs fed a diet based on corn and soybean meal. Asian-Aust J Anim Sci. 2010;23(12):1632–8.

16. Kaput J, Rodriguez RL. Nutritional genomics: the next frontier in the postgenomic era. Physiol Genomics. 2004;16(2):166–77.

17. Astle J, Ferguson JT, German JB, Harrigan GG, Kelleher NL, Kodadek T, et al. Characterization of proteomic and metabolomic responses to dietary factors and supplements. J Nutr. 2007;137(12):2787–93.

18. Zhang JZ, Li DF. Effect of conjugated linoleic acid on inhibition of prolyl hydroxylase 1 in hearts of mice. Lipids Health Dis. 2012;11:22.

19. Wang JJ, Li DF, Dangott LJ, Wu GY. Proteomics and its role in nutrition research. J Nutr. 2006;136(7):1759–62.

20. Kitteringham NR, Abdullah A, Walsh J, Randle L, Jenkins RE, Sison R, et al. Proteomic analysis of Nrf2 deficient transgenic mice reveals cellular defence and lipid metabolism as primary Nrf2-dependent pathways in the liver. J Proteomics. 2010;73(8):1612–31.

21. Luo J, Zheng A, Meng K, Chang W, Bai Y, Li K, et al. Proteome changes in the intestinal mucosa of broiler (Gallus gallus) activated by probiotic Enterococcus faecium. J Proteomics. 2013;91:226–41.

22. Reichardt F, Habold C, Chaumande B, Ackermann A, Ehret-Sabatier L, Le Maho Y, et al. Interactions between ingested kaolinite and the intestinal mucosa in rat: proteomic and cellular evidences. Fundam Clin Pharmacol. 2009;23(1):69–79.

23. Lowe SE, Theodorou MK, Trinci AP. Cellulase and xylanase of an anaerobic rumen fugus grown on wheat straw, wheat straw holocellulose, cellulose, xylan. Appl Environ Microbiol. 1987;53(6):1216–23.

24. Gao Y, Zhou X, Yu JX, Jin YC, Li C, Liu JY, et al. Effects of non-starch polysaccharide enzymes addition on growth performance, carcass traits and meat quality of growing-finishing pigs. Chin J Vet Sci. 2014;34(5): 820–4 (In Chinese).

25. Wang X, Yang F, Liu C, Zhou H, Wu G, Qiao S, et al. Dietary supplementation with the probiotic Lactobacillus fermentum I5007 and the antibiotic aureomycin differentially affects the small intestinal proteomes of weanling piglets. J Nutr. 2012;142(1):7–13.

26. Olsen JV, Blagoev B, Gnad F, Macek B, Kumar C, Mortensen P, et al. Global, in vivo, and site-specific phosphorylation dynamics in signaling networks. Cell. 2006;127(3):635–48.

27. Su L, Cao L, Zhou R, Jiang Z, Xiao K, Kong W, et al. Identification of novel biomarkers for sepsis prognosis via urinary proteomic analysis using iTRAQ labeling and 2D-LC-MS/MS. PLoS One. 2013;8(1):e54237.

28. Hakimov HA, Walters S, Wright TC, Meidinger RG, Verschoor CP, Gadish M, et al. Application of iTRAQ to catalogue the skeletal muscle proteome in pigs and assessment of effects of gender and diet dephytinization. Proteomics. 2009;9(16):4000–16.

29. Long B, Yin C, Fan Q, Yan G, Wang Z, Li X, et al. Global liver proteome analysis using iTRAQ reveals AMPK-mTOR-autophagy signaling is altered by intrauterine growth restriction in newborn piglets. J Proteome Res. 2016;15(4):1262–73.

30. Zhang LZ, Yan WY, Wang ZL, Guo YH, Yi Y, Zhang SW. Differential protein expression analysis following olfactory learning in Apis cerana. J Comp Physiol A. 2015;201(11):1053–61.

31. Zhang J, Li C, Tang X, Lu Q, Sa R, Zhang H. High concentrations of atmospheric ammonia induce alterations in the hepatic proteome of broilers (Gallus gallus): an iTRAQ-based quantitative proteomic analysis. PLoS One. 2015;10(4):e0123596.

32. Kiarie E, Romero LF, Nyachoti CM. The role of added feed enzymes in promoting gut health in swine and poultry. Nutr Res Rev. 2013;26(1): 71–88.

33. Khadem A, Lourenço M, Delezie E, Maertens L, Goderis A, Mombaerts R, et al. Does release of encapsulated nutrients have an important role in the efficacy of xylanase in broilers? Poult Sci. 2016;95(5):1066–76.

34. Willamil J, Badiola JI, Torrallardona D, Geraert PA, Devillard E. Effect of enzyme supplementation on nutrient digestibility and microbial metabolite concentrations in ileal and caecal digesta of growing pigs. Book of abstracts of 11th International Symposium on Digestive Physiology of Pigs; 2009.

35. Zijlstra RT, Owusu-Asiedu A, Simmins PH. Future of NSP-degrading enzymes to improve nutrient utilization of co-products and gut health in pigs. Livest Sci. 2010;134(1–3):255–7.

36. Bakke-McKellep AM, Penn MH, Salas PM, Refstie S, Sperstad S, Landsverk T, et al. Effects of dietary soyabean meal, inulin and oxytetracycline on intestinal microbiota and epithelial cell stress, apoptosis and proliferation in the teleost Atlantic salmon (Salmo salar L.). Br J Nutr. 2007;97(4):699–713.

37. Dehghan-Kooshkghazi M, Mathers JC. Starch digestion, large-bowel fermentation and intestinal mucosal cell proliferation in rats treated with the alpha-glucosidase inhibitor acarbose. Br J Nutr. 2004;91(3):357–65.

38. Klejnot M, Gabrielsen M, Cameron J, Mleczak A, Talapatra SK, Kozielski F, et al. Analysis of the human cofilin 1 structure reveals conformational changes required for actin binding. Acta Crystallogr D Biol Crystallogr. 2013;69(Pt 9): 1780–8.

39. Nakano K, Kanai-Azuma M, Kanai Y, Moriyama K, Yazaki K, Hayashi Y, et al. Cofilin phosphorylation and actin polymerization by NRK/NESK, a member of the germinal center kinase family. Exp Cell Res. 2003;287(2):219–27.

40. Insall R, Müller-Taubenberger A, Machesky L, Köhler J, Simmeth E, Atkinson SJ, et al. Dynamics of the Dictyostelium Arp2/3 complex in endocytosis, cytokinesis, and chemotaxis. Cell Motil Cytoskeleton. 2001;50(3):115–28.

41. Hewitt EW. The MHC, class I antigen presentation pathway: strategies for viral immune evasion. Immunology. 2003;110(2):163–9.

42. Tomatsu S, Orii KO, Vogler C, Nakayama J, Levy B, Grubb JH, et al. Mouse model of N-acetylgalactosamine-6-sulfate sulfatase deficiency (Galns-/-) produced by targeted disruption of the gene defective in Morquio A disease. Hum Mol Genet. 2003;12(24):3349–58.

43. Settembre C, Fraldi A, Medina DL, Ballabio A. Signals from the lysosome: a control centre for cellular clearance and energy metabolism. Nat Rev Mol Cell Biol. 2013;14(5):283–96.

44. Takaoka A, Wang Z, Choi MK, Yanai H, Negishi H, Ban T, et al. DAI (DLM-1/ ZBP1) is a cytosolic DNA sensor and an activator of innate immune response. Nature. 2007;448(7152):501–5.

45. Hayashi T, Nishitsuji H, Takamori A, Hasegawa A, Masuda T, Kannagi M. DNA-dependent activator of IFN-regulatory factors enhances the transcription of HIV-1 through NF-κB. Microbes Infect. 2010;12(12–13):937–47.

46. Levy M, Thaiss CA, Zeevi D, Dohnalová L, Zilberman-Schapira G, Mahdi JA, et al. Microbiota-modulated metabolites shape the intestinal microenvironment by regulating NLRP6 inflammasome signaling. Cell. 2015;163(6):1428–43.

47. Jia M, Jing Y, Ai Q, Jiang R, Wan J, Lin L, et al. Potential role of catalase in mice with lipopolysaccharide/D-galactosamine-induced fulminant liver injury. Hepatol Res. 2014;44(11):1151–8.

48. Spolarics Z, Wu JX. Role of glutathione and catalase in H_2O_2 detoxification in LPS-activated hepatic endothelial and Kupffer cells. Am J Physiol. 1997; 273(6 Pt 1):G1304–11.

49. Lillig CH, Berndt C, Holmgren A. Glutaredoxin systems. Biochim Biophys Acta. 2008;1780(11):1304–17.

50. Arnér ES, Holmgren A. Physiological functions of thioredoxin and thioredoxin reductase. Eur J Biochem. 2000;267(20):6102–9.

51. Khadem Ansari MH, Karimipour M, Salami S, Shirpoor A. The effect of ginger (Zingiber officinale) on oxidative stress status in the small intestine of diabetic rats. Int J Endocrinol Metab. 2008;6(3):144–50.

52. Fattman CL, Schaefer LM, Oury TD. Extracellular superoxide dismutase in biology and medicine. Free Radic Biol Med. 2003;35(3):236–56.

53. Cashman JR, Zhang J. Human flavin-containing monooxygenases. Annu Rev Pharmacol Toxicol. 2006;46:65–100.

54. Schurig H, Beaucamp N, Ostendorp R, Jaenicke R, Adler E, Knowles JR. Phosphoglycerate kinase and triosephosphate isomerase from the hyperthermophilic bacterium Thermotoga maritima form a covalent bifunctional enzyme complex. EMBO J. 1995;14(3):442–51.

55. Costa E, Guidotti A. Diazepam binding inhibitor (DBI): a peptide with multiple biological actions. Life Sci. 1991;49(5):325–44.

56. Mori T, Kondo H, Hase T, Tokimitsu I, Murase T. Dietary fish oil upregulates intestinal lipid metabolism and reduces body weight gain in C57BL/6 J mice. J Nutr. 2007;137(12):2629–34.

57. Wang F, Kohan AB, Dong HH, Yang Q, Xu M, Huesman S, et al. Overexpression of apolipoprotein C-III decreases secretion of dietary triglyceride into lymph. Physiological Rep. 2014;2(3):e00247.

58. Shimizu M. Interaction between food substances and the intestinal epithelium. Biosci Biotechnol Biochem. 2010;74(2):232–41.

59. Tsukita S, Furuse M, Itoh M. Multifunctional strands in tight junctions. Nat Rev Mol Cell Biol. 2001;2(4):285–93.

60. Karbach U. Paracellular calcium transport across the small intestine. J Nutr. 1992;122 Suppl 3:672–7.

61. Hisada N, Satsu H, Mori A, Totsuka M, Kamei J, Nozawa T, et al. Low-molecular-weight hyaluronan permeates through human intestinal Caco-2 cell monolayers via the paracellular pathway. Biosci Biotechnol Biochem. 2008;72(4):1111–4.

10

An interactomics overview of the human and bovine milk proteome over lactation

Lina Zhang[1], Aalt D. J. van Dijk[2,3,4] and Kasper Hettinga[1]* (ORCID)

Abstract

Background: Milk is the most important food for growth and development of the neonate, because of its nutrient composition and presence of many bioactive proteins. Differences between human and bovine milk in low abundant proteins have not been extensively studied. To better understand the differences between human and bovine milk, the qualitative and quantitative differences in the milk proteome as well as their changes over lactation were compared using both label-free and labelled proteomics techniques. These datasets were analysed and compared, to better understand the role of milk proteins in development of the newborn.

Methods: Human and bovine milk samples were prepared by using filter-aided sample preparation (FASP) combined with dimethyl labelling and analysed by nano LC LTQ-Orbitrap XL mass spectrometry.

Results: The human and bovine milk proteome show similarities with regard to the distribution over biological functions, especially the dominant presence of enzymes, transport and immune-related proteins. At a quantitative level, the human and bovine milk proteome differed not only between species but also over lactation within species. Dominant enzymes that differed between species were those assisting in nutrient digestion, with bile salt-activated lipase being abundant in human milk and pancreatic ribonuclease being abundant in bovine milk. As lactation advances, immune-related proteins decreased slower in human milk compared to bovine milk. Notwithstanding these quantitative differences, analysis of human and bovine co-expression networks and protein-protein interaction networks indicated that a subset of milk proteins displayed highly similar interactions in each of the different networks, which may be related to the general importance of milk in nutrition and healthy development of the newborn.

Conclusions: Our findings promote a better understanding of the differences and similarities in dynamics of human and bovine milk proteins, thereby also providing guidance for further improvement of infant formula.

Keywords: Proteomics, Protein interaction networks, Lactation, Human milk, Bovine milk

Background

Milk is one of the richest foods, as it provides complete nutrition and bioactive components for healthy development of the newborn. These nutritional and bioactive components are essential for the neonate, for example for cognitive development, pathogen prevention, intestinal microflora modulation, and development of the immune system [1, 2]. Of these bioactive components, proteins have attracted great attention because of their importance in the protection of the neonate. With the development of proteomics techniques, more and more

proteins, including both high and low abundant proteins, were characterized in the last few decades [3–5].

However, milk proteins are variable in presence and concentration due to many factors. One of the most obvious factors causing differences in protein concentration is species differences [6]. Caseins accounts for 80% (w/w) of the bovine milk proteins, and for 50% of human milk proteins [4]. In addition, β-lactoglobulin exists in bovine milk but cannot be found in human milk [6, 7]. Human and bovine milk diverge not only in their high abundant protein composition, but also in their low abundant protein composition. A total of 268 and 269 proteins were previously identified in human and bovine milk, respectively, in our previous study [8]. Of these

* Correspondence: kasper.hettinga@wur.nl
[1]Dairy Science and Technology, Food Quality and Design Group, Wageningen University, Postbox 8129, 6700EV Wageningen, The Netherlands
Full list of author information is available at the end of the article

proteins, 44 from human milk and 51 from bovine milk were related to the host defense system. Specifically, the concentration of proteins involved in the mucosal immune system, immunoglobulin A, CD14, lactoferrin, and lysozyme, were present in much higher concentration in human milk than bovine milk [8].

Furthermore, milk proteins also differ in concentration over lactation. Immunoglobulins have been reported to change rapidly in concentration from colostrum to mature milk in both human [9, 10] and bovine milk [11–13]. Moreover, the low abundant proteins, such as complement proteins, lipid synthesis and transport proteins, and enzymes were also reported to change as lactation advances [14, 15]. However, the differences in changes of proteins over lactation has not been reported between human milk and bovine milk directly, although we reported the changes in the species separately [13, 16–18].

As human milk is used as reference and bovine milk is used as protein source for producing infant formula [19], the differences in the health outcomes between breastfed and formula-fed infants could be related to the differences in the nutrient intake [6]. Breastfed infants were reported to have fewer infections (gastrointestinal infections, acute otitis media), reduced risk for celiac disease, obesity, and diabetes compared to formula-fed infants [19]. Therefore, the aim of this study is to better understand the role of different proteins, especially those involved in immune activity, in both human milk and bovine milk through elaborating the existing data in qualitative and quantitative proteome [8] and their changes over lactation [13, 16–18]. Separate interactomics studies of human and bovine milk proteins have previously been performed, using published data collected from many different sources [20, 21]. In this study, the analysis is a comparative data analysis on both species simultaneously, where data has been collected on a single instrument [8, 13, 16–18], throughout lactation, allowing a better comparison between species.

In the current study, the human and bovine milk data in Data set 1 [8] was reanalysed by Maxquant to give a more precise comparison in the quantitative differences between human and bovine milk proteins. The changes of both human and bovine milk proteome over lactation in Data set 2 [13, 16–18] were reanalysed using a co-expression (expression meaning the relative abundance) network approach and integrated with protein-protein interaction network data. The additional analysis enhances the comparison between human and bovine milk proteome from both qualitative and quantitative differences in milk proteome and their differences in changes over lactation. This should contribute to better understanding of the differences and similarities in biological functions networks of proteins, especially with regard to immune activity, in both the human and bovine milk proteome.

Result

A total of 379 proteins were quantified through reanalyzing the human and bovine milk of data set 1 prepared by filter-aided sample preparation (FASP) and LC-MS/MS. The specific number of identified proteins in milk fat globule membrane (MFGM) and milk serum proteins for both human and bovine species are shown in Fig. 1. Of these quantified proteins, 93 proteins present in both species. Figure 2 shows that both human milk and bovine milk have similar distribution over biological functions in quantified MFGM and milk serum proteins. Transport proteins, enzymes, and immune-related proteins were the three dominant biological function groups in both human and bovine milk (Fig. 2). The biological enrichment of these three protein groups were shown in Additional file 1: Table S1. However, the number of proteins in these three dominant groups was different between human and bovine milk. Bovine milk contained a higher number of transport proteins than human milk (Fig. 2), which was dominated by lipid and protein transporters. Although the number of enzymes were similar, they were quite different in the type between human and bovine milk. The enzymes assisting nutrient digestion were bile salt-activated lipase (CEL), and lipoprotein lipase (LPL) alpha-trypsin chain 1 (PRSS1) in human milk (Table 1) [16, 18], whereas pancreatic ribonuclease 1 (RNASE1), LPL, and ribonuclease 4 (RNASE4) were dominant in bovine milk [13, 17].

Tables 1 and 2 show the quantitative differences of common MFGM and milk serum proteins between human and bovine milk. Lipid synthesis and transport proteins, including fatty acid-binding protein, heart (FABP3), perilipin-2 (PLIN2), butyrophilin subfamily 1 member A1 (BTN1A1), lactadherin (MFGE8), and platelet glycoprotein 4 (CD36), were present at approximately 10–100 times higher abundance in bovine MFGM ($p < 0.05$). Serum albumin (ALB), monocyte differentiation antigen CD14 (CD14), alpha-lactalbumin (LALBA), lactoferrin (LTF), toll-like receptor 2 (TLR2), alpha-1-antitrypsin (SERPINA1), alpha-1-antichymotrypsin (SERPINA3), clusterin (CLU), and polymeric immunoglobulin receptor (PIGR) showed higher concentrations in human milk, especially for ALB, LTF, SERPINA3, and CD14, which were around 20–100 times higher in human milk serum ($p < 0.05$).

Since milk serum protein content is far higher than MFGM protein content [20], the quantitative changes over lactation were only determined for milk serum. A total of 299 proteins were quantified in bovine milk serum [13, 17] and 247 in human milk serum [16, 18] by FASP and dimethyl labelling combined with LC-MS/MS. There were 71 common proteins quantified in human and bovine milk serum, with 34 of them quantified in every time point over lactation. In addition to the high number of transport proteins in bovine milk serum, the

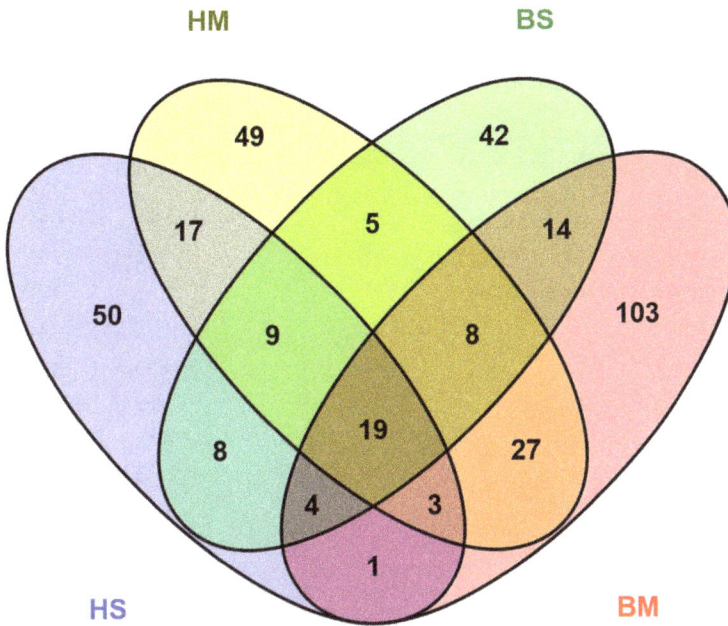

Fig. 1 The number of identified proteins in human milk and bovine milk (HS is human serum protein; HM is human milk fat globule membrane (MFGM) protein; BS is bovine serum protein; BM is bovine MFGM; identified number of proteins in HS, HM, BS, BM are 111. 137, 109, 179)

concentration of the transport proteins (calculated based on the summed intensity based absolute quantification (iBAQ values)) was higher in bovine milk serum than human milk serum, whereas enzymes were higher in human milk serum (Figs. 2 and 3).

Although the biological function distribution were similar in the identified proteins between human and bovine (Fig. 2), the quantitative changes of these protein groups differed over lactation (Fig. 3). Immune-related protein group decreased during the course of lactation, whereas transport protein and enzymes increased (Fig. 3). Moreover, the changing rate of the protein with the same functionality differed between species (Fig. 3); for

instance, immune-related proteins, LTF, complement C3 (C3), PIGR, and osteopontin (SPP1) decreased much faster in bovine milk serum compared to human milk serum (Fig. 4). The changes in immune-related proteins over lactation are important for two reasons. Firstly, immune-related proteins had relatively higher concentration in human milk than bovine milk. Secondly, these proteins play important roles in the protection of the neonate, which may therefore be proteins of interest for application in infant formula. Hierarchical clustering (Fig. 4) shows that these immune-related proteins are correlated to each other. In addition to the correlation of proteins related to complement and coagulation

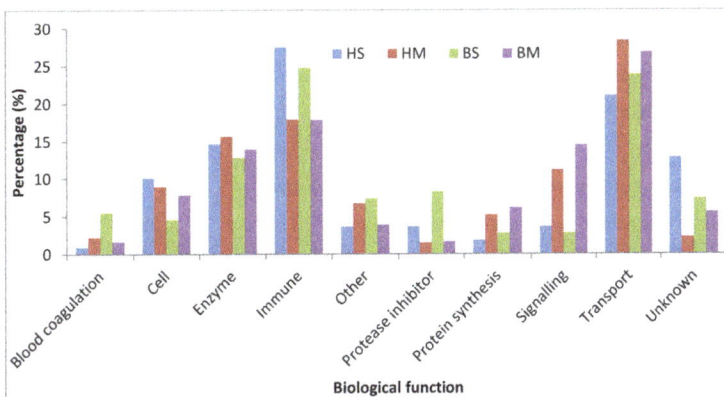

Fig. 2 The distribution of biological functions found in human and bovine milk ((HS is human serum protein; HM is human MFGM protein; BS is bovine serum protein; BM is bovine MFGM)

Table 1 Quantitative comparison of overlap human milk MFGM and bovine milk MFGM (Bold proteins are significantly different proteins by one-way ANOVA; the values are log10 average iBAQ values of proteins; Human milk samples were collected from 10 healthy mothers that were between 3 and 10 months in lactation; Bovine milk samples were collected from 30 clinically healthy cows that were between 3 weeks and 10 months in lactation; data was normally distributed)

Gene name	Protein name	HM	CM	Log_2 (HM/CM)	P value
LTF	**Lactotransferrin**	6.87	5.03	6.13	0.000
ALB	**Serum albumin**	6.02	4.46	5.18	0.000
FOLR1	**Folate receptor alpha**	5.16	3.85	4.35	0.000
CD14	**Monocyte differentiation antigen CD14**	5.37	4.51	2.87	0.000
LALBA	**Alpha-lactalbumin**	7.16	6.44	2.38	0.001
TLR2	**Toll-like receptor 2**	4.30	3.70	2.01	0.008
KRT79	Keratin, type II cytoskeletal 79	4.47	4.10	1.23	0.586
ATP2B2	Plasma membrane calcium-transporting ATPase 2	3.38	3.10	0.95	0.164
YWHAZ	14-3-3 protein zeta/delta	4.29	4.02	0.89	0.084
FASN	Fatty acid synthase	4.02	3.81	0.70	0.125
RRAS	Related RAS viral (R-ras) oncogene homolog	3.67	3.46	0.67	0.591
CSN2	Beta-casein	7.23	7.08	0.48	0.729
SPP1	Osteopontin	5.06	5.09	−0.09	0.585
RAB10	Ras-related protein Rab-10	4.55	4.61	−0.20	0.720
EEF1A1	Elongation factor 1-alpha 1	4.17	4.23	−0.22	0.686
SAR1A	GTP-binding protein SAR1a	4.35	4.43	−0.26	0.737
LSS	Lanosterol synthase	4.27	4.39	−0.38	0.387
STX3	Syntaxin-3	4.64	4.76	−0.39	0.300
RAB5C	Ras-related protein Rab-5C	3.58	3.73	−0.51	0.850
CD9	CD9 antigen	5.79	5.99	−0.67	0.245
XDH	**Xanthine dehydrogenase/oxidase**	6.11	6.34	−0.76	0.047
ANXA2	Annexin A2	4.20	4.43	−0.76	0.081
STOM	**Erythrocyte band 7 integral membrane protein**	5.28	5.55	−0.88	0.027
ACTG1	**Actin, cytoplasmic 2**	4.71	4.98	−0.90	0.026
CD59	CD59 molecule, complement regulatory protein	6.12	6.40	−0.94	0.057
FGFBP1	**Fibroblast growth factor-binding protein 1**	4.60	4.90	−1.00	0.017
CIDEA	Cell death activator CIDE-A	4.57	4.87	−1.02	0.077
SAR1B	GTP-binding protein SAR1b	3.41	3.72	−1.03	0.280
HSP90AA1	Heat shock protein HSP 90-alpha	2.95	3.31	−1.19	0.510
RAB1A	**Ras-related protein Rab-1A**	4.68	5.06	−1.24	0.021
EHD4	EH domain-containing protein 4	3.53	3.91	−1.26	0.077
BTN1A1	**Butyrophilin subfamily 1 member A1**	6.80	7.32	−1.70	0.001
PLIN2	**Perilipin-2**	6.24	6.80	−1.87	0.001
GNB1	**Guanine nucleotide-binding protein G(I)/G(S)/G(T) subunit beta-1**	3.79	4.36	−1.87	0.006
VAT1	**Synaptic vesicle membrane protein VAT-1 homolog**	3.98	4.55	−1.88	0.003
YWHAB	14-3-3 protein beta/alpha	3.03	3.60	−1.91	0.432
UBC	**Polyubiquitin-C**	3.56	4.16	−2.02	0.001
RAB18	**Ras-related protein Rab-18**	5.29	5.93	−2.11	0.001
MUC1	**Mucin-1**	3.79	4.47	−2.26	0.046
RAC1	**Ras-related C3 botulinum toxin substrate 1**	3.85	4.60	−2.48	0.017
RAB2A	**Ras-related protein Rab-2A**	3.93	4.70	−2.57	0.000

Table 1 Quantitative comparison of overlap human milk MFGM and bovine milk MFGM (Bold proteins are significantly different proteins by one-way ANOVA; the values are log10 average iBAQ values of proteins; Human milk samples were collected from 10 healthy mothers that were between 3 and 10 months in lactation; Bovine milk samples were collected from 30 clinically healthy cows that were between 3 weeks and 10 months in lactation; data was normally distributed) *(Continued)*

PIGR	Polymeric immunoglobulin receptor	5.17	5.99	−2.73	0.000
NUCB1	Nucleobindin-1	2.58	3.54	−3.22	0.067
YKT6	Synaptobrevin homolog YKT6	3.68	4.64	−3.22	0.000
FABP3	Fatty acid-binding protein, heart	5.09	6.09	−3.33	0.001
CSN1S1	Alpha-S1-casein	6.69	7.79	−3.66	0.000
ABCG2	ATP-binding cassette, sub-family G, member 2	4.92	6.05	−3.75	0.000
ACSL1	Acyl-CoA synthetase long-chain family member 1	3.64	4.81	−3.89	0.000
HSPA8	Heat shock cognate 71 kDa protein	3.19	4.41	−4.06	0.000
DHRS1	Dehydrogenase/reductase (SDR family) member 1	3.77	5.03	−4.17	0.000
CD36	Platelet glycoprotein 4	4.78	6.25	−4.90	0.000
GNB2	Guanine nucleotide-binding protein G(I)/G(S)/G(T) subunit beta-2	3.40	5.34	−6.43	0.000
IGL@	IGL@ protein	4.39	6.49	−6.96	0.000
MFGE8	Lactadherin	4.46	6.72	−7.53	0.000

cascades, such as C3, complement factor I (CFI), complement factor B (CFB), SERPINA1, antithrombin-III (SERPINC1), and alpha-2-HS-glycoprotein (AHSG) discussed before [13], CLU, alpha-1-acid glycoprotein 1 (ORM1), actin, cytoplasmic 1 (ACTB), LTF, SPP1, and PIGR also showed close interactions in both human and bovine milk serum (Fig. 4).

In order to compare the common human and bovine milk serum proteome at the network level, we converted our expression data to co-expression networks, and obtained available protein-protein interaction data for both species. Analysis of protein-protein interaction data indicated that the milk serum proteins quantified in our study are highly connected. For example, 310 interactions were observed for 66 human milk serum proteins, which is roughly 50 times higher than the number of interactions expected for randomly chosen proteins. The observed high interaction density was statistically significant according to the statistical test provided by STRING ($p < 10^{-6}$).

Comparing the co-expression networks to each other, for 34 proteins quantified in every time point in both human and bovine milk serum, 18 were aligned to the equivalent protein in the other species. For these proteins, if they have expression similarity with another protein in human milk, it is likely that they also have expression similarity with that protein in bovine milk, and vice versa. For the other 16 proteins, network alignment indicated that this was not the case. In other words, these proteins have expression similarities with different proteins in human milk than in bovine milk, and are indicative of changes in the expression network between the two species (Fig. S1). The similarity between

the human and bovine expression networks was also quantified using the correlation between the expression correlation coefficients. This resulted in a Pearson correlation coefficient of R = 0.23 ($p < 10^{-7}$) between the expression Pearson correlation coefficients in human and bovine milk serum proteome. Comparing the human co-expression network with the protein interaction network, for 34 proteins, 17 were aligned to themselves. For these proteins, if they have expression similarity with another protein, it is likely that they also have protein interaction with that protein. Out of these, 13 proteins were among the above-mentioned 18 proteins which were aligned to the equivalent protein in the human-bovine co-expression network alignment. This indicates a common core of 13 proteins with relatively highly conserved interaction in each of the networks (Fig. 5). These include the immune-related C3, CLU, ACTB, SERPINA1, SPP1, PIGR, and LTF.

The large agreement between co-expression networks and protein interaction networks observed based on the network alignment (Additional file 2: Figure S1 and Additional file 3: Table S2) was confirmed by analysing the relation between interaction status in the protein-protein interaction network, and expression correlation (both in human and bovine milk, Additional file 4: Table S3). The average expression correlation coefficient of non-interacting proteins is −0.06 +/−0.37, whereas for interacting proteins it is 0.18+/−0.37 (human) and 0.14+/−0.51 (bovine) respectively (Fig. 6). According to a Kolmogorov-Smirnov test, the differences between the distribution of correlation coefficients for interacting and for non-interacting proteins is significant: $p \sim 10^{-5}$ (human interacting vs non-interacting) and $p \sim 10^{-3}$ (bovine interacting vs

Table 2 Quantitative comparison of overlap human milk serum and bovine milk serum (Bold proteins are significantly different proteins by one-way ANOVA; the values are log10 average iBAQ values of proteins; Human milk samples were collected from 10 healthy mothers that were between 3 and 10 months in lactation; Bovine milk samples were collected from 30 clinically healthy cows that were between 3 weeks and 10 months in lactation; data was normally distributed)

Gene name	Protein name	HS	CS	Log2 (HS/CS)	P value
ALB	**Serum albumin**	7.30	5.27	6.75	0.000
FASN	**Fatty acid synthase**	4.38	2.40	6.59	0.000
XDH	**Xanthine dehydrogenase**	5.36	3.59	5.86	0.000
SERPINA3	**Alpha-1-antichymotrypsin**	5.44	4.02	4.72	0.000
LTF	**Lactoferrin**	7.17	5.84	4.44	0.002
PRSS1	**Alpha-trypsin chain 1**	6.74	5.42	4.39	0.000
CD14	**Monocyte differentiation antigen CD14**	5.52	4.24	4.23	0.000
CLU	**Clusterin**	5.74	4.64	3.66	0.002
CSN2	**Beta-casein**	7.80	6.79	3.34	0.002
LDHB	L-lactate dehydrogenase B chain	4.25	3.30	3.16	0.124
SERPINA1	**Alpha-1-antitrypsin**	5.38	4.62	2.52	0.007
PIGR	**Polymeric immunoglobulin receptor**	6.47	5.84	2.08	0.003
GC	Vitamin D-binding protein	5.20	4.64	1.89	0.026
LPL	Lipoprotein lipase	4.05	3.56	1.65	0.043
SPP1	osteopontin	6.35	5.95	1.34	0.107
HSPA8	Heat shock 70 kDa protein 8	3.89	3.51	1.25	0.277
LALBA	Alpha-lactalbumin	7.79	7.49	1.00	0.073
SERPINF2	Alpha-2-plasmin inhibitor	3.89	3.60	0.95	0.302
FABP3	Fatty acid-binding protein 3	5.33	5.13	0.66	0.359
B2M	Beta-2-microglobulin	5.47	5.36	0.37	0.779
BTN1A1	Butyrophilin subfamily 1 member A1	5.60	5.53	0.23	0.679
NUCB2	Nucleobindin 2	4.44	4.40	0.12	0.543
PLIN2	Perilipin-2	4.49	4.46	0.10	0.943
CSN1S1	Alpha-S1-casein	7.33	7.31	0.10	0.836
APOE	Apolipoprotein E	3.81	3.99	−0.59	0.285
SERPINC1	Antithrombin-III	3.86	4.05	−0.64	0.677
RAB18	Ras-related protein Rab-18	3.70	3.96	−0.86	0.725
MFGE8	Lactadherin	4.88	5.16	−0.94	0.402
NPC2	Epididymal secretory protein E1	4.49	4.89	−1.32	0.513
C3	Complement C3	4.56	4.96	−1.33	0.074
AZGP1	Zinc-alpha-2-glycoprotein	4.88	5.32	−1.47	0.197
AHSG	Alpha-2-HS-glycoprotein	4.32	4.83	−1.69	0.149
IGL@	**IGL@ protein**	6.19	6.76	−1.88	0.025
CFB	Complement factor B	2.66	3.27	−2.03	0.138
ORM1	Alpha-1-acid glycoprotein 1	4.94	5.56	−2.04	0.298
FGFBP1	**Fibroblast growth factor-binding protein 1**	3.69	4.64	−3.16	0.002
LPO	**Leucine-rich alpha-2-glycoprotein**	4.25	5.23	−3.26	0.004
NUCB1	**Nucleobindin 1**	4.05	5.35	−4.30	0.008
IDH1	**Isocitrate dehydrogenase 1**	3.00	4.48	−4.92	0.007

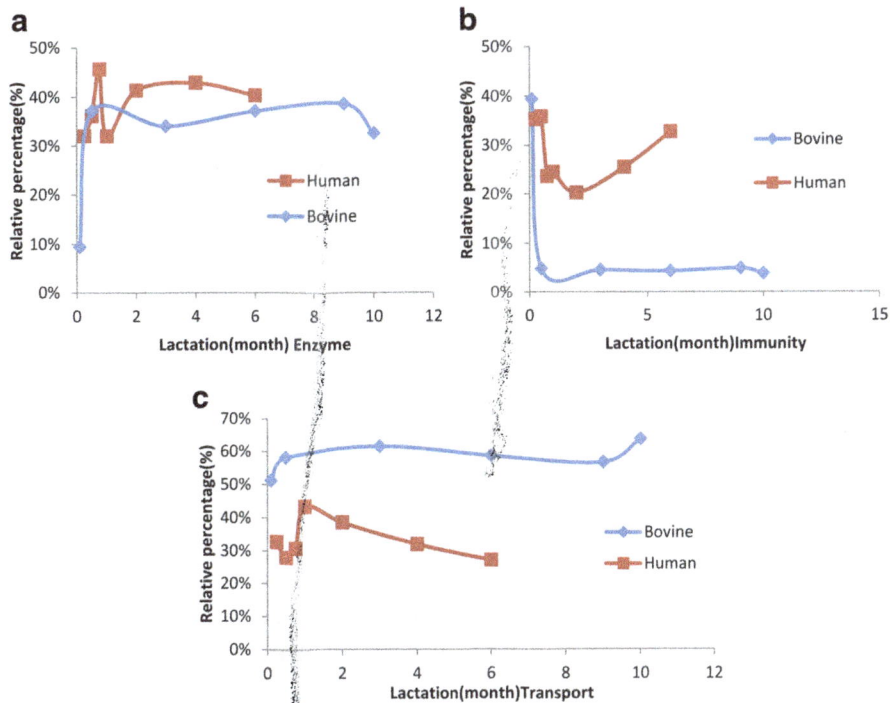

Fig. 3 The relative changes of enzyme (**a**), immunity (**b**) and transport proteins (**c**) over lactation between human and bovine milk. The percentage is calculated through the total iBAQ value of proteins in each biological function group divided by the total iBAQ value of proteins belonging to these three groups

non-interacting), respectively. Similarly, a Mann–Whitney U Test indicated that the means are significantly different ($p \sim 10^{-5}$ for human interacting vs non-interacting and $p \sim 0.005$ for bovine interacting vs non-interacting).

Discussion

Previous studies described some comparisons of the milk proteome between species [20–22]; however, they only used single samples, either mature milk collected at certain lactation stages or a pooled samples from different lactation stage. Also some reviews [23, 24] on milk proteome were based on single species, with no comparisons between different species. This is because the data they used are from different studies. Differences in lactation stage, differences in sample preparation methods, and differences in instruments make it difficult to compare the proteome between species at the same time points over lactation. This study was the first one to compare the changes of milk protein profile between human and bovine species at the same time points from colostrum to 6 months lactation by using the same sample preparation method and the same instrument. Our comparative analysis between the human and bovine lactation proteome was performed by reanalysing data from several of our previous studies [8, 13, 16–18]. The time-based comparison between human and bovine milk proteins, may help us to know better the differences in the

needs between infants and calves. This may also provide guidance on the improvement of infant formula composition on different stages. Although the data interpretation of the lactation stage studies is limited by the small sample size ($n = 4$) for both species, the separate results for bovine and human milk are similar to previously published studies on the biological functions of bovine and human milk protein, with many proteins in both species contributing to nutrient transport and immune protection [23, 24]. The annotation in this study gives a first insight in the comparison in the milk proteomes between human and bovine and their changes over lactation. The network analysis indicates that both the biological functions and the concentration of proteins have similarities between human and bovine milk. The reanalysed results in the current study should contribute to better understanding of the differences and similarities in the biological functions and micronutrients between human and bovine milk proteome.

A total of 390 proteins were quantified using Maxquant in both human and bovine milk (Fig. 1), which is higher compared to our previous study [8]. However, the number of identified proteins were lower than that reported in previous studies [10, 20, 21, 23, 24]. First, this comparison is based on one study not on a large number of reviewed studies [23, 24]. Second, the lower number of identified proteins can be related to both the

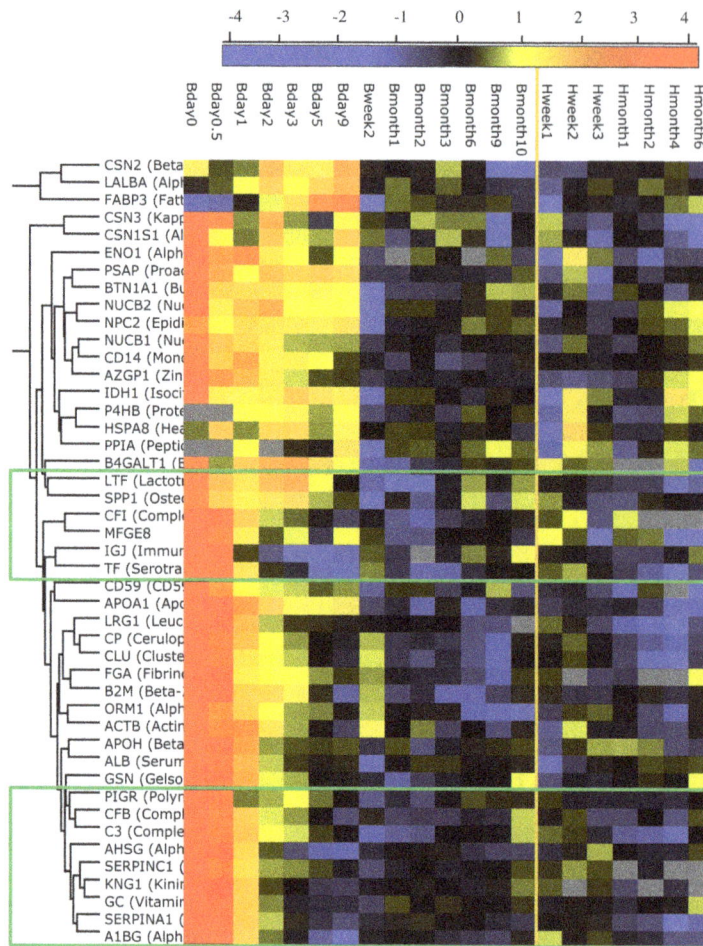

Fig. 4 The changes in the protein concentration from human and bovine milk over lactation (B-bovine milk; H-human milk)

identification criteria (reducing identification confidence) and the extensive protein fractionation (increasing the proteome coverage but decreasing the precision of protein quantification), as discussed in our previous paper [16]. Moreover, Maxquant was time cost-efficient in protein quantification. This indicates the advantages of Maxquant in quantifying milk proteins. The higher number of quantified proteins in data set 1 than data set 2 can be related to the differences in the preparation methods. Label free was used for dataset 1 and dimethyl labelling was used for dataset 2. The shift from label free to dimethyl labelling in two studies is because dimethyl labelling is much more sensitive and precise to pick up small differences between two samples [25]. The lower number of quantified proteins in our studies compared with previous studies (e.g. 573 proteins from bovine milk [23], 1606 [22] and 976 [15] proteins from human milk) can be related to the extensive protein fractionation in these previous studies and less strict identification criteria as discussed in our previous paper [16].

The higher number of quantified MFGM proteins than milk serum proteins in both human and bovine (data set 1) is consistent with the numbers of identified proteins reported previously [8]. It is not surprising, as MFGM represent the epithelial cell, the place where the milk fat is synthesised and secreted [26, 27]. The low amount of transport proteins in human milk can be mainly related to the absence of the major transport protein β-lactoglobulin (LGB) in human milk [28], which is the most abundant protein in bovine milk serum. In addition, the lower concentration of lipid synthesis and secretion proteins in human milk (Table 1 and 2) also contributes to the relatively low amount of transport proteins in human milk.

The relative high amount of enzymes (Fig. 3) and the high biological enrichment (Additional file 1: Table S1) in human milk can probably be attributed to the immature gastrointestinal tract of infants at birth. Although the development of the gastrointestinal tract starts from the fetal stage, the maturation of the gastrointestinal digestive function is not complete at birth [29]. It

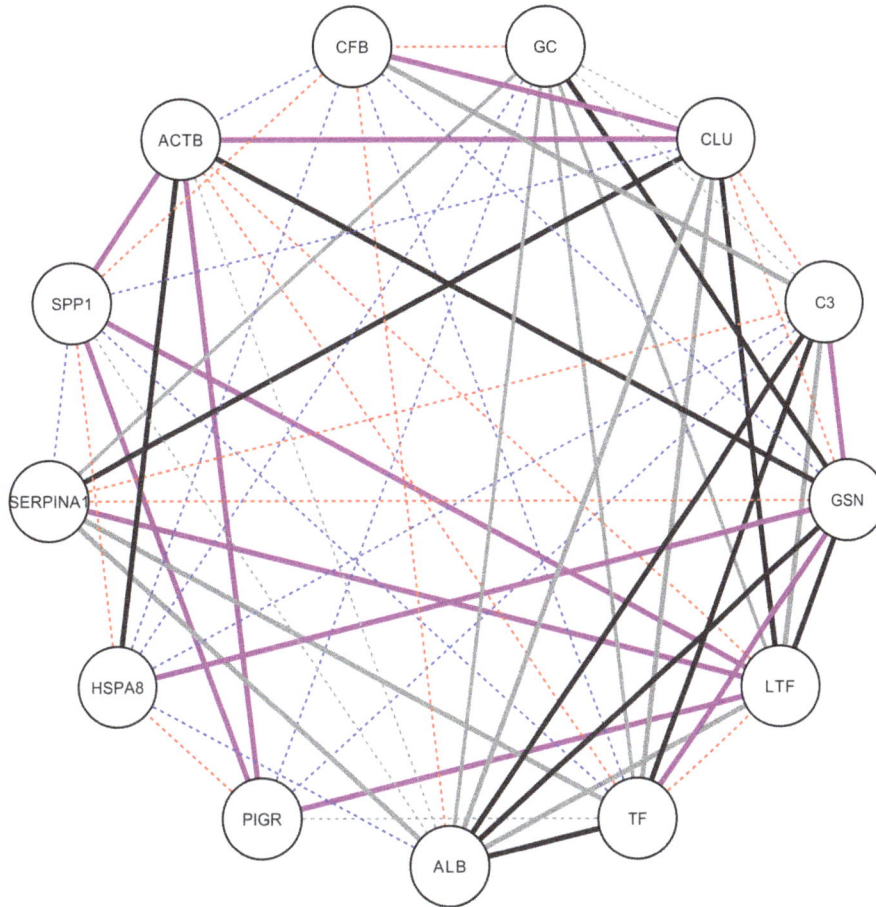

Fig. 5 Common conserved core observed in network alignment between protein interaction network and bovine and human co-expression networks. Edge colors indicate in which of the networks interaction occur: only in human co-expression network (*red*), only in bovine co-expression network (*blue*), in both human and bovine co-expression network (*purple*), in human co-expression network and in protein interaction network (*black*), or other combinations of networks (*grey*). In addition, line type differentiates interactions occurring only in one network (*dashed lines*) from interactions occurring in multiple networks (*straight lines*)

experiences a dramatic switch in the nutrients from amniotic fluid before birth to colostrum after birth and the energy supply switches from glucose-dominated to lipid-dominated [30]. This transition requires the digestion of lipids and proteins prior to their absorption in the gastrointestinal tract [30]. The high abundant enzymes related to lipid and protein degradation in human milk, such as bile salt-activated lipase, lipoprotein lipase, trypsin, and cathepsin D [31], suggests that human milk itself contributes to the digestive capacity, thereby being able to more effectively deal with immature luminal digestion [32]. The differences in the dominant digestive enzymes between human milk (bile salt-activated lipase) and bovine milk (ribonuclease pancreatic), which have been discussed in our previous papers [17] may thus reflect the differences in the needs for support of the digestion system between infants and calves.

Previous studies have reported that calves develop their own immune system in a few weeks [33], whereas infants produce their own immunoglobulins only after 2 or 3 months [15]. The relatively higher amount and slower decrease of immune-related proteins in human milk (Fig. 3) may be related to the slower maturation of immune system in infants than calves, as hypothesized before [8]. This hypothesis is consistent with the in-depth comparison between human and bovine milk proteome (Tables 1 and 2, Figs. 3 and 4).

However, the common proteins present in human and bovine milk (Fig. 1) suggest the similarity in the milk proteome between human and bovine. Several common immune-related proteins in the network analysis of both biological functions and co-expression levels (Fig. 5) indicate the comparable immunological functions of milk proteins in protecting the neonate. In addition to the importance of dominant immune-related proteins, such as LTF and immunoglobulins discussed previously [14, 15], the low abundant immune-related proteins, including C3, CFB, SERPINA1, ACTB, and SPP1 (Fig. 5), play

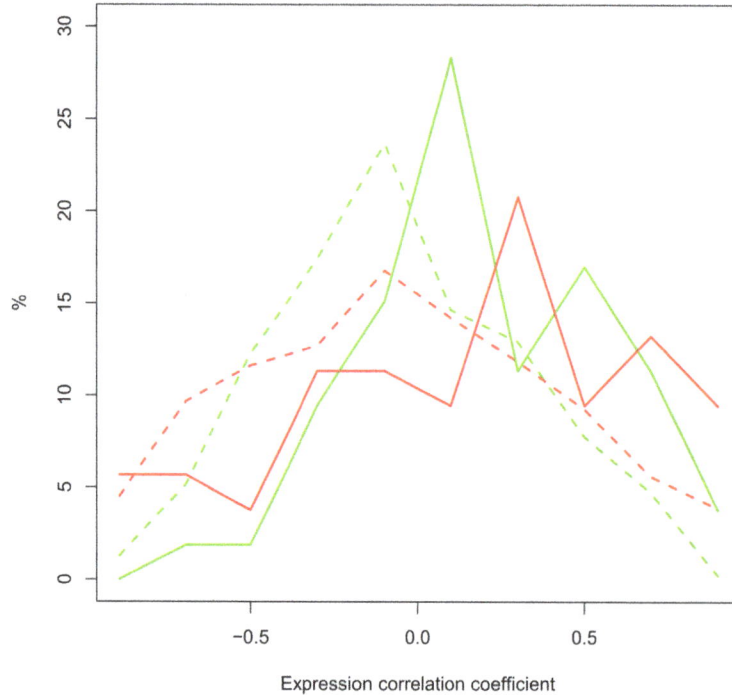

Fig. 6 Proteins interacting in the protein-protein interaction network have higher expression correlation than proteins not interacting. Histogram of expression Pearson correlation coefficients for human (*green*) and bovine (*red*) protein pairs, separately for interacting (*straight lines*) and for non-interacting (*dashed lines*) protein pairs

important roles in the immune system, especially innate immune system [10, 15]. The high abundance of innate immune-related proteins in early lactation (Fig. 4) may be due to its rapid reaction against broad groups of pathogens in the gastrointestinal tract of the neonate [8, 34], especially just after birth. SERPINA1 plays a dual role in regulating the complement and coagulation pathway [35], but also protecting the immune-related proteins against degradation during digestion. ACTB not only plays a role in the cell cytoskeleton but is also involved in innate immune response, according to research using a mice model [36]. SPP1 could protect the intestinal tract of infants against pathogens or bacteria, due to its cytokine-like properties and it being a key factor in the initiation of T helper 1 immune responses [37]. PIGR is the receptor of immunoglobulins A and M, facilitating their secretion in the mammary gland. The high correlation between SERPINA1, LTF, C3, ACTB, SPP1, and PIGR (Fig. 5) in both human and bovine milk reflects the interactions between innate and adaptive immune system and the complex nature of biological interrelationships between milk proteins in protecting the neonate.

The other common proteins in Fig. 5, LTF, TF, ALB, vitamin D-binding (GC), play roles in transport and delivery of nutrients through binding minerals, vitamins, fatty acid, steroids, glucocorticoid/progestin, and heme

derivatives, and thus facilitate their uptake in the intestinal tract [38]. The correlation of these proteins in both human and bovine milk (Fig. 5) could be related to need for providing this range of micronutrients that are necessary for the growth of the neonate.

The distribution of expression correlation coefficients (Fig. 6) over lactation in both human and bovine milk proteome for protein pairs not interacting in the protein interaction network is shifted towards negative values compared to the distribution for protein pairs that are interacting. This suggests an interplay between protein-protein interactions and expression similarity. Such similarity between these different types of networks was also observed based on network alignment. In all mammals, milk provision is a complex process with changes in milk composition and interactions between parent and young beyond the straightforward nutritional function [39]. The similarity in the milk proteome may be related to their main functions in providing nutrients and protection to the neonate. The differences in the milk proteome between species may be due to their unique lactation strategies to accommodate reproductive success and adapt to the specific environment. This suggests an interplay between protein-protein interactions and expression similarity.

The comparison of the milk proteome between human and bovine over lactation provides more information on the similarity and differences of milk protein profile over lactation. This study can be used as a start point for further biological function investigation of proteins discussed in the paper. Proteins differing between human and bovine are interesting from an infant nutrition point-of-view. Further evaluation of the biological significance of these proteins, and on the feasibility of the application of such proteins in infant formula can be conducted. With respect to the proteins with high similarity based on the network alignment, they may still differ in digestibility or have different nutritional values due to the differences in amino acid sequence and post-translation modifications between species. Further studying this will contribute to a better understanding of protein functionality in human and bovine milk, and may provide guidance on the improvement of infant formula.

Conclusions

The qualitative and quantitative differences between human and bovine milk proteome as well as the differences in the concentration changes over lactation help us to better understand the role of milk proteins in the development of the digestive and immune system of the neonates in general, including differences between infants and calves. The similarities in both protein-protein interaction network and expression correlation between human and bovine milk proteome indicates the importance of milk proteins in providing nutrients and protection to the neonate. This in-depth comparison between human and bovine milk contributes to a better understanding on the biological functions, especially immunological functions, of milk proteins between human and bovine.

Methods
Materials
In this study, we reanalysed the data collected on a single instrument [8, 13, 16–18] from both human and bovine milk proteome for an in-depth comparison throughout lactation.

Data set 1-Qualitative and quantitative differences between human and bovine milk proteome study
This data is based on the study of Hettinga, et al. [8]. Human milk was collected from 10 healthy mothers between 3 and 10 months in lactation. Samples of 10 mL were collected and frozen for later analysis. After thawing, the 10 samples were pooled. One bovine tank milk sample was collected from the university farm "De Ossekampen" in Wageningen, The Netherlands, which was milk from 30 clinically healthy cows which were between 3 weeks and 10 months in lactation.

Data set 2-The comparison in the changes of human and bovine milk proteome over lactation
This data set is based on our previous studies [13, 16–18]. Human milk samples were collected from women who gave birth at the obstetric department in VU medical center (VUmc) in Amsterdam. All women who delivered singleton term infants (gestational age 37–42 weeks) were eligible for this study. Women with haemolysis elevated liver enzymes, low platelet syndrome, history of breast surgery, and (gestational) diabetes mellitus were excluded. The samples collected at week 1, 2, 3, 4, 8, 16, 24 were used for this study. Approximately 5–10 mL was collected in a polypropylene bottle after 1 min of pumping for every sample. and stored at −18 °C immediately afterwards.

Bovine milk was collected from four healthy cows in a farm in Zaffelaere, Belgium. The cows were milked using an automatic milking system. Samples were collected from day 0 to the end of lactation. Samples collected at day 0, 0.5, 1, 2, 3, 5, 9, 14, month 1, 2, 3, 6, 9 and the latest time point of the lactation (10 months for cow 1, 11 months for cow 2 and 12 months for cow 3, the latest time point was missed for cow 4) were used for this study. The samples were frozen immediately at −20 °C after collection and transferred frozen to the laboratory for further analysis.

Methods
Milk serum separation
The separation of milk serum was performed according to a previous study [8]. The samples were centrifuged at 1,500 × g for 10 min at 10 °C (Beckman coulter Avanti J-26 XP centrifuge, rotor JA-25.15). The milk fat was removed and the obtained supernatant was transferred to the ultracentrifuge tubes followed by ultracentrifugation at 100,000 × g for 90 min at 4 °C (Beckman L-60, rotor 70 Ti). After ultracentrifugation, samples were separated into three phases. The top layer was remaining milk fat, the middle layer was milk serum (with some free soluble caseins), and the bottom layer (pellet) was casein. Milk serum was used for filter aided sample preparation as described below after the measurement of protein content by the BCA protein assay (Fisher Scientific).

Proteomic techniques
Filter aided sample preparation Filter aided sample preparation (FASP) was performed as previously described [40]. Milk serum samples (20 μL), including samples of each time point and pooled samples of each included woman, were diluted in 100 mM Tris/HCl pH 8.0 + 4% SDS + 0.1 M Dithiotreitol (SDT-lysis buffer) to get a 1 μg/μL protein solution. Samples were then incubated for 10 min at 95 °C, and centrifuged at 18407 g for 10 min, after cooling down to room temperature.

Twenty μL of each sample were directly added to the middle of 180 μL 0.05 M iodoacetamide/100 mM Tris/HCl pH 8.0 + 8 M urea (UT) in a low binding Eppendorf tube and incubated for 10 min while mildly shaking at room temperature. The sample was transferred to a Pall 3 K omega filter (10–20 kDa cutoff, OD003C34; Pall, Washington, NY, USA) and centrifuged at 15871 g for 30 min. Three repeated centrifugations at 15871 g for 30 min were carried out after adding three times 100 μL UT. After that, 110 μL 0.05 M NH₄HCO₃ in water (ABC) were added to the filter unit and the samples were centrifuged again at 15871 g for 30 min. Then, the filter was transferred to a new low-binding Eppendorf tube. One hundred μL ABC containing 0.5 μg trypsin were added followed by overnight incubation at room temperature. Finally, the sample was centrifuged at 15871 g for 30 min, and 3.5 μL 10% trifluoroacetic acid (TFA) were added to the filtrate to adjust the pH value of the sample to around 2. These samples were ready for dimethyl labeling.

Dimethyl labeling The dimethyl labeling was carried out by on-column dimethyl labeling according to [22]. The trypsin digested samples of pooled milk serum from each individual mothers and cows collected at the different time points were labeled with light reagent (the mix of CH_2O and cyanoborohydride), whereas trypsin digested milk serum samples of the individual mothers and cows at each time point were labeled with heavy reagent (the mix of CD_2O and cyanoborohydride). Stage tips containing 2 mg Lichroprep C18 (25 um particles) column material (C18+ Stage tip) were made in-house. The C18+ Stage tip column was washed 2 times with 200 μL methanol. The column was conditioned with 100 μL of 1 mL/L formic acid in water (HCOOH) after which samples were loaded on the C18+ Stage tip column. The column was washed with 100 μL 1 mL/L HCOOH, and then slowly flushed with 100 μL labeling reagent (0.2% CH_2O or CD_2O and 30 mM cyanoborohydride in 50 mM phosphate buffer pH 7.5) in about 10 min. The column was washed again with 200 μL 1 mL/L HCOOH. Finally, the labeled peptides were eluted with 50 μL of 70% acetonitrile/30% 1 mL/L HCOOH from the C18+ Stage tip columns. The samples were then dried in a vacuum concentrator (Eppendorf Vacufuge®) at 45 °C for 20 to 30 min until the volume of each sample decreased to 15 μL or less. The pairs of light dimethyl label and heavy dimethyl label were then mixed up and the volume was adjusted to exactly 100 μL by adding 1 mL/L HCOOH. These samples were ready for analysis by LC-MS/MS.

LC-MS/MS Eighteen μL of the trypsin digested and dimethyl labeled milk fractions were injected on a 0.10 × 30 mm Magic C18AQ 200A 5 μm beads (Michrom Bioresources Inc., USA) pre-concentration column (prepared in house) at a maximum pressure of 270 bar. Peptides were eluted from the pre-concentration column onto a 0.10 × 200 mm Prontosil 300-3-C18H Magic C18AQ 200A 3 μm analytical column with an acetonitrile gradient at a flow of 0.5 μL/min, using gradient elution from 8 to 33% acetonitrile in water with 0.5 v/v% acetic acid in 50 min. The column was washed using an increase in the percentage acetonitrile to 80% (with 20% water and 0.5 v/v% acetic acid in the acetonitrile and the water) in 3 min. A P777 Upchurch microcross was positioned between the pre-concentration and analytical column. An electrospray potential of 3.5 kV was applied directly to the eluent via a stainless steel needle fitted into the waste line of the microcross. Full scan positive mode FTMS spectra were measured between m/z 380 and 1400 on a LTQ-Orbitrap XL (Thermo electron, San Jose, CA, USA). CID fragmented MS/MS scans of the four most abundant doubly- and triply-charged peaks in the FTMS scan were recorded in data-dependent mode in the linear trap (MS/MS threshold = 5.000).

Data analysis
The acquired datasets were analyzed by using MaxQuant (Version 1.5.2.8, http://www.maxquant.org/) and the built-in Andromeda search engine with a UniProt human and bovine database (http://www.uniprot.org/; accessed March 2012). The search parameters were as follows: variable modifications of protein N-terminal acetylation and methionine oxidation, and fixed modification of cysteine carbamidomethylation. The minimum peptide length was set to 7 amino acids and a maximum of 2 missed cleavages was allowed for the search. Trypsin/P was selected as the semi-specific proteolytic enzyme. The global false discovery rate (FDR) cut off used for both peptides and proteins was 0.01 [41]. Label-free quantitation was performed in MaxQuant. To further improve the quantification accuracy, only the razor/unique peptides were used for quantitative calculations. The other parameters used were the default settings in MaxQuant software for processing MS/MS data.

All known contaminants (i.e. keratins, trypsin), and proteins detected in less than half of the samples, were removed from each sample set of proteins identified. The origin and function of the identified proteins was taken from UniProtKB (http://www.uniprot.org/; accessed March 2012) for recommended protein name, gene name, and protein function. It was verified that the human and bovine proteins with the same protein name were orthologous using a reciprocal best BLAST hit approach. DAVID Bioinformatics Resource 6.7 (https://david.ncifcrf.gov/) was used for protein biological function classification and protein group enrichment. Protein concentrations were calculated as the average of all peptide peak intensities from five replicates divided by the number of theoretically observable tryptic peptides (intensity based absolute quantification, or iBAQ, [42, 43]). Perseus software v.1.2.0.16 (Martinsried, Germany) was used to test for hierarchical clustering and significant differences between species. Hierarchical clustering in Perseus software was used for clustering proteins identified in both human and bovine milk based on their relative abundance. This procedure is performing hierarchical clustering of rows (proteins) and columns (samples) and produces a visual heat map representation of the clustered matrix. The ratios between the concentration found in human milk (milk fat globulin membrane-MFGM and serum) and bovine milk (MFGM and serum) were calculated as the difference (on ^{10}log scale) of the iBAQ value of the human MFGM versus the bovine MFGM and human serum versus bovine serum. ANOVA was applied to compare MFGM and serum in both species, and the p-values obtained were adjusted with false discovery rate (FDR)-based correction in order to account for the effect of multiple comparisons.

Protein-protein interactions for proteins in both human and bovine milk proteome were obtained from STRING [44]. In order to interpret the interaction density (number of observed interactions divided by total possible number of interactions) of milk proteins, this density was compared with the interaction density of all human/bovine STRING proteins. A statistical test for the significance of the observed high density in the milk proteome was performed using the approach provided by STRING [45].

For co-expression network analysis, a cutoff of 0.3 on the absolute value of the Pearson correlation was applied, in order to get a number of interactions in the co-expression networks that would be comparable to that in the STRING interaction networks. Pinalog [46] was used to align different networks to each other, taking into account both sequence similarity between proteins and topological similarity (i.e. similarity of interaction partners for each protein). For visualization, VANLO [47] and Cytoscape [48] were applied. Comparison of distributions with Kolmogorov-Smirnov test was performed using the R-function ks.test.

Additional files

Additional file 1: Table S1. The biological functional enrichment of immunity, transport and enzyme protein groups in both human and bovine milk. (DOCX 13 kb)

Additional file 2: Figure S1. Network alignment between bovine (red) and human (green) co-expression networks. Equivalent nodes are connected by thin straight lines and are at comparable positions in the two networks. (TIF 18148 kb)

Additional file 3: Table S2. Alignment between bovine and human co-expression networks. (DOCX 14 kb)

Additional file 4: Table S3. Alignment between protein interaction network and human co-expression network. (DOCX 15 kb)

Abbreviations
ACTB: Actin, cytoplasmic 1; AHSG: Alpha-2-HS-glycoprotein; ALB: Serum albumin; BTN1A1: Butyrophilin subfamily 1 member A1; CD14: Monocyte differentiation antigen CD14; CD36: Platelet glycoprotein 4; CEL: Bile salt-activated lipase; CFB: Complement factor B; CFI: Complement factor I; CLU: Clusterin; FABP3: Fatty acid-binding protein, heart; FASP: Filter-aided sample preparation; FDR: False discovery rate; GC: Vitamin D-binding; GO: Gene ontology; iBAQ Value: intensity based absolute quantification; LALBA: Alpha-lactalbumin; LGB: β-lactoglobulin; LPL: Lipoprotein lipase; LTF: Lactoferrin; MFGE8: Lactadherin; MFGM: Milk fat globule membrane; ORM1: Alpha-1-acid glycoprotein 1; PIGR: Polymeric immunoglobulin receptor; PLIN2: Perilipin-2; PRSS1: Alpha-trypsin chain 1; RNASE1: Pancreatic ribonuclease 1; RNASE4: Ribonuclease 4; SERPINA1: Alpha-1-antitrypsin; SERPINA3: Alpha-1-antichymotrypsin; SERPINC1: Antithrombin-III; SPP1: Osteopontin; TLR2: Toll-like receptor 2

Acknowledgements
We thank Marita de Waard, Hester Verheijen and Hans van Goudoever for collecting the human milk samples. We thank Jeroen Heck for collecting the bovine milk samples. We thank Jos A. Hageman for the statistical support in the primary data analysis. We thank Sjef Boeren for performing the LC-MS/MS on all samples.

Funding
Not applicable.

Authors' contribution
LZ designed the experiment, performed sample preparation and preliminary data analysis, and drafted the manuscript. AvD performed the interactomics data analysis, participated in discussion on result interpretation, and wrote the data analysis section and part of the results and discussion of the manuscript. KH participated in the experiment design and data interpretation discussion, and revised the manuscript. All authors read and approved the final manuscript.

Competing interests
The authors declare that they have no competing interests.

Author details
[1]Dairy Science and Technology, Food Quality and Design Group, Wageningen University, Postbox 8129, 6700EV Wageningen, The Netherlands. [2]Biometris, Wageningen University and Research Centre, P.O. Box 166700 AA Wageningen, The Netherlands. [3]Bioinformatics Group, Wageningen University Droevendaalsesteeg 1, 6708 PB Wageningen, The Netherlands. [4]Bioscience, cluster Applied Bioinformatics, Wageningen University and Research, Droevendaalsesteeg 1, 6708 PB Wageningen, The Netherlands.

References
1. Casado B, Affolter M, Kussmann M. OMICS-rooted studies of milk proteins, oligosaccharides and lipids. J Proteomics. 2009;73(2):196–208.
2. German JB, Dillard CJ, Ward RE. Bioactive components in milk. Curr Opin Clin Nutr Metab Care. 2002;5(6):653–8.
3. Reinhardt TA, Lippolis JD. Bovine milk fat globule membrane proteome. J Dairy Res. 2006;73(4):406–16.
4. Séverin S, Wenshui X. Milk biologically active components as nutraceuticals: Review. Crit Rev Food Sci Nutr. 2005;45(7–8):645–56.
5. Smolenski G, Haines S, Kwan FYS, Bond J, Farr V, Davis SR, Stelwagen K, Wheeler TT. Characterisation of host defence proteins in milk using a proteomic approach. J Proteome Res. 2007;6(1):207–15.
6. D'Auria E, Agostoni C, Giovannini M, Riva E, Zetterstrom R, Fortin R, Greppi GF, Bonizzi L, Roncada P. Proteomic evaluation of milk from different mammalian species as a substitute for breast milk. Acta Paediatr Int J Paediatr. 2005;94(12):1708–13.
7. Mercier JC, Vilotte JL. Structure and function of milk protein genes. J Dairy Sci. 1993;76(10):3079–98.
8. Hettinga K, van Valenberg H, de Vries S, Boeren S, van Hooijdonk T, van Arendonk J, Vervoort J. The host defense proteome of human and bovine milk. PLoS. One 2011;6(4):e19433.

9.   Politis I, Chronopoulou R. Milk peptides and immune response in the neonate. Adv Exp Med Biol. 2008;606:253–69.

10.   Zhang Q, Cundiff J, Maria S, McMahon R, Woo J, Davidson B, Morrow A. Quantitative analysis of the human milk whey proteome reveals developing milk and mammary-gland functions across the first year of lactation. Proteomes. 2013;1(2):128–58.

11.   Stelwagen K, Carpenter E, Haigh B, Hodgkinson A, Wheeler TT. Immune components of bovine colostrum and milk. J Anim Sci. 2009;87(13 Suppl):3–9.

12.   Senda A, Fukuda K, Ishii T, Urashima T. Changes in the bovine whey proteome during the early lactation period. Anim Sci J. 2011;82(5):698–706.

13.   Zhang L, Boeren S, Hageman JA, van Hooijdonk T, Vervoort J, Hettinga K. Bovine milk proteome in the first 9 days: protein interactions in maturation of the immune and digestive system of the newborn. PLoS One. 2015;10(2): e0116710.

14.   Liao Y, Alvarado R, Phinney B, Lonnerdal B. Proteomic characterization of human milk whey proteins during a twelve-month lactation period. J Proteome Res. 2011;10(4):1746–54.

15.   Gao X, McMahon RJ, Woo JG, Davidson BS, Morrow AL, Zhang Q. Temporal changes in milk proteomes reveal developing milk functions. J Proteome Res. 2012;11(7):3897–907.

16.   Zhang L, de Waard M, Verheijen H, Boeren S, Hageman JA, van Hooijdonk T, Vervoort J, van Goudoever JB, Hettinga K. Changes over lactation in breast milk serum proteins involved in the maturation of immune and digestive system of the infant. J Proteomics. 2016;147:40–7. http://dx.doi.org/10.1016/ j.jprot.2016.02.005.

17.   Zhang L, Boeren S, Hageman JA, van Hooijdonk T, Vervoort J, Hettinga K. Perspective on calf and mammary gland development through changes in the bovine milk proteome over a complete lactation. J Dairy Sci. 2015;98(8): 5362–73.

18.   Zhang L, de Waard M, Verheijen H, et al. Changes over lactation in breast milk serum proteins involved in the maturation of immune and digestive system of the infant. Data Brief. 2016;7:362–5. doi:10.1016/j.dib.2016.02.046.

19.   Hernell O. Human milk vs. cow's milk and the evolution of infant formulas. In: Nestle Nutrition Workshop Series: Pediatric Program. 67th ed. 2011. p. 17–28.

20.   Reinhardt TA, Lippolis JD, Nonnecke BJ, Sacco RE. Bovine milk exosome proteome. J Proteomics. 2012;75(5):1486–92.

21.   Reinhardt TA, Lippolis JD. Developmental changes in the milk fat globule membrane proteome during the transition from colostrum to milk. J Dairy Sci. 2008;91(6):2307–18.

22.   Beck KL, Weber D, Phinney BS, Smilowitz JT, Hinde K, Lonnerdal B, Korf I, Lemay DG. Comparative proteomics of human and macaque milk reveals species-specific nutrition during postnatal development. J Proteome Res. 2015;14:2143–57.

23.   D'Alessandro A, Scaloni A, Zolla L. Human milk proteins: an interactomics and updated functional overview. J Proteome Res. 2010;9(7):3339–73.

24.   D'Alessandro A, Zolla L, Scaloni A. The bovine milk proteome: cherishing, nourishing and fostering molecular complexity. An interactomics and functional overview. Mol BioSyst. 2011;7:579–97.

25.   Lu J, Boeren S, de Vries SC, van Valenberg HJ, Vervoort J, Hettinga K. Filter-aided sample preparation with dimethyl labeling to identify and quantify milk fat globule membrane proteins. J Proteomics. 2011;75(1):34–43.

26.   McManaman JL, Neville MC. Mammary physiology and milk secretion. Adv Drug Deliv Rev. 2003;55(5):629–41.

27.   Lu J, van Hooijdonk T, Boeren S, Vervoort J, Hettinga K. Identification of lipid synthesis and secretion proteins in bovine milk. J Dairy Res. 2014;81(1):65–72.

28.   Hinz K, O'Connor PM, Huppertz T, Ross RP, Kelly AL. Comparison of the principal proteins in bovine, caprine, buffalo, equine and camel milk. J Dairy Res. 2012;79(2):185–91.

29.   Lindquist S, Hernell O. Lipid digestion and absorption in early life: an update. Curr Opin Clin Nutr Metab Care. 2010;13(3):314–20.

30.   Abrahamse E, Minekus M, van Aken GA, van de Heijning B, Knol J, Bartke N, Oozeer R, van der Beek EM, Ludwig T. Development of the digestive system-experimental challenges and approaches of infant lipid digestion. Food Digestion. 2012;3(1–3):63–77.

31.   Khaldi N, Vijayakumar V, Dallas DC, Guerrero A, Wickramasinghe S, Smilowitz JT, Medrano JF, Lebrilla CB, Shields DC, German JB. Predicting the important enzymes in human breast milk digestion. J Agric Food Chem. 2014;62(29): 7225–32.

32.   Dallas DC, Smink CJ, Robinson RC, Tian T, Guerrero A, Parker EA, Smilowitz JT, Hettinga KA, Underwood MA, Lebrilla CB, et al. Endogenous human milk peptide release is greater after preterm birth than term birth. J Nutr. 2015; 145(3):425–33.

33.   Chase CC, Hurley DJ, Reber AJ. Neonatal immune development in the calf and its impact on vaccine response. Vet Clin North Am Food Anim Pract. 2008;24(1):87–104.

34.   Jensen GS, Patel D, Benson KF. A novel extract from bovine colostrum whey supports innate immune functions. II. Rapid changes in cellular immune function in humans. Prev Med. 2012;54:124–9.

35.   Law RH, Zhang Q, McGowan S, Buckle AM, Silverman GA, Wong W, Rosado CJ, Langendorf CG, Pike RN, Bird PI, Whisstock JC. An overview of the serpin superfamily. Genome Biol. 2006;7(5):216.

36.   Man SM, Ekpenyong A, Tourlomousis P, Achouri S, Cammarota E, Hughes K, Rizzo A, Ng G, Wright JA, Cicuta P, et al. Actin polymerization as a key innate immune effector mechanism to control Salmonella infection. Proc Natl Acad Sci U S A. 2014;111(49):17588–93.

37.   Schack L, Lange A, Kelsen J, Agnholt J, Christensen B, Petersen TE, Sørensen ES. Considerable variation in the concentration of osteopontin in human milk, bovine milk, and infant formulas. J Dairy Sci. 2009;92(11):5378–85.

38.   Lonnerdal B. Nutritional and physiologic significance of human milk proteins. Am J Clin Nutr. 2003;77(6):1537s–43s.

39.   Lefèvre CM, Sharp JA, Nicholas KR. Evolution of lactation: Ancient origin and extreme adaptations of the lactation system. In: Annual Review of Genomics and Human Genetics. 11th ed. 2010. p. 219–38.

40.   Wisniewski JR, Zougman A, Nagaraj N, Mann M. Universal sample preparation method for proteome analysis. Nat Methods. 2009;6:359–62.

41.   Michalski A, Cox J, Mann M. More than 100,000 detectable peptide species elute in single shotgun proteomics runs but the majority is inaccessible to data-dependent LC-MS/MS. J Proteome Res. 2011;10:1785–93.

42.   Malmström J, Beck M, Schmidt A, Lange V, Deutsch EW, Aebersold R. Proteome-wide cellular protein concentrations of the human pathogen Leptospira interrogans. Nature. 2009;460(7256):762–5.

43.   Schwanhausser B, Busse D, Li N, Dittmar G, Schuchhardt J, Wolf J, Chen W, Selbach M. Global quantification of mammalian gene expression control. Nature. 2011;473(7347):337–42.

44.   Szklarczyk D, Franceschini A, Wyder S, Forslund K, Heller D, Huerta-Cepas J, Simonovic M, Roth A, Santos A, Tsafou KP, et al. STRING v10: protein-protein interaction networks, integrated over the tree of life. Nucleic Acids Res. 2015;43(Database issue):D447–452.

45.   Franceschini A, Szklarczyk D, Frankild S, Kuhn M, Simonovic M, Roth A, Lin J, Minguez P, Bork P, von Mering C, Jensen LJ. STRING v9.1: protein-protein interaction networks, with increased coverage and integration. Nucleic Acids Res. 2013;41(Database issue):D808–15.

46.   Phan HT, Sternberg MJ. PINALOG: a novel approach to align protein interaction networks–implications for complex detection and function prediction. Bioinformatics. 2012;28(9):1239–45.

47.   Brasch S, Linsen L, Fuellen G. VANLO–interactive visual exploration of aligned biological networks. BMC bioinformatics. 2009;10:327.

48.   Shannon P, Markiel A, Ozier O, Baliga NS, Wang JT, Ramage D, Amin N, Schwikowski B, Ideker T. Cytoscape: a software environment for integrated models of biomolecular interaction networks. Genome Res. 2003;13(11):2498–504.

11

Overgrazing induces alterations in the hepatic proteome of sheep (*Ovis aries*): an iTRAQ-based quantitative proteomic analysis

Weibo Ren[1], Xiangyang Hou[1], Yuqing Wang[1], Warwick Badgery[2], Xiliang Li[1], Yong Ding[1], Huiqin Guo[3], Zinian Wu[1], Ningning Hu[1], Lingqi Kong[1], Chun Chang[1], Chao Jiang[1] and Jize Zhang[1*]

Abstract

Background: The degradation of the steppe of Inner Mongolia, due to overgrazing, has resulted in ecosystem damage as well as extensive reductions in sheep production. The growth performance of sheep is greatly reduced because of overgrazing, which triggers massive economic losses every year. The liver is an essential organ that has very important roles in multiple functions, such as nutrient metabolism, immunity and others, which are closely related to animal growth. However, to our knowledge, no detailed studies have evaluated hepatic metabolism adaption in sheep due to overgrazing. The molecular mechanisms that underlie these effects remain unclear.

Methods: In the present study, our group applied isobaric tags for relative and absolute quantitation (iTRAQ)-based quantitative proteomic analysis to investigate changes in the protein profiles of sheep hepatic tissues when nutrition was reduced due to overgrazing (12.0 sheep/ha), with the goal of characterizing the molecular mechanisms of hepatic metabolism adaption in sheep in an overgrazing condition.

Results: The body weight daily gain of sheep was greatly decreased due to overgrazing. Overall, 41 proteins were found to be differentially abundant in the hepatic tissue between a light grazing group and an overgrazing group. Most of the differentially expressed proteins identified are involved in protein metabolism, transcriptional and translational regulation, and immune response. In particular, the altered abundance of kynureninase (KYNU) and HAL (histidine ammonia-lyase) involved in protein metabolic function, integrated with the changes of serum levels of blood urea nitrogen (BUN) and glucose (GLU), suggest that overgrazing triggers a shift in energy resources from carbohydrates to proteins, causing poorer nitrogen utilization efficiency. Altogether, these results suggest that the reductions in animal growth induced by overgrazing are associated with liver proteomic changes, especially the proteins involved in nitrogen compounds metabolism and immunity.

Conclusions: This provides new information that can be used for nutritional supplementation to improve the growth performance of sheep in an overgrazing condition.

Keywords: Overgrazing, Liver, proteomics, Sheep

* Correspondence: jzz2006@126.com
[1]Key Laboratory of Forage Grass, Ministry of Agriculture, Institute of Grassland Research, Chinese Academy of Agricultural Sciences, Hohhot 010010, Inner Mongolia, China
Full list of author information is available at the end of the article

Background

The Inner Mongolian steppe is the most important region for mutton and milk production in China [1]. However, this natural grassland has been severely damaged by overgrazing in recent decades [2]. The degradation of the Inner Mongolian steppe due to overgrazing also damages the ecosystem and has extensively reduced animal productions. Increasing evidence shows that the quantity and quality of herbage and growth of sheep were substantially reduced due to overgrazing [3, 4]. Increasing grazing intensity elevated the odour source of volatile organic compounds in grassland plants and altered the morphological response of the host plants [5]. A previous study demonstrated that stocking rate rather than management system determined the ecological sustainability of pastoral livestock system [6]. Most importantly, the body weight gain was significantly decreased (up to 55%) due to overgrazing during the grazing season in multiple years studies [7, 8]. However, few studies have been conducted on the effects of overgrazing on the metabolic alterations related to sheep growth due to the research gap between animal nutritional metabolism and grazing [9, 10]. Furthermore, the molecular mechanism of the growth reduction induced by overgrazing in sheep is unknown.

The liver plays a central role in the regulation of the metabolism of carbohydrates, proteins, lipids and other nutrients in animals. Additionally, the liver has multiple other physiological functions in the body including immune response, regulating inflammation and the removal of xenobiotics [11, 12]. A previous gene array study on beef cows demonstrated marked hepatic responses to high or low grazing herbage allowances of native grasslands, including genes associated with glucogenesis, fatty acid oxidation, cell growth, DNA replication and transcription [13]. The proteomic profile of hepatic tissue in goats fed a high-grain diet demonstrated that an altered expression of hepatic proteins was related to amino acids metabolism [14]. To date, most studies on grazing animal have simply focused on body weight gain relating to intake based on the quality and quantity of herbage. However, few detailed studies have been conducted on the hepatic response to overgrazing in sheep.

Studies have shown lack of correlation between mRNA and protein expression abundance due to RNA editing and posttranslational modifications [15, 16]. Thus, the elucidation of protein expression is more imperative [17]. Previous research demonstrated that changes in animal growth performance are closely related to alterations in the protein expression in the hepatic tissue [18, 19]. A number of enzymes or functional proteins in the liver participate in physiological processes relevant to immunity, detoxification, nutrient metabolism, and others [20]. It is not practical to simultaneously measure all protein expressions of hepatic tissue using classical biotechnologies such as western blot, immunohistochemical staining or ELISA.

Thus, we hypothesized that overgrazing can confer negative effects on the hepatic protein expression. Therefore, the objective of this study was to identify and characterize candidate proteins that were differentially induced in the livers of sheep from an overgrazed pasture during the grazing season using a label-based iTRAQ procedure (isobaric tags for relative and absolute quantitation) followed by LC-MS/MS.

Results

Effect of overgrazing on herbage and animals

The herbage crude protein (CP) and acid detergent lignin (ADL) contents in the overgrazing (OG) group were 34.4 and 19.6% greater, respectively, than those in the light grazing (LG) group over the grazing period ($P = 0.003$; $P = 0.049$). However, the gross energy and nitrogen free extract (NFE) contents were significantly decreased in the OG group than the LG group (16.42 ± 0.37 kJ/g vs. 17.53 ± 0.12 kJ/g, $P = 0.008$; 40.2 ± 1.5 g/kg vs. 46.9 ± 4.2 g/kg, $P = 0.048$). There was no significant effect of overgrazing on neutral detergent fibre (NDF) or acid detergent fibre (ADF). However, NDF tended to decrease in the OG group ($P = 0.096$). A full description of the effects of overgrazing on herbage and animal growth performance are given in Additional file 1: Table S1 and Table 1, respectively.

In this study, all sheep had similar body weights (31.5 ± 4.5 kg) at the beginning of the grazing experiment. Throughout the experimental grazing period (total 90 d), the OG sheep had a 21.4% reduction in daily weight gain (27 g) ($P = 0.042$) and 14.0% reduction in carcass weight (2.8 kg) ($P = 0.047$) (Table 1). Additionally, two most important organ indices (calculated based on the weight of the spleen and liver) in the OG group were significantly lower than those in the LG group ($P = 0.004$; $P = 0.010$) (Table 1).

Table 1 Effect of overgrazing on the growth and immune organ indexes of sheep

	Groups		
	LG[c]	OG[d]	P value
Daily gain (g)	153 ± 13^a	126 ± 9^b	0.042
Carcass weight cold (kg)	22.8 ± 1.2^a	20.0 ± 1.2^b	0.047
Index of spleen (%)[c]	0.43 ± 0.02^a	0.34 ± 0.03^b	0.004
Index of liver (%)[c]	3.36 ± 0.11^a	2.98 ± 0.03^b	0.010

[a, b] Values within a row not sharing a common superscript letter indicate significant difference at $P < 0.05$. Numbers are means $\pm$ SD. (Daily gain: $n = 12$ for LG, $n = 48$ for OG; $n = 3$ for indices of immune organs). Immune organ indexes were calculated as the ration of organ weight to body weight
[c] LG light grazing
[d] OG overgrazing

Effect of overgrazing on biochemical parameters of serum

The effects of overgrazing on the biochemical parameters of serum in sheep are shown in Table 2. The alanine aminotransferase (ALT) and aspartate aminotransferase (AST) activities in serum were significantly increased in response to overgrazing compared with light grazing ($P = 0.038$; $P = 0.009$). Furthermore, the serum concentration of blood urea nitrogen (BUN) was significantly higher in the OG group than in the LG group ($P = 0.046$). In contrast, the levels of total protein (TP), glucose (GLU), non-esterified fatty acid (NEFA) and insulin-like growth factor 1 (IGF-1) in the serum were significantly decreased in response to overgrazing compared with light grazing ($P = 0.044$; $P = 0.017$; $P = 0.032$; $P = 0.018$). However, the levels of triglyceride (TG) and cholesterol (CHOL) were similar between the two groups ($P > 0.05$).

Identification and comparison of proteins of differential abundance

Using iTRAQ analysis, a total of 27,287 peptide spectral matches were found, and 2,153 proteins were identified within the FDR (false discover rate) of 1% (Additional file 2: Table S2). Following the statistical analysis, 45 proteins were found to be differentially expressed in hepatic tissue between the LG and OG groups, with 8 being up-regulated and 37 down-regulated (Additional file 3: Table S3).

Table 2 Effect of overgrazing on serum biochemical parameters of sheep

	Groups		
	LG[c]	OG[d]	P value
ALT (IU/L)[e]	20.73 ± 9.41[b]	45.97 ± 10.86[a]	0.038
AST (IU/L)[f]	164.93 ± 28.94[b]	386.37 ± 75.32[a]	0.009
TP (mmol/L)[g]	73.03 ± 4.05[a]	64.23 ± 3.32[b]	0.044
BUN (mmol/L)[h]	5.03 ± 0.42[b]	6.77 ± 0.96[a]	0.046
GLU (mmol/L)[i]	7.80 ± 0.72[a]	5.71 ± 0.58[b]	0.017
TG (mmol/L)[j]	0.36 ± 0.06	0.35 ± 0.07	0.835
CHOL (μmol/l)[k]	1.88 ± 0.84	1.44 ± 0.11	0.417
NEFA (mmol/L)[l]	0.61 ± 0.01[a]	0.44 ± 0.09[b]	0.032
IGF-1 (ng/mL)[m]	30.76 ± 4.01[a]	21.56 ± 4.18[b]	0.018

[a, b] Values within a column not sharing a common superscript letter indicate significant difference at $P < 0.05$. Numbers are means ± SD. ($n = 6$)
[c] LG light grazing
[d] OG overgrazing
[e] ALT alanine aminotransferase
[f] AST aspartate aminotransferase
[g] TP total protein
[h] BUN blood urea nitrogen
[i] GLU glucose
[j] TG triglyceride
[k] CHOL cholesterol
[l] NEFA non-esterified fatty acid
[m] IGF-1 insulin-like growth factor 1

A total of 41 proteins of differential abundance were grouped into nine classes based on putative functions: protein modification and metabolism (14.6%), transcriptional and translational regulation (14.6%), immune response, apoptosis and inflammation (14.6%), energy metabolism (12.2%), miscellaneous (12.2%), lipid metabolism (9.8%), stress response and detoxification (9.8%), cell cytoskeleton (7.3%) and cell growth and proliferation (4.9%) (Fig. 1). Those related to protein modification and metabolism, transcriptional and translational regulation, immune response, apoptosis and inflammation, and energy metabolism were predominant and accounted for approximately 55% of the differentially-expressed proteins. A comparison of proteins of differential abundance with functional grouping between the two grazing intensities indicated that more protein species were down- regulated in the overgrazing sheep (34 versus 7, respectively) (Table 3). Most importantly, the protein species that participated in energy metabolism, lipid metabolism, cell cytoskeleton, and cell growth and proliferation were found to be down-regulated in OG sheep in the present study.

GO annotations and pathway analysis

In the cellular component group, the differentially expressed proteins are concentrated in the cell part and membrane-bounded organelle (Fig. 2). In the molecular functional group, the differentially expressed proteins that were binding proteins (protein binding, ion binding and heterocyclic compound binding), metabolic enzymes (hydrolase activity, transferase activity and oxidoreductase activity) and structural molecules (structural constituents of ribosomes) were ranked at the top of the category occupancy, suggesting that their related functions were important in the livers of the sheep (Fig. 2). In the biological process category, the proteins that participate in cellular processes (macromolecule metabolism), single-organism process and metabolism (nitrogen compound metabolism, heterocycle metabolism and lipid metabolism) were at the highest ratios among the differentially expressed proteins (Fig. 2), suggesting that overgrazing primarily results in changes to nutrient metabolism.

Furthermore, GO annotation and a KEGG pathway enrichment analysis were performed to determine the over-represented biological events and to provide a primary overview of the hepatic proteome impacted by overgrazing. The DAVID 6.7 software identified highly overrepresented GO classes including cellular components, molecular functions and biological processes (Additional file 4: Table S4). For these identified differentially expressed proteins, the cellular component classifications were enriched in the cytoplasmic membrane-bounded vesicle, membrane-bounded vesicle, cytoplasmic vesicle and vesicle. According to the molecular function classifications, the differentially expressed proteins were enriched in vitamin binding,

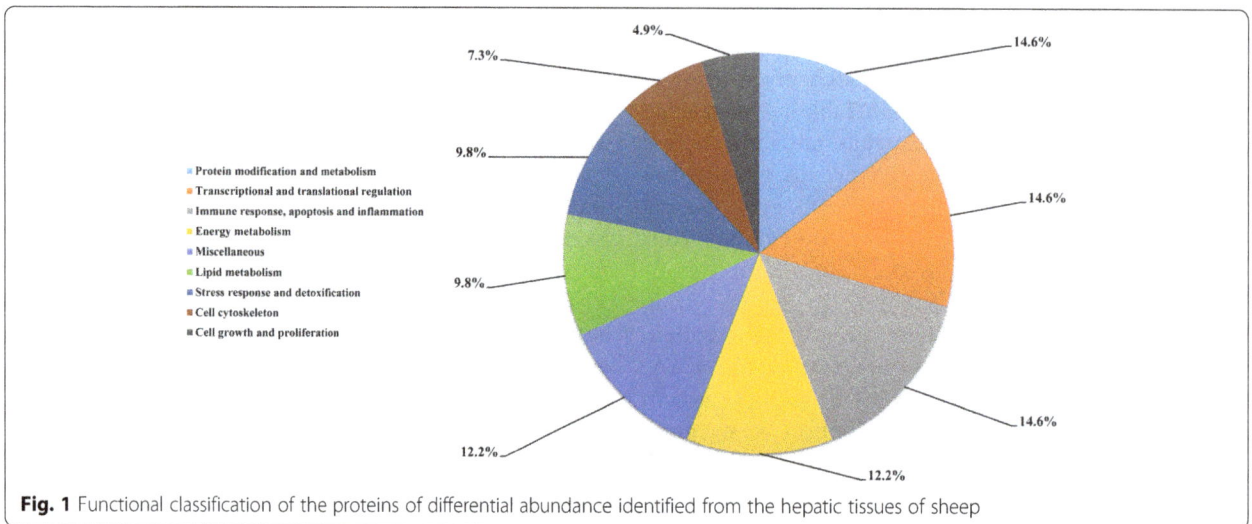

Fig. 1 Functional classification of the proteins of differential abundance identified from the hepatic tissues of sheep

cofactor binding, vitamin B6 binding and pyridoxal phosphate binding. Enriched biological process classifications of the differentially expressed proteins included the heterocycle catabolic process, response to unfolded proteins, the cellular amino acid catabolic process and response to protein stimulus, which implies that the protein metabolic process was greatly influenced in the livers of sheep in an overgrazing condition. A KEGG analysis showed that insulin signaling was significantly enriched ($P = 0.047$) in the identified pathways, which indicates that overgrazing had affected glucose and energy metabolism in the sheep.

Target proteomics validation of important proteins of differential abundance

Three differentially expressed proteins (KYNU involved in amino acid catabolism, FASN involved in fatty acids synthesis, and ARSD as disease marker) were selected for validation of proteomic data at the protein level using parallel reaction monitoring (PRM). The results of PRM analysis indicated that candidate proteins show similar trends as the iTRAQ results, which implied the credibility of the proteomics data (Fig. 3).

Effect of overgrazing on immune response and inflammatory indexes of serum

As shown in the list of differentially expressed protein species involved, those related to immune response and inflammation were ranked at one of the top putative protein functions. To further evaluate the effect of overgrazing on the immune response and inflammatory parameters of sheep, we measured the levels of primary immunoglobulins (Ig) and inflammatory cytokines in serum (Table 4). The levels of primary inflammatory cytokines, including interleukin-1β (IL-1β), interleukin-4 (IL-4) and interleukin-6 (IL-6), were greatly increased in the OG group compared with the LG group ($P = 0.004$; $P = 0.007$; $P = 0.001$). While,

the levels of IgA, IgG and interferon-γ (IFN-γ) were significantly lower in the OG group than in the LG group ($P = 0.050$; $P = 0.022$; $P = 0.036$).

Discussion

The liver is a vital organ that plays important roles in multiple physiological functions such as nutrient metabolism, detoxification, immune response and others. Any change in dietary components can be closely related to animal growth, which is reflected in variations of hepatic protein expressions in most cases. Whether overgrazing has a direct effect on the functional hepatocytes (thereby causing synthetic and metabolic changes in the sheep's liver) remains unclear. Substantially less is understood regarding the molecular mechanism of hepatic response in sheep with reduced nutrition due to overgrazing. In the present study, an iTRAQ-based quantitative proteomic analysis integrated with biochemical and immune analyses was applied to investigate the hepatic response in overgrazed sheep. When integrated with the proteomic profiles, biochemical and immune detections, these data suggest that overgrazing triggers a shift in energy resources from carbohydrates to proteins, which results in the impairment of nutrient metabolism (protein and lipid) and immunity, which may be the reasons for reduced growth in sheep.

Serum levels of ALT and AST serve as important indicators of hepatic health status and were found to be greatly elevated in OG sheep, indicating that overgrazing had a severely negative effect on hepatic function [21]. Additionally, a higher level of BUN in the serum of the OG group suggested that overgrazing was associated with poorer nitrogen utilization efficiency [22]. Furthermore, a decreased serum level of TP and an increased serum level of GLU in OG sheep during the period may have been caused by a shift in energy resources from carbohydrates to proteins

Table 3 List of differentially expressed proteins in hepatic samples from overgrazing group and light grazing group

Accession[a]	Description[b]	Gene symbol	Fold change	p-value	Biological process GO term
Energy metabolism					
W5NYL7	Uncharacterized protein OS = Ovis aries GN = MTHFD1L PE = 3 SV = 1 - [W5NYL7_SHEEP]	MTHFD1L	0.69	0.0403	ATP binding
Q5MIB6	Glycogen phosphorylase, brain form OS = Ovis aries GN = PYGB PE = 2 SV = 3 - [PYGB_SHEEP]	PYGB	0.82	0.0397	Carbohydrate metabolic process
W5PM30	Uncharacterized protein OS = Ovis aries GN = CREB3L3 PE = 4 SV = 1 - [W5PM30_SHEEP]	CREB3L3	0.72	0.0223	Maintaining systemic energy homeostasis
W5NUJ4	Uncharacterized protein OS = Ovis aries GN = NEK9 PE = 4 SV = 1 - [W5NUJ4_SHEEP]	NEK9	0.82	0.0185	ATP binding
B9VXW5	Mammalian target of rapamycin OS = Ovis aries PE = 2 SV = 1 - [B9VXW5_SHEEP]	mTOR	0.83	0.0051	ATP binding
Protein modification and metabolism					
W5PJN6	Kynureninase OS = Ovis aries GN = KYNU PE = 3 SV = 1 - [W5PJN6_SHEEP]	KYNU	1.21	0.0148	Tryptophan catabolic process
W5Q678	Uncharacterized protein OS = Ovis aries GN = SEC24D PE = 4 SV = 1 - [W5Q678_SHEEP]	SEC24D	0.81	0.0369	Intracellular protein transport
W5QHW3	Uncharacterized protein OS = Ovis aries GN = THNSL2 PE = 4 SV = 1 - [W5QHW3_SHEEP]	THNSL2	0.60	0.0349	Serine binding
W5PFC9	Uncharacterized protein (Fragment) OS = Ovis aries GN = LOC101117129 PE = 4 SV = 1 - [W5PFC9_SHEEP]	LOC101117129	0.78	0.0099	Enzyme inhibitor activity
W5QF53	Uncharacterized protein OS = Ovis aries GN = ZFAND2B PE = 4 SV = 1 - [W5QF53_SHEEP]	ZFAND2B	0.67	0.0095	Maintain cellular folding capacity
W5PRG9	Histidine ammonia-lyase OS = Ovis aries GN = HAL PE = 3 SV = 1 - [W5PRG9_SHEEP]	HAL	0.83	0.0062	Elimination of ammonia from the substrate
Accession[a]	Description[b]	Gene symbol	Fold change	p-value	Biological process GO term
Lipid metabolism					
W5PMR6	Uncharacterized protein OS = Ovis aries GN = DERL1 PE = 4 SV = 1 - [W5PMR6_SHEEP]	DERL1	0.73	0.0362	ApoB secretion
W5P8F9	Uncharacterized protein OS = Ovis aries PE = 4 SV = 1 - [W5P8F9_SHEEP]	None	0.74	0.0203	Lipid binding
W5PKK8	Uncharacterized protein OS = Ovis aries GN = ESYT1 PE = 4 SV = 1 - [W5PKK8_SHEEP]	ESYT1	0.81	0.0344	Lipid binding
W5Q6U0	Uncharacterized protein OS = Ovis aries GN = FASN PE = 4 SV = 1 - [W5Q6U0_SHEEP]	FASN	0.79	0.0147	De novo synthesis of fatty acids
Transcriptional and translational regulation					
W5P328	Uncharacterized protein OS = Ovis aries GN = EIF2A PE = 4 SV = 1 - [W5P328_SHEEP]	EIF2A	1.26	0.0487	Regulation of translation
W5P2A1	Uncharacterized protein OS = Ovis aries PE = 3 SV = 1 - [W5P2A1_SHEEP]	None	0.81	0.0482	Translational elongation
W5PLU3	Uncharacterized protein OS = Ovis aries GN = ZNF207 PE = 4 SV = 1 - [W5PLU3_SHEEP]	ZNF207	0.68	0.0389	Transcription regulation and chromatin organization
W5PHI1	Uncharacterized protein (Fragment) OS = Ovis aries GN = MRPL3 PE = 3 SV = 1 - [W5PHI1_SHEEP]	MRPL3	0.59	0.0325	Structural constituent of ribosome
W5PTA9	Uncharacterized protein OS = Ovis aries GN = DCPS PE = 4 SV = 1 - [W5PTA9_SHEEP]	DCPS	0.69	0.0039	Regulation of RNA stability
B0FZM0	Ribosomal protein L14-like protein (Fragment) OS = Ovis aries PE = 2 SV = 1 - [B0FZM0_SHEEP]	None	0.75	0.0028	Structural constituent of ribosome
Accession[a]	Description[b]	Gene symbol	Fold change	p-value	Biological process GO term
Immune response, apoptosis and inflammation					
W5QAE0		TMBIM6	1.31	0.0247	

Table 3 List of differentially expressed proteins in hepatic samples from overgrazing group and light grazing group *(Continued)*

Accession[a]	Description[b]	Gene symbol	Fold change	p-value	Biological process GO term
	Uncharacterized protein OS = Ovis aries GN = TMBIM6 PE = 3 SV = 1 - [W5QAE0_SHEEP]				Intrinsic apoptotic signaling pathway
W5PKK4	Uncharacterized protein OS = Ovis aries GN = CCAR2 PE = 4 SV = 1 - [W5PKK4_SHEEP]	CCAR2	1.51	0.0402	Positive regulation of apoptotic process
W5PUV3	Uncharacterized protein OS = Ovis aries GN = NT5E PE = 3 SV = 1 - [W5PUV3_SHEEP]	NT5E	0.83	0.0387	Marker of lymphocyte differentiation
W5P8T5	Uncharacterized protein OS = Ovis aries GN = PPM1B PE = 3 SV = 1 - [W5P8T5_SHEEP]	PPM1B	0.73	0.0358	Protein serine/threonine phosphatase activity
W5NZ57	Proteasome subunit beta type OS = Ovis aries GN = PSMB10 PE = 3 SV = 1 - [W5NZ57_SHEEP]	PSMB10	0.78	0.0344	T cell proliferation
W5Q0Z8	Uncharacterized protein OS = Ovis aries GN = TRIM56 PE = 4 SV = 1 - [W5Q0Z8_SHEEP]	TRIM56	0.29	0.0011	Regulator of host innate immunity
Stress response and detoxification					
W5PGA7	UDP-glucuronosyltransferase OS = Ovis aries GN = UGT2B7 PE = 3 SV = 1 - [W5PGA7_SHEEP]	UGT2B7	0.79	0.0376	Elimination of potentially toxic xenobiotics and endogenous compounds
W5PWA2	Uncharacterized protein OS = Ovis aries PE = 3 SV = 1 - [W5PWA2_SHEEP]	None	1.24	0.0140	Oxidoreductase activity
W5PKE4	Uncharacterized protein OS = Ovis aries PE = 4 SV = 1 - [W5PKE4_SHEEP]	None	0.56	0.0024	Oxidation reduction process
W5P214	Transgelin OS = Ovis aries GN = TAGLN PE = 3 SV = 1 - [W5P214_SHEEP]	TAGLN	0.82	0.0006	Stress response related
Cell growth and proliferation					
W5Q8Q6	Uncharacterized protein OS = Ovis aries GN = KANK2 PE = 4 SV = 1 - [W5Q8Q6_SHEEP]	KANK2	0.81	0.0474	Promotion of cell proliferation
W5P0W0	Uncharacterized protein (Fragment) OS = Ovis aries GN = IST1 PE = 4 SV = 1 - [W5P0W0_SHEEP]	IST1	0.72	0.0015	Cytokinesis
Accession[a]	Description[b]	Gene symbol	Fold change	p-value	Biological process GO term
Cell cytoskeleton					
W5PK38	Uncharacterized protein (Fragment) OS = Ovis aries GN = VASP PE = 4 SV = 1 - [W5PK38_SHEEP]	VASP	0.76	0.0135	Actin cytoskeleton organization
W5Q5K3	Uncharacterized protein (Fragment) OS = Ovis aries GN = NCKAP1 PE = 4 SV = 1 - [W5Q5K3_SHEEP]	NCKAP1	0.69	0.0022	Regulation of actin cytoskeleton
W5Q8P7	Uncharacterized protein (Fragment) OS = Ovis aries GN = SNTB1 PE = 4 SV = 1 - [W5Q8P7_SHEEP]	SNTB1	0.59	0.0004	Structural molecule activity
Miscellaneous					
W5PEV3	Uncharacterized protein OS = Ovis aries GN = ARSD PE = 4 SV = 1 - [W5PEV3_SHEEP]	ARSD	1.24	0.0138	Disease marker
W5NRU6	Uncharacterized protein OS = Ovis aries GN = MYADM PE = 4 SV = 1 - [W5NRU6_SHEEP]	MYADM	1.36	0.0102	Disease marker
W5QGB6	Uncharacterized protein OS = Ovis aries GN = GCHFR PE = 4 SV = 1 - [W5QGB6_SHEEP]	GCHFR	0.79	0.0386	Negative regulation of biosynthetic process
W5P2J9	Uncharacterized protein OS = Ovis aries PE = 3 SV = 1 - [W5P2J9_SHEEP]	RUSC1	0.77	0.0342	Transferase activity
W5PMB1	Uncharacterized protein OS = Ovis aries GN = SNX3 PE = 4 SV = 1 - [W5PMB1_SHEEP]	SNX3	0.82	0.0197	Iron homeostasis

[a] Uniprot_Ovis aries _27110_20151123 database accession number
[b] The name of the protein exclusive of the identifier that appears in the database

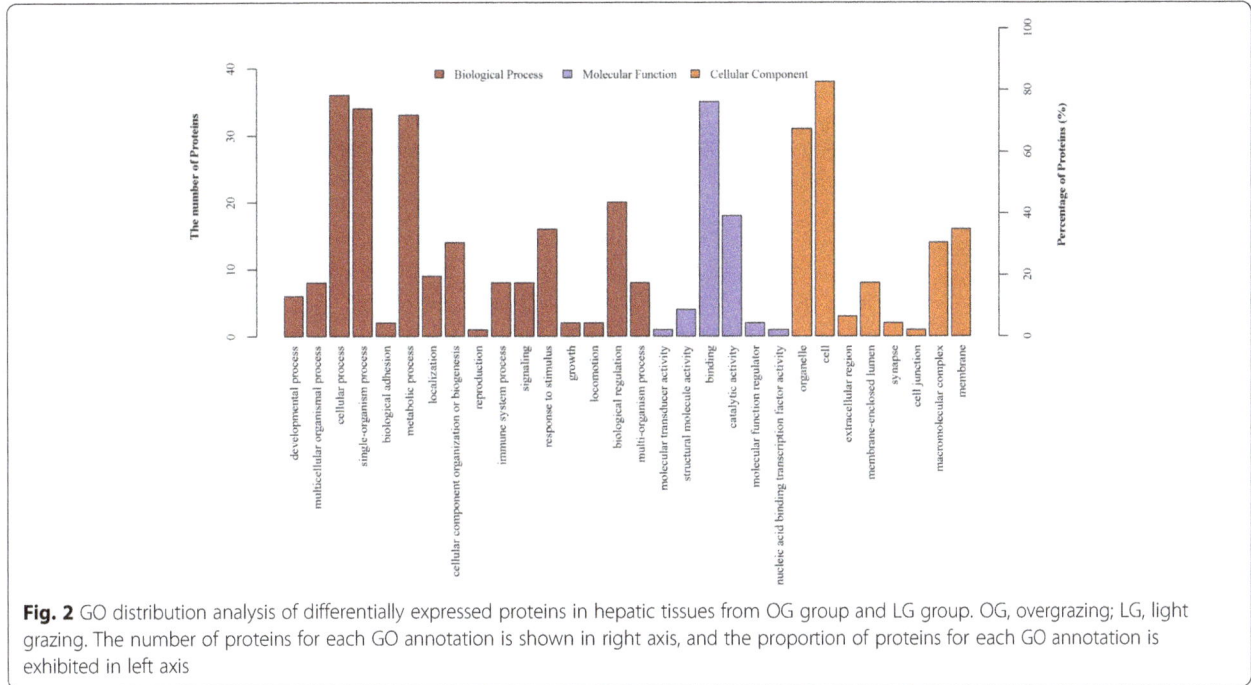

Fig. 2 GO distribution analysis of differentially expressed proteins in hepatic tissues from OG group and LG group. OG, overgrazing; LG, light grazing. The number of proteins for each GO annotation is shown in right axis, and the proportion of proteins for each GO annotation is exhibited in left axis

due to the imbalance in protein and fermentable carbohydrates in the herbage [20]. Moreover, lower levels of NEFA and IGF-1 were observed in the OG group, which indicated that fatty acid biosynthesis and anabolic effects in sheep were reduced due to overgrazing [20]. Both above nutrients are very closely related to animal growth.

The liver is an essential nutrient metabolism organ in which, nitrogen compounds metabolism takes a central role. Nitrogen compounds metabolism, including protein metabolism, amino acid metabolism, ammonia toxic elimination and others, can thus be an important indicator of hepatocyte health. [23]. In this study, nearly all six differentially expressed protein species related to protein

modification and metabolism were decreased in OG sheep with the exception of KYNU (kynureninase). Among these proteins, KYNU is an enzyme within the tryptophan metabolism pathway [24]. A previous study demonstrated that tryptophan was the third most limiting amino acid in growing lambs, and its inadequate supply or increased catabolism *in vivo* can trigger limited protein deposition and the elevation of urinary N excretion [25]. The up-regulation of KYNU was observed in the livers of OG sheep and is consistent with the higher serum level of BUN in the present study. HAL (histidine ammonia-lyase) catalyzes the elimination of ammonia from the substrate to form (*E*)-urocanate

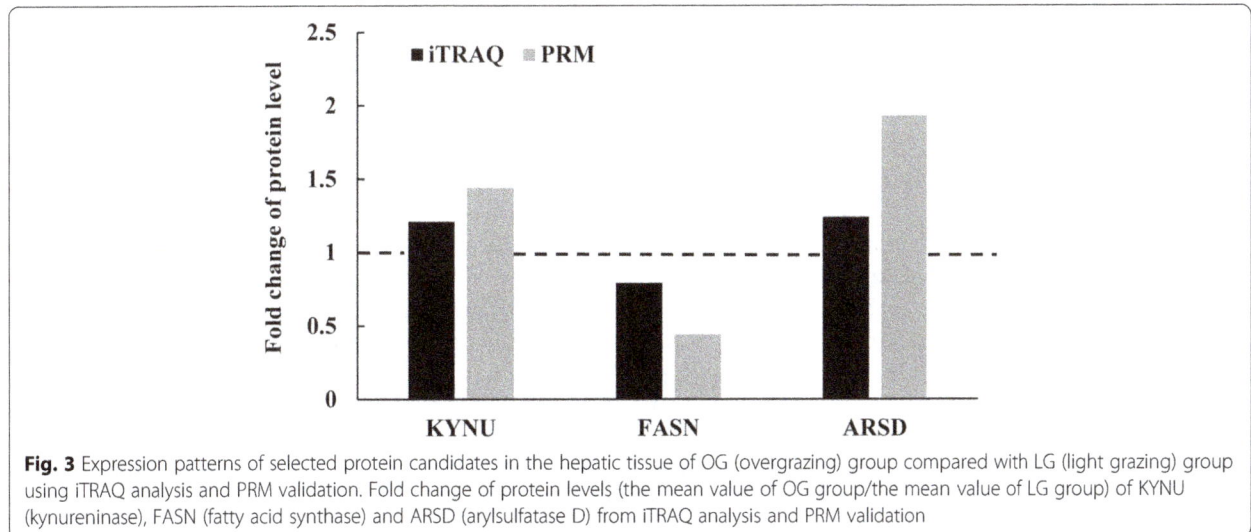

Fig. 3 Expression patterns of selected protein candidates in the hepatic tissue of OG (overgrazing) group compared with LG (light grazing) group using iTRAQ analysis and PRM validation. Fold change of protein levels (the mean value of OG group/the mean value of LG group) of KYNU (kynureninase), FASN (fatty acid synthase) and ARSD (arylsulfatase D) from iTRAQ analysis and PRM validation

Table 4 Effect of overgrazing on the immune responses and inflammatory cytokines of sheep

	Groups		
	LG[c]	OG[d]	P value
IL-1β (pg/mL)[e]	10.34 ± 0.72[b]	22.90 ± 3.68[a]	0.004
IL-4 (pg/mL)[f]	21.38 ± 2.19[b]	29.06 ± 1.43[a]	0.007
IL-6 (pg/mL)[g]	90.61 ± 22.03[b]	222.50 ± 10.81[a]	0.001
IFN-γ (pg/mL)[h]	30.76 ± 4.01[a]	21.56 ± 4.18[b]	0.050
IgA (g/L)[i]	0.73 ± 0.06[a]	0.54 ± 0.06[b]	0.022
IgG (g/L)[j]	19.85 ± 1.95[a]	15.19 ± 1.73[b]	0.036

[a, b] Values within a column not sharing a common superscript letter indicate significant difference at $P < 0.05$. Numbers are means ± SD. ($n = 6$)
[c] LG light grazing
[d] OG overgrazing
[e] IL-1β interleukin-1β
[f] IL-4 interleukin-4
[g] IL-6 interleukin-6
[h] IFN-γ interferon-γ
[i] IgA immunoglobulin A
[j] IgG immunoglobulin G

[26]. The down-regulation of this protein indicates that overgrazing may result in perturbations in ammonia detoxification in hepatocytes, which can be another reason for the increased concentration of BUN in the OG group. Taken together, the expression changes of KYNU and HAL integrated with the results of serum levels of BUN and GLU indicate that overgrazing triggers a shift in energy resources from carbohydrates to proteins causing poorer nitrogen utilization efficiency. A GO annotation enrichment analysis done in the present study also showed that cellular amino acid catabolic process is overrepresented in the hepatic proteome of sheep due to overgrazing. The down-regulation of the other four proteins involved in protein metabolism, including SEC24D (SEC24 homologue D, COPII coat complex component), THNSL2 (threonine Synthase-Like 2), LOC101117129 (UniProt database accession W5PFC9) and ZFAND2B (Zinc finger, AN1-Type Domain 2B), may reflect reduced protein transport, binding and folding capacity [27–29].

Energy production is one of the key functions of the liver, which participates in a number of physiological processes. However, energy production and mitochondrial function are usually found to be impaired in hepatic dysfunction [30]. PYGB (glycogen phosphorylase, brain form) serves as a glucose metabolism protein and contributes to the regulation of carbohydrate metabolism [31]. CREB3L3 (cAMP-responsive element-binding protein 3-like 3) maintains systemic energy homeostasis throughout the entire body [32]. mTOR (mammalian target of rapamycin) is a central signaling molecule that impacts most cellular functions including energy metabolism promotion [33]. MTHFD1L (methylenetetrahydrofolate dehydrogenase (NADP+ dependent) 1-like) and NEK9 (NIMA-related kinase 9) are ATP binding

proteins that are involved in mitochondrial function and DNA replication, respectively [34, 35]. These proteins were both down-regulated in OG sheep, suggesting that overgrazing interferes with energy production and metabolism; this is consistent with the decreased serum GLU level in OG sheep in the present study.

The down-regulated proteins DERL1, a lipid binding protein (UniProt database accession W5P8F9), ESYT1 (extended synaptotagmin-like protein 1) and FASN (fatty acid synthase) are classified as lipid metabolism proteins based on their primary function. Of these proteins, DERL1 is a putative dislocon component in the ER (endoplasmic reticulum) membrane that plays an important role in ApoB secretion [36]; FASN stimulates the de novo synthesis of fatty acids [37]. The down-regulation of these proteins implies decreased lipid metabolism in the OG group and is consistent with the finding in this study of a lower serum level of NEFA observed in OG sheep.

The liver is an important immune organ in the body that plays indispensable roles in immune response, apoptosis and inflammatory reactions [38]. In this study, two protein species related to apoptosis were up-regulated in the OG group. TMBIM6 (transmembrane BAX inhibitor motif containing 6) promotes apoptosis in prolonged stress or severe conditions [39]; CCAR2 (cell cycle and apoptosis regulator 2) leads to increased level of p53-mediated apoptosis and DNA damage [40]. The up-regulation of TMBIM6 and CCAR2 indicates that overgrazing may lead to a long-term stressful situation and trigger apoptosis in sheep hepatocytes. Other proteins involved in the immune response, including NT5E (5'-nucleotidase, ecto), PPM1B (protein phosphatase, Mg^{2+}/Mn^{2+} dependent, 1B), PSMB10 (proteasome subunit beta 1) and TRIM56 (tripartite motif containing 56), are down-regulated, which suggests that the sheep immunity is suppressed during overgrazing, thus increasing the likelihood of bacterial or viral infection and reduced growth performance [41, 42]. These observations were consistent with the reduced immune organ indexes and serum levels of immunoglobulins and the increased concentrations of inflammatory cytokines in OG sheep. Moreover, the reduced abundance of proteins relevant to stress response and detoxification, including UGT2B7 (UDP-glucuronosyltransferase), an oxidoreductase (UniProt database accession W5PKE4), and TAGLN (transgelin), were observed in the OG group, which indicates that long-term overgrazing may lead to oxidative stress and that it inhibits detoxification in the sheep liver [43, 44].

Cytoskeletal proteins play a crucial role in liver protection and maintaining both the cellular structure and integrity of hepatocytes [45]. In this study, three differential protein species related to the cytoskeleton were down-regulated in the livers of OG sheep. VASP (vasodilator-stimulated phosphoprotein) is associated with filamentous actin formation and plays a widespread role in cell adhesion and motility [46]. NCKAP1 (NCK-associated protein 1) is an integral membrane protein that regulates actin cytoskeleton organization [47]. SNTB1 (beta-1-syntrophin) is a peripheral membrane protein that is associated with mediating high-density lipoprotein (HDL) metabolism in the liver [48]. This is consistent with the elevated AST and ALT serum levels in the OG group of this study, which may interfere with liver protection and normal hepatocytes structure. Other proteins relevant to cell growth and proliferation, including KANK2 (KN motif and ankyrin repeat domain 2) and IST1 (increased sodium tolerance 1), are also down-regulated and may harm the regeneration of hepatocytes in OG sheep due to overgrazing [49, 50]. Furthermore, the reduced abundance of proteins involved in transcriptional and translational regulation, including a ribosomal protein (UniProt database accession W5P2A1), ZNF207 (zinc finger protein 207), MRPL3 (mitochondrial ribosomal protein L3), DCPS (decapping enzyme, scavenger) and ribosomal protein L14-like protein, was observed in the OG group, which indicates a decreased capacity for protein metabolism to provide sufficient nutritional nitrogen compounds for growth [51–53].

Conclusions

In summary, the present proteomic analysis demonstrated that overgrazing leads to differential abundances of a number of hepatic proteins in sheep. The functional groupings of those altered proteins are primarily related to protein metabolism, transcriptional and translational regulation, and immune response. Some of the other proteins are involved in nutrient metabolism (energy and lipid metabolism), stress response, and cellular functions (cell cytoskeleton, cell growth and proliferation). Additionally, biochemical and immune analyses provided sufficient physiological evidence. All results obtained from the present study suggest that overgrazing induces a shift in energy resources from carbohydrates to proteins, causing the impairment of nutrient metabolism (protein and lipid) and immunity, which may be the reasons for the reduced growth in sheep. Future studies will investigate the application of nutritional supplementation to improve the growth performance of sheep in overgrazing conditions.

Methods

Study area and experimental design

This study was conducted by the Institute of Grassland Research, Chinese Academy of Agricultural Science, Hohhot in the Xilin River Basin, Inner Mongolia Autonomous Region, China (116°32' E, 44°15' N). There was 8 ha of an experimental site with pasture dominated by three grass species: *Leymus chinensis*, *Stipa krylovii* and S. grandis.

A total of 60 Uzhumchin wethers were used in the present study. The sheep were born in summer 2013 and at the initiation of the grazing experiment in June were approximately 24 months old with an average live weight of 31.5 ± 4.5 kg. Water and minerals in lick stones were provided *ad libitum* during the grazing experiment.

The duration of the grazing experiment was 90 days from June 10th 2015 to September 5th 2015. There were a total of 6 plots (1.33 ha per plot) in the experimental site (3 plots per each grazing intensity), comprising a light grazing group (4 sheep per plot) and an overgrazing group (16 sheep per plot). Therefore, two different grazing intensities were realized: 3.0 (light grazing, LG) and 12.0 sheep/ha (overgrazing, OG).

Herbage samples were obtained monthly and combined at the end of the grazing experiment for chemical composition analysis. The herbage sample collection method was followed a previous study [6]. All details are described in Additional file 5. The content of CP, gross energy, NFE, NDF, ADF and ADL were determined in the herbage sample following the protocol of a previous study [7].

Data and sample collection

The live weight of all animals was measured at day 0 and day 90 (end of experiment, after 12 h of fasting) of the grazing season, and the mean daily gain was calculated. At the termination of the experiment, 2 sheep per plot in both groups ($n = 6$) were randomly chosen for blood collection. Each blood sample was collected from the cervical vein using a sterilized syringe. Sera samples were obtained via centrifugation at 2000 g for 30 min at 4 °C, then at 400 g for 10 min at 4 °C, and all sera samples were stored at –80 °C for further analysis. After blood collection, one sheep per plot ($n = 3$) was slaughtered using standard procedures established by the Chinese Academy of Agricultural Sciences. To calculate the indices of immune organs, the spleen and liver were excised and weighted, respectively. Immune organ indices were calculated as the ratio of organ weight to body weight. Hepatic samples (~2 g) were then washed with ice cold sterilized PBS, frozen in liquid nitrogen, and stored at –80 °C for further proteomic analysis.

Biochemical, immune response and inflammation analyses

For biochemical, immune response and inflammatory parameters of the serum, the concentration of ALT, AST, TP, BUN, GLU, TG, CHOL, NEFA, IgA and IgG were measured using a fully automatic biochemistry analyser (Hitachi 7020, Tokyo, Japan); IL-1β, IL-4, IL-6, IGF-1 and IFN-γ were determined using a corresponding diagnostic kit (Nanjing Jiancheng Bioengineering Institute, Nanjing, China) according to the instructions of the manufacturer.

Hepatic sample preparation and protein extraction

A total of six hepatic samples (one sheep per plot, three biological replications per group) were collected for protein extraction and subsequent proteomic analysis. Each hepatic sample (~0.5 g) was ground after being frozen in liquid nitrogen in a Dounce glass grinder. The grinded powder was precipitated with 10% trichloroacetic acid (TCA) (w/v) and 90% ice-cold acetone at –20 °C for 2 h. The precipitate in the sample was obtained via centrifugation at 20000 g for 30 min at 4 °C and subsequently washed with ice-cold acetone. The precipitate was then lysed in the lysis buffer [8 M urea, 30 mM 4-(2-hydroxyethyl)-1-piperazineethanesulfonic acid (HEPES), 1 mM phenylmethanesulfonyl fluoride (PMSF), 2 mM ethylene diamine tetraacetic acid (EDTA), and 10 mM dithiothreitol (DTT)]. The crude tissue extracts were centrifuged to remove the remaining debris. The tissue lysates were reduced for 1 h at 37 °C in a water bath through the addition of 10 mM DTT and then alkylated with for 1 h with the addition of 55 mM iodoacetamide in the dark. Afterwards, the lysates were precipitated by adding 4 volumes of pre-chilled acetone. The pellets were then washed three times with pre-chilled pure acetone and resuspended in the buffer (50% TEAB and 0.1% SDS). The centrifugation was repeated to remove the undissolved pellets. Subsequently, protein quantitation was performed using a Bio-Rad Bradford Protein Assay Kit (Hercules, CA, USA).

iTRAQ experiments

Each sample was digested with modified sequence grade trypsin (Promega Corporation, Madison, WI) at a 1:30 ratio (3.3 µg trypsin: 100 µg target) overnight at 37 °C. Each isobaric tag (114, 115, 116, 117, 118 and 119) (AB Sciex, Framingham, MA, USA) was solubilized in 70 µL isopropanol and then added to its respective sample (three biological replications per group). Incubation continued for 2 h at room temperature. All labelled peptides were pooled together and separated using SCX chromatography. The eluted peptides were dried under a vacuum and analyzed via LC-MS/MS based on Q-Exactive (Thermo Scientific). All detailed procedures are described in Additional file 5.

MS data processing and bioinformatics analysis

Peptide and protein identifications were analyzed using Proteome Discover 1.4 (Thermo Fisher Scientific) and searched with the Mascot search engine (Matrix Science, London, U.K.; version 2.3.0) against the database Uniprot_Ovis aries _27110_20151123.fasta (Nov 23rd, 2015 with 27,110 protein sequences) with the following parameters: enzyme: trypsin; variable modifications: oxidation (M), gln-pyro-glu (N-term Q), and iTRAQ8plex (Y); fixed modification: carbamidomethyl (C), iTRAQ8plex (N-term), and iTRAQ8plex (K); peptide mass tolerance: ± 20 ppm; fragment mass tolerance: 0.1 Da; maximum missed cleavages: 2. Identified peptides had an ion score above the threshold of peptide identity in Mascot, and protein identifications were accepted as the false discovery rate (FDR) ≤ 0.01 in which at least one such unique peptide match was specific for the protein. Median ratio normalization was performed to obtain the quantitative protein ratios. Proteins with a 1.2-fold change or greater (P-values < 0.05) were considered significant.

The Gene Ontology (GO) distribution for all proteins that were differentially expressed in the hepatic tissue of overgrazed sheep was classified using Blast2GO software (http://www.blast2go.com/). The Database for Annotation, Visualization and Integrated Discovery (DAVID) 6.7 (http://david.abcc.ncifcrf.gov/) and the Kyoto encyclopedia of genes and genomes (KEGG) data base (http://www.genome.jp/kegg/), were used to classify differentially altered proteins in significantly overrepresented pathways and GO terms.

Target analysis by PRM

Three representing proteins of interest were selected to perform the targeted quantification and verification among all proteins of differential abundance under LG and OG conditions from sheep livers. Samples were analyzed using PRM (Additional file 6: Table S5 for settings), which is applied for proteomic data verification in a number of studies [54, 55]. Details of this method are provided in the online Additional file 5.

Statistical analysis

Statistical analyses were performed with SPSS Statistics 17.0 (SPSS, Inc., Chicago, IL, USA). All results are expressed as the means ± SD. A group difference was analyzed using Student's t test, and P < 0.05 was considered significant.

Additional files

Additional file 1: Table S1. Effect of overgrazing on primary nutritional indexes of herbage. (DOCX 14 kb)

Additional file 2: Table S2. List of all proteins (n = 2153) identified in the study. (XLSX 148 kb)

Additional file 3: Table S3. List of all differentially expressed proteins (n = 45) identified in the study. (XLSX 11 kb)

Additional file 4: Table S4. Enrichment analysis of differentially expressed proteins in the hepatic tissue of sheep. (DOCX 15 kb)

Additional file 5: The detailed description of the experiment methods, including herbage sample collection, SCX chromatography, mass spectrometry, targeted protein quantitation. (DOCX 15 kb)

Additional file 6: Table S5. Parallel reaction monitoring conditions for targeted natural abundance peptides and isotopically labeled peptide standards. (DOCX 14 kb)

Abbreviations

ADF: Acid detergent fibre; ADL: Acid detergent lignin; ALT: Alanine aminotransferase; AST: Aspartate aminotransferase; BUN: Blood urea nitrogen;

CCAR2: Cell cycle and apoptosis regulator 2; CHOL: Cholesterol; CP: Crude protein; CREB3L3: cAMP-responsive element-binding protein 3-like 3; ESYT1: Extended synaptotagmin-like protein 1; FASN: Fatty acid synthase; GLU: Glucose; HAL: Histidine ammonia-lyase; IGF-1: Insulin-like growth factor 1; iTRAQ: Isobaric tags for relative and absolute quantitation; KYNU: Kynureninase; mTOR: Mammalian target of rapamycin; NDF: Neutral detergent fibre; NEFA: Non-esterified fatty acid; NEK9: NIMA-related kinase 9; NFE: Nitrogen free extract; PRM: Parallel reaction monitoring; PYGB: Glycogen phosphorylase; TG: Triglyceride; TMBIM6: Transmembrane BAX inhibitor motif containing 6; TP: Total protein

Acknowledgements
This research was funded by the Chinese National Key Basic Research and Development Program (2014CB138804) and the Chinese National Key technology research and development program (2012BAD13B07). Shanghai Applied Protein Technology provided the technical support in iTRAQ proteomic analysis and target protein verification.

Funding
This research was supported by the Chinese National Key Basic Research and Development Program (No: 2014CB138804) and the Chinese National Key technology research and development program (No: 2012BAD13B07).

Authors' contributions
WR, XH and JZ developed and framed the research questions. WR and JZ analyzed proteomics data. XL and YD were involved in the analysis of serum biochemical, immune and inflammatory data. YW, ZW, NH, LK, CC and CJ were involved in sample collection. WB and HG were involved in revising the manuscript. WR and JZ drafted the manuscript. All authors read and approved the final manuscript.

Competing interests
The authors declare that they have no competing interests.

Author details
[1]Key Laboratory of Forage Grass, Ministry of Agriculture, Institute of Grassland Research, Chinese Academy of Agricultural Sciences, Hohhot 010010, Inner Mongolia, China. [2]NSW Department of Primary Industries, Orange Agricultural Institute, Orange, NSW 2800, Australia. [3]College of Life Sciences, Inner Mongolia Agricultural University, Hohhot 010019, Inner Mongolia, China.

References
1. Lin L, Dickhoefer U, Müller K, Wang C, Glindemann T, Hao J, et al. Growth of sheep as affected by grazing system and grazing intensity in the steppe of Inner Mongolia, China. Livest Sci. 2012;144(1–2):140–7.
2. Zuo Z, Zhang M, Gao J, Wen S, Hou P, Gao Y. Allelopathic effects of Artemisia frigida Willd. on growth of pasture grasses in Inner Mongolia, China. Biochem Syst Ecol. 2011;39(4–6):377–83.
3. Lin L, Dickhoefer U, Müller K, Susenbeth A. Grazing behavior of sheep at different stocking rates in the Inner Mongolian steppe, China. Appl Anim Behav Sci. 2011;129(1):36–42.
4. Wang CJ, Tas BM, Glindemann T, Mueller K, Schiborra A, Schoenbach P, et al. Rotational and continuous grazing of sheep in the inner Mongolian steppe of China. J Anim Physiol Anim Nutr. 2009;93(2):245–52.
5. Zhang R, Zhang W, Zuo Z, Li R, Wu J, Gao Y. Inhibition effects of volatile organic compounds from Artemisia frigida Willd. on the pasture grass intake by lambs. Small Ruminant Res. 2014;121(2–3):248–54.
6. Hao J, Dickhoefer U, Lin L, Müller K, Glindemann T, Schönbach P, et al. Effects of rotational and continuous grazing on herbage quality, feed intake and performance of sheep on a semi-arid grassland steppe. Arch Anim Nutr. 2013;67(1):62–76.
7. Glindemanna T, Wang C, Tas B, Schiborrab A, Gierusb M, Taubeb F, et al. Impact of grazing intensity on herbage intake, composition, and digestibility and on live weight gain of sheep on the inner Mongolian steppe. Livest Sci. 2009;124(1–3):142–7.
8. Dickhoefer U, Bösing BM, Hasler M, Hao J, Lin L, Müller K, et al. Animal responses to herbage allowance: forage intake and body weight gain of sheep grazing the inner Mongolian steppe - results of a six-year study. J Anim Sci. 2016;94(5):2059–71.
9. Garcia F, Carrère P, Soussana F, Baumont R. The ability of sheep at different stocking rates to maintain the quality and quantity of their diet during the grazing season. J Agr Sci. 2003;140(1):113–24.
10. Lachica M, Aguilera F. Energy expenditure of walk in grassland for small ruminants. Small Ruminant Res. 2005;59(2–3):105–21.
11. Nemeth E, Baird W, O'Farrelly C. Microanatomy of the liver immune system. Semin Immunopathol. 2009;31(3):333–43.
12. Trauner M, Meier PJ, Boyer JL. Molecular pathogenesis of cholestasis. N Engl J Med. 1998;339(17):1217–27.
13. Laporta J, Rosa GJ, Naya H, Carriquiry M. Liver functional genomics in beef cows on grazing systems: novel genes and pathways revealed. Physiol Genomics. 2014;46(4):138–47.
14. Jiang X, Zeng T, Zhang S, Zhang Y. Comparative proteomic and bioinformatic analysis of the effects of a high-grain diet on the hepatic metabolism in lactating dairy goats. PLoS One. 2013;8(11):e80698.
15. Astle J, Ferguson JT, German JB, Harrigan GG, Kelleher NL, Kodadek T, et al. Characterization of proteomic and metabolomic responses to dietary factors and supplements. J Nutr. 2007;137(12):2787–93.
16. Zhang J, Li D. Effect of conjugated linoleic acid on inhibition of prolyl hydroxylase 1 in hearts of mice. Lipids Health Dis. 2012;11:22.
17. Wang J, Li F, Dangott J, Wu Y. Proteomics and its role in nutrition research. J Nutr. 2006;136(7):1759–62.
18. Sherlock L, McKeegan DE, Cheng Z, Wathes CM, Wathes DC. Effects of contact dermatitis on hepatic gene expression in broilers. Br Poult Sci. 2012;53(4):439–52.
19. Zhang J, Li C, Tang X, Lu Q, Sa R, Zhang H. High concentrations of atmospheric ammonia induce alterations in the hepatic proteome of broilers (Gallus gallus): an iTRAQ-based quantitative proteomic analysis. PLoS One. 2015;10(4):e0123596.
20. Kaufmann LD, Dohme-Meier F, Münger A, Bruckmaier RM, van Dorland HA. Metabolism of grazed vs. zero-grazed dairy cows throughout the vegetation period: hepatic and blood plasma parameters. J Anim Physiol Anim Nutr. 2012;96(2):228–36.
21. Akbar H, Grala TM, Vailati Riboni M, Cardoso FC, Verkerk G, McGowan J, et al. Body condition score at calving affects systemic and hepatic transcriptome indicators of inflammation and nutrient metabolism in grazing dairy cows. J Dairy Sci. 2015;98(2):1019–32.
22. Kohn RA, Dinneen MM, Russek-Cohen E. Using blood urea nitrogen to predict nitrogen excretion and efficiency of nitrogen utilization in cattle, sheep, goats, horses, pigs, and rats. J Anim Sci. 2005;83(4):879–89.
23. de Oliveiraa S, Camposb S, Oliveirac C, Britod F, Valadares Filhob C, Detmannb E, et al. Nutrient digestibility, nitrogen metabolism and hepatic function of sheep fed diets containing solvent or expeller castorseed meal treated with calcium hydroxide. Anim Feed Sci Tech. 2010;158(1–2):15–28.
24. Harden JL, Lewis SM, Lish SR, Suárez-Fariñas M, Gareau D, Lentini T, et al. The tryptophan metabolism enzyme L-kynureninase is a novel inflammatory factor in psoriasis and other inflammatory diseases. J Allergy Clin Immunol. 2016;137(6):1830–40.
25. Van E Nolte J, Löest CA, Ferreira AV, Waggoner JW, Mathis CP. Limiting amino acids for growing lambs fed a diet low in ruminally undegradable protein. J Anim Sci. 2008;86(10):2627–41.
26. Katona A, Toşa MI, Paizs C, Rétey J. Inhibition of histidine ammonia lyase by heteroaryl-alanines and acrylates. Chem Biodivers. 2006;3(5):502–8.
27. Garbes L, Kim K, Rieß A, Hoyer-Kuhn H, Beleggia F, Bevot A, et al. Mutations in SEC24D, encoding a component of the COPII machinery, cause a syndromic form of osteogenesis imperfecta. Am J Hum Genet. 2015;96(3):432–9.
28. Mason CC, Hanson RL, Ossowski V, Bian L, Baier LJ, Krakoff J, et al. Bimodal distribution of RNA expression levels in human skeletal muscle tissue. BMC Genomics. 2011;12:98.
29. Glinka T, Alter J, Braunstein I, Tzach L, Wei Sheng C, Geifman S, et al. Signal-peptide-mediated translocation is regulated by a p97-AIRAPL complex. Biochem J. 2014;457(2):253–61.
30. Young TA, Bailey SM, Van Horn CG, Cunningham CC. Chronic ethanol consumption decreases mitochondrial and glycolytic production of ATP in liver. Alcohol Alcohol. 2006;41(3):254–60.
31. Das A, Durrant D, Koka S, Salloum FN, Xi L, Kukreja RC. Mammalian target of rapamycin (mTOR) inhibition with rapamycin improves cardiac function in

type 2 diabetic mice: potential role of attenuated oxidative stress and altered contractile protein expression. J Biol Chem. 2014;289(7):4145–60.

32. Nakagawa Y, Satoh A, Yabe S, Furusawa M, Tokushige N, Tezuka H, et al. Hepatic CREB3L3 controls whole-body energy homeostasis and improves obesity and diabetes. Endocrinology. 2014;155(12):4706–19.

33. Laplante M, Sabatini DM. mTOR signaling in growth control and disease. Cell. 2012;149(2):274–93.

34. Momb J, Lewandowski JP, Bryant JD, Fitch R, Surman DR, Vokes SA, et al. Deletion of Mthfd1l causes embryonic lethality and neural tube and craniofacial defects in mice. Proc Natl Acad Sci U S A. 2013;110(2):549–54.

35. Smith SC, Petrova AV, Madden MZ, Wang H, Pan Y, Warren MD, et al. A gemcitabine sensitivity screen identifies a role for NEK9 in the replication stress response. Nucleic Acids Res. 2014;42(18):11517–27.

36. Suzuki M, Otsuka T, Ohsaki Y, Cheng J, Taniguchi T, Hashimoto H, et al. Derlin-1 and UBXD8 are engaged in dislocation and degradation of lipidated ApoB-100 at lipid droplets. Mol Biol Cell. 2012;23(5):800–10.

37. Wang Z, Li Q, Chamba Y, Zhang B, Shang P, Zhang H, et al. Identification of genes related to growth and lipid deposition from transcriptome profiles of pig muscle tissue. PLoS One. 2015;10(10):e0141138.

38. Racanelli V, Rehermann B. The liver as an immunological organ. Hepatology. 2006;43(2 Suppl 1):S54–62.

39. Robinson KS, Clements A, Williams AC, Berger CN, Frankel G. Bax inhibitor 1 in apoptosis and disease. Oncogene. 2011;30(21):2391–400.

40. Magni M, Ruscica V, Buscemi G, Kim JE, Nachimuthu BT, Fontanella E, et al. Chk2 and REGγ-dependent DBC1 regulation in DNA damage induced apoptosis. Nucleic Acids Res. 2014;42(21):13150–60.

41. Stremitzer S, Sunakawa Y, Zhang W, Yang D, Ning Y, Stintzing S, et al. Variations in genes involved in immune response checkpoints and association with outcomes in patients with resected colorectal liver metastases. Pharmacogenomics J. 2015;15(6):521–9.

42. Kasthuri SR, Umasuthan N, Whang I, Lim BS, Jung HB, Oh MJ, et al. Molecular characterization and expressional affirmation of the beta proteasome subunit cluster in rock bream immune defense. Mol Biol Rep. 2014;41(8):5413–27.

43. Jeong HU, Kim JH, Lee DY, Shim HJ, Lee HS. In vitro metabolic pathways of the new anti-diabetic drug evogliptin in human liver preparations. Molecules. 2015;20(12):21802–15.

44. Wang SH, Cheng CY, Tang PC, Chen CF, Chen HH, Lee YP, et al. Differential gene expressions in testes of L2 strain Taiwan country chicken in response to acute heat stress. Theriogenology. 2013;79(2):374–82.

45. Sergi C, Abdualmjid R, Abuetabh Y. Canine liver transplantation model and the intermediate filaments of the cytoskeleton of the hepatocytes. J Biomed Biotechnol. 2012;2012:131324.

46. Tu WW, Ji LD, Qian HX, Zhou M, Zhao JS, Xu J. Tributyltin induces disruption of microfilament in HL7702 cells via MAPK-mediated hyperphosphorylation of VASP. Environ Toxicol. 2016;31(11):1530–8.

47. Loo CS, Chen CW, Wang PJ, Chen PY, Lin SY, Khoo KH, et al. Quantitative apical membrane proteomics reveals vasopressin-induced actin dynamics in collecting duct cells. Proc Natl Acad Sci U S A. 2013;110(42):17119–24.

48. Hebel T, Eisinger K, Neumeier M, Rein-Fischboeck L, Pohl R, Meier EM, et al. Lipid abnormalities in alpha/beta2-syntrophin null mice are independent from ABCA1. Biochim Biophys Acta. 2015;1851(5):527–36.

49. Gee HY, Zhang F, Ashraf S, Kohl S, Sadowski CE, Vega-Warner V, et al. KANK deficiency leads to podocyte dysfunction and nephrotic syndrome. J Clin Invest. 2015;125(6):2375–84.

50. Guo EZ, Xu Z. Distinct mechanisms of recognizing endosomal sorting complex required for transport III (ESCRT-III) protein IST1 by different microtubule interacting and trafficking (MIT) domains. J Biol Chem. 2015; 290(13):8396–408.

51. Russo J, Santucci-Pereira J, Russo IH. The genomic signature of breast cancer prevention. Genes. 2014;5(1):65–83.

52. Galmiche L, Serre V, Beinat M, Assouline Z, Lebre AS, Chretien D, et al. Exome sequencing identifies MRPL3 mutation in mitochondrial cardiomyopathy. Hum Mutat. 2011;32(11):1225–31.

53. Zhou M, Bail S, Plasterer HL, Rusche J, Kiledjian M. DcpS is a transcript-specific modulator of RNA in mammalian cells. RNA. 2015;21(7):1306–12.

54. Xue T, Liu P, Zhou Y, Liu K, Yang L, Moritz RL, et al. Interleukin-6 induced "Acute" phenotypic microenvironment promotes Th1 anti-tumor immunity in cryo-thermal therapy revealed by shotgun and parallel reaction monitoring proteomics. Theranostics. 2016;6(6):773–94.

55. Mackey KR, Post AF, McIlvin MR, Cutter GA, John SG, Saito MA. Divergent responses of Atlantic coastal and oceanic Synechococcus to iron limitation. Proc Natl Acad Sci U S A. 2015;112(32):9944–9.

Protein profiles of hatchery egg shell membrane

N. C. Rath[1*], R. Liyanage[2], S. K. Makkar[3] and J. O. Lay Jr.[2]

Abstract

Background: Eggshells which consist largely of calcareous outer shell and shell membranes, constitute a significant part of poultry hatchery waste. The shell membranes (ESM) not only contain proteins that originate from egg whites but also from the developing embryos and different contaminants of microbial and environmental origins. As feed supplements, during post hatch growth, the hatchery egg shell membranes (HESM) have shown potential for imparting resistance of chickens to endotoxin stress and exert positive health effects. Considering that these effects are mediated by the bioactive proteins and peptides present in the membrane, the objective of the study was to identify the protein profiles of hatchery eggshell membranes (HESM).

Methods: Hatchery egg shell membranes were extracted with acidified methanol and a guanidine hydrochloride buffer then subjected to reduction/alkylation, and trypsin digestion. The methanol extract was additionally analyzed by matrix assisted laser desorption ionization-time of flight mass spectrometry (MALDI-TOF-MS). The tryptic digests were analyzed by liquid chromatography and tandem mass spectrometry (LC-MS-MS) to identify the proteins.

Results: Our results showed the presence of several proteins that are inherent and abundant in egg white such as, ovalbumin, ovotransferrin, ovocleidin-116, and lysozyme, and several proteins associated with cytoskeletal, cell signaling, antimicrobial, and catalytic functions involving carbohydrate, nucleic acid, and protein metabolisms. There were some blood derived proteins most likely originating from the embryos and several other proteins identified with different aerobic, anaerobic, gram positive, gram negative, soil, and marine bacterial species some commensals and others zoonotic.

Conclusion: The variety of bioactive proteins, particularly the cell signaling and enzymatic proteins along with the diverse microbial proteins, make the HESM suitable for nutritional and biological application to improve post hatch immunity of poultry.

Keywords: Hatchery egg shell membrane, Proteins, Mass spectrometry

Background

Egg shells that constitute a significant part of poultry hatchery waste, comprise largely of calcium carbonate crusts and underlying shell membranes laced with proteins originating from egg whites as well as contaminants of microbial and environmental origins [1]. The shell membranes have been considered for various engineering, agricultural, and biomedical applications, spurring many studies of their biochemical compositions [2–9]. Most such studies however, have used membranes harvested from fresh eggs and shown to contain proteins

and peptides of structural, antimicrobial, cell regulatory, and stress protective categories which likely contribute to their some of their health beneficial and therapeutic effects [2, 10–19]. Recently, we reported that the hatchery shell membranes (HESM), used as nutritional supplement, in post hatch chicks, promote growth, show protection against bacterial lipopolysaccharide (LPS) induced inflammatory response and diminish stress markers such as plasma levels of corticosterone [20]. Presuming that these effects could be mediated through the bioactive proteins present in the HESM, we were interested to analyze the proteomic composition of HESM that has never been investigated. Besides, the protein population of HESM can be qualitatively different from the fresh

* Correspondence: Narayan.rath@ars.usda.gov
[1]USDA/Agricultural Research Service, Poultry Production and Product Safety Research Unit, Fayetteville, AR 72701, USA
Full list of author information is available at the end of the article

egg shell membrane because the embryo development not only induce dynamic changes in egg proteome [21, 22] but also, alter the membrane structure to facilitate the hatching process. Hence, the current study was conducted utilizing the HESM that were dried at room temperature in a chemical hood without additional cleanup for 2 month period in the room temperature.

Methods
Chemicals and reagents
The reagents and devices such as C18 Nu tips (Glysci.com), Spectra/Por membranes (Spectrumlabs.com), bicinchoninic acid (BCA) protein assay kit, Pierce C18 spin columns, MS grade trypsin (Fisher Scientific.com), peptide and protein calibration standards (m/z 500–16000, Bruker Daltonics, Bremen, Germany), and 2-iodoacetamide (IAA) (MP Biomedical, OH) were purchased from their respective vendors. Most other chemicals and supplies including 1, 4-dithiothreitol (DTT), 2, 5-dihydroxybenzoic acid (DHB), were purchased from Sigma Aldrich (St. Louis, MO).

HESM preparation
Empty eggshells were obtained from the University of Arkansas poultry hatcheries, dried at room temperature under the hood for 2 months, the membranes separated manually then ground to coarse powder using an IKA mill (Cole Palmer). Duplicate samples of membrane powder from two separate preparations, each made from >50 eggs, were extracted (a) with 0.1% acetic acidified 70% methanol for soluble proteins, peptides, and their degradation products, and (b) with 4 M guanidine hydrochloride (GuHCl) containing 20 mM EDTA, and 50 mM Na-acetate, pH 5.8 for all soluble proteins [17, 23]. The membrane powder were extracted with 20 volumes of respective solutions overnight at 4 °C by constant stirring then centrifuged at 20,000 g for 15 min and the supernatants containing all soluble factors were dialyzed against excess volumes of 50 mM ammonium bicarbonate with three successive changes using 1000 Da Spectra/Por dialysis membranes. The solutions containing approximately 20 µg of protein from each extract, measured by BCA protein assay, were reduced with 10 mM dithiothreitol (DTT) for 1 h at 60 °C and alkylated with 40 mM iodoacetamide (IAA) then digested with 0.4 µg trypsin for 48 h at 37 °C [23, 24]. The tryptic digests were desalted using Pierce C18 spin columns according to the manufacturer suggested protocol prior to liquid chromatography-tandem mass spectrometry (LC-MS/MS).

LC-MS/MS analysis
Following desalting with C18 columns, the peptides were eluted, dried and re-suspended in 0.1% formic acid (FA) then subjected to LC-MS/MS using an Agilent 1200 series micro flow HPLC coupled to a Bruker Amazon-SL quadrupole ion trap mass spectrometer, and a captive spray ionization source. Tryptic peptides were separated at a solvent flow rate of 1.6 µL/min with 0 to 40% gradients of 0.1% FA (solvent A) and acetonitrile (ACN) in 0.1% FA (solvent B) over a period of 300 min each using a C_{18} (150 × 0.1 mm, 3.5 µm particle size, 300 Å pore size, Zorbax SB) capillary column. The captive electrospray source was operated in a positive ion mode with a dry gas temperature of 150 °C, dry nitrogen flow 3 L/minute, and capillary voltage of 1500 V. The data were acquired in the auto MS (n) mode with optimized trapping condition for the ions at m/z 1000. MS scans were performed in the enhanced scanning mode (8100 m/z/second), while the collision-induced dissociation or the MS/MS fragmentation scans performed automatically for top ten precursor ions for 1 min each in the UltraScan mode (32,500 m/z/second). The samples were run three times as technical repeats.

MALDI-TOF analysis
For direct matrix assisted laser desorption ionization-time of flight mass spectrometry (MALDI-TOF-MS), aliquots of methanol extracts were subjected to reduction/ alkylation using DTT and iodoacetamide (IAA) as described previously [25] or mock treated omitting DTT from the reaction for respective controls then spotted on a MALDI target plate mixing at 1:1 ratio with two dihydroxybenzoic acid. In all cases, samples were first subjected C18 nu tip cleaning up process using manufacturer recommended protocol before spotting on the MALDI target plate. The mass spectra were acquired with a Bruker Ultraflex II MALDI-TOF mass spectrometer (Bruker Daltonics GMBH, Bremen, Germany) operated in the positive-ion reflectron mode with the TOF analyzer calibrated with Bruker peptide standard II. Accurate mono isotopic protonated, intact peptide, masses were determined by MALDI-TOF-MS using combinations of external and internal calibration procedures, spotting the samples with equal volumes of α-cyano-4-hydroxycinnamic acid matrix, prepared in 0.1% FA, and 50:50 water/ACN. The MALDI-TOF-MS data were processed and some selective peaks were fragmented using LIFT-TOF/TOF MS to obtain their identities [17].

Data analysis
Peaks were picked from the LC-MS/MS chromatogram using instrument's default settings and the mzML files created using Bruker Data Analysis 4.0 software. The mzML files created as such were used in Bruker Protein-Scape 3.1 server with MASCOT data base search tool to perform MS/MS data search against the UniProt Gallus and NCBI bacterial data bases to identify the proteins.

All proteins were identified with a 95% confidence limit and <1% false discovery rate with at least one unique peptide. The parent ion mass- and fragment ion mass tolerance were both, set at 0.6 Da with cysteine carbamidomethylation and methionine oxidation as fixed and variable modifications. MASCOT.dat data base search files and.mzML raw data files were then exported into Skyline v3.1 software (http://proteome.gs.washington.edu/software/Skyline) [26, 27] to further refine the tryptic peptide identification based on their relative hydrophobicity and retention time correlations. Protein identities were also refined based on the uniqueness of the identified tryptic peptides by comparing them against respective background proteomes (either UniProt Gallus or the NCBI Bacteria data bases). The MALDI-TOF-MS data were processed using Bruker Flex Analysis 3.3 and Bruker BioTools 3.1 software. Peptides were identified by MASCOT MS/MS ion search tool with a peptide mass tolerance of 200 ppm and MS/MS ion tolerance of 0.6 Da. Functional annotation and classification of the proteins identified by LC-MS/MS with two or more unique peptides were performed using gene ontology (GO) based PANTHER classification system (http://www.pantherdb.org) [28] and proteins refined by Skyline from both guanidine and methanol extracts. Also, the same proteins were used in STRING functional protein association networks, version 10 (http://string-db.org) [29], under a high confidence setting to identify protein co-occurrence and possible functional relations.

Results

Table 1 shows the list of 41 proteins identified in the methanol extract using LC-MS/MS analyses and the Additional file 1: Table S1 shows the identities of 167 proteins in the GuHCl extracts of which 11 were common in both extracts which occur in egg white such as ovalbumin, ovomucoid, ovocleidin-116, and lysozyme (Table 1, names underlined). Several identified proteins on the basis of a single peptide were considered tentative although they were identified repeatedly in replicate samples and by Skyline retention time correlation refinements. These identifications were largely occurred in methanol extract which may relate to degraded proteins. There were 12 uncharacterized proteins including one in methanol extract. Several structural proteins that included collagens, keratin, proteoglycans (lumican, decorin) were identified in the HESM. Different keratins that included both cytoskeletal type 1 and non-cytoskeletal were abundant in HESM. Many cytoskeletal and their cognate proteins such as actin, vinculin, gelsolin, tubulin, vimentin, thymosin β4, and some blood associated proteins (annexins, hemoglobin, epsilon globin, and serum albumin) predominated in the guanidine extract. Cell signaling and enzyme proteins largely, associated with energy, protein, and nucleic acid metabolisms, and several antimicrobial proteins that included ovotransferrin, lysozyme, ovocleidin-116, ovocalyxin-36, keratin peptides, and gallinacin 9 and 10, were present. Some of the other proteins that secured high scores were embryonic and development associated proteins such as zona pellucida sperm binding proteins, elongation factors, and histone cluster proteins. Functional annotation with PANTHER using combined protein IDs from both extracts (methanol and guanidine) and GO-slim molecular function sorted the proteins into six major categories using 98 gene products with the binding, structural, and catalytic proteins topping the list (Fig. 1). A classification of protein functionality showed two dominant associations, one, with nucleic acid binding and the other with cytoskeletal function, followed by structural, transfer/ carrier, and oxidoreductase genre of proteins (Fig. 2). Some proteins uniquely present in methanol extract of HESM, such as gallinacin 9, thymosin β4, and septin, were not detected in the guanidine HCl extract. A protein interaction and co-occurrence map, using STRING bioinformatics, showed a major cluster of proteins associated with nuclear and metabolic activities (eg: ribosomal proteins) and two others associated with carbohydrate metabolism and actin/ actin-cognate function, respectively (Fig. 3). The MALDI-TOF MS of methanol extract showed several peaks of which 2, corresponding to m/z 4643 and 4773 (Fig. 4a), showed mass shifts by 348 Da upon reduction/ alkylation (Fig. 4b), suggestive of carbamidomethylation of three disulfides, whereas, few other peaks such as m/z 2682, 2624, 2835, and 2126 did not show any change. Based on our previous identification in ESM [17], we surmised that the m/z 4773 was the mature gallinacin 10 peptide with a corresponding sequence of "DPLFPDTVACRTQGNFCRAGACPPTFTIS GQCHGGLLNCCAKIPAQ" and the m/z 4643 was the same peptide with the truncation of glutamine (Q) from the N-terminal. Fragmentation of m/z 2682 peptide by LIFT-MS/MS (Fig. 5) yielded a partial sequence corresponding to "AGGSVPGRPLPNEAL" which upon protein blast (BLASTP) (http://blast.ncbi.nlm.nih.gov/Blast.cgi? PAGE=Proteins), in non-redundant protein data base showed >90% identity with one Apicomplexan specific protein associated with Eimeria, a protozoa that causes coccidiosis in poultry. Using NCBI bacteria data base search, the proteins in GuHCl extract, showed 50 proteins belonging to different species of gram positive, gram negative, aerobic and anaerobic families of bacteria that included Candidatus, Enterococci, Yersinia, and Butyrivibrio, and several phytobacteria (Sphingomonas taxi), Nocardioide (soil bacteria), and a marine bacterium (hyphomonas, marinobacter adhaerens) (Additional file 1: Table S2). Many of these identifications were tentative based on the matches with their respective enzyme proteins with

Table 1 Proteins identified in methanol extract by LC-MS/MS. Proteins with a single peptide are considered "tentatively identified"

	Accession	Protein	MW [kDa]	Scores	# Peptides
1	ENSGALP00000036403	Ovalbumin	42.9	629.9	11
2	ENSGALP00000005544	Ovomucoid	22.6	338.1	7
3	ENSGALP00000017755	Ovocleidin-116	76.8	225	2
4	ENSGALP00000008163	Orosomucoid 1 (ovoglycoprotein) precursor	22.3	167	3
5	ENSGALP00000042635	peptidyl-prolyl cis-trans isomerase FKBP1A	8.9	163.5	3
6	ENSGALP00000016177	Lysozyme C	16.2	129.2	2
7	ENSGALP00000016632	keratin 8, type II	42.1	119.1	2
8	ENSGALP00000031725	SH3 domain binding glutamic acid-rich protein like (SH3BGRL), mRNA	12.9	109.6	2
9	ENSGALP00000027483	Ubiquitin-fold modifier 1	9	102.1	1
10	ENSGALP00000035930	Gallinacin-9	7.3	85.7	2
11	ENSGALP00000010763	Ovocalyxin-36 precursor	58.3	83.7	1
12	ENSGALP00000019758	Diazepam binding inhibitor (GABA receptor modulator, acyl-CoA binding protein) (DBI), mRNA.	9.6	69.8	2
13	ENSGALP00000026846	Gallinacin-10	7.1	62.6	2
14	ENSGALP00000006093	Keratin, type I cytoskeletal 19	46	62.3	3
15	ENSGALP00000009976	Usher syndrome 1C	100.1	51.9	1
16	ENSGALP00000012729	Uncharacterized protein	21	50.9	1
17	ENSGALP00000040654	Thymosin, beta 4	5	47.8	2
18	ENSGALP00000043411	Collagen, type XVI, alpha 1	156.6	42	1
19	ENSGALP00000002523	Signal peptidase complex subunit 1 homolog	27.2	37.9	1
20	ENSGALP00000013908	Zinc finger BED domain-containing protein 4	132.4	37.8	2
21	ENSGALP00000038912	Alpha-D-globin (HBAD), mRNA.	15.7	37.3	1
22	ENSGALP00000019988	Utrophin	398.6	36.5	2
23	ENSGALP00000000876	Fatty acid-binding protein, heart	14.8	34.6	1
24	ENSGALP00000026777	Nociceptin precursor	21.4	34.4	1
25	ENSGALP00000026863	Polycystic kidney and hepatic disease 1 (autosomal recessive)	440	32.7	1
26	ENSGALP00000018601	Serine peptidase inhibitor, Kazal type 2 (acrosin-trypsin inhibitor)	6	26.7	1
27	ENSGALP00000025439	Probable arginyl-tRNA synthetase, mitochondrial	65.3	25.2	1
28	ENSGALP00000039913	Transforming, acidic coiled-coil containing protein 1	86.4	24.5	2
29	ENSMGAP00000001722	elaC ribonuclease Z 2	94.2	24.2	1
30	ENSGALP00000031519	Fibroblast growth factor 2	16.2	22.8	1
31	ENSGALP00000039675	Nuclear receptor coactivator 2	44.7	21.5	1
32	ENSGALP00000000325	WD repeat-containing protein 36	98.2	21.2	1
33	ENSGALP00000019412	Septin 3	40	20.5	1
34	ENSGALP00000027541	High mobility group protein B1	24.9	20	1
35	ENSGALP00000014919	tRNA (adenine-N(1)-)-methyltransferase non-catalytic subunit TRM6	54.2	19.8	1
36	ENSGALP00000043133	Fibrinogen silencer binding protein	36.4	19	1
37	ENSGALP00000000726	SH3 domain binding glutamate-rich protein like 3	10.5	18.8	1
38	ENSGALP00000038283	UPF3 regulator of nonsense transcripts homolog B (yeast)	56.9	18.4	1
39	ENSGALP00000000275	Mutated in colorectal cancers	112.6	18.3	1
40	ENSGALP00000038735	Polyubiquitin-B Ubiquitin	109.6	15.5	1
41	ENSGALP00000040476	Neuregulin 2	65.5	15	1

The underlined proteins also occur in guanidine HCl extract (Additional file 1: Table S1)

PANTHER GO-Slim Molecular Function
Total # Genes: 98 Total # function hits: 82

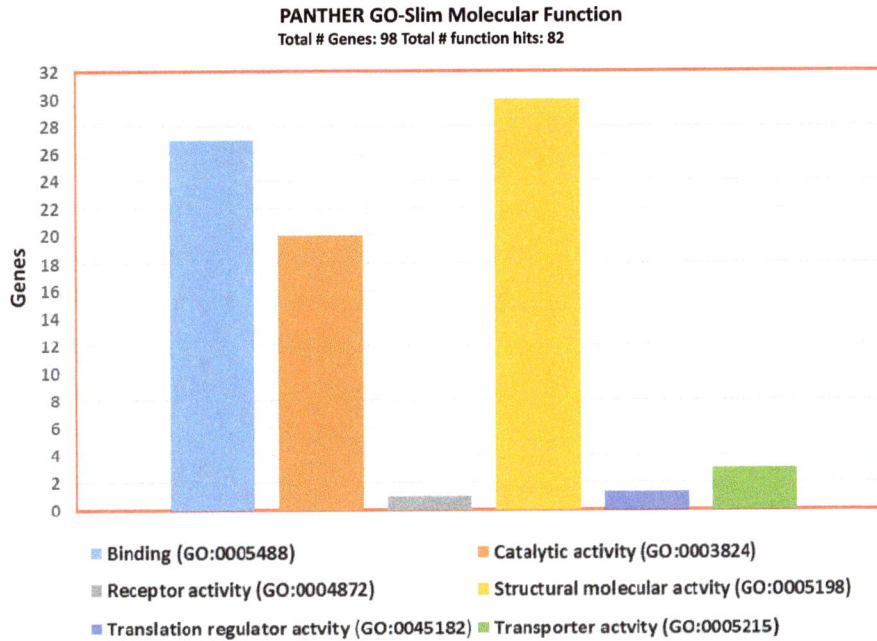

- Binding (GO:0005488)
- Catalytic activity (GO:0003824)
- Receptor activity (GO:0004872)
- Structural molecular actvity (GO:0005198)
- Translation regulator actvity (GO:0045182)
- Transporter actvity (GO:0005215)

Fig. 1 Molecular function annotation of proteins, identified with two or more unique peptides, using Protein Analysis through Evolutionary Relationships (PANTHER)

PANTHER Protein Class
Total # Genes: 98 Total # protein class hits: 102

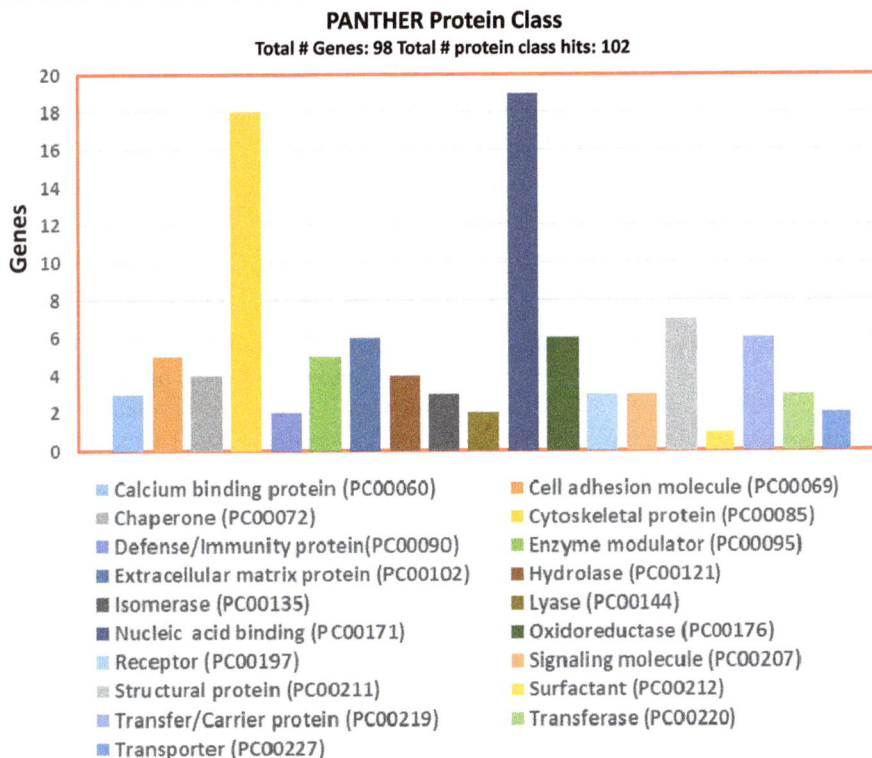

- Calcium binding protein (PC00060)
- Cell adhesion molecule (PC00069)
- Chaperone (PC00072)
- Cytoskeletal protein (PC00085)
- Defense/Immunity protein(PC00090)
- Enzyme modulator (PC00095)
- Extracellular matrix protein (PC00102)
- Hydrolase (PC00121)
- Isomerase (PC00135)
- Lyase (PC00144)
- Nucleic acid binding (P C00171)
- Oxidoreductase (PC00176)
- Receptor (PC00197)
- Signaling molecule (PC00207)
- Structural protein (PC00211)
- Surfactant (PC00212)
- Transfer/Carrier protein (PC00219)
- Transferase (PC00220)
- Transporter (PC00227)

Fig. 2 Functional annotation by protein class of HESM proteins, identified with two or more unique peptides, using Protein Analysis through Evolutionary Relationships (PANTHER)

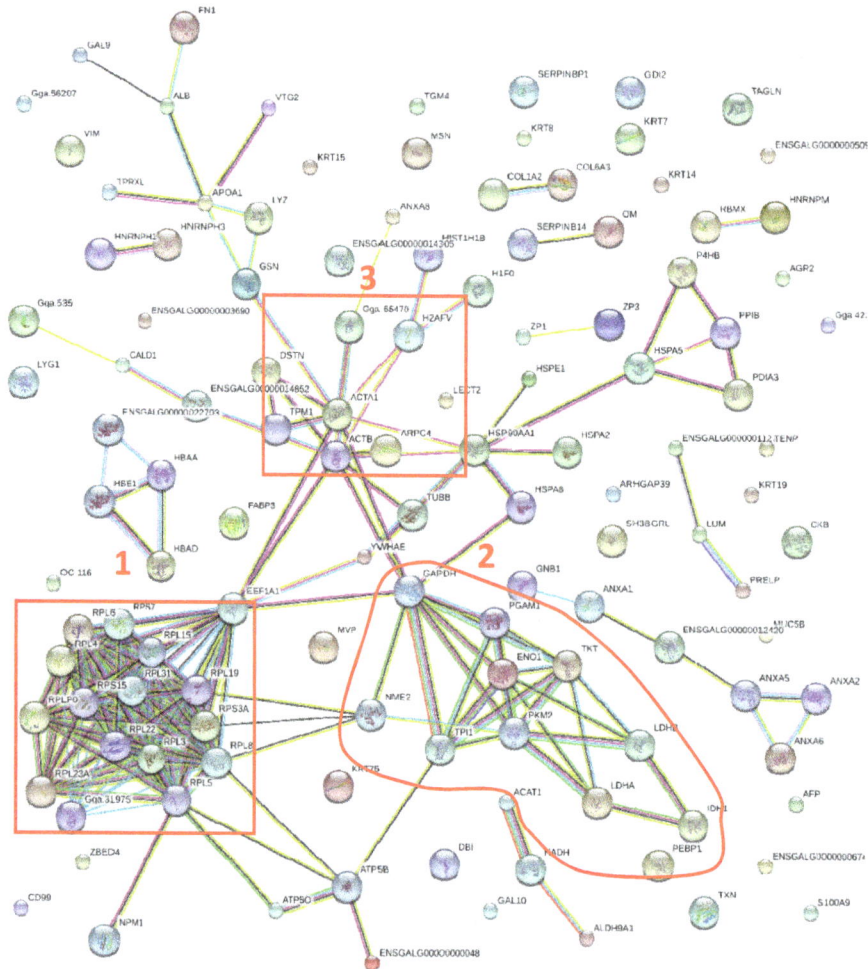

Fig. 3 The protein-protein interaction network of proteins identified with two or more unique peptides. Rectangles showing, 1. Actin node, 2. Ubiquitin node, and 3. Ribosomal and nuclear protein clusters

one unique peptide occurring in replicate sample. The maximum number of bacterial proteins identified belonged to Pseudomonas M10, a family of gram negative bacteria followed by Candidatus, Hyphomonas, and Sphingomonas.

Discussion

As expected the HESM contained many proteins such as ovalbumin, ovomucoid, ovocleidin-116, and lysozyme that are inherent to egg proteome [8, 30–32]. Structural proteins such as keratins, collagens, decorin, lumican, and tropomyosin, and cytoskeletal proteins, such as actin, tubulin, vimentin, and their interacting proteins, were present along with several blood (hemoglobin, α-D-globin, annexin, fibronectin) and embryo associated proteins (vitellogenin, zona pellucida sperm-binding proteins, and vitelline membrane outer layer protein). These results are consistent with the reports of Cordeiro and Hincke [32] who showed the occurrence of fibronectin, vitellogenin, and apolipoproteins in the shell membranes

of fertilized eggs. The HESM also contained many different enzymes, protease inhibitors, and signaling proteins similar to fresh ESM. However, the HESM seemed to be differentially populated with several proteins associated with glycolysis, carbohydrate, protein, and nucleic acid metabolisms. Many ribosomal, histone, and ribonuclear proteins, and enzymes, associated with protein folding and translocations, were also identified in the HESM extract. However, there was a conspicuous absence of clusterin, a glycoprotein which we and others have found present in the fresh ESM and has been reported as a major component of egg proteome [17, 19, 21, 33–36]. Clusterin is implicated in the maintenance of cellular homeostasis, developmental remodeling, and in the egg, it is correlated with the changes in viscosity of egg white [36–39]. In our previous study with fresh ESM we identified several clusterin peptide fragments in the methanol extract of ESM. However, the absence of any clusterin peptide in HESM under similar extraction

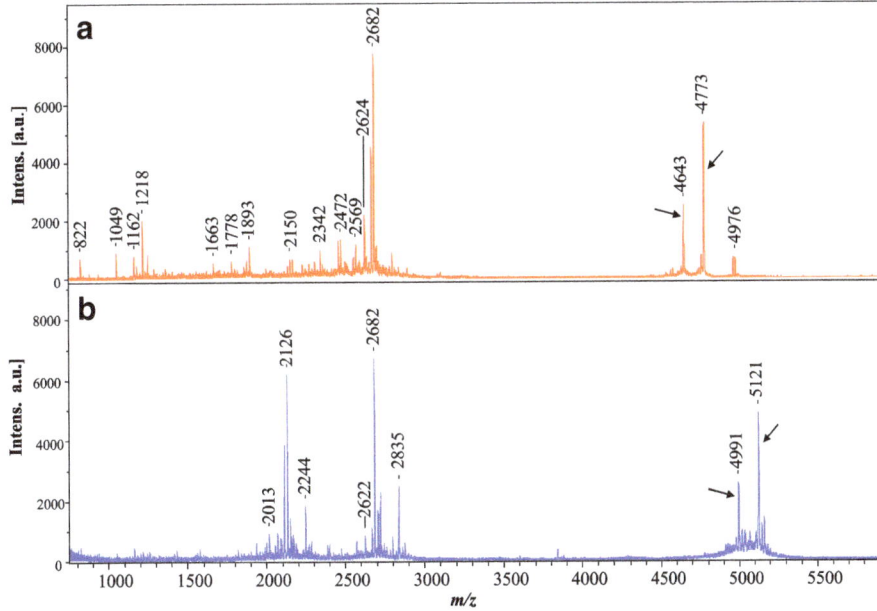

Fig. 4 MALDI-TOF profile of HESM methanol extract before (**a**) and after reduction and alkylation (**b**). Two peaks corresponding to m/z 4643 and 4773 (**a**) show mass shifts by 348 Da corresponding to m/z 4991 and 5121 (**b**), respectively (*arrows*)

condition was intriguing. It is possible that clusterin, in shell membrane, is involved in early embryo development and utilized completely, as the development reaches to its termination. Similarly, there was also a conspicuous absence of gallin, an antimicrobial peptide that occurs uniquely in egg proteome of fresh eggs [17, 33, 40] whereas two other avian defensins, gallinacin 9 and 10, were identified along with several other antimicrobial proteins including ovotransferrin, lysozyme, keratin, and ovomucoid. Because the protein profiles of egg undergo qualitative and quantitative changes both, during incubation and passive storage [21, 22, 32, 38,

41, 42], the differences in protein profiles of membranes between fresh and hatchery shells appear logical and consistent.

A functional association bioinformatics showed co-occurrence of proteins associated with ubiquitination and apoptosis activities, indicative of large scale protein degradation, expected, with the embryo development and hatching. The other notable protein clusters were carbohydrate and energy metabolizing enzymes (glyceraldehyde phosphate dehydrogenase, α-enolase, isocitrate dehydrogenase, aldolase, triose phosphate isomerase) as well as the proteins associated to nucleic acid metabolism, and

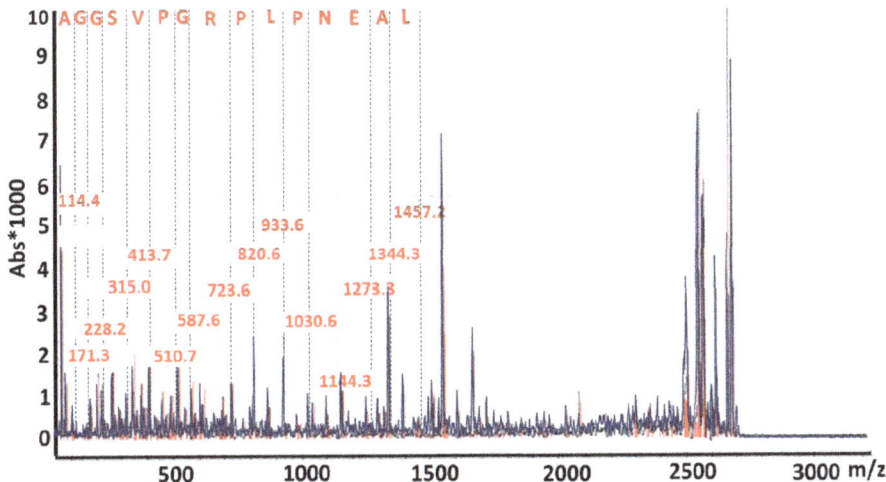

Fig. 5 Fragmentation of m/z 2682 showing a partial sequence 'AGGSVPGRPLPNEAL' corresponding to Eimeria Apicomplexan specific protein matching by >92% identity

stress protection functions such as heat shock and heat shock cognate proteins (HSP). Besides the endogenous proteins and peptides, the HESM also contained several proteins identified with multi species of both gram positive and gram negative bacteria some belonging to families of phytobacteria, soil bacteria even, marine bacteria. Proteins identified with bacteria such as Candidatus, Enterococci, Yersinia, Butyrivibrio Pseudomonas, Clostridium, and Vibrio are zoonotic that are implicated in human gastrointestinal and respiratory illness including one peptide that belonged to a Eimeria, a protozoa, that causes coccidiosis and enteritis in poultry.

The objective of the study was to gain a better understanding of the potential of egg membrane as an adjuvant that likely provides epigenetic conditioning influencing the post hatch immunity through its proteomic constituents [20]. Based on the above results, it appears that the HESM with its plethora of bioactive proteins, peptides, enzymes, and multi species microbial protein factors, may be a suitable modulator of immunity of post hatch chickens with its muco-adhesive properties and action on gastrointestinal mucosa. Besides, the HESM protein profile also provides a reference to compare and identify similar other allogeneic and xenogeneic sources of material that can improve immunity of post hatch poultry reducing the dependence on antibiotics growth promoters. The identification of diverse bacterial proteins in the HESM makes it appealing since numerous body of evidence has shown that the immunity and disease resistance is greatly influenced by biodiverse factors that include various bacterial and viral antigens [43–45].

Conclusion

The HESM is a rich source of a variety of bioactive proteins particularly belonging to signaling enzymatic, and regulatory varieties that along with many microbial proteins make it uniquely suitable for use as a modulator to improve post hatch immunity of poultry.

Additional file

Additional file 1: Table S1. Proteins identified in guanidine HCL extract of eggshell membranes by LC-MS/MS. Proteins identified with a single peptide are considered "**tentatively identified**". **Table S2.** Bacterial proteins identified in Guanidine HCl extracts of HESM. Proteins identified with one only peptide are considered "**tentatively identified**". (DOCX 68 kb)

Abbreviations
ACN: Acetonitrile; BCA: Bicinchoninic acid; DTT: Dithiothreitol; ESM: Egg shell membrane; FA: Formic acid; GuHCl: Guanidine hydrochloride; HESM: Hatchery egg shell membrane; IAA: Iodoacetamide; LC-MS/MS: Liquid chromatography-tandem mass spectrometry; MALDI-TOF: Matrix assisted laser desorption ionization time of flight; MS: Mass spectrometry

Acknowledgment
We thank Sonia Tsai and Scott Zornes for assistance.

Funding
This work was carried out as in house project and partly conducted in the State wide Mass spectrometry Facility supported by a grant P30 GM103450 to the University of Arkansas.

Authors' contributions
NCR and RL conceived and designed the experiments; experiments performed by SKM, RL and NCR; NCR, RL, SKM wrote the manuscript, discussed and approved by NCR, RL, and JL.

Competing interests
The authors declare that they have no competing interests.

Author details
[1]USDA/Agricultural Research Service, Poultry Production and Product Safety Research Unit, Fayetteville, AR 72701, USA. [2]Statewide Mass Spectrometry Facility, University of Arkansas, Fayetteville, AR 72701, USA. [3]Department of Poultry Science, University of Arkansas, Fayetteville, AR 72701, USA.

References
1. Glatz P, Miao Z, Rodda B. Handling and treatment of poultry hatchery waste: a review. Sustainability. 2011;3:216–37.
2. Cordeiro CM, Hincke MT. Recent patents on eggshell: shell and membrane applications. Recent Pat Food Nutr Agric. 2011;3:1–8.
3. Vlad V, Biova LLC. Eggshell membrane separation method. (USPTO ed., vol. US7954733; 2009.
4. Kovacs-Nolan J, Phillips M, Mine Y. Advances in the value of eggs and egg components for human health. J Agric Food Chem. 2005;53:8421–31.
5. Baláž M. Eggshell membrane biomaterial as a platform for applications in materials science. Acta Biomater. 2014;10:3827–43.
6. King'ori AM. A review of the uses of poultry eggshells and shell membranes. Int J Poult Sci. 2011;10:908–12.
7. Miksík I, Eckhardt A, Sedláková P, Mikulikova K. Proteins of insoluble matrix of avian (gallus gallus) eggshell. Connect Tissue Res. 2007;48:1–8.
8. Mikšík I, Ergang P, Pácha J. Proteomic analysis of chicken eggshell cuticle membrane layer. Anal Bioanal Chem. 2014;406:7633–40.
9. Kozuka M, Murao S, Yamane T, Inoue T, Ohkubo I, Ariga H. Rapid and simple purification of lysozyme from the egg shell membrane. J Nutr Sci Vitaminol (Tokyo). 2015;61:101–3.
10. Makkar S, Rath NC, Packialakshmi B, Huff WE, Huff GR. Nutritional effects of egg shell membrane supplements on chicken performance and immunity. Poult Sci. 2015;94:1184–9.
11. Ruff KJ, DeVore DP, Leu MD, Robinson MA. Eggshell membrane: a possible new natural therapeutic for joint and connective tissue disorders. Results from two open-label human clinical studies. Clin Interv Aging. 2009;4:235–40.
12. Ruff KJ, Winkler A, Jackson RW, DeVore DP, Ritz BW. Eggshell membrane in the treatment of pain and stiffness from osteoarthritis of the knee: a randomized, multicenter, double-blind, placebo-controlled clinical study. Clin Rheumatol. 2009;28:907–14.
13. Shi Y, Rupa P, Jiang B, Mine Y. Hydrolysate from eggshell membrane ameliorates intestinal inflammation in mice. Int J Mol Sci. 2014;15:22728–42.
14. Hincke MT, Nys Y, Gautron J. The role of matrix proteins in eggshell formation. Journal of poultry science. 2010;47:208–19. 208–219.
15. Rose ML, Hincke MT. Protein constituents of the eggshell: eggshell-specific matrix proteins. Cell Mol Life Sci. 2009;66:2707–19.
16. Mann K, Olsen JV, Macek B, Gnad F, Mann M. Identification of new chicken egg proteins by mass spectrometry-based proteomic analysis. World's Poultry Science J. 2008;64:209–18. 209–218.
17. Makkar S, Liyanage R, Kannan L, Packialakshmi B, Lay JO, Rath NC. Chicken egg shell membrane associated proteins and peptides. J Agric Food Chem. 2015;63:9888–98.
18. Ahlborn GJ, Clare DA, Sheldon BW, Kelly RW. Identification of eggshell membrane proteins and purification of ovotransferrin and beta-NAGase from hen egg white. Protein J. 2006;25:71–81.

19. Huopalahti R, López-Fandiño R, Anton M, editors. Bioactive egg compounds. Berlin: Springer; 2007.

20. Makkar SK, Rath NC, Packialakshmi B, Zhou ZY, Huff GR, Donoghue AM. Nutritional supplement of hatchery eggshell membrane improves poultry performance and provides resistance against endotoxin stress. PLoS One. 2016;11, e0159433.

21. Liu Y, Qiu N, Ma M. Comparative proteomic analysis of hen egg white proteins during early phase of embryonic development by combinatorial peptide ligand library and matrix-assisted laser desorption ionization-time of flight. Poult Sci. 2013;92:1897–904.

22. Kaweewong K, Garnjanagoonchorn W, Jirapakkul W, Roytrakul S. Solubilization and identification of hen eggshell membrane proteins during different times of chicken embryo development using the proteomic approach. Protein J. 2013;32:297–308.

23. Kannan L, Liyanage R, Lay Jr J, Packialakshmi B, Anthony N. Identification and structural characterization of avian beta-defensin 2 peptides from pheasant and quail. J Proteomics Bioinform. 2013;6:031–7.

24. Kannan L, Rath NC, Liyanage R, Lay JO. Effect of toll-like receptor activation on thymosin beta-4 production by chicken macrophages. Mol Cell Biochem. 2010;344:55–63.

25. Kannan L, Rath NC, Liyanage R, Lay Jr JO. Evaluation of beta defensin 2 production by chicken heterophils using direct MALDI mass spectrometry. Molec Immunol. 2009;46:3151–6.

26. Schilling B, Rardin MJ, MacLean BX, Zawadzka AM, Frewen BE, Cusack MP, Sorensen DJ, Bereman MS, Jing E, Wu CC, et al. Platform-independent and label-free quantitation of proteomic data using MS1 extracted ion chromatograms in sSkyline: application to protein acetylation and phosphorylation. Mol Cell Proteomics. 2012;11:202–14.

27. D'Souza A, Schilling B, Chytrowski J, MacLean B, Broudy D, Shulman NJ, MacCoss MJ, Gibson BW. MS1Probe-Implementation of a Statistical Tool for MS1-based Quantitation in Skyline for High Throughput Quantitative Analysis. In: Proceedings of the 61st Annual ASMS Conference on Mass Spectrometry & Allied Topics Minneapolis, MN. 2013.

28. Mi H, Poudel S, Muruganujan A, Casagrande JT, Thomas PD. PANTHER version 10: expanded protein families and functions, and analysis tools. Nucleic Acids Res. 2016;44:D336–42.

29. Szklarczyk D, Franceschini A, Wyder S, Forslund K, Heller D, Huerta-Cepas J, Simonovic M, Roth A, Santos A, Tsafou KP, et al. STRING v10: protein–protein interaction networks, integrated over the tree of life. Nucleic Acids Res. 2015;43:D447–52.

30. Mann K, Mann M. In-depth analysis of the chicken egg white proteome using an LTQ Orbitrap Velos. Proteome Sci. 2011;9:7.

31. Farinazzo A, Restuccia U, Bachi A, Guerrier L, Fortis F, Boschetti E, Fasoli E, Citterio A, Righetti PG. Chicken egg yolk cytoplasmic proteome, mined via combinatorial peptide ligand libraries. J of Chromatography A. 2009; 1216:1241–52.

32. Cordeiro CM, Hincke MT. Quantitative proteomics analysis of eggshell membrane proteins during chick embryonic development. J Proteomics. 2016;130:11–25.

33. Mann K. The chicken egg white proteome. Proteomics. 2007;7:3558–68.

34. Guérin-Dubiard C, Pasco M, Mollé D, Désert C, Croguennec T, Nau F. Proteomic analysis of hen egg white. J Agric Food Chem. 2006;54:3901–10.

35. Mann K, Gautron J, Nys Y, McKee MD, Bajari T, Schneider WJ, Hincke MT. Disulfide-linked heterodimeric clusterin is a component of the chicken eggshell matrix and egg white. Matrix Biol. 2003;22:397–407.

36. Wang J, Wu J. Proteomic analysis of fertilized egg white during early incubation. EuPA Open Proteom. 2014;2:38–59.

37. Jones SE, Jomary C. Clusterin. Int J Biochem Cell Biol. 2002;34:427–31.

38. Omana DA, Liang Y, Kav NN, Wu J. Proteomic analysis of egg white proteins during storage. Proteomics. 2011;11:144–53.

39. Rohne P, Prochnow H, Koch-Brandt C. The CLU-files: disentanglement of a mystery. Biomol Concepts. 2016;1:1–15.

40. Gong D, Wilson PW, Bain MM, McDade K, Kalina J, Herve-Grepinet V, Nys Y, Dunn IC. Gallin; an antimicrobial peptide member of a new avian defensin family, the ovodefensins, has been subject to recent gene duplication. BMC Immunol. 2010;11:12.

41. Qiu N, Ma M, Zhao L, Liu W, Li Y, Mine Y. Comparative proteomic analysis of egg white proteins under various storage temperatures. J Agric Food Chem. 2012;60:7746–53.

42. Qiu N, Liu W, Ma M, Zhao L, Li Y. Differences between fertilized and unfertilized chicken egg white proteins revealed by 2-dimensional gel electrophoresis-based proteomic analysis. Poult Sci. 2013;92:782–6.

43. Pi C, Allott EH, Ren D, Poulton S, Lee SY, Perkins S, Everett ML, Holzknecht ZE, Lin SS, Parker W. Increased biodiversity in the environment improves the humoral response of rats. PLoS One. 2015;10, e0120255.

44. Hanson MA, Gluckman PD. Early developmental conditioning of later health and disease: physiology or pathophysiology? Physiol Rev. 2014;94:1027–76.

45. Morris JA, Harrison LM, Lauder RM, Telford DR, Neary R. Low dose, early mucosal exposure will minimize the risk of microbial disease. Med Hypotheses. 2012;79:630–4.

13

Proteomic analysis of microparticles isolated from malaria positive blood samples

Samuel Antwi-Baffour[1*], Jonathan Kofi Adjei[1,2], Francis Agyemang-Yeboah[2], Max Annani-Akollor[2], Ransford Kyeremeh[1], George Awuku Asare[1] and Ben Gyan[3]

Abstract

Background: Malaria continues to be a great public health concern due to the significant mortality and morbidity associated with the disease especially in developing countries. Microparticles (MPs), also called plasma membrane derived extracellular vesicles (PMEVs) are subcellular structures that are generated when they bud off the plasma membrane. They can be found in healthy individuals but the numbers tend to increase in pathological conditions including malaria. Although, various studies have been carried out on the protein content of specific cellular derived MPs, there seems to be paucity of information on the protein content of circulating MPs in malaria and their association with the various signs and symptoms of the disease. The aim of this study was therefore to carry out proteomic analyses of MPs isolated from malaria positive samples and compare them with proteins of MPs from malaria parasite culture supernatant and healthy controls in order to ascertain the role of MPs in malaria infection.

Methods: Plasma samples were obtained from forty-three (43) malaria diagnosed patients (cases) and ten (10) healthy individuals (controls). Malaria parasite culture supernatant was obtained from our laboratory and MPs were isolated from them and confirmed using flow cytometry. 2D LC-MS was done to obtain their protein content. Resultant data were analyzed using SPSS Ver. 21.0 statistical software, Kruskal Wallis test and Spearman's correlation coefficient *r*.

Results: In all, 1806 proteins were isolated from the samples. The MPs from malaria positive samples recorded 1729 proteins, those from culture supernatant were 333 while the control samples recorded 234 proteins. The mean number of proteins in MPs of malaria positive samples was significantly higher than that in the control samples. Significantly, higher quantities of haemoglobin subunits were seen in MPs from malaria samples and culture supernatant compared to control samples.

Conclusion: A great number of proteins were observed to be carried in the microparticles (MPs) from malaria samples and culture supernatant compared to controls. The greater loss of haemoglobin from erythrocytes via MPs from malaria patients could serve as the initiation and progression of anaemia in *P.falciparum* infection. Also while some proteins were upregulated in circulating MPs in malaria samples, others were down regulated.

Keywords: Proteomics, Microparticles, Malaria, Plasmodium, Plasma membrane

* Correspondence: s.antwi-baffour@chs.edu.gh
[1]Department of Medical Laboratory Sciences, School of Biomedical and Allied Health Sciences, College of Health Sciences, University of Ghana, P. O. Box KB 143Korle-Bu, Accra, Ghana
Full list of author information is available at the end of the article

Background

Microparticles (MPs) also called plasma membrane derived extracellular vesicles (PMEVs) are a heterogeneous group of small sub-membrane fragments or membrane coated vesicles shed from the plasma membrane of various cells during normal cellular activities like growth, senescence, proliferation and apoptosis [1]. MPs carry proteins, lipids and nucleic acids from host cells and are means of intercellular communication and it has been shown that analysis of MPs from blood samples can provide information about the state and progression of a particular disease or condition [2].

Malaria is caused by five species of the genus *Plasmodium* which is a unicellular protozoan parasite. The disease is a major cause of mortality and morbidity in many developing countries especially in Sub-Saharan Africa. It is estimated 3.4 billion people worldwide risk being infected with malaria in 104 countries [3]. Complications associated with malaria infection particularly in severe malaria include fever/chills, coagulopathy and anaemia among other symptoms. At the molecular level, up-regulation of certain cytokines is also thought to relate to malaria associated high fever [4].

The degree of anaemia experienced in malaria does not always correspond to the parasitaemia level [5]. This is partly caused by a mild bone marrow suppression of erythrocyte production and the collection of complement containing complexes on erythrocyte surfaces after infection which promotes splenic removal of these erythrocytes. Lysis of both infected and uninfected erythrocytes is also considered to be a contributing factor [5]. Severe malaria caused by *Plasmodium falciparum* is considered to be associated with the dysregulation of the coagulation system which include endothelial damage, lower levels of anticoagulation and the release of procoagulant MPs [6].

To this end, malaria has been associated with an increase in the level of circulating plasma MPs and plasma concentrations of endothelial MPs (EMPs) which may be proportional to disease severity [7, 8]. The role of infected erythrocyte-derived MPs in cellular communication has been investigated but the protein content (proteomic analysis) of MPs isolated in malaria is yet to be explored [9]. Proteomic analysis on circulating MPs obtained from plasma of malaria positive blood samples once explored will give a general idea of the protein and protein groups borne by these MPs that may influence the pathophysiology of malaria infection. This study therefore sought to examine the protein composition of plasma MPs of malaria samples and comparing them with proteins of MPs from healthy controls in order to explore their effect on the pathogenesis of malaria and the possible linkage of circulating plasma MPs to malaria anaemia.

Existing literature indicate that elevated MPs levels have been seen in cancer, sepsis, pulmonary hypertension, idiopathic thrombocytopenic purpura and atherosclerosis [10]. Researchers also contend that increased endothelial microparticle level correlating with disease severity has been seen in malaria [11]. Again studies in mice models indicate that microparticles contributed to induction of systemic inflammation [12, 13]. Others have shown that MPs released after malaria infection which are primarily erythrocyte-derived are capable of activating macrophage through toll-like receptors (TLR) and may enhance infectivity as their count elevates and investigations show they contain parasite components some of which promote pathogen invasion of erythrocytes [14].

Regev-Rudzki et al. (2013) postulated that MPs released in malaria are also capable of activating the blood–brain barrier which exacerbates inflammation [15]. The perplexing feature of malarial anemia which is increased clearance of uninfected erythrocytes can also be attributed to the release of parasite antigens in MPs during entry in erythrocytes. These erythrocyte-adhesive proteins probably adhere to erythrocytes resulting in IgG and complement binding which promotes their elimination from peripheral circulation [16]. Furthermore, Schorey et al. stated that MPs released from P. falciparum-infected cells are able to modulate hosts immune response thereby impairing surveillance [17].

Proteomic analysis on circulating plasma MPs obtained from plasma of malaria positive blood samples once explored will give a general idea of the protein and protein groups borne by these MPs thereby influencing the pathophysiology of malaria infection. This study seeks to examine the protein composition of plasma MPs of malaria samples in order to explore their effect on the pathogenesis of malaria and the possible linkage of circulating plasma MPs to the anaemia.

Methods

Aim

The aim of this study was to carry out proteomic analyses of MPs isolated from malaria positive samples and compare them with proteins of MPs from healthy controls in order to ascertain the role of MPs in malaria infection.

Design and setting

The study was a cross-sectional study conducted over a 2-year period, from May 2014 to June, 2016. The samples were collected at the Sunyani Regional Hospital, Ghana and all the laboratory work-up were carried out at the Department of Molecular Medicine, Kwame Nkrumah University of Science and Technology (KNUST), Noguchi Memorial Institute for Medical

Research (NMIMR) and the Proteomic and Flow cytometry Core Facilities of the Indiana University School of Medicine.

The characteristics of participants/materials

In all, fifty three (53) participants who were out-patients were recruited into the study. There were 43 (23 males and 20 females) cases and 10 (5 males and 5 females) controls. The age range for controls was 28–47 years while the age range for patients was 1 week to 62 years.

Processes

Sample collection

Convenience sampling was employed for this study. Three milliliters (3 ml) of blood was collected from 43 laboratory diagnosed malaria patients. The parasite density was confirmed using independently prepared thick film. Control samples were obtained from 10 apparently healthy individuals. The samples were categorized into three (3): mild, moderate and high based on the level of parasitaemia.

Thick film preparation

Briefly, 6 microlitres (6 µl) of blood was placed in the middle of a glass slide for a thick film preparation using the WHO standardized template. The slides were then air-dried and subsequently stained as below.

Thick film staining and examination

The films were stained with 1:10 diluted Paskem® Giemsa for 10 min, air dried and examined using the Olympus °CX 31 microscope (Tokyo, Japan). Parasite count was calculated using the formula

$$Parasite\,count = \frac{Number\,of\,parasite\,counted}{Number\,of\,WBC\,counted} \times \frac{6000}{\mu l}$$

Where 6000 is the assumed WBC count per 1 µl of blood.

Purification of MPs from plasma

Briefly, EDTA-anticoagulated blood (3 mL) was centrifuged at $160 \times g$ for 5 min, and plasma was stored at $-20\,°C$ until further processing. During processing, samples were thawed and centrifuged at 4000 x g for 60 min to obtain platelet-free plasma. The resultant supernatant was then spun for 120 min at $19,000 \times g$ to obtain the microparticle pellet after the supernatant had been discarded [18]. The MP pellet was then dissolved in 50 µl of PBS and stored at $-80\,°C$.

Flow cytometry analysis

All reagents used in the flow cytometry experiments were from BD Biosciences unless otherwise stated. 50 µl

of phosphate buffered saline (PBS) was added to the thawed MP pellet. Equal volume of Annexin V binding buffer was added to label the MPs. Labeled plasma samples were analyzed on a BD FACSAria™ flow cytometer. MPs isolated from plasma were gated (Annexin V+) based on their forward (FSC) and side (SSC) scatter distribution as compared to the distribution of synthetic 0.7–0.9 µm SPHERO™ Amino Fluorescent Particles (Spherotech Inc. Libertyville, Illinois, US) (Fig. 1a). Taking into account the presence of phophatidylserine (PS) residues in MPs surface, events present in Annexin V+ region were accessed for their positive staining with annexin V (BD Bioscience). The flow cytometry analysis done in this study was to verify the presence of Annexin V+ MPs in the pellet.

Proteomic analysis

One hundred microlitres (100 µl) of frozen MPs fractions were thawed and resuspended in 60% methanol and 0.1 M ammonium bicarbonate pH 8.0. Incubation with diothiothreitol followed by cysteine alkylation with iodoacetamide reduced the proteins. The solution was incubated overnight with mass spectrometry grade trypsin (Pierce, USA) at 37 °C to digest the proteins. The obtained tryptic peptides underwent desalting using reversed phase cartridge by washing in 0.5% acetic acid. The peptides were then eluted with 95% acetonitrile and 0.5% acetic acid. The eluted peptides were lyophilized and re-suspended in 0.5% acetic acid and directly analyzed using 2-dimensional liquid chromatography tandem mass spectrometry (2D-LC-MS/MS). The samples were loaded onto a microcapillary strong cation exchange column and fractions were collected using an increasing salt step elution gradient. Afterwards, each fraction was analyzed using the LTQ ion trap mass spectrometer (Thermo-Fisher Scientific, Waltham, MA). The

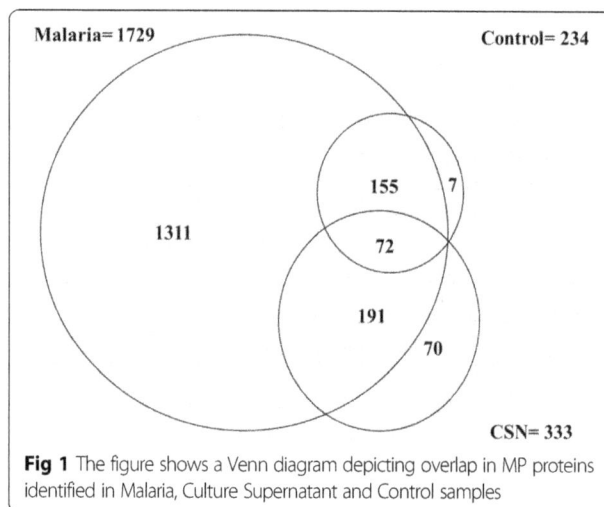

Fig 1 The figure shows a Venn diagram depicting overlap in MP proteins identified in Malaria, Culture Supernatant and Control samples

acquired MS/MS spectra were searched against protein database available at www.uniprot.org [19]. The protein interactions were acquired using the string database available at http://string-db.org/cgi/.

Western blotting

Experimental Procedures: MPs were thawed and lyzed in sample buffer and were loaded on SDS-PAGE gels. The presence of the identified proteins in the samples was analyzed by Western blot. Bands at the predicted molecular weight for the proteins were observed [20].

All experiments (both independent and technical) were repeated 2–4 times.

Data analysis

Proteins identified were analyzed under molecular function, biological processes, subcellular location and cellular component using the uniprot database. Resultant data were further analyzed using SPSS 21.0 statistical software. Differences between the medians of the various groups were analyzed by Kruska Willis test. Pairwise correlations were evaluated with Spearman's correlation coefficient r. A p-value < 0.05 was considered to be statistically significant. For illustration purposes the samples were categorized into 3 sub-groups according to the parasitaemia based on the old system of classification (1 +, 2+ and 3+ which corresponded to parasite counts <5000/μl, >5000/μl but <10,000/μl and parasite count > 10,000/μl respectively in this study) to analyze their association with proteins released in MPs.

Results

Participant characteristics

The characteristics of subjects of this study are presented in Table 1. All patients were reporting on an outpatient basis. The age range for controls was 28–47 years while the age range for patients was 1 week to 62 years.

Plasmodium (malaria) species

The predominant plasmodium specie found in Ghana is *falciparum*. This phenomenon was seen in our study where all the patients tested had *P. falciparum* malaria.

Table 1 Patient Characteristics

	Control	Patients
Number	10	43
Female (%)	50	47.5
Male (%)	50	53.5
Median Age (years)	37.5 (28–47)	22 (0.003–62)
Number below 5 years	0	9
Number pregnant	0	4

Subsequently the MPs types seen also carried *P. falciparum* markers as well as those of other activated cells (endothelial cells, erythrocytes, leukocytes and platelets).

Red blood cell indices

The red blood cell indices showed significant variation between the patients and the control. For example there was significance in haemoglobin concentration, MCV, MCH and RDW as seen in Table 2.

Proteomics analysis

A total of 1729 proteins were identified in the malaria positive samples. The culture supernatant (CSN) showed 333 proteins while a total of 234 proteins were identified in the control samples. Seventy-two (72) proteins were common to the malaria positive samples, CSN and the control. The details are shown in Fig. 1 below.

The mean number of proteins obtained from MPs in malaria samples was 517.58 ± 56.58 while that of control was 191.50 ± 5.20. The independent sample t-test showed significantly higher number of proteins released in MPs from malaria samples. Twenty-nine *P. falciparum* proteins were identified in at least one malaria positive sample (Table 3).

Twenty three (23) RAB proteins were identified in MPs from malaria samples exclusively without any being found in the control samples as presented in Table 4. Out of the 1729 proteins found in the MPs from malaria positive samples, 653 were identified in plasma MPs from 1 category of samples only while 1076 proteins were identified in at least two categories and 697 proteins were found to be common to all 3 categories. The details of the relationship between the proteins released in plasma MPs from the 3 categories of malaria positive samples are shown in Fig. 2 below.

Plasma MPs and inflammation

Some of the proteins identified showed strong link to inflammation. Inflammatory proteins seen in malaria MPs but not control sample MPs include Heat shock protein

Table 2 Red Blood Cell Indices

Parameter (Unit)	Control	Patients	P-value
	Mean ± SD	Mean ± SD	
Hb (g/dl)	13.20 ± 0.94	10.35 ± 2.61	**0.037**
Hct (%)	36.30 ± 3.35	30.74 ± 8.10	0.184
MCV (fl)	92.95 ± 2.37	80.68 ± 10.75	**0.029**
MCH (pg)	33.78 ± 0.35	26.82 ± 3.66	**0.000**
MCHC (g/dl)	35.75 ± 0.64	33.28 ± 2.64	0.072
RBC count ($\times 10^{12}$/l)	3.83 ± 0.26	3.86 ± 0.83	0.934
RDW (%)	12.05 ± 0.29	15.80 ± 3.35	**0.032**

A p-value of < 0.05 was deemed to be statistically significant

Table 3 *P. falciparum* proteins identified in MPs from malaria positive samples

Accession	Protein Name	Protein MW	Species
P04934	Merozoite surface protein 1	196199.8	PLAFC
P08569	Merozoite surface protein 1	193722.4	PLAFM
P13819	Merozoite surface protein 1	193721.5	PLAFF
P19598	Merozoite surface protein 1	192465	PLAF3
P86287	Actin-1	41871	PLAFX
Q8I4X0	Actin-1	41871	PLAF7
P10988	Actin-1	41843	PLAFO
P11144	Heat shock 70 kDa protein	74288	PLAFA
Q00080	Elongation factor 1-alpha	49041.2	PLAFK
P06719	Knob-associated histidine-rich protein	71941.5	PLAFN
P14643	Tubulin beta chain	49751.4	PLAFK
Q7KQL5	Tubulin beta chain	49751.4	PLAF7
P38545	GTP-binding nuclear protein Ran	24875.5	PLAFA
Q27727	Enolase	48704.2	PLAFA
Q8IJN7	Enolase	48678.1	PLAF7
Q9UAL5	Enolase	48662.1	PLAFG
P14140	Tubulin beta chain	49814.4	PLAFA
P13830	Ring-infected erythrocyte surface antigen	124907.8	PLAFF
Q25761	ADP-ribosylation factor 1	20840	PLAFO
Q7KQL3	ADP-ribosylation factor 1	20912	PLAF7
Q94650	ADP-ribosylation factor 1	20912	PLAFA
P12078	Heat shock 70 kDa protein PPF203 (Fragment)	23057.9	PLAFA
P13816	Glutamic acid-rich protein	80551.3	PLAFF
P19260	Merozoite surface antigen 2, allelic form 2	28555.5	PLAFG
P19599	Merozoite surface antigen 2	27890.3	PLAFF
P50490	Apical membrane antigen 1	71968.2	PLAFG
Q03498	V-type proton ATPase catalytic subunit A	68577	PLAFA
P06916	300 kDa antigen AG231 (Fragment)	33968.1	PLAFF
P04928	S-antigen protein	33695.1	PLAFN

(HSP) 90-alpha, HSP 90-beta, 60 kDa HSP mitochondrial, 70 kDa HSP, HSP beta-1, Transforming growth factor beta-1 induced transcript 1 protein, Macrophage migration inhibitory factor and Transforming growth factor beta-1. Again, inflammatory proteins released in malaria MPs but in reduced quantities included: Heat shock cognate 71 kDa protein, Heat shock 70 kDa protein 1A/1B, Heat shock 70 kDa protein 6, Heat shock 70 kDa protein 1-like.

Plasma MPs and the complement system
A good number of complement associated proteins were seen in the MPs from the malaria samples. It was observed that the levels of Complement C3 and Complement C4-B, Complement C5, Complement C9, Complement factor B and Complement C1q subcomponent subunit C were significantly elevated in malaria plasma MPs compared to controls (Table 5).

Plasma MPs and haemostasis
The following proteins associated with coagulation were release in malaria plasma MPs but not in control plasma. They are; Thrombospondin-1, Thrombospondin-4, Coagulation factor V, Coagulation factor XII and Coagulation factor XIII A chain. Again, Fibrinogen beta chain and plasminogen released in the malaria plasma MPs were significantly higher compared to control plasma. There was however no significant difference in levels of Fibrinogen alpha chain, Fibrinogen gamma chain and von Willebrand factor between the 2 groups. Antithrombin-III in plasma MPs was however significantly reduced in the patients compared to the control group. The summary is presented in Table 6.

Table 4 List of Rab Proteins identified in MPs isolated from malaria samples

Accession	Protein Name	Protein MW	Species
Q9H0U4	Ras-related protein Rab-1B	22171.4	HUMAN
O00194	Ras-related protein Rab-27B	24608.1	HUMAN
P62820	Ras-related protein Rab-1A	22678	HUMAN
P61006	Ras-related protein Rab-8A	23668.4	HUMAN
P61106	Ras-related protein Rab-14	23897.2	HUMAN
P61026	Ras-related protein Rab-10	22541.1	HUMAN
P31150	Rab GDP dissociation inhibitor alpha	50583.2	HUMAN
P51149	Ras-related protein Rab-7a	23490	HUMAN
Q92930	Ras-related protein Rab-8B	23584.3	HUMAN
Q9NRW1	Ras-related protein Rab-6B	23461.9	HUMAN
P62491	Ras-related protein Rab-11A	24393.7	HUMAN
P50395	Rab GDP dissociation inhibitor beta	50663.7	HUMAN
Q15286	Ras-related protein Rab-35	23025.4	HUMAN
P20340	Ras-related protein Rab-6A	23593	HUMAN
P51148	Ras-related protein Rab-5C	23482.8	HUMAN
P51153	Ras-related protein Rab-13	22774.3	HUMAN
Q13637	Ras-related protein Rab-32	24997.5	HUMAN
Q96AX2	Ras-related protein Rab-37	24815.4	HUMAN
Q96E17	Ras-related protein Rab-3C	25952.4	HUMAN
Q9UL25	Ras-related protein Rab-21	24347.8	HUMAN
P61020	Ras-related protein Rab-5B	23707	HUMAN
P61019	Ras-related protein Rab-2A	23545.8	HUMAN
Q9NP72	Ras-related protein Rab-18	22977.3	HUMAN

Plasma MPs and haemoglobin sub-units
Some haemoglobin subunit proteins were seen in the circulating MPs from both malaria positive samples and controls. The quantities of haemoglobin subunits in plasma MPs from malaria positive samples were

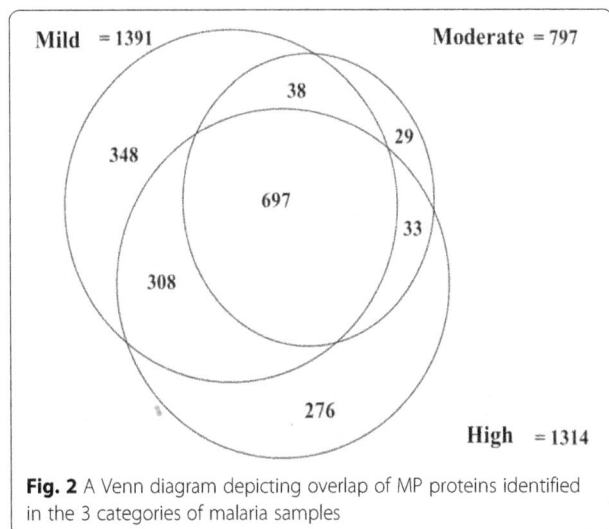

Fig. 2 A Venn diagram depicting overlap of MP proteins identified in the 3 categories of malaria samples

however significantly higher as compared to controls (Table 7).

Plasma MPs and adhesive proteins
Some adhesive and receptor proteins were identified in the plasma MPs from the malaria samples but they were absent in the control samples. They include: Merozoite surface protein 1, Knob-associated histidine-rich protein, Disintegrin and metalloproteinase domain-containing protein-10,-12,-17, Complement receptor type 1, Cysteine and glycine-rich protein 1, Ring-infected erythrocyte surface antigen, Glycophorin-binding protein, Intercellular adhesion molecule 2, A disintegrin and metalloproteinase with thrombospondin motif 13, Merozoite surface antigen 2, Endothelial cell-selective adhesion molecule.

Plasma MPs and cytoskeletal proteins
Cytoskeletal proteins were identified in the malaria samples but they were absent in the control samples. Spectrin alpha chain, erythrocytic 1, Spectrin beta chain erythrocytic 1, Coronin 1A, Coronin 1C, Myosin 9, Actin-1, Actin-related protein 3, Actin-related protein 2, Actin-related protein 2/3 complex subunit 3, Actin-related protein 2/3 complex subunit 4 and Actin-related protein 2/3 complex subunit 5.

Flow cytometer distribution for MPs show a typical forward-/side-scatter distribution
MPs can be identified by their forward-/side- scatter appearance on flow cytometer. As can be seen from Fig. 3, the forward/side scatter distribution of the MPs from the samples shows a similar pattern in line with the classical appearance of MPs on flow cytometer and comparable to those obtained by others in similar experiments. This, in combination with other properties such as the proteins detected give credence to the fact that throughout our experiments, MPs were being obtained.

Protein enrichment of the isolated MPs
Figure 4 shows protein enrichment of MPs (A) and proteins differentially expressed by MPs (B) by Western blot. The proteins were loaded according to the sample categories of mild, moderate and high malaria as well as no-malaria and bands at the predicted molecular weight for each of the proteins were observed.

Correlation coefficient of clinical variables
Tables 8, 9 and 10 show the Spearman correlation coefficients of various groups of proteins identified in MPs from malaria samples against the parasite count and peripheral haemoglobin concentration. There was no significant correlation of all the haemoglobin subunits against either the peripheral haemoglobin concentration or parasite count.

Table 5 Analysis of complement proteins from samples

Parameter	Patients	Control	P-value
Non-Parametric	Median (Q1-Q3)	Median (Q1-Q3)	
Complement C3	192.5 (181–204)	134 (88–183)	0.083
Complement C4-B	103 (96–110)	38 (23–60)	0.000
C4b-binding protein alpha chain	18.5 (18–19)	15 (10.25–19.75)	0.192
Complement C1r subcomponent	15.5 (12–19)	16.5 (6.75–24.25)	0.895
Complement C1s subcomponent	9 (5–13)	6 (4–11.75)	0.311
Complement C5	11 (9–13)	4 (2–8)	0.620
Complement C1q subcomponent subunit B	3.5 (3–4)	5 (3–7)	0.142
Complement component C9	23 (21–25)	11 (7–18)	0.000
Complement factor B	11.5 (8–15)	5 (3–9.75)	0.025
Complement factor H	12.5 (5–20)	5 (2–10)	0.150
Complement C1q subcomponent subunit A	2 (1–4)	3 (2–5)	0.315
Complement C1q subcomponent subunit C	4 (2–6)	2 (1–3)	0.035
Complement component C7	6 (4–8)	2 (1.75–4)	0.061
Parametric	Mean ± SD	Mean ± SD	
Complement component C6	4.27 ± 0.43	1.33 ± 0.52	**0.000**

A p-value of < 0.05 was deemed to be statistically significant

Peripheral haemoglobin however negatively correlated with parasite count ($r = -0.415$, $p < 0.001$) and Mannose-binding lectin serine protease 2 ($r = -0.664$, $p < 0.001$) while it positively correlated with Complement C3 ($r = 0.457$, $p < 0.001$). Parasite count correlated negatively with Complement C3 ($r = -0.359$, $p < 0.05$) and correlated positively with Complement C1s subcomponent ($r = 0.407$, $p < 0.001$) and C4b-binding protein beta chain ($r = 0.845$, $p < 0.05$).

Discussion

The study identified an array of proteins from the MPs isolated from the malaria positive samples, culture supernatant and healthy controls. The mean haemoglobin (Hb) of the studied subjects was significantly lower than that of the control group ($p = 0.037$). This finding agrees with a study in the Brazilian Amazon in which out of a total of 7831 people studied, individuals with malaria were seen to have the lowest Hb compared to community controls and malaria-negative febrile patients [21]. The results of our study also collaborate with the results from a study carried out in Nigeria [22]. These observations are expected because, as far back as in 1947 the plasmodium was reported to "consume" haemoglobin during its growth within erythrocytes [23]. *P. falciparum* for instance is capable of digesting more than 80% of the erythrocyte haemoglobin [24].

Table 6 Analysis of haemostatic proteins from samples

Parameter	Control	Patient	P-value
Non-Parametric	Median (Q1-Q3)	Median (Q1-Q3)	
von Willebrand factor	11.5 (11–12)	15 (5.5–27)	0.493
Antithrombin-III	17 (13–21)	5.5 (3–8)	0.001
Thrombospondin-1	39 (17.25–71.25)	0	0.000
Coagulation factor V	16.50 (5.25–36.50)	0	0.000
Coagulation factor XIII A chain	8 (1–19)	0	0.000
Thrombospondin-4	0.00 (0–3.75)	0	0.043
Coagulation factor XII	1 (0–2)	0	0.004
Plasminogen	30.0 (10.25–43.75)	0	0.000
Parametric data	Mean ± SD	Mean ± SD	
Fibrinogen alpha chain	88.5 ± 0.58	59.01 ± 28.35	0.045
Fibrinogen beta chain	74.13 ± 6.93	56.79 ± 27.62	0.225
Fibrinogen gamma chain	53.37 ± 9.24	37.56 ± 16.05	0.066

Table 7 Analysis of haemoglobin subunit proteins in samples

Parameter	Control	Patients	P-value
	Median (Q1-Q3)	Median (Q1-Q3)	
Haemoglobin subunit gamma-2	1 (1–1)	23 (10–63)	0.001
Haemoglobin subunit beta	13 (11–15)	143 (69–274)	0.000
Haemoglobin subunit gamma-1	1 (1–1)	28 (10.5–64.5)	0.000
Haemoglobin subunit alpha	11 (9–13)	125 (52–284)	0.000
Haemoglobin subunit delta	3.5 (3–4)	72 (40–138)	0.000
Haemoglobin subunit epsilon	1 (1–1)	21 (8–51)	0.000

This goes to explain the significant negative correlation of Hb and parasite count ($r = -0.415$, $p < 0.001$) indicating that the greater the parasite density in a patient the lower the Hb, as was also reported in a study in Thailand [25].

Again, there was a reduction of the haematocrit (Hct) of the patients (Table 2) as compare to controls which was however not significant. Similar results were reported in patients with uncomplicated malaria where the authors indicated that, uncomplicated malaria was

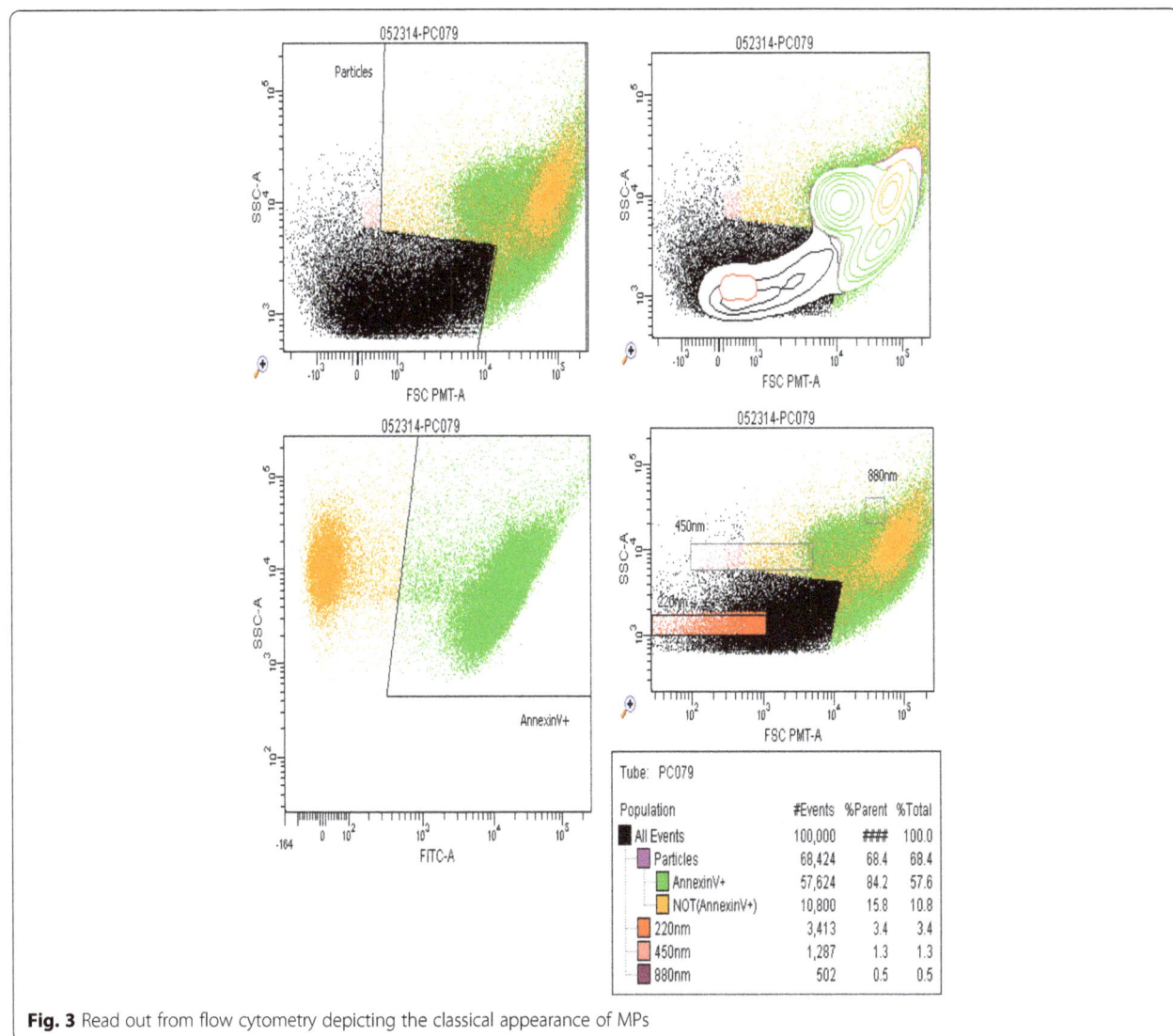

Fig. 3 Read out from flow cytometry depicting the classical appearance of MPs

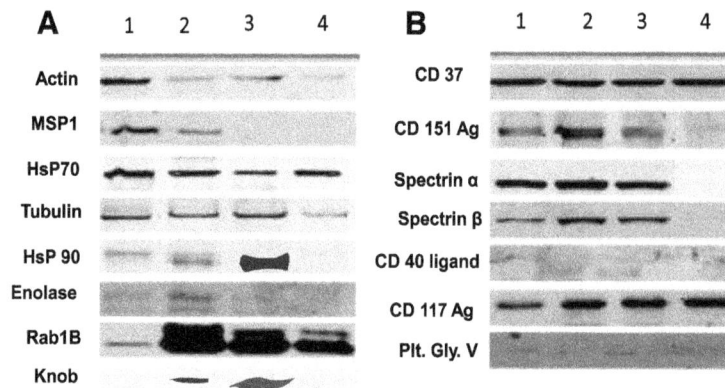

Fig. 4 A figure showing protein enrichment of MPs **a** and proteins differentially expressed by MPs **b** through western blotting. The proteins were loaded according to the sample categories of mild, moderate and high malaria as well as no malaria and bands at the predicted molecular weight for each of the proteins were observed. The figure shows representative gels of 2–4 experiments

associated with milder biochemical alteration and haemolysis as opposed to complicated/severe malaria [26]. A study done at ECWA Community Health Centre, Bukuru, Jos in Nigeria however showed significant reduction of the haematocrit of malaria samples compared to healthy controls [22].

The mean cell volume (MCV) of the patients was significantly lower than that of the control group (Table 2). This result was unexpected because previous studies in a population near Thailand-Myanmar border indicated the

Table 8 Spearman's rank correlation coefficients of Haemoglobin subunits and coagulation proteins released in MPs against peripheral haemoglobin concentration and parasite count

Microparticle Protein	HB	Parasite Count
HB		−0.415**
Haemoglobin subunit epsilon	0.2	−0.25
Haemoglobin subunit gamma-2	0.201	−0.227
Haemoglobin subunit beta	−0.063	0.085
Haemoglobin subunit gamma-1	0.197	−0.17
Haemoglobin subunit alpha	−0.09	0.212
Haemoglobin subunit delta	−0.059	0.13
Haemoglobin subunit mu	−0.5	0.5
Alpha-Haemoglobin-stabilizing protein	0.5	−0.5
Fibrinogen beta chain	0.211	−0.254
Fibrinogen alpha chain	0.029	0.121
Fibrinogen gamma chain	0.285	−0.131
Coagulation factor V	0.177	−0.151
Coagulation factor XIII A chain	0.053	0.037
von Willebrand factor	0.036	0.047*
Antithrombin-III	0.175	−0.079
Coagulation factor XII	0.2	−0.006

*Correlation significant at $p < 0.05$
**Correlation significant at $p < 0.001$

Table 9 Spearman's rank correlation coefficients of Complement proteins released in MPs and peripheral haemoglobin concentration and parasite count

Complement Proteins	HB	Parasite Count
Complement C4-B	0.21	−0.161
Complement component C6	−0.207	−0.414
Complement C3	0.457**	−0.359*
C4b-binding protein alpha chain	0.095	0.103
Complement C1r subcomponent	0.006	0.045
Complement C1s subcomponent	−0.081	0.407**
Complement C5	−0.017	0.025
Complement C1q subcomponent subunit B	0.147	−0.191
Complement component C9	0.045	−0.111
Complement factor B	0.007	0.144
Complement factor H	0.078	−0.004
Complement C1q subcomponent subunit A	−0.071	0.352
Complement C1q subcomponent subunit C	0.124	−0.224
Complement component C8 beta chain	0.202	0.251
Complement component C7	0.025	0.350
Complement factor I	−0.059	−0.736
Complement component C8 alpha chain	0.421	0.089
Complement component C8 gamma chain	−0.344	0.465
Complement decay-accelerating factor	−0.514	0.062
C4b-binding protein beta chain	−0.338	0.845*
Complement factor H-related protein 2	−0.258	0.775
Complement factor I	−0.059	−0.736
Complement component C8 alpha chain	0.421	0.089
Complement component C8 gamma chain	−0.344	0.465

*Correlation significant at $p < 0.05$
**Correlation significant at $p < 0.001$

Table 10 Spearman's rank correlation coefficients of selected microparticle proteins against Parasite count and Haemoglobin

Microparticle Protein	HB	Parasite Count
Leukocyte elastase inhibitor	−0.106	−0.206
P-selectin	0.078	−0.023
Mannan-binding lectin serine protease 2	−0.664**	0.176
Intercellular adhesion molecule 1	−0.686	0.169
Intercellular adhesion molecule 2	−0.093	−0.109
Intercellular adhesion molecule 3	−0.726*	0.676
Mannan-binding lectin serine protease 1	−0.157	0.258

*Correlation significant at $p < 0.05$
**Correlation significant at $p < 0.0$

opposite finding usually because as erythrocytes are being destroyed in malaria infection, the bone marrow is stimulated to push more new erythrocytes whose mean cell volume are higher than older erythrocytes [25]. Also, the mean cell haemoglobin (MCH) of patients was significantly lower than that of the control group. This result however contradicts with the results of the population near the Thailand-Myanmar border where the predominant malaria parasite is *P. vivax* [25]. Therefore the results seen in this study can probably be explained by the ability of *P. falciparum* (predominant in Ghana) to consume haemoglobin during its growth in the erythrocyte [23] and the propensity of the erythrocytes in malaria to release more haemoglobin-bearing MPs into circulation leaving the erythrocytes with less haemoglobin compared with erythrocytes from healthy controls [23].

The red cell distribution width (RDW) was significantly higher ($p = 0.032$) in the patients compared to the control group. It was however consistent with results from a *P. vivax* malaria study [27]. The mean RDW was 15.80 and RDW greater than 15 has been shown to be a predictive value of malaria infection, although some researchers contest this [28, 29]. In *P.vivax* malaria however, the elevated RDW results from the initial size increase of parasitized erythrocytes which is followed by erythrocyte rapture. The same mechanism however cannot be said to be associated with RDW increase in *P. falciparum* infection since parasitized cells maintain their sizes [29]. The elevated RDW seen in this study could result from the bone marrow's effort to balance erythrocytes loss in malaria by pushing more new erythrocytes into circulation which also increases the macrocyte percentage [30, 31].

Again, our data indicated that a significantly higher amount of all haemoglobin subunits released came from MPs of malaria positive samples. This point to an interesting phenomenon where the erythrocytes from malaria patients lose their haemoglobin to vesiculation and the vesicles that result may be eliminated by the body

thereby contributing to anaemia. Table *3* shows a list of 24 RAB proteins found in the malaria samples. They include proteins which play pivotal roles in relation to MP docking, fusion and appropriate targeting of various cellular compartments [32].

Also, among the proteomics data were some erythrocyte cytoskeletal proteins which have been referred to as low abundance cytoskeletal proteins [33]. They include myosin chains, moesin, ezrin and F-actin capping proteins. Furthermore, some Pf proteins which have been described as potential drug targets were identified in the MPs, they include: enolase, Hsp 90, hypoxanthine guanine phosphoribosyl transferase, L-lactate dehydrogenase and Phosphoglycerate kinase. This is because the parasite through these proteins in their pathways starts processes such as glycolysis, haemoglobin digestion and salvage of purines [34]. Tubulin Beta which is a brain antigen that can discriminate cerebral malaria from other forms of the disease was identified in MPs from 38 samples [35]. This protein chain is an antimalarial target [36].

Enolase is a glycolytic enzyme and especially essential in ATP generation in organisms that are devoid of the Krebs cycle. In malaria, surface enolase assists in parasite invasion by binding to plasminogen [37]. Enolase was found in MPs of 38 of the samples. It is known to play a crucial role in the parasite invasion of the midgut of the mosquito [38]. A study has shown that its levels are elevated in infected red blood cells compared to uninfected cells. Antibodies against merozoite surface enolase have the ability to interfere with *P. falciparum* invasion of the red cell thereby conferring a partial protection against malaria [39]. Knob-associated histidine-rich protein (KAHRP) was identified in only 1 of the 43 malaria samples. KAHRP is a major component of the knob which is an electron-dense protrusion located on the membrane of infected erythrocytes. The knob is the site of adhesive interaction between infected erythrocytes and vascular endothelial cells [40].

Plasmodium falciparum Merozoite surface protein 1 (MSP-1) which is also called precursor to major merozoite surface antigens (PMMSA) or merozoite surface antigen 1 (MSA-1) was identified in MPs from 2 of the malaria samples [41]. This protein is the most widely studied of proteins of *Plasmodium falciparum*. MSP-1 is well conserved among *P. falciparum* isolates and hence it was identified in the MP isolated from the culture supernatant [42]. This protein binds to erythrocytes in a sialic dependent manner suggesting it is a receptor to a ligand on erythrocyte surface permitting adhesion of *Plasmodium falciparum* [41].

Also, merozoite surface antigen 2 (MSP2) which is a surface coat protein essential for the survival of the blood-stages of *P. falciparum* was identified [43]. This protein and its allelic form 2 were identified in MPs from one sample. Also identified in one sample was *Plasmodium falciparum* Apical membrane antigen 1 (PfAMA1) which is synthesized during schizogony and transported to micronemes. PfAMA1 translocates onto the merozoite surface prior to erythrocyte invasion when it serves as an adhesion molecule playing a central role in the invasion process [44]. Ring-infected erythrocyte surface antigen (RESA), a protein expressed in early stage gametocytes, final stages of schizont and stored in dense granules within the merozoites was found in one sample. RESA binds to spectrin, its primary site on erythrocyte and is associated with the membrane of recently invaded erythrocyte but it is only evident in the cell up to 24 h after parasite invasion [45].

Glutamic acid-rich protein (GLURP) found in one sample is a molecule located on the surface of merozoites and also in the hepatic stage [46]. This protein is an antigen and its epitopes defined by non-repetitive sequence are thought of to be more effective antibody-dependent cellular inhibition process and is believed to be involved in acquired protective immunity to malaria. These epitopes are therefore proposed to be preferred in vaccine formulations against malaria [47, 48]. Also, *P. falciparum* actin 1(PfACT1) was expressed in the MPs of 35 samples. PfACT1 is a ubiquitously expressed protein and it is expressed throughout the lifecycle of the *Plasmodium*. However, its isoform PfACT 2 which is expressed only in the sexual stages of the parasite was not isolated in any of the samples [49].

Heat shock 70 kDa protein (Hsp70) was expressed in 35 samples. *P. falciparum* has six Hsp70s [50]. Hsp70 proteins have molecular chaperone functions, and are involved in a number of processes including protein degradation and controlling of the activity of regulatory proteins, protein folding and protein translocation across membranes. Hsp 70 members are present in almost all cellular compartments [51] and this could account for its identification in a good number of our samples. Also, elongation factor-1 alpha was found in the MPs from 7 of the malaria positive samples. It is an abundant protein that is an essential element in eukaryotic protein translation, in species of *Plasmodium* [52]. It is key in the proliferation of the blood stages of *Plasmodium* [53].

Precursors of serine-repeat antigen (SERA) proteins are synthesized in the late trophozoites [54]. Among organisms in the apicomplexan phylum and with the exception of *Theileria* found in cattle, species of *Plasmodium* are the only organisms that the gene family translating into Serine-repeat antigen (SERA) protein has been found [55]. This protein was found in 1 sample. Investigation with anti-SERA antibody has established SERA protein as a target for antibody dependent cellular inhibition of *P. falciparum* development [56].

Also identified was *Plasmodium falciparum* ADP-ribosylation factor 1 (PfARF1) which is activated after binding to GTP. In the secretary pathway, PfARF1 regulates vesicular biogenesis and trafficking processes [57]. It is also thought of to be involved in the transport of MSP-1 from the endoplasmic reticulum and plays a role in the activation of a calcium-signalling mechanism in the parasite [58]. Hypothetical protein identified 300 kDa antigen AG231 (Fragment) was also found in 1 sample. This protein was discovered in about 93% of 65 patients living in a malaria endemic area in Papua Guinea. It is found in schizonts and trophozoites. Again, *P. falciparum* S-antigen proteins were found in the MPs of 2 samples. They are heat-stable antigens that are serologically diverse among varied isolates of the parasite [59].

There was significant difference in the levels of the following complement system proteins released in plasma circulating MPs between the malaria samples and that of control: complement component C6, complement C4-B, complement component C9, complement factor B and Complement C1q subcomponent subunit C. The difference between the patients and the control group in complement component C3 was however not significant but correlation analysis indicated a significant negative correlation between C3 released in MPs and parasite count in the malaria samples. There was significant positive correlation between C3 and haemoglobin ($r = 0.457$). It is known that MPs are able to bind to complement component C1q and cause the C3 to be fixed on the MPs after exposure to normal human serum through the classical pathway. These complement-fixed MPs are capable of binding to erythrocytes and remove them from circulation [60]. It could therefore be presumed that complement-fixed MPs bound to erythrocyte might be a mechanism that explains the increase erythrocyte clearance in malaria leading to reduced haemoglobin concentration. This presumption can be related to a similar work that explains the role of MPs in complement activation relating to the pathogenesis of rheumatoid arthritis [61].

Also, Hb negatively correlated with Mannan-binding lectin serine protease 2 released in MPs (Table *10*). This may be explained by the ability of this protein to activate complement and interact with phagocytes and also serves as an opsonin to Plasmodium infected erythrocytes facilitating the elimination of such erythrocytes from circulation and invariably causing reduction in Hb [62].

C4 binding protein beta chain in MPs from malaria samples correlated positively with parasite count. C4 binding protein is an abundant plasma protein whose natural function is to inhibit the classical and lectin pathways of complement activation. Research has shown that the fusion of oligomerization domains of C4 binding

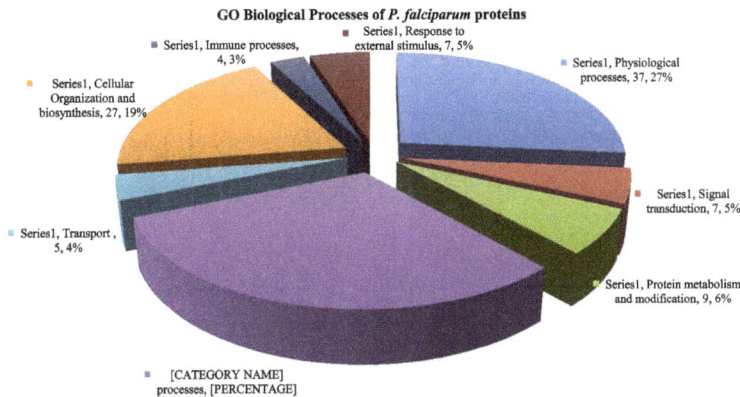

Fig. 5 3D pie chart of the Gene Ontology of all *P. falciparum* proteins isolated from samples that are involved in biological processes

protein to MSP-1 from *Plasmodium yoelii* improved its immunogenicity [63]. C4 binding protein beta chain released in MPs could hence be acting as an adjuvant to MSP-1 antigens thereby eliciting the appropriate immune response against the parasite. Antithrombin concentration with the MPs in the patients was significantly low compared with the controls [64]. Serum concentration of antithrombin III particularly in severe malaria is reduced [65, 66]. This is because generally malaria is associated with the consumption of Antithrombin [67]. Figures 5 and 6 represent isolated proteins that are involved in biological and molecular functions, Fig. 7 represent isolated proteins that are cellular components and Fig. 8 a network model delineating the *P. falciparum* proteins isolated from MPs of malaria positive samples and their associated pathways.

In terms of limitations, because convenience sampling was used in this study further stratification of the patients and control was not done to assess the association of parameters like age category, sex and duration of onset of symptoms on the proteomic profile of subjects. If possible, a further prospective sampling involving sex and age matched controls should be employed in future

study so that data could be stratified to address the afore-mentioned limitation.

Conclusion

Our study has clearly established that there is an up regulation of protein content of circulating plasma MPs from malaria patients than in healthy controls. A great variety of protein content can be seen in malaria patients although a large number of proteins identified were common to the 3 categories of malaria samples. Furthermore, higher amounts of haemoglobin subunits are released into circulating MPs of malaria patients. These MPs when finally eliminated from circulation have the propensity to cause anaemia. Interaction enrichment indicates that the *P. falciparum* proteins isolated from the malaria samples are least partially connected as a group.

The protein-protein interaction (PPI) of the interaction of the *P. falciparum* indicated that the proteins have more interactions among themselves than what is expected for a random set of proteins of similar size drawn from the *P. falciparum* genome. The enrichment indicates the proteins are at least partially connected as a group.

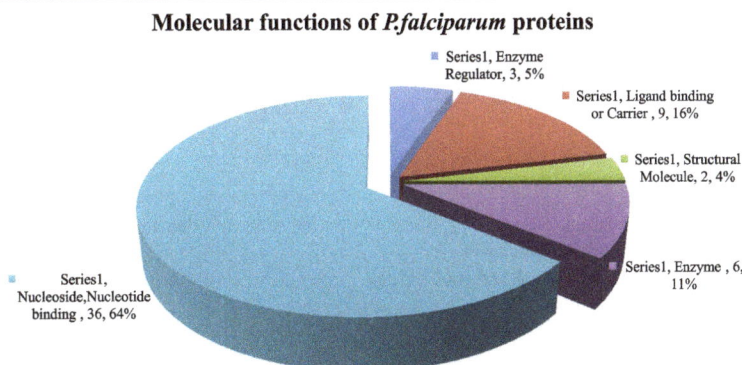

Fig. 6 3D pie chart of Gene Ontology of the *P. falciparum* proteins isolated from the malaria samples that are involved in molecular functions

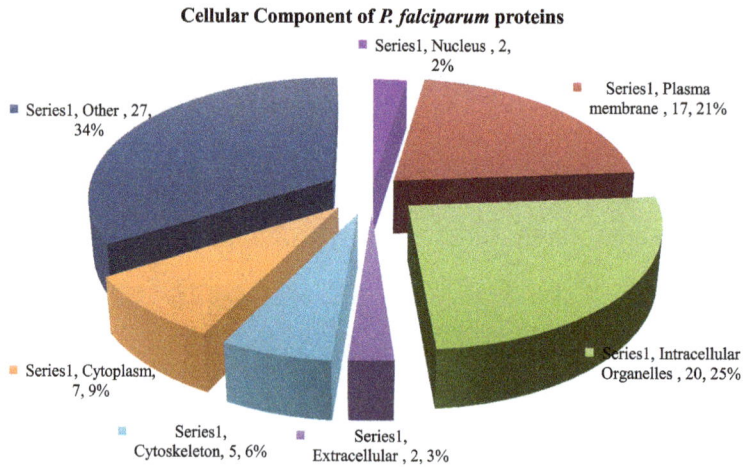

Fig. 7 3D pie chart of Gene Ontology of *P. falciparum* proteins isolated in the malaria samples that are cellular components

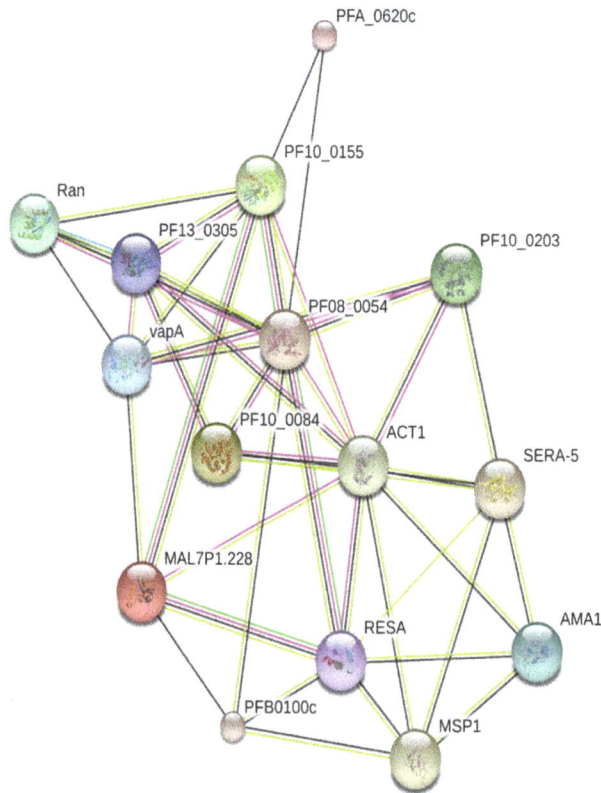

Fig. 8 A network model delineating the *P. falciparum* proteins isolated from the MPs in malaria positive plasma and their associated pathways. The node colours represent MP proteins and their interactions (colour coded basis for a particular interaction)

Abbreviations

ATP: Adenosine triphosphate; cryo-TEM: Cryogenic transmitting electron microscopy; DNA: Deoxyribonucleic acid; EC: Ectosomes; EDTA: Ethylene diamine tetra acetic acid; ELISA: Enzyme linked immunosorbent assay; EMP: Endothelial microparticles; EPCR: Endothelial protein C receptor; ErMP: Erythrocyte microparticles; EV: Extracellular vesicles; Hb: Haemoglobin; HbF: Fetal haemoglobin; Ig: Immunoglobulin; IL: Interleukin; LMP: Leucocyte derived-MP; MBL: Mannose-binding lectin; MCH: Mean cell haemoglobin; MCHC: Mean cell haemoglobin concentration; MonoMP: Monocyte derived MP; MP: Microparticles; MV: Microvesicles; MW: Molecular weight; nano-HPLC-MS/MS: nano-high performance liquid chromatography combined with tandem mass spectrometry; NeuMP: Neutrophil Microparticle; *P.falciparum*: *Plasmodium falciparum*; PBS: Phosphate buffered saline; PC: Phosphatidylcholine; PE: Phosphatidylethanolamine; Plt. Gly. V: Platelet glycoprotein V; PMEV: Plasma membrane derived extracellular vesicles; PMP: Platelet microparticles; PNH: Paroxysmal nocturnal haemoglobinuria; PS: Phosphatidylserine; RNA: Ribonucleic acid; SAH: Subarachnoid haemorrhage; SCD: Sickle cell disease; SM: Sphingomyelin; TF: Tissue factor; TNF-α: Tumour necrosis factor alpha.

Acknowledgements

We are grateful to the Ke Hu Laboratory, Department of Biology and the Proteomic and Flow cytometry Core Facilities all of the Indiana University School of Medicine for their assistance in analyzing the samples.

Funding

This study was funded by the Office of Research, Innovation and Development (ORID) of the University of Ghana, Legon. ORID did not play any role in the design of the study and collection, analysis, and interpretation of data and in writing the manuscript.

Authors' contributions

SAB is the PI of the study, participated in the design, co-supervised the research and drafted the manuscript. JKA and RK participated in the design of the study and carried out the experimental work. MA participated in the design and co-supervised the work. BG, GAA and FAY participated in the supervision of the work and proof reading of the manuscript. All authors read and approved the final manuscript.

Competing interests

The authors declare that they have no competing interests.

Author details

[1]Department of Medical Laboratory Sciences, School of Biomedical and Allied Health Sciences, College of Health Sciences, University of Ghana, P. O. Box KB 143Korle-Bu, Accra, Ghana. [2]Department of Molecular Medicine, School of Medical Sciences Kwame Nkrumah University of Science and Technology, Kumasi, Ghana. [3]Noguchi Memorial Institute of Medical Research, University of Ghana, Legon, Ghana.

References

1. Antwi-Baffour SS. Molecular characterisation of plasma membrane-derived vesicles. J Biomed Sci. 2015;22(1):1–7.
2. Ratajczak J, Wysoczynski M, Hayek F, Janowska-Wieczorek A, Ratajczak M. Membrane-derived microvesicles: important and underappreciated mediators of cell-to-cell communication. Leukemia. 2006;20(9):1487–95.
3. WHO. World malaria report 2013. Zurich: World Health Organization; 2014.
4. Hu WC. Microarray analysis of PBMC after Plasmodium falciparum infection: Molecular insights into disease pathogenesis. Asian Pac J of Trop Med. 2016;9(4):313–23.
5. Buffet PA, Safeukui I, Milon G, Mercereau-Puijalon O, David PH. Retention of erythrocytes in the spleen: a double-edged process in human malaria. Curr Opin Hematol. 2009;16(3):157–64.
6. Moxon CA, Heyderman RS, Wassmer SC. Dysregulation of coagulation in cerebral malaria. Mol Biochem Parasit. 2009;166(2):99–108.
7. Upasana S, Prakash KS, Shantanu KK, Biranchi NM, Manoranjan R. Association of TNF level with production of circulating cellular microparticles during clinical manifestation of human cerebral malaria. Hum Immunol. 2013;74(6):713–21.
8. Combes V, Taylor TE, Juhan-Vague I, Mège J-L, Mwenechanya J, Tembo M, et al. Circulating endothelial microparticles in malawian children with severe falciparum malaria complicated with coma. JAMA. 2004;291(21):2542–4.
9. Mantel PY, Hoang AN, Goldowitz I, Potashnikova D, Hamza B, Vorobjev I, et al. Malaria-infected erythrocyte-derived microvesicles mediate cellular communication within the parasite population and with the host immune system. Cell Host Microbe. 2013;13(5):521–34.
10. Owens III AP, Mackman N. Microparticles in Hemostasis and Thrombosis. Circ Res. 2011;108(10):1284–97.
11. Mfonkeu JBP, Gouado I, Fotso Kuaté H, Zambou O, Amvam Zollo PH, Grau GER, et al. Elevated Cell-Specific Microparticles Are a Biological Marker for Cerebral Dysfunctions in Human Severe Malaria. PLoS One. 2010;5(10):e13415.
12. Nantakomol D, Dondorp AM, Krudsood S, Udomsangpetch R, Pattanapanyasat K, Combes V, et al. Circulating Red Cell–derived Microparticles in Human Malaria. J Infect Dis. 2011;203(5):700–6.
13. Marcilla A, Martin-Jaular L, Trelis M, de Menezes-Neto A, Osuna A, Bernal D, et al. Extracellular vesicles in parasitic diseases. J Extracell Vesicles. 2014;310: 3402/jev.v3403.25040.
14. Taraschi TF, Trelka D, Martinez S, Schneider T, O'Donnell ME. Vesicle-mediated trafficking of parasite proteins to the host cell cytosol and erythrocyte surface membrane in Plasmodium falciparum infected erythrocytes. Int J Parasitol. 2001;31(12):1381–91.
15. Regev-Rudzki N, Wilson DW, Carvalho TG, Sisquella X, Coleman BM, Rug M, et al. Cell-cell communication between malaria-infected red blood cells via exosome-like vesicles. Cell. 2013;153(5):1120–33.
16. Haldar K, Mohandas N. Malaria, erythrocytic infection, and anemia. ASH Education Program Book. 2009;1:87–93.
17. Schorey JS, Cheng Y, Singh PP, Smith VL. Exosomes and other extracellular vesicles in host–pathogen interactions. EMBO Rep. 2015;16(1):24–43.
18. Grant R, Ansa-Addo E, Stratton D, Antwi-Baffour S, Jorfi S, Kholia S, et al. A filtration-based protocol to isolate human plasma membrane-derived vesicles and exosomes from blood plasma. J Immunol Methods. 2011;371(1–2):143–51.
19. Dean WL, Lee MJ, Cummins TD, Schultz DJ, Powell DW. Proteomic and functional characterisation of platelet microparticle size classes. Thromb Haemost. 2009;102(4):711–8.
20. Sandvig K, Llorente A. Proteomic Analysis of Microvesicles Released by the Human Prostate Cancer Cell Line PC-3. Mol Cell Proteomics. 2012;11(7):M111.012914.
21. Siqueira AM, Cavalcante JA, Vítor-Silva S, Reyes-Lecca RC, Alencar AC, Monteiro WM, et al. Influence of age on the haemoglobin concentration of malaria-infected patients in a reference centre in the Brazilian Amazon. Mem Inst Oswaldo Cruz. 2014;109(5):569–76.
22. Meraiyebu A, Akintayo C, Nenchi D. Evaluation of Pcv and Hemoglobin Variations among Malaria Positive and Malaria Negative Patients, At the Ecwa Community Health Centre Bukuru, Jos. Nigeria. IOSR-PHR. 2012;2(6):65–9.
23. Black RH. The Consumption of Haemoglobin by Malaria Parasites. Ann Trop Med Parasit. 1947;41(2):215–7.
24. Lazarus MD, Schneider TG, Taraschi TF. A new model for hemoglobin ingestion and transport by the human malaria parasite Plasmodium falciparum. J Cell Sci. 2008;121(11):1937–49.
25. Kotepui M, Phunphuech B, Phiwklam N, Chupeerach C, Duangmano S. Effect of malarial infection on haematological parameters in population near Thailand-Myanmar border. Malaria J. 2014;13:218.
26. Muwonge H, Kikomeko S, Sembajjwe LF, Seguya A, Namugwanya C. How Reliable Are Hematological Parameters in Predicting Uncomplicated Plasmodium falciparum Malaria in an Endemic Region? ISRN Trop Med. 2013;2013:1–9.
27. Koltas IS, Demirhindi H, Hazar S, Ozcan K. Supportive presumptive diagnosis of Plasmodium vivax malaria. Thrombocytopenia and red cell distribution width. Saudi Med J. 2007;28(4):535–9.
28. Lathia TB, Joshi R. Can hematological parameters discriminate malaria from nonmalarious acute febrile illness in the tropics? Indian J Med Sci. 2004;58(6):239–44.
29. Jairajpuri ZS, Rana S, Hassan MJ, Nabi F, Jetley S. An Analysis of Hematological Parameters as a Diagnostic test for Malaria in Patients with Acute Febrile Illness: An Institutional Experience. Oman Med J. 2014;29(1):12–7.
30. Bunyaratvej A, Butthep P, Bunyaratvej P. Cytometric analysis of blood cells from malaria-infected patients and in vitro infected blood. Cytometry. 1993;14(1):81–5.
31. Caicedo O, Ramirez O, Mourão MPG, Ziadec J, Perez P, Santos JB, et al. Comparative Hematologic Analysis of Uncomplicated Malaria in Uniquely

Different Regions of Unstable Transmission in Brazil and Colombia. Am J Trop Med Hyg. 2009;80(1):146–51.

32. Kim H-S, Choi D-Y, Yun SJ, Choi S-M, Kang JW, Jung JW, et al. Proteomic analysis of microvesicles derived from human mesenchymal stem cells. J Proteome Res. 2011;11(2):839–49.

33. Pasini EM, Thomas AW, Mann M. Red blood cells: proteomics, physiology and metabolism. In ESH handbook on disorders of erythropoiesis, erythrocytes and iron metabolism. Paris: European School of Haematology; 2008. pp. 96.

34. Acharya P, Pallavi R, Chandran S, Chakravarti H, Middha S, Acharya J, et al. A glimpse into the clinical proteome of human malaria parasites Plasmodium falciparum and Plasmodium vivax. Proteom Clin Appl. 2009;3(11):1314–25.

35. Bansal D, Herbert F, Lim P, Deshpande P, Bécavin C, Guiyedi V, et al. IgG Autoantibody to Brain Beta Tubulin III Associated with Cytokine Cluster-II Discriminate Cerebral Malaria in Central India. PLoS One. 2009;4(12):e8245.

36. Fennell BJ, Naughton JA, Dempsey E, Bell A. Cellular and molecular actions of dinitroaniline and phosphorothioamidate herbicides on Plasmodium falciparum: tubulin as a specific antimalarial target. Mol Biochem Parasitol. 2006;145(2):226–38.

37. Ghosh AK, Jacobs-Lorena M. Surface-expressed enolases of Plasmodium and other pathogens. Mem Inst Oswaldo Cruz. 2011;106(1):85–90.

38. Alam A, Neyaz MK, Ikramul HS. Exploiting unique structural and functional properties of malarial glycolytic enzymes for antimalarial drug development. Malar Res Treat. 2014;2014:451065.

39. Pal Bhowmick I, Kumar N, Sharma S, Coppens I, Jarori GK. Plasmodium falciparum enolase: stage-specific expression and sub-cellular localization. Malaria J. 2009;8(1):1–16.

40. Weng H, Guo X, Papoin J, Wang J, Coppel R, Mohandas N, et al. Interaction of Plasmodium falciparum knob-associated histidine-rich protein (KAHRP) with erythrocyte ankyrin R is required for its attachment to the erythrocyte membrane. BBA – Biomembranes. 2014;1838(1, Part B):185–92.

41. Mazumdar S, Mukherjee P, Yazdani SS, Jain SK, Mohmmed A, Chauhan VS. Plasmodium falciparum Merozoite Surface Protein 1 (MSP-1)-MSP-3 Chimeric Protein: Immunogenicity Determined with Human-Compatible Adjuvants and Induction of Protective Immune Response. Infect Immun. 2010;78(2):872–83.

42. Holder AA, Blackman MJ, Burghaus PA, Chappel JA, Ling IT, McCallum-Deighton N, et al. A malaria merozoite surface protein (MSP1)-structure, processing and function. Mem Inst Oswaldo Cruz. 1992;87(Suppl):337–42.

43. Krishnarjuna B, Andrew D, MacRaild CA, Morales RAV, Beeson JG, Anders RF, et al. Strain-transcending immune response generated by chimeras of the malaria vaccine candidate merozoite surface protein 2. Sci Rep. 2016;6:20613.

44. Triglia T, Healer J, Caruana SR, Hodder AN, Anders RF, Crabb BS, et al. Apical membrane antigen 1 plays a central role in erythrocyte invasion by Plasmodium species. Mol Microbiol. 2000;38(4):706–18.

45. Tibúrcio M, Dixon MWA, Looker O, Younis SY, Tilley L, Alano P. Specific expression and export of the Plasmodium falciparum Gametocyte EXported Protein-5 marks the gametocyte ring stage. Malaria J. 2015;14(1):1–12.

46. Borre MB, Dziegiel M, Høgh B, Petersen E, Rieneck K, Riley E, et al. Primary structure and localization of a conserved immunogenicPlasmodium falciparum glutamate rich protein (GLURP) expressed in both the preerythrocytic and erythrocytic stages of the vertebrate life cycle. Mol Biochem Parasit. 1991;49(1):119–31.

47. Theisen M, Soe S, Oeuvray C, Thomas AW, Vuust J, Danielsen S, et al. The Glutamate-Rich Protein (GLURP) of Plasmodium falciparum Is a Target for Antibody-Dependent Monocyte-Mediated Inhibition of Parasite Growth In Vitro. Infect Immun. 1998;66(1):11–7.

48. Dodoo D, Theisen M, Kurtzhals JAL, Akanmori BD, Koram KA, Jepsen S, et al. Naturally Acquired Antibodies to the Glutamate-Rich Protein Are Associated with Protection against Plasmodium falciparum Malaria. J Infect Dis. 2000;181(3):1202–5.

49. Schüler H, Mueller A-K, Matuschewski K. Unusual properties of Plasmodium falciparum actin: new insights into microfilament dynamics of apicomplexan parasites. FEBS Lett. 2005;579(3):655–60.

50. Hatherley R, Blatch GL, Bishop ÖT. Plasmodium falciparum Hsp70-x: a heat shock protein at the host–parasite interface. J Biomol Struct Dyn. 2014;32(11):1766–79.

51. Nyakundi DO, Vuko LAM, Bentley SJ, Hoppe H, Blatch GL, Boshoff A. Plasmodium falciparum Hep1 is Required to Prevent the Self Aggregation of PfHsp70-3. PLoS One. 2016;11(6):e0156446.

52. Suarez CE, McElwain TF. Transfection systems for Babesia bovis: a review of methods for the transient and stable expression of exogenous genes. Vet Parasitol. 2010;167(2 4):205 15.

53. Costa RM, Nogueira F, de Sousa KP, Vitorino R, Silva MS. Immunoproteomic analysis of Plasmodium falciparum antigens using sera from patients with clinical history of imported malaria. Malaria J. 2013;12(1):1–7.

54. Fairlie WD, Spurck TP, McCoubrie JE, Gilson PR, Miller SK, McFadden GI, et al. Inhibition of Malaria Parasite Development by a Cyclic Peptide That Targets the Vital Parasite Protein SERA5. Infect Immun. 2008;76(9):4332–44.

55. Huang X, Liew K, Natalang O, Siau A, Zhang N, Preiser PR. The Role of Serine-Type Serine Repeat Antigen in Plasmodium yoelii Blood Stage Development. PLoS One. 2013;8(4):e60723.

56. Soe S, Singh S, Camus D, Horii T, Druilhe P. Plasmodium falciparum Serine Repeat Protein, a New Target of Monocyte-Dependent Antibody-Mediated Parasite Killing. Infect Immun. 2001;70(12):7182–4.

57. Thavayogarajah T, Gangopadhyay P, Rahlfs S, Becker K, Lingelbach K, Przyborski JM, et al. Alternative Protein Secretion in the Malaria Parasite Plasmodium falciparum. PLoS One. 2015;10(4):e0125191.

58. Cook WJ, Smith CD, Senkovich O, Holder AA, Chattopadhyay D. Structure of Plasmodium falciparum ADP-ribosylation factor. Acta Crystallogr Sect F: Struct Biol Cryst Commun. 2010;66(Pt 11):1426–31.

59. Anders RF, Brown GV, Edwards A. Characterization of an S antigen synthesized by several isolates of Plasmodium falciparum. P Natl Acad Sci USA. 1983;80(21):6652–6.

60. Gasser O, Schifferli JA. Microparticles released by human neutrophils adhere to erythrocytes in the presence of complement. Exp Cell Res. 2005;307(2):381–7.

61. Biró É, Nieuwland R, Tak PP, Pronk LM, Schaap MCL, Sturk A, et al. Activated complement components and complement activator molecules on the surface of cell-derived microparticles in patients with rheumatoid arthritis and healthy individuals. Ann Rheum Dis. 2007;66(8):1085–92.

62. Garred P, Nielsen MA, Kurtzhals JAL, Malhotra R, Madsen HO, Goka BQ, et al. Mannose-Binding Lectin Is a Disease Modifier in Clinical Malaria and May Function as Opsonin for Plasmodium falciparum- Infected Erythrocytes. Infect Immun. 2003;71(9):5245–53.

63. Ogun SA, Dumon-Seignovert L, Marchand J-B, Holder AA, Hill F. The Oligomerization Domain of C4-Binding Protein (C4bp) Acts as an Adjuvant, and the Fusion Protein Comprised of the 19-Kilodalton Merozoite Surface Protein 1 Fused with the Murine C4bp Domain Protects Mice against Malaria. Infect Immun. 2008;76(8):3817–23.

64. Clemens R, Pramoolsinsap C, Lorenz R, Pukrittayakamee S, Bock HL, White NJ. Activation of the coagulation cascade in severe falciparum malaria through the intrinsic pathway. Brit J Haematol. 1994;87(1):100–5.

65. Mohanty D, Ghosh K, Nandwani SK, Shetty S, Phillips C, Rizvi S, Parmar BD. Fibrinolysis, inhibitors of blood coagulation, and monocyte derived coagulant activity in acute malaria. Am J Hematol. 1997;54(1):23–9.

66. Mostafa AG, Bilal NE, Abass AE, Elhassan EM, Mohmmed AA, Adam I. Coagulation and Fibrinolysis Indicators and Placental Malaria Infection in an Area Characterized by Unstable Malaria Transmission in Central Sudan. Malar Res Treat. 2015;2015:369237.

67. Dondorp AM, Pongponratn E, White NJ. Reduced microcirculatory flow in severe falciparum malaria: pathophysiology and electron-microscopic pathology. Acta Trop. 2004;89(3):309–17.

14

Proteomic profiling of HBV infected liver biopsies with different fibrotic stages

Seyma Katrinli[1], Kamil Ozdil[2], Abdurrahman Sahin[2], Oguzhan Ozturk[2], Gozde Kir[3], Ahmet Tarik Baykal[4], Emel Akgun[4], Omer Sinan Sarac[5], Mehmet Sokmen[2], H. Levent Doğanay[2]* and Gizem Dinler Doğanay[1]*

Abstract

Background: Hepatitis B virus (HBV) is a global health problem, and infected patients if left untreated may develop cirrhosis and eventually hepatocellular carcinoma. This study aims to enlighten pathways associated with HBV related liver fibrosis for delineation of potential new therapeutic targets and biomarkers.

Methods: Tissue samples from 47 HBV infected patients with different fibrotic stages (F1 to F6) were enrolled for 2D-DIGE proteomic screening. Differentially expressed proteins were identified by mass spectrometry and verified by western blotting. Functional proteomic associations were analyzed by EnrichNet application.

Results: Fibrotic stage variations were observed for apolipoprotein A1 (APOA1), pyruvate kinase PKM (KPYM), glyceraldehyde 3-phospahate dehydrogenase (GAPDH), glutamate dehydrogenase (DHE3), aldehyde dehydrogenase (ALDH2), alcohol dehydrogenase (ALDH1A1), transferrin (TRFE), peroxiredoxin 3 (PRDX3), phenazine biosynthesis-like domain-containing protein (PBLD), immuglobulin kappa chain C region (IGKC), annexin A4 (ANXA4), keratin 5 (KRT5). Enrichment analysis with Reactome and Kegg databases highlighted the possible involvement of platelet release, glycolysis and HDL mediated lipid transport pathways. Moreover, string analysis revealed that HIF-1α (Hypoxia-inducible factor 1-alpha), one of the interacting partners of HBx (Hepatitis B X protein), may play a role in the altered glycolytic response and oxidative stress observed in liver fibrosis.

Conclusions: To our knowledge, this is the first protomic research that studies HBV infected fibrotic human liver tissues to investigate alterations in protein levels and affected pathways among different fibrotic stages. Observed changes in the glycolytic pathway caused by HBx presence and therefore its interactions with HIF-1α can be a target pathway for novel therapeutic purposes.

Keywords: Liver fibrosis, Chronic hepatitis B, Two-dimensional difference gel electrophoresis, Proteomics, Glycolysis

Background

Chronic hepatitis B (CHB) is a global health care problem, around two billion people have been infected with the virus and annually 800.000 people lose their lives due to consequences of hepatitis B infection [1]. The virus itself is non-cytopathic, and ongoing inappropriate inflammatory response against hepatitis B virus (HBV) causes the liver damage [2]. Inflammation in liver triggers activation of effector cells, mainly hepatic stellate cells, and this yields to deposition of extracellular matrix

causing fibrosis [3]. Approximately out of 10% of HBV infected patients, this ongoing fibrogenesis culminates in cirrhosis [4] and, if those with cirrhosis left untreated, the five-year survival drops to 50% [5]. Furthermore, annually 2% of cirrhotic patients develops hepatocellular carcinoma (HCC) and in total half of all liver cancers are due to HBV, worldwide [1, 6]. To estimate which patient with CHB will develop fibrosis and/or cirrhosis rapidly, is yet a controversial issue. Although many pathological and epidemiological research have suggested that several elements, including host genetic factors, viral factors, inflammation and alcohol consumption are involved in the development of fibrosis in chronic HBV, hepatic cellular pathways causing liver fibrosis in HBV infection is not yet fully understood [7–10]. Advances in molecular and

* Correspondence: levent.doganay@ueh.gov.tr; gddoganay@itu.edu.tr
[2]Gastroenterology, Umraniye Teaching and Research Hospital, Umraniye, Istanbul, Turkey
[1]Molecular Biology Biotechnology and Genetics Research Center (MOBGAM), Istanbul Technical University, Sariyer, Istanbul, Turkey
Full list of author information is available at the end of the article

biochemical technologies are expected to clarify the pathways involved in development of HBV associated liver fibrosis, and efforts to do so might yield to the discovery of potential targets for novel anti-fibrotic treatments.

Proteomic studies on liver fibrosis mainly focused on HCC or hepatitis C driven cirrhosis with limited samples. One of the earlier studies has compared tumor tissue and surrounding non-tumor tissue from eight HCC patients and has showed overexpression of 14-3-3γ protein in HCC [11]. Another study investigated the proteomic differences between tumor and adjacent nontumor tissue samples of 12 HBV-associated HCC patients and found out upregulation of members of the heat shock protein 70 and 90 families and downregulation of metabolism-associated and mitochondrial and peroxisomal proteins in HCC [12]. Molleken et al., also has analyzed cirrhotic septa and liver parenchyme of seven cirrhotic patients and has discovered an increase in cell structure associated proteins which are actin, prolyl 4-hydroxylase, tropomyosin, calponin, transgelin and human microfibril-associated protein 4 (MFAP-4) [13]. However, all these studies investigate the alterations occuring at the very late stage of fibrosis and did not give information about the proteomic changes during fibrosis progression. To understand the pathways related to fibrosis, the proteomic changes between different fibrotic stages should be investigated. Proteomic studies evaluating serum or plasma proteins corresponding to different fibrotic stages of HCV and HBV infected patients exist in literature [14–17]. Although these studies are promising for determining serum biomarkers for non-invasive diagnosis of liver fibrosis, they are insufficient for exposing pathways related to fibrosis progression. Such a study should involve analysis of liver tissue samples with different fibrotic stages and so far, there are a few research available. In one of these research, HCV-infected human liver tissue from 15 patients at different stages of fibrosis were analyzed by $^{16}O/^{18}O$ stable isotope labeling in combination with the accurate mass and time (AMT) tag approach and demonstrated association of oxidative stress and hepatic mitochondrial dysfunction with HCV pathogenesis [18]. A recent study performed differential label-free proteomics approach using 27 biopsies from patients with HCV-associated hepatic fibrosis and reported alterations in lumican (LUM), fibulin 5 (FBLN5), cysteine and glycine-rich protein 2 (CSRP2), calponin 2 (CNN2), transgelin (TAGLN), collagen alpha-1(XIV) chain (COL14A1), and MFAP-4, then verify the expression of these proteins on a transcriptional level and with targeted proteomic approach in different cohorts composed by a total of 77 and 68 HBV or HCV infected patients with liver fibrosis, respectively [19]. These mentioned studies focuses on proteomics of HCV associated liver fibrosis, only the latter

offered a partial targeted proteomic information about limited group corresponding to different fibrotic stages in HBV infection. Hereby, a broad fibrotic stage specific proteomic profiling study of HBV associated liver fibrosis is still missing in literature.

In this study, to obtain an insight into the pathogenesis of HBV related liver fibrosis, we firstly performed differential proteome analysis of the human liver tissue specimens with different fibrotic stages classified according to Ishak, using two dimensional difference gel electrophoresis (2D-DIGE) followed by tandem mass spectrometry. Findings of this study will highlight cellular pathways associated with liver fibrosis in HBV infection for future development of new treatments through detection of target proteins for drug development, and will offer putative stage specific biomarkers.

Methods
Patients
Human liver tissues from 47 chronic HBV patients were collected between 2014 and 2015 from Gastroenterology and Hepatology Clinic at Umraniye Teaching and Research Hospital with the prior approval of the Ethical Committee at the teaching hospital and with prior written informed consents obtained from patients. All patients had been diagnosed CHB and were followed up at the hepatology clinic. All were HBsAg positive with detectable HBV-DNA. All had been considered as a candidate for antiviral treatment and had been scheduled for liver biopsy. Patients co-infected with other hepatitis virus (hepatitis delta virus, HCV) or HIV infection, patients with other liver diseases (hemochromatosis, alpha 1 anti-trypsin deficiency, auto immune hepatitis, primary biliary cirrhosis, sclerosing cholangitis, Wilson's disease, obstructing biliary disease, malignancy, non-alcoholic steatohepatitis) or patients with alcohol intake more than 20 mg/day or patients taking antiviral treatment or immunosuppressive treatment were all excluded. All tissues were obtained before antiviral therapy. Biopsies were obtained under ultrasonographic guidance by 16G Hepafix needle. The length of the biopsies was not shorter than 2.5 cm. All biopsy specimens were stained with haematoxyin-eosin and Masson's trichrome. Classification of the fibrotic stages and inflammation grades were done according to Ishak's classification by an experienced pathologist who was blinded to proteomic analysis [20].

The study was conducted in accord with the ethical principles originating in the Declaration of Helsinki.

Sample preparation of tissue samples
Fresh frozen liver tissue specimens were thawed and each were homogenized with 120 μL of T-PER™ Tissue Protein Extraction Reagent (Pierce Chemical, Rockford,

IL) containing 0,1% (v/v) protease inhibitor cocktail (Roche, Mannheim, Germany) using hand homogenizer. The homogenates were incubated 5 min on ice and then centrifuged at 10.000×g for 5 min. The resulting supernatant was collected, and the protein concentration was measured by BCA assay with BSA as the standard. The ReadyPrep™ 2-D Cleanup Kit (Bio-Rad, Hercules, CA, USA) was further used to purify the homogenate according to the manufacturer's protocol. Before processing to 2D-DIGE, samples in the same group were pooled to decrease individual differences.

Two dimension-difference gel electrophoresis (2D-DIGE)

For tissue samples, the same group of sample was labelled with either Cy3 or Cy5 and ran in two different gels to eliminate the effect of dyes. 60 μg protein of each group was labelled with 400 pmol of Cy3 or Cy5 and 60 μg protein from internal standard (10 μg from each group) was labelled with Cy2. After 30 min incubation in dark, the labeling reaction was stopped by adding 1 μL of 10 mM lysine. Two different groups (Cy3 and Cy5) and the internal standard (Cy2) were run per gel. The three labelled samples were mixed and the volume was adjusted to 150 μL with rehydration buffer (8 M urea, 2% CHAPS, 50 mM dithiothreitol (DTT), 0.2% (w/v) Bio-Lyte® 3/10 ampholytes, and trace amount of Bromophenol Blue). All gels, ten in total, were processed and analyzed simultaneously. The first dimension was performed on Protean IEF Cell (Bio-Rad) using 7 cm, pH 5-8 IPG gel strips. Isoelectric focusing (IEF) was carried on at 20 °C under the following conditions: 14 h at 50 V; 20 min at 250 V; 2 h at 4000 V and held at 4000 V until total Vh reached 20000 Vh. After IEF, the IPG strips were equilibrated for 15 min in a reduction buffer (6 M urea, 2% SDS, 0.375 M Tris-HCl (pH 8.8), 20% glycerol, and 2% (w/v) DTT) and subsequently alkylated for 10 min in an alkylation buffer (6 M urea, 2% SDS, 0.375 M Tris-HCl (pH 8.8), 20% glycerol and 2.5% iodoacetamide). After equilibration, the strips were overlaid on individual 12% polyacrylamide gels and added 0.5% agarose to immobilize the strips. The second dimensional separation was carried out in the Bio-Rad Mini Protean system and the electrophoresis was run at 125 V for 1.5 h.

Gel image and data analysis

The gels were scanned using ChemiDoc™ MP Imaging System (Bio-Rad) at three different settings (Cy2, blue laser 488 nm and 520 bp 40 filter; Cy3, green laser 532 nm and 580 bp 30 filter; Cy5, red laser 633 nm and 670 bp 30 filter). The scanned gels were semi-automatically matched and analyzed by the PD-Quest software v.8.0.1 (Bio-Rad). Spot volumes (pixels*spot size) were assigned to each protein spot and normalized

to internal standard. In our experimental design, besides Cy2 labeled internal pooled standard, we also run Cy3 or Cy5 labeled F1 pool in each gel. Therefore, we had two internal controls which enabled better alignment of gels for more accurate spot analysis. For proteome profile comparison, proteome profile of each six Ishak fibrotic stage (F1-F6) was compared to each other by Student's t-test in search for differentially expressed spots between stages. The quantitative difference in percent volume of spots more than two fold was considered as differential expression change. Competent spots with a p-value ≤0.05 were considered statistically significant and withheld for protein identification by mass spectrometry.

In-gel digestion and mass spectrometric analysis

Spot picking was carried out with preparative gels. Two-dimensional electrophoresis was performed as described under "Two dimension-difference gel electrophoresis" except that the IPG strips were loaded with 300 μg of protein consisting of equal amounts from each stage, and gels were stained with colloidal Coomassie Brillant Blue. Protein spots of interest were matched between the images of analytical and preparative gels and manually cut from preparative gels; then and transferred to a 1.5 mL siliconikzed Eppendorf tube. Subsequently, the transferred gel spots were destained in a destaining solution (100 μL acetonitrile (ACN): 100 mM ammonium bicarbonate (ABC), v/v, 1:1). After prereduction at 50 °C using 100% acetonitrile, gel pieces were reduced (10 mM DTT at 80 °C for 30 min) and then alkylated (20 mM iodoacetamide at room temperature and dark for 45 min). After a subsequent washing step (250 mM ABC and 50% ACN), the pieces were dehydrated in 100% ACN and dried in at 50 °C. A total volume of 30 μL trypsin (20 ng/μL in 50 mM ABC) added to the samples incubated overnight at 37 °C.

Peptide samples were dissolved in 1% formic acid and 2% ACN, then peptide sequences were analysed by tandem mass spectrometry (MS/MS) on a SYNAPT G2-Si mass spectrometer (Waters, Milford, MA, USA). Peak list files containing mass spectral data were processed using the ProteinLynxGlobalServer software (PLGS V3.02; Waters) and were searched against the reviewed Uniprot homosapiens database. Database search parameters were set according to previously published data [21].

Validation with western blotting

Fifty microgram of tissue proteins from each fibrotic stage pool were mixed with NuPAGE® LDS Sample Buffer (4X) and NuPAGE® Reducing Agent (10X) (Invitrogen, CA, USA) and were heated at 95 °C for 5 min, prior to loading to 12% sodium dodecyl sulfate-polyacrylamide gel electrophoresis (SDS-PAGE). The iblot2 system (Invitrogen, CA,

USA) was used for transferring the proteins to polyvinylidene fluoride (PVDF) membrane. After using 0.5% skim milk to blot the membrane for 2 h at room temperature, APOA11, KPYM, GAPDH, DHE3, ALDH1, ALDH1A1, TRFE, PRDX3, PBLD were detected by mouse anti-APOA11 antibody (Cell Signaling Technology, CST, 1:250), rabbit anti-KPYM antibody (CST, 1:250), rabbit anti-GAPDH antibody (CST, 1:1000), rabbit anti-DHE3 antibody (CST, 1:500), mouse anti-ALDH2 antibody (Thermo Scientific, 1:500), rabbit anti-ALDH1A1 antibody (CST, 1:500), mouse anti-TRFE (Thermo Scientific, 1:500), rabbit anti-PRDX3 (Thermo Scientific, 1:250), mouse anti-PBLD (Novus, 1:1000) at 4 °C overnight. After washing three times in TBS-T buffer (137 mM Sodium Chloride, 20 mM Tris, 0.1% Tween-20, pH 7.6), the second antibody conjugated with horseradish peroxidase [anti-mouse (CST, 1:5000) or anti-rabbit (CST, 1:5000), depending on applied primer antibody]. Protein signals were detected by ECL detection system (Bio-Rad, Hercules, CA, USA) and images were acquired by ChemiDoc™ MP Imaging System (Bio-Rad). Quantitative analysis of western blotting images was performed by ImageLab software v5.2.1 (Bio-Rad) and statistical analysis was done by Student's t-test. Normalization could not be done by a cytosolic housekeeping protein (i.e. β-actin, β-tubulin, GAPDH) due to their expressional changes within the different fibrotic stages, therefore, both rabbit anti-histoneH3 antibody (CST, 1:500) and ponceau statining were used for loading control and normalization.

Associations between identified proteins and cellular pathways in liver

Functional associations between identified proteins from liver tissue and cellular pathways were analyzed by processing the protein list with on-line EnrichNet application [22] using KEGG [23] and Reactome [24] databases. The significance of overlap between protein sets was decided by using a combination of one-side Fisher's exact test ($q < 0.05$) and network similarity scores (XD-scores). The threshold values were assessed by EnrichNet with a regression fit equivalent to a Fisher q value of 0.05 and an upper boundary of 95% confidence for linear fitting. The trend of these identified pathways among development of fibrotic stages was determined by Reactome expression analysis. Interation networks for the differentially expressed genes were conducted using the STRING tool [25].

Results and discussion

Selection of study group

Human liver tissues from 47 chronic HBV patients [male = 30, female = 17; mean age (range) = 41.3 (19–65) years] were collected between 2014 and 2015 from Gastroenterology and Hepatology Clinic at Umraniye

Teaching and Research Hospital. Fibrosis stages according to Ishak F-score of these 47 subjects were as followed: F1 ($n = 7$), F2 ($n = 20$), F3 ($n = 12$), F4 ($n = 3$), F5 ($n = 2$), F6 ($n = 3$). In addition, patient's characteristics and laboratory results within 10 days of biopsy were given on Table 1. A more detailed demographic data of patient's according to fibrotic stages is also presented in Additional file 1: Table S1.

Quantitative comparison and identification of protein spots

Tissue samples from individuals in the same group were pooled together for analysis to flatten intrinsic individual differences and augment common characteristic traits only related to fibrotic stages. Each tissue sample group was labelled with either Cy3 or Cy5 and ran in two different gels (Additional file 1: Figure S1). For tissue samples, ten separate 2D-PAGE gels were semi-automatically matched and resulted in a combined reference image with

Table 1 Patient's characteristics of the proteome study ($n = 47$)

		Statistics ($N = 47$)
Age (mean ± sem)		40.98 ± 2.063 (19–66)
Sex	Male	30 (63.8%)
	Female	17 (36.2%)
F-score	1	7
	2	20
	3	12
	4	3
	5	2
	6	3
HAI (mean ± sem)		5.36 ± 0.366 (2–11)
logHBVDNA (mean ± sem)		6.04 ± 0.263 (1.77–8.94)
AST (mean ± sem)		54.59 ± 7.697 (16–309)
ALT (mean ± sem)		88.89 ± 14.569 (14–441)
PLT (x10³) (mean ± sem)		218 ± 8.4 (119–370)
PT (mean ± sem)		13.23 ± 0.13 (11.4–16.1)
INR (mean ± sem)		1.03 ± 0.0178 (0.85–1.39)
HDL (mean ± sem)		46.65 ± 2.29 (27–72)
LDL (mean ± sem)		116 ± 5.27 (60–167)
Triglyceride (mean ± sem)		105.17 ± 9.59 (36–359)
Glucose (mean ± sem)		103.02 ± 6.1 (76–295)
Waist (cm) (mean ± sem)		92.48 ± 2.1 (62–120)
BMI (mean ± sem)		27 ± 0.7 (19–38)
HBeAg	Positive	11 (23.4%)
	Negative	36 (76.6%)

SEM standard error of mean, *F* fibrosis, *HAI* hepatic activity index, *HBV* hepatitis B virus, *AST* aspartate transaminase, *ALT* alanine transaminase, *PLT* platelet, *PT* prothrombin time, *INR* international normalized ratio, *HDL* high density lipoprotein, *LDL* low density lipoprotein, *BMI* body mass index, *HBeAg* hepatitis B early antigen

222 protein spots (Fig. 1). According to univariate analyses and trend test between individual fibrotic stages 25 spots were obtained and cut from preparative gel for mass fingerprinting (Additional file 1: Figure S2). Mass spectrometry analysis resulted in 12 different protein identities, summarized in Table 2.

Differential expressed proteins were found to be APOA1, KPYM, GAPDH, DHE3, ALDH1, ALDH1A1, TRFE, PRDX3, PBLD, IGKC, ANXA4, KRT5. To confirm stage-dependent changes observed in 2D-DIGE, western blotting analysis was carried out for selected proteins (Fig. 2). The trends of alterations for analyzed proteins were consistent with that of 2D-DIGE results. When each protein expression was compared to that of F1: ALDH1A1 and ALDH2 showed a significant increase in F6 ($p < 0.05$); APOA1 has significantly reduced in F4 and F5 ($p < 0.01$) and F6 ($p < 0.001$); DHE3 is significantly increased in F2 ($p < 0.01$); GAPDH was significantly increased in F3 ($p < 0.05$) and decreased in F6 ($p < 0.05$); PBLD expression is significantly increased in F3 ($p < 0.05$); PKM is significantly increased in F3 ($p < 0.01$) and F4 ($p < 0.001$), also showed an increased trend in F5 and F6; PRDX3 showed an increased trend in F5 and TRFE showed an increased trend in F2 which is consistent with 2D-DIGE results.

Inter-stage analysis of tissue proteins

APOA1 (SSP0303) expression was significantly decreased during F3 ($p < 0.05$) and F4, F5, F6 compared to that of F1 ($p < 0.001$) (Fig. 3). Apolipoprotein A, which is included in noninvasive FibroTest is a putative biomarker of HCV associated fibrosis [26]. This well-known biomarker, for the first time with our study, was shown to be downregulated in human liver in a stage specific manner.

DHE3 (SSP7603) expression was significantly increased in F2 compared to that of F1 ($p < 0.05$) (Fig. 3). Pyruvate kinase PKM (KPYM) (SSP7802) expression was significantly increased in F5 as opposed to F1 ($p < 0.001$) (Fig. 3). GAPDH (SSP9201) was significantly increased in F3 ($p < 0.05$) and decreased in F6 ($p < 0.05$) compared to that of F1 (Fig. 3). All of these proteins are members of the glycolysis pathway, justifying the existence of abnormalities in glycolytic mechanism of patients having HBV related liver fibrosis.

ALDH2 (SSP2402) expression was significantly increased in F6 when compared to the results of F1 ($p < 0.05$) (Fig. 3). Aldehyde dehydrogenase is involved in ethanol metabolism along with alcohol dehydrogenase. ADH1A1 (SSP3303) expression was also significantly increased during F3, F5 and F6 ($p < 0.001$) compared to that of F1 (Fig. 3).

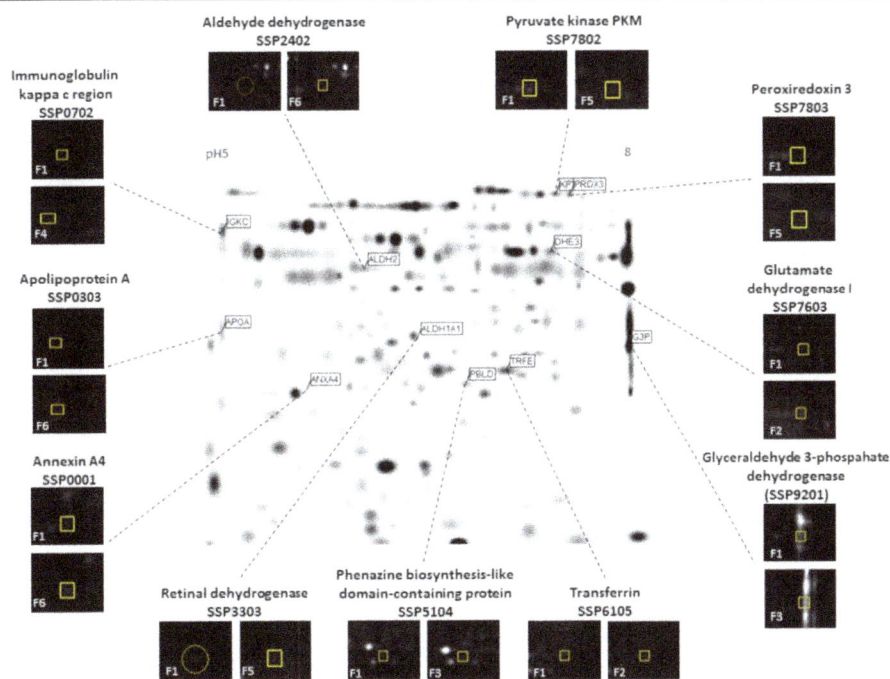

Fig. 1 Reference gel image and selected spots. Reference gel image built up from 10 separate 2D-PAGE gels from HBV-infected patient's liver tissue samples formed by PDQuest, containing 222 valid protein spots. The eleven selected proteins are visualized on reference gel: annexin A4 (SSP0001), apolipoprotein A (SSP0303), immuglobulin kappa chain C region (SSP0702), aldehyde dehydrogenase (SSP2402), retinal dehydrogenase (SSP3303), phenazine biosynthesis-like domain-containing protein (SSP5104), transferrin (SSP6105), glutamate dehydrogenase I (SSP7603), pyruvate kinase PKM (SSP7802), peroxiredoxin 3 (SSP7803), glyceraldehyde 3-phosphate dehydrogenase (SSP9201). The enlarged images display differences in protein expression (spot volume = pixels*spot size) according to F1 stage

Table 2 List of identified proteins in tissue samples

Spot No (SSP)	Protein ID	Procession No	PLGS score	Coverage	MW (kDa)	pI	Biological function
0001	annexin A4	P09525	1876	46.7	35860	5.7	epithelial cell differentiation
0303	apolipoprotein A	P02647	2211	46.8	30758	5.4	Carrier (cholesterol)
0702	immuglobulin kappa chain C region	P01834	11563	30.2	11601	5.5	Adaptive immunity
2402	aldehyde dehydrogenase	P05091	4205	46.4	56345	6.7	Alcohol metabolism
3303	alcohol dehydrogenase	P07327	550	22.7	39832	7.9	Alcohol metabolism
5104	phenazine biosynthesis-like domain-containing protein	P30039	1633	32.3	31765	6.1	Epithelial maintenance
6105	transferrin	P02787	922	43.1	77013	6.8	Carrier (Fe)
7603	glutamate dehydrogenase I	P04406	509	17.6	36030	8.7	Glycolysis
7802	pyruvate kinase PKM	P14618	896	25.0	57900	7.8	Glycolysis
7803	peroxiredoxin 3	P30048	804	34.4	27675	7.7	Oxidative stress
9201	glyceraldehyde 3-phospahate dehydrogenase	P04406	509	17.6	36030	8.7	Glycolysis
7804	keratin, type II cytoskeletal 5	P13647	197	15.6	62339	7.8	Cytoskeleton

TRFE (SSP6105) expression is significantly increased in F2 ($p < 0.001$) and decreased in F4 ($p < 0.05$) when compared to that of F1 (Fig. 3). PRDX3 (SSP7803) expression was significantly increased in F5 and F6 ($p < 0.001$) compared to that of F1 (Fig. 3).

IGKC (SSP0702) was significantly increased during F2, F4 and F6 ($p < 0.001$) compared to that of F1 (Fig. 3). PBLD (SSP5104) expression is significantly increased in F3 compared to that of F1 ($p < 0.05$) (Fig. 3). Altered expression of these inflammatory proteins, IGKC and PBLD, depends on occasionally fluctuating inflammatory activity in liver due to fibrosis and HBV infection; hence, this situation may not be directly associated with patient's fibrotic stages.

ANXA4 (SSP0001) expression was significantly increased during F4 ($p < 0.01$), F5 ($p < 0.001$) and F6 ($p < 0.01$) compared to that of F1 (Fig. 3). KRT5 (SSP7804) expression was significantly increased during F2 ($p < 0.05$) compared to that of F1 (Fig. 3).

Associations between identified proteins and cellular pathways in liver

A heat map graphic was created to classify proteins that share the same expressional alterations and cluster together (Fig. 4a). We observed two main clusters that include sub-clusters. First cluster has three sub-clusters comprising first with ANXA4 and IGKC partnering and having DHE3 as a neighbor; second with APOA1 and

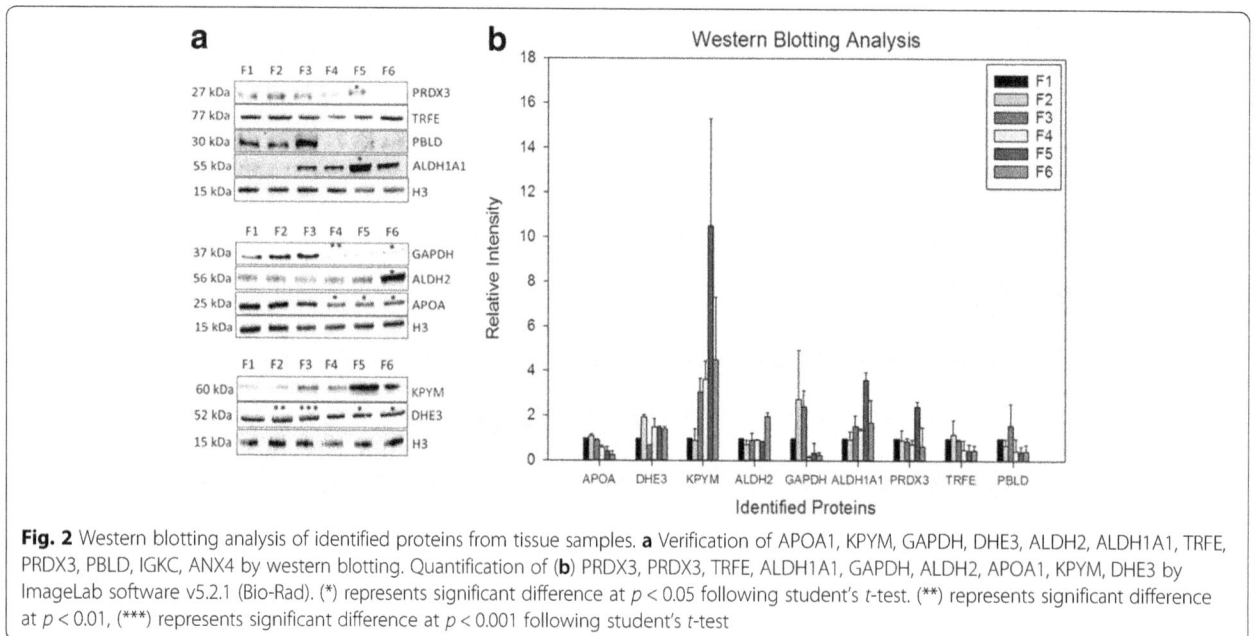

Fig. 2 Western blotting analysis of identified proteins from tissue samples. **a** Verification of APOA1, KPYM, GAPDH, DHE3, ALDH2, ALDH1A1, TRFE, PRDX3, PBLD, IGKC, ANX4 by western blotting. Quantification of (**b**) PRDX3, PRDX3, TRFE, ALDH1A1, GAPDH, ALDH2, APOA1, KPYM, DHE3 by ImageLab software v5.2.1 (Bio-Rad). (*) represents significant difference at $p < 0.05$ following student's t-test. (**) represents significant difference at $p < 0.01$, (***) represents significant difference at $p < 0.001$ following student's t-test

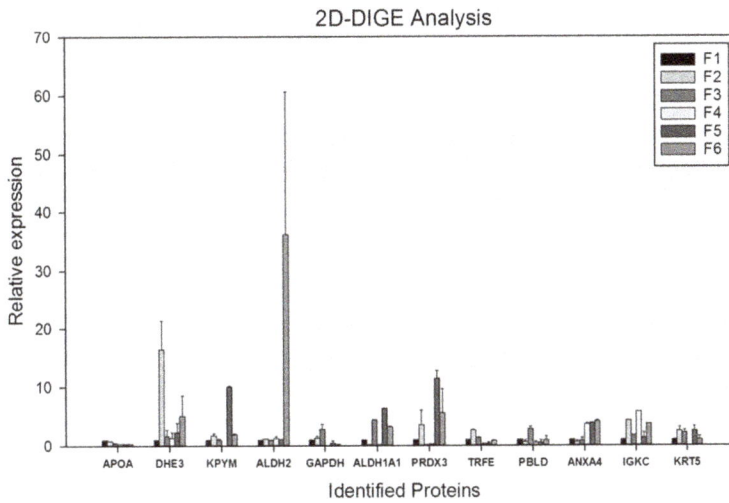

Fig. 3 Selected proteins in liver tissue proteome (inter-stage analysis). Protein expression is represented by relative spot volume (pixels*spot size) normalized according to internal standard and are grouped according to Ishak's classification (F1-F2-F3-F4-F5-F6)

TRFE together, also having DHE3 as a neighbor; third with KPYM and GAPDH. The second cluster includes ALDH1A1 having two separate sub-clusters with PRDX3 and ALDH2, respectively. From this heat map, we can observe close association between APOA1 and TRFE, clustering of PRDX3 with ethanol metabolism proteins, ALDH1 and ALDH2, suggesting the role of PRDX3 in alcohol metabolism in HBV-associated liver fibrosis.

For the prediction of possible involvement of the identified proteins in the fibrosis development process in HBV infection and thus reveal the pathways influenced by HBV related fibrosis pathogenesis, we processed the protein dataset in EnrichNet. Enrichment analysis using KEGG and Reactome databases exposed implications of

the identified proteins in several cellular pathways (Additional file 1: Table S2 and S3). XD-scores were regarded to be more than 0.49 and 0.80 threshold values, as estimated by the application for the KEGG and Reactome databases, respectively.

To further investigate the alterations in these identified pathways (further platelet release, glycolysis, HDL mediated lipid transport) along with fibrosis progression, the rate trend of these pathways were assessed by Reactome expression analysis (Fig. 4b). This analysis showed a decreasing trend in further platelet release along with liver fibrosis. Glycolysis showed fluctuations along with the fibrosis development. HDL mediated lipid transport was determined by APOA1 levels. As APOA1 levels

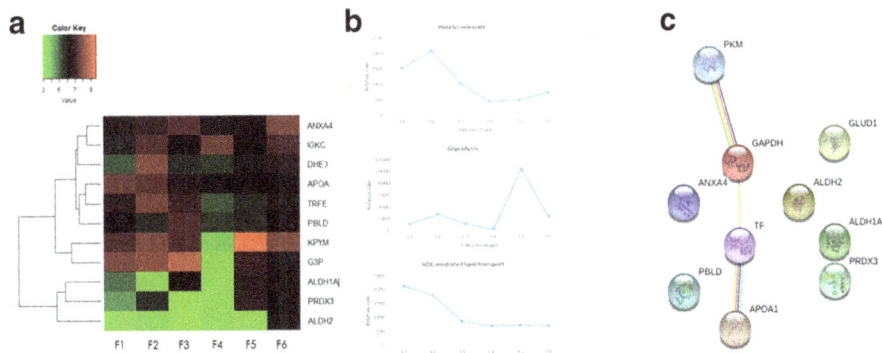

Fig. 4 Bioinformatic analysis. **a** Clustering of identified proteins from tissue samples according to tissue fibrotic stages. The degree of similarity in profiles is shown by colour (base-two logarithmical scale is above the figure). The average protein abundance of the 11 identified proteins (labelled with Cy3 and Cy5 dyes) is calculated relative to internal standard (labelled with Cy2 dye). **b** Alterations in identified cellular pathways along with fibrosis progression. Alterations in pathways were determined by using expression analysis of Reactome database. **c** Evidence view of identified liver tissue proteins in String database. (Confidence score > 0.700). Different line colors represent the types of evidence for the association. *Green* indicates neighborhood. *Gray* indicates co-expression. *Cyan* indicates association in current databases. *Magenta* indicates experimental data. *Yellow* indicates text mining

were decreasing toward cirrhosis, HDL mediated lipid transport activity was also showed a decreasing trend toward cirrhosis.

Highest confident string analysis (confidence score > 0.9), showed a functional link between PKM and GAPDH (combined confidence score = 0.946), TRFE and APOA11 (combined confidence score = 0.957), TRFE and GAPDH (confidence score = 0.941) (Fig. 4c). Evidence suggest that the co-expression of PKM and GAPDH are not yet shown in *Homo sapiens* but they are shown to be co-expressed in homologues species closest to *Homo sapiens* (Additional file 1: Table S4). Experimental research showed an altered expression level in HCC for PKM and GAPDH [27] and also revealed an intact protein-protein interaction between PKM and GAPDH in nonmalignant skin cancer [28], suggesting the importance of this interaction in tumor supression and apoptosis.

Evidence suggest that the co-expression of TRFE and APOA1 are not yet shown in *Homo sapiens* but they are shown to be co-expressed in close homologues species (Additional file 1: Table S5). Experimental researches also showed an intact protein-protein interaction between TRFE and APOA1 in human sera [29]. TRFE and APOA1 interaction is also present in "Release of platelet secretory granule components" pathway as found out in our Reactome database research.

We also performed string analysis with high confidence (0.900) to all proteins that come up from interstage analysis of tissue samples to search for specific interacting partners that can be related with liver fibrosis. As a result, we observed that there is a significant link between PKM and HIF-1α which is an important interacting partner of Hepatitis virus X protein (HBx) (Additional file 1: Figure S3).

Conclusions

Liver fibrosis currently represents one of the severe global health problems and chronic HBV infection is one of the major risk factors of hepatic fibrogenesis. Although, liver biopsy is currently accepted as the gold standard for diagnosis of liver fibrosis, the tissue specific biomarkers would be invaluable to overcome inter-observer variability and to offer an unbiased clinical result. The first step in the development of such biomarkers is the knowledge of disease-associated proteins and pathways. Although various proteomic studies that investigated altered protein levels in HCV infected liver tissue samples [13, 18, 30] are present in literature, there are no available study concerning proteomic changes in HBV infected tissue samples.

In this study, we identified cellular and metabolic pathways that are associated with liver fibrosis in HBV infection to highlight critical players in the development of fibrosis. Major cellular pathways like glycolysis,

alcohol metabolism, oxidative stress regulation, platelet release and HDL mediated lipid transport get affected by HBV infection. Apart from these observed effects on specific pathways, some individual protein expression alterations were observed. Changes in these protein levels can also play a critical role in various cellular pathways, especially in cell motility, immunity, cellular transport, alcohol metabolism and oxidative stress, mediating stage dependent liver fibrogenesis.

One of the affected mechanisms in liver fibrosis is cell motility, and our results demonstrated also a change in this mechanism by revealing enhanced expression of Keratin 5, which is member of intermediate filaments constitute the cytoskeletal scaffold within epithelial cells, assists cell architecture and provides the cells with the ability to withstand mechanical and non-mechanical stresses [31]. During fibrosis, epidermal growth factor (EGF) expression in the liver increases and the activation of epidermal growth factor receptor (EGFR) by increased EGF levels results in upregulation of KRT5 [32]. ANXA4 which is responsible for membrane fusion of exocytotic vesicles was upregulated in our study. There are body of evidence suggesting an increase in ANXA4 expression in terminally differentiated hepatocytes and HCC [33]. Therefore, increased ANXA4 expression at advanced fibrosis stages might be an indicator of ongoing carcinogenesis. Altogether, these alterations may suggest the role of cell motility in liver fibrosis.

Observation of transferrin level changes in our study, highlight the importance of iron metabolism in cirrhosis since transferrin produced by hepatocytes is a crucial member of iron metabolism [34]. A recent study demonstrated a significant decrease in serum transferrin levels in advanced fibrosis than in mild fibrosis [35]. In combination with our findings, it is tempting to propose that transferrin synthesis might increase during the early stage of liver disease, and thereafter it decreases as fibrosis advances as a result of reduced capacity of transferrin synthesis.

We also observed a downregulation in APOA1 protein that plays a role in HDL mediated lipid transport. The relation of HDL mediated lipid transport with liver fibrosis was well-known and noninvasive FibroTest involving serum apolipoprotein-A1 have already been in clinical use for diagnosis of liver fibrosis [26]. APOA1 which is the carrier of (HDL), is secreted by hepatocytes and was shown to carry anti-inflammatory properties [36]. Experiments have shown that the level of serum APOA1 was downregulated in CHB and CHB-related HCC [37, 38]. There are studies revealing deregulation of lipid metabolism and oxidative stress in HCV infected fibrotic liver tissue samples [18]. Increased ROS levels and dysregulated hepatic lipid metabolism, altogether, causes lipotoxicity leading to more and more ROS accumulation and eventually induction of hepatocyte apoptosis and fibrogenesis [39]. In our study, we also showed similar dysfunctioning in lipid metabolism and oxidative stress in HBV infected liver tissues.

Liver is known to be as a central organ in lipogenesis, gluconeogenesis and cholesterol metabolism. Therefore, alterations in insulin response, β-oxidation, lipid storage and transport, autophagy and an imbalance in chemokines and nuclear receptor signaling are all responsible for liver fibrogenesis [39]. In our study, we identified glycolysis as an altered pathway in HBV associated liver fibrosis due to changes in the DHE3, GAPDH and KPYM expression and also demonstrated altered expression of ALDH2 and ALDH1A1 which are the key members of alcohol metabolism contributing Acetyl Coenzyme A (Ac-CoA) production and ROS accumulation. Recent data has demonstrated that the transdifferentiation of quiescent HSCs into myofibroblasts activates glycolysis and results in lactate accumulation through the induction of HIF1-α for metabolic reprogramming of HSCs [40]. One of the main contributors in HBV infection is HBx protein that has been shown to interact with various host proteins [41, 42]. HIF-1α which is one of the interaction partners of HBx [43], activates the transcription of PKM [44]. HIF-1α also trans-activates the gene encoding pyruvate dehydrogenase kinase 1 (PDK1) which inactivates the TCA cycle enzyme, pyruvate dehydrogenase (PDH) responsible for conversion of pyruvate to acetyl-CoA thereby inhibiting TCA cycle [45]. We observed an increase in PKM expression in advanced fibrosis and cirrhosis, probably due to increased HIF-1α production. Moreover, we observed increased ALDH2 and ADH1A1 expression which are the enzymes responsible for metabolizing alcohol to acetate in cirrhosis. Therefore, we propose a model in which inhibition of PDH

by HIF-1α triggers an alternative method for acetyl-coA production which is in our case through alcohol metabolism. However, upregulation of this pathway along with decreased APOA1 levels disrupts hepatic lipid metabolism and results in ROS accumulation. Increased ROS accumulation therefore induces peroxiredoxins which are responsible for reducing intracellular H_2O_2, and it can be the reason of increased PRDX3 levels observed in our study. As a result, observed activation of PKM and possible inhibition of PDH by HIF-1α may upregulate glycolysis pathway to overcome the TCA cycle inhibition (Fig. 5).

It is worthy to mention that all samples that were used in this study were needle biopsies that can only represent a very small fraction of the liver (approximately 1/50000). Since fibrotic tissue is not distributed homogeneously inside the organ, this form of liver biopsy may bring on sampling error [46]. By taking these facts into account, in our study, we pooled the samples from each stage to minimize the sampling errors resulting from biopsy procedures.

In conclusion, our proteomic approach characterized various pathways associated with HBV infection dependent-stage specific liver fibrosis. Our results suggest HIF-1α, through its interactions with HBx, as a critical modulator of HBV associated liver fibrosis, yielding variations mainly in glycolysis, oxidative stress, ethanol and lipid metabolism within the hepatic cells causing a fibrotic environment by the rearrangement of extra cellular matrix. The study of Moon et al. showed reduced liver fibrosis in HIF-1α deficient mice and therefore pointed HIF-1α as a crucial regulator of profibrotic mediator produced during fibrogenesis, correlating well with our findings [47].

For future studies, validation of determined pathways from our study in large datasets and multiple cohorts would further allow the usage of protein candidates as diagnostic and therapeutic targets in HBV-related fibrosis.

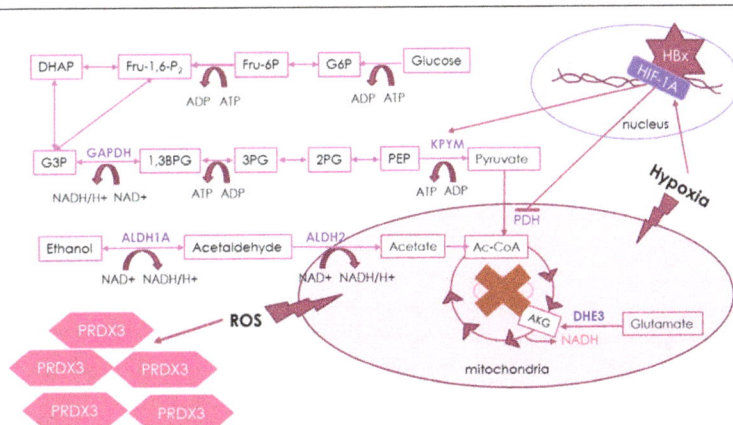

Fig. 5 Proposed model for HBV related liver fibrosis. HBx interacts with HIF-1A that activates PKM expresstion and inhibits PDH expression. This causes and alternate route for Ac-CoA production that results in increased ALDH1A1 and ALDH2 expression and also ROS accumulation, leading to elevated PRDX3 level

Acknowledgements
Not applicable.

Funding
Gilead Turkey Fellowship 2013
Turkish Viral Hepatitis Association (VHSD)
Istanbul Technical University Internal Funds

Authors' contributions
SK performed 2D-DIGE experiments, statistical analysis and drafted the manuscript. KO participated in the design of the study. AS and OO carried on the collection of samples and participated in grading of fibrosis. GK carried on the grading of tissue samples. ATB and EA performed mass spectrometry analysis. OSS participated in statistical analysis. MS participated in the design of the study. GDD and LD conceived of the study, and participated in its design and coordination and helped to draft the manuscript. All authors read and approved the final manuscript.

Competing interests
The authors declare that they have no competing interests.

Author details
[1]Molecular Biology Biotechnology and Genetics Research Center (MOBGAM), Istanbul Technical University, Sariyer, Istanbul, Turkey. [2]Gastroenterology, Umraniye Teaching and Research Hospital, Umraniye, Istanbul, Turkey. [3]Pathology, Umraniye Teaching and Research Hospital, Umraniye, Istanbul, Turkey. [4]Department of Medical Biochemistry, School of Medicine, Acibadem University, Istanbul, Turkey. [5]Computer Engineering, Istanbul Technical University, Sariyer, Istanbul, Turkey.

References
1. Lozano R, Naghavi M, Foreman K, Lim S, Shibuya K, Aboyans V, et al. Global and regional mortality from 235 causes of death for 20 age groups in 1990 and 2010: a systematic analysis for the Global Burden of Disease Study 2010. Lancet. 2012;380(9859):2095–128.
2. Bertoletti A, Ferrari C. Innate and adaptive immune responses in chronic hepatitis B virus infections: towards restoration of immune control of viral infection. Gut. 2012;61(12):1754–64.
3. Rockey DC, Bell PD, Hill JA. Fibrosis–A Common Pathway to Organ Injury and Failure. N Engl J Med. 2015;373(1):96.
4. McMahon BJ. The natural history of chronic hepatitis B virus infection. Hepatology. 2009;49(5 Suppl):S45–55.
5. Weissberg JI, Andres LL, Smith CI, Weick S, Nichols JE, Garcia G, et al. Survival in chronic hepatitis B. An analysis of 379 patients. Ann Intern Med. 1984;101(5):613–6.
6. Trepo C, Chan HL, Lok A. Hepatitis B virus infection. Lancet. 2014;384(9959): 2053–63.
7. Gutierrez-Reyes G, Gutierrez-Ruiz MC, Kershenobich D. Liver fibrosis and chronic viral hepatitis. Arch Med Res. 2007;38(6):644–51.
8. Doganay L, Fejzullahu A, Katrinli S, Yilmaz Enc F, Ozturk O, Colak Y, et al. Association of human leukocyte antigen DQB1 and DRB1 alleles with chronic hepatitis B. World J Gastroenterol. 2014;20(25):8179–86.
9. Jiang DK, Sun J, Cao G, Liu Y, Lin D, Gao YZ, et al. Genetic variants in STAT4 and HLA-DQ genes confer risk of hepatitis B virus-related hepatocellular carcinoma. Nat Genet. 2013;45(1):72–5.
10. Sarin SK, Kumar M, Lau GK, Abbas Z, Chan HL, Chen CJ, et al. Asian-Pacific clinical practice guidelines on the management of hepatitis B: a 2015 update. Hepatol Int. 2016;10(1):1–98.
11. Lee IN, Chen CH, Sheu JC, Lee HS, Huang GT, Yu CY, et al. Identification of human hepatocellular carcinoma-related biomarkers by two-dimensional difference gel electrophoresis and mass spectrometry. J Proteome Res. 2005;4(6):2062–9.
12. Sun W, Xing B, Sun Y, Du X, Lu M, Hao C, et al. Proteome analysis of hepatocellular carcinoma by two-dimensional difference gel electrophoresis: novel protein markers in hepatocellular carcinoma tissues. Mol Cell Proteomics. 2007;6(10):1798–808.
13. Molleken C, Sitek B, Henkel C, Poschmann G, Sipos B, Wiese S, et al. Detection of novel biomarkers of liver cirrhosis by proteomic analysis. Hepatology. 2009;49(4):1257–66.
14. Cheung KJ, Tilleman K, Deforce D, Colle I, Van Vlierberghe H. The HCV serum proteome: a search for fibrosis protein markers. J Viral Hepat. 2009; 16(6):418–29.
15. Ho AS, Cheng CC, Lee SC, Liu ML, Lee JY, Wang WM, et al. Novel biomarkers predict liver fibrosis in hepatitis C patients: alpha 2 macroglobulin, vitamin D binding protein and apolipoprotein AI. J Biomed Sci. 2010;17:58.
16. Lu Y, Liu J, Lin C, Wang H, Jiang Y, Wang J, et al. Peroxiredoxin 2: a potential biomarker for early diagnosis of hepatitis B virus related liver fibrosis identified by proteomic analysis of the plasma. BMC Gastroenterol. 2010;10:115.
17. Mohamadkhani A, Jazii FR, Sayehmiri K, Jafari-Nejad S, Montaser-Kouhsari L, Poustchi H, et al. Plasma myeloperoxidase activity and apolipoprotein A-1 expression in chronic hepatitis B patients. Arch Iran Med. 2011;14(4):254–8.
18. Diamond DL, Jacobs JM, Paeper B, Proll SC, Gritsenko MA, Carithers Jr RL, et al. Proteomic profiling of human liver biopsies: hepatitis C virus-induced fibrosis and mitochondrial dysfunction. Hepatology. 2007;46(3):649–57.
19. Bracht T, Schweinsberg V, Trippler M, Kohl M, Ahrens M, Padden J, et al. Analysis of disease-associated protein expression using quantitative proteomics-fibulin-5 is expressed in association with hepatic fibrosis. J Proteome Res. 2015;14(5):2278–86.
20. Ishak K, Baptista A, Bianchi L, Callea F, De Groote J, Gudat F, et al. Histological grading and staging of chronic hepatitis. J Hepatol. 1995;22(6): 696–9.
21. Serhatli M, Baysal K, Acilan C, Tuncer E, Bekpinar S, Baykal AT. Proteomic study of the microdissected aortic media in human thoracic aortic aneurysms. J Proteome Res. 2014;13(11):5071–80.
22. Glaab E, Baudot A, Krasnogor N, Schneider R, Valencia A. EnrichNet: network-based gene set enrichment analysis. Bioinformatics. 2012;28(18):i451–7.
23. Kanehisa M, Goto S, Sato Y, Kawashima M, Furumichi M, Tanabe M. Data, information, knowledge and principle: back to metabolism in KEGG. Nucleic Acids Res. 2014;42(Database issue):D199–205.
24. Croft D, Mundo AF, Haw R, Milacic M, Weiser J, Wu G, et al. The Reactome pathway knowledgebase. Nucleic Acids Res. 2014;42(Database issue):D472–7.
25. Jensen LJ, Kuhn M, Stark M, Chaffron S, Creevey C, Muller J, et al. STRING 8-a global view on proteins and their functional interactions in 630 organisms. Nucleic Acids Res. 2009;37(Database issue):D412–6.
26. European Association for Study of L, Asociacion Latinoamericana para el Estudio del H. EASL-ALEH Clinical Practice Guidelines: Non-invasive tests for evaluation of liver disease severity and prognosis. J Hepatol. 2015;63(1):237–64.
27. Megger DA, Bracht T, Kohl M, Ahrens M, Naboulsi W, Weber F, et al. Proteomic differences between hepatocellular carcinoma and nontumorous liver tissue investigated by a combined gel-based and label-free quantitative proteomics study. Mol Cell Proteomics. 2013;12(7):2006–20.
28. Lau E, Kluger H, Varsano T, Lee K, Scheffler I, Rimm DL, et al. PKCepsilon promotes oncogenic functions of ATF2 in the nucleus while blocking its apoptotic function at mitochondria. Cell. 2012;148(3):543–55.
29. Zhou M, Lucas DA, Chan KC, Issaq HJ, Petricoin 3rd EF, Liotta LA, et al. An investigation into the human serum "interactome". Electrophoresis. 2004; 25(9):1289–98.
30. Diamond DL, Krasnoselsky AL, Burnum KE, Monroe ME, Webb-Robertson BJ, McDermott JE, et al. Proteome and computational analyses reveal new insights into the mechanisms of hepatitis C virus-mediated liver disease posttransplantation. Hepatology. 2012;56(1):28–38.
31. Coulombe PA, Omary MB. 'Hard' and 'soft' principles defining the structure, function and regulation of keratin intermediate filaments. Curr Opin Cell Biol. 2002;14(1):110–22.
32. Komuves LG, Feren A, Jones AL, Fodor E. Expression of epidermal growth factor and its receptor in cirrhotic liver disease. J Histochem Cytochem. 2000;48(6):821–30.
33. Li T, LI S, Qin X. Effects of ANXA4 on cell adhesive ability and expression of adhesion-related genes in hepatocellular carcinoma MHCC97H cell line. Chin J Clin Lab Sci. 2013;31(8):588–91.
34. Fleming RE, Ponka P. Iron overload in human disease. N Engl J Med. 2012; 366(4):348–59.
35. Cho HJ, Kim SS, Ahn SJ, Park JH, Kim DJ, Kim YB, et al. Serum transferrin as a liver fibrosis biomarker in patients with chronic hepatitis B. Clin Mol Hepatol. 2014;20(4):347–54.
36. Yang F, Yan S, He Y, Wang F, Song S, Guo Y, et al. Expression of hepatitis B

virus proteins in transgenic mice alters lipid metabolism and induces oxidative stress in the liver. J Hepatol. 2008;48(1):12–9.

37. Norton PA, Gong Q, Mehta AS, Lu X, Block TM. Hepatitis B virus-mediated changes of apolipoprotein mRNA abundance in cultured hepatoma cells. J Virol. 2003;77(9):5503–6.

38. Ai J, Tan Y, Ying W, Hong Y, Liu S, Wu M, et al. Proteome analysis of hepatocellular carcinoma by laser capture microdissection. Proteomics. 2006;6(2):538–46.

39. Bechmann LP, Hannivoort RA, Gerken G, Hotamisligil GS, Trauner M, Canbay A. The interaction of hepatic lipid and glucose metabolism in liver diseases. J Hepat. 2012;56(4):952–64.

40. Chen Y, Choi SS, Michelotti GA, Chan IS, Swiderska-Syn M, Karaca GF, et al. Hedgehog controls hepatic stellate cell fate by regulating metabolism. Gastroenterology. 2012; 143(5): 1319-29 e1-11.

41. Xie N, Chen X, Zhang T, Liu B, Huang C. Using proteomics to identify the HBx interactome in hepatitis B virus: how can this inform the clinic? Expert Rev Proteomics. 2014;11(1):59–74.

42. Decorsiere A, Mueller H, van Breugel PC, Abdul F, Gerossier L, Beran RK, et al. Hepatitis B virus X protein identifies the Smc5/6 complex as a host restriction factor. Nature. 2016;531(7594):386–9.

43. Yoo YG, Cho S, Park S, Lee MO. The carboxy-terminus of the hepatitis B virus X protein is necessary and sufficient for the activation of hypoxia-inducible factor-1alpha. FEBS Lett. 2004;577(1-2):121–6.

44. Luo W, Hu H, Chang R, Zhong J, Knabel M, O'Meally R, et al. Pyruvate kinase M2 is a PHD3-stimulated coactivator for hypoxia-inducible factor 1. Cell. 2011;145(5):732–44.

45. Kim JW, Tchernyshyov I, Semenza GL, Dang CV. HIF-1-mediated expression of pyruvate dehydrogenase kinase: a metabolic switch required for cellular adaptation to hypoxia. Cell Metab. 2006;3(3):177–85.

46. Bedossa P, Dargere D, Paradis V. Sampling variability of liver fibrosis in chronic hepatitis C. Hepatology. 2003;38(6):1449–57.

47. Moon J-O, Welch TP, Gonzalez FJ, Copple BL. Reduced liver fibrosis in hypoxia-inducible factor-1α-deficient mice. Am J Physiol Gastrointest Liver Physiol. 2009;296(3):G582–92.

15

Interaction of peroxiredoxin V with dihydrolipoamide branched chain transacylase E2 (DBT) in mouse kidney under hypoxia

Sun Hee Ahn[1†], Hee-Young Yang[1†], Gia Buu Tran[2†], Joseph Kwon[3], Kyu-Yeol Son[4], Suhee Kim[1], Quoc Thuong Dinh[2], Seunggon Jung[5], Ha-Mi Lee[4], Kyoung-Oh Cho[4] and Tae-Hoon Lee[1,2*]

Abstract

Background: Peroxiredoxin V (Prdx V) plays a major role in preventing oxidative damage as an effective antioxidant protein within a variety of cells through peroxidase activity. However, the function of Prdx V is not limited to peroxidase enzymatic activity per se. It appears to have unique function in regulating cellular response to external stimuli by directing interaction with signaling protein. In this study, we identified Prdx V interacting partners in mouse kidney under hypoxic stress using immunoprecipation and shotgun proteomic analysis (LC-MS/MS).

Results: Immunoprecipitation coupled with nano-UPLC-MS[E] shotgun proteomics was employed to identify putative interacting partners of Prdx V in mouse kidney in the setting of hypoxia. A total of 17 proteins were identified as potential interacting partners of Prdx V by a comparative interactomics analysis in kidney under normoxia versus hypoxia. Dihydrolipoamide branched chain transacylase E2 (DBT) appeared to be a prominent candidate protein displaying enhanced interaction with Prdx V under hypoxic stress. Moreover, hypoxic kidney exhibited altered DBT enzymatic activity compared to normoxia. An enhanced colocalization of these two proteins under hypoxic stress was successfully observed in vitro. Furthermore, peroxidatic cysteine residue (Cys48) of Prdx V is likely to be responsible for interacting with DBT.

Conclusions: We identified several proteins interacting with Prdx V under hypoxic condition known to induce renal oxidative stress. In hypoxic condition, we observed an enhanced interaction of Prdx V and DBT protein as well as increased DBT enzymatic activity. The results from this study will contribute to enhance our understanding of Prdx V's role in hypoxic stress and may suggest new directions for future research.

Keywords: Peroxiredoxin V, Hypoxia, Kidney, Dihydrolipoamide branched-chain transacylase, Peroxidatic cysteine

Background

Peroxiredoxins (Prdxs) are known to include six isoforms with different cellular localizations, and constitute 0.1–1% of total soluble protein in most human cells [1]. Prdxs play major roles in preventing oxidative damage as the effective antioxidant proteins within a variety of cells through peroxidase activity [2]. In mammalian cells, Prdx V was identified originally as a transcriptional corepressor [3], and a peroxisomal and mitochondrial antioxidant protein [4]. Recently, Prdx V was found to be a stress-inducible factor under specialized oxidative stress conditions, especially hypoxic stress [5]. This protein appears to be multifunctional, and the full spectrum of cellular functions of Prdx V remains unknown.

Hypoxia is one of the most important factors influencing clinical outcomes in the renal environment [6]. A growing body of literature has implicated hypoxia in the pathogenesis of both acute and chronic renal disease [7-9]. The partial pressure of oxygen (pO$_2$) in the kidney is ordinarily sustained at well-balanced levels by complex

---

* Correspondence: thlee83@chonnam.ac.kr
†Equal contributors
[1]Department of Oral Biochemistry, Dental Science Research Institute, Medical Research Center for Biomineralization Disorders, School of Dentistry, Chonnam National University, 300 Yongbong-Dong, Buk-Ku, Gwangju 500-757, Republic of Korea
[2]Department of Molecular Medicine, Graduate School, Chonnam National University, Gwangju, Republic of Korea
Full list of author information is available at the end of the article

functional interactions among renal blood flow, glomerular filtration rate, O_2 consumption, and arteriovenous O_2 shunting [10]. Kidney is particularly susceptible to hypoxic damage depending on the delicacy of these complex functional interactions. Renal cells use various molecular pathways that allow them to respond and adapt to changes in renal oxygenation [11]. The efficient metabolic acclimation to low pO_2 is thus a vital factor for maintenance of renal transport ability and essential for cell survival. For example, the prolyl-4-hydroxylase domain (PHD)/hypoxia-inducible factor (HIF) pathway has a primal role in metabolic reprogramming under low pO_2 conditions so that it regulates cellular energy and glucose metabolism at multiple levels. HIFs alter metabolic conditions from oxidative phosphorylation to anaerobic glycolysis, and inhibit mitochondrial respiration and ROS generation. It does this by increasing the expression of glycolytic enzymes, by blocking the conversion of pyruvate to acetyl CoA through transcriptional upregulation of pyruvate dehydrogenase kinase, and by regulating the expression of proteins that compose the mitochondrial respiratory chain [11]. Also, mitochondrial biogenesis and its consequent processes enhance metabolic pathways such as fatty acid oxidation and boost antioxidant defense mechanisms that remediate injury from tissue hypoxia, and glucose or fatty acid overburden, all of which would otherwise contribute to the pathogenesis of acute and chronic kidney disease [12].

Our previous study indicated that Prdx V exerted protective effects in the hypoxic kidney by regulating a variety of individual proteins in a protein network. Using shotgun proteomic analysis, it has been shown that knocking down Prdx V influences the expression of a variety of protein groups associated with oxidative stress, mitochondrial transport, fatty acid metabolism, amino acid/nucleic acid metabolism, glycolysis/gluconeogenesis, and the cytoskeleton. Additionally, the hypoxic kidneys in Prdx V knock-down mice (Prdx V^{si}) showed insufficient activities of mitochondrial metabolic enzymes, especially aconitase 2 (Aco2), acyl-CoA dehydrogenase C-4 to C-12 straight chain (Acadm), and acyl-CoA oxidase 1 (Acox1) [13]. Taken together, these findings suggest that Prdx V may be involved in the coupling of a broad range of cellular signaling cascades to maintain renal homeostasis under hypoxic conditions. To gain further insights into the mechanisms regulated by Prdx V in hypoxic conditions, we employed an approach to compare the interacting partners in the kidneys under normoxia versus hypoxia. Here, we report Prdx V interactome using a strategy of immunoprecipitating Prdx V-protein complexes in the hypoxic kidney. Our data reveal a promising target protein interacting with Prdx V in the hypoxic kidney and provide a potential therapeutic target for chronic kidney disease.

Results

Identification of Prdx V interactome in normoxic and hypoxic mouse kidney

Previously, we reported that Prdx V was a novel regulator of renal homeostasis under hypoxic stress, altering protein networks associated with oxidative stress, fatty acid metabolism, and mitochondrial dysfunction [13]. However, it is still unclear which protein/molecules interact with Prdx V in regulating kidney homeostasis. To capture potential interaction partners of mouse Prdx V, we used an anti-Prdx V antibody to immunoprecipitate Prdx V and its interacting proteins from normoxic and hypoxic kidneys. The immunoprecipitates were then subjected to gel-assisted digestion followed by nano-UPLC-MS analysis. According to Han et al., to maximize protein digestion efficiency and recovery (>90%) [14], we employed the gel-assisted digestion method. We next subjected the digested protein to a nano-UPLC-MS^E proteomic analysis to identify proteins interacting with Prdx V. We compared the proteomic data from three independent experiments to determine meaningful targets with high reproducibility. In detail, 27 (149 spectra) and 33 (276 spectra) proteins were identified as Prdx V's interaction proteins under normoxic and hypoxic conditions, respectively. Table 1 summarizes the potential interacting partners of Prdx V, identified under normoxic and hypoxic conditions. Among them, 13 proteins showed increased interaction with Prdx V in the hypoxic versus the normoxic kidney: Rab43, DBT, Alb, Pcca, Krt76, Krt14, Krt17, Krt84, Krt72, Krt74, Krt77, Krt42, and Pccb. On the other hand, four proteins showed decreased interaction with Prdx V in the hypoxic versus the normoxic kidney: Gba2, Txn1, Krt78, and Krt32 (Table 1).

To confirm our proteomics analysis for identifying Prdx V interaction partners, coprecipitation experiments were performed, targeting some representative proteins. As shown in Additional file 1: Figure S1, Rab43, Alb, and Pccb were shown to strongly coprecipitate with Prdx V in hypoxia, consistent with the proteomics analysis in Table 1. Taken together, these findings suggest that Prdx V could act as a direct regulator in hypoxia and be involved in maintaining kidney homeostasis.

Interaction of Prdx V and DBT in hypoxic mouse kidney

Of the Prdx V-interacting proteins, DBT appeared to be an interesting target. As shown in Table 1, DBT was identified two of three times in hypoxia, whereas it was not detected in normoxia. In our previous study, DBT protein appeared to be associated with Prdx V in hypoxic stress [13]. DBT is known as a transacylase component of the mitochondrial multienzyme BCKDH complex and it regulates branched-chain amino acid (BCCA) degradation. This raises the questions as to whether hypoxia enhances the Prdx V-DBT interaction and/or influences DBT

Table 1 Protein showing altered interaction by Prdx V immunoprecipitation during hypoxic stress[a]

Accession	Description	Gene	Score	MS/MS spectra	Frequency[b]	
					Normoxia	Hypoxia
IPI00130467	Ras related protein Rab 43 isoform b	Rab43	228.7	6	ND	1/3
IPI00130535	Lipoamide acyltransferase component of branched chain α-keto acid dehydrogenase complex, mitochondial	Dbt	302.6	8	ND	2/3
IPI00131695	Serum albumin	Alb	325.3	3	1/3	3/3
IPI00225123	Non-lysosomal glucosylceramidase	Gba2	191.4	9	1/3	ND
IPI00226993	Thioredoxin	Txn1	1155.1	3	3/3	2/3
IPI00330523	Propionyl CoA carboxylase alpha chain, mitochondrial	Pcca	474.0	21	2/3	3/3
IPI00346834	Keratin type II cytoskeletal 2, oral	Krt76	184.2	7	ND	1/3
IPI00348328	Keratin Kb40	Krt78	580.5	19	1/3	ND
IPI00122281	Keratin type I cuticular Ha2	Krt32	260.3	8	1/3	ND
IPI00227140	Keratin type I cytoskeletal 14	Krt14	1399.7	16	ND	1/3
IPI00230365	Keratin type I cytoskeletal 17	Krt17	1322.1	16	ND	1/3
IPI00347019	Keratin type II cuticular Hb4	Krt84	330.8	10	ND	1/3
IPI00347096	Keratin type II cytoskeletal 72	Krt72	337.2	8	ND	1/3
IPI00420970	Keratin type II cytoskeletal 74	Krt74	1999.7	4	1/3	3/3
IPI00462140	Keratin type II cytoskeletal 1b	Krt77	2014.2	7	ND	1/3
IPI00468696	Keratin type I cytoskeletal 42	Krt42	798.6	12	ND	1/3
IPI00606510	Propionyl CoA carboxylase beta chain, mitochondrial	Pccb	266.6	10	2/3	3/3

[a]Proteins were affinity-purified from mouse kidneys under both normoxic and hypoxic conditions as bound interactors with Prdx V immunoprecipitation. The purified immunoprecipitates were applied to acrylamide gel-associated tryptic digestion and subjected to nano-UPLC-MS/MS for protein identification.
[b]Frequency represents the number of times that the interactors were observed in three independent experiments. ND, not detected.

enzymatic activity. Thus, we first confirmed the protein interaction using a reverse immunoprecipitation (Figure 1A). As a result, we found a four-fold higher interaction between Prdx V and DBT in hypoxia than normoxia (Figure 1B). To our knowledge, this is the first report showing that Prdx V could bind directly to DBT and this interaction was enhanced by hypoxic stress, which might suggest a strong linkage to amino acid metabolism and oxidative stress.

To examine the effect of hypoxia on DBT enzymatic activity, we assayed DBT enzymatic activities in mouse kidneys under normoxic and hypoxic conditions. DBT enzymatic activities in hypoxic and normoxic mouse kidneys were 1.20 ± 0.05 and 0.88 ± 0.04 mU/mg, respectively ($n = 10$ mice for each group; Figure 2). This indicates that hypoxic treatment resulted in a significant increase in DBT activity (about 1.5-fold) versus normoxic conditions (Figure 2, inset bar graph).

We next performed co-immunostaining of Prdx V and DBT to determine their subcellular co-localization in normoxic and hypoxic conditions. HEK293 cells were cotransfected with a myc-tagged DBT and a HA-tagged Prdx V. As shown in Figure 3, DBT (green) was colocalized with Prdx V proteins (red) in hypoxic HEK293 cells, while the complex was weakly detectable in normoxic cell. Colocalized pixels were quantified as the percentage of overlapping selected red (Prdx V) and green

pixels (DBT) and revealed that co-localization of these proteins increased 3.8-fold in hypoxia versus normoxia (Figure 3, inset bar graph).

Cysteine 48 of Prdx V is required for the interaction with DBT

Human Prdx V has one catalytic cysteine residue, Cys 48, which is highly conserved in all other peroxiredoxins (Prdx I-IV and VI), and this cysteine residue is involved in Prdx V's peroxidase activity [15]. Most recently, we demonstrated that Prdx V interacts with Jak2 via the catalytic Cys 48 to modulate the Jak2-Stat5 pathway [16]. Based on our previous observations, we next asked whether the catalytic Cys 48 plays a role in the Prdx V-DBT interaction. To test this, we transiently cotransfected HEK293 cells with a myc-tagged DBT and a HA-tagged wild-type (WT) or a cysteine mutant Prdx V (C48S, C73S, C152S, and C48/152S) expression construct. Then, total cell lysate proteins from transfected-HEK293 cells under normoxia and hypoxia were used for immunoprecipitations with anti-myc-conjugated agarose. The cellular expression of DBT and PrdxV for each transfected construct showed the same background for every Prdx V mutant (Figure 4A). As shown in Figure 4B, WT Prdx V bound weakly to DBT under normoxic conditions, but this interaction showed a significant increase under hypoxic stress. In the case of the Prdx V C152S mutant, the

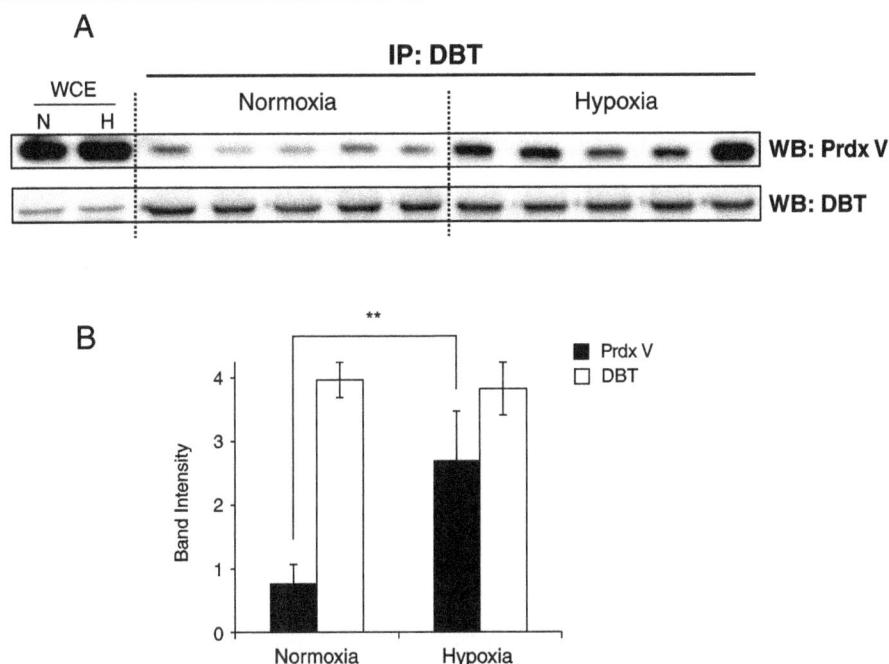

Figure 1 Coprecipitation of endogenous Prdx V and DBT in normoxic and hypoxic mouse kidney. (A) Immunoprecipitates identified by DBT from normoxic and hypoxic (H, lanes 3, 4) mouse kidney were probed with a Prdx V antibody (n = 5). **(B)** Histogram showing numerical data obtained by densitometry analysis of **(A)**. The histogram graph is expressed as mean values ± standard deviation (**p < 0.05). WCE, whole cell lysate of mouse kidney.

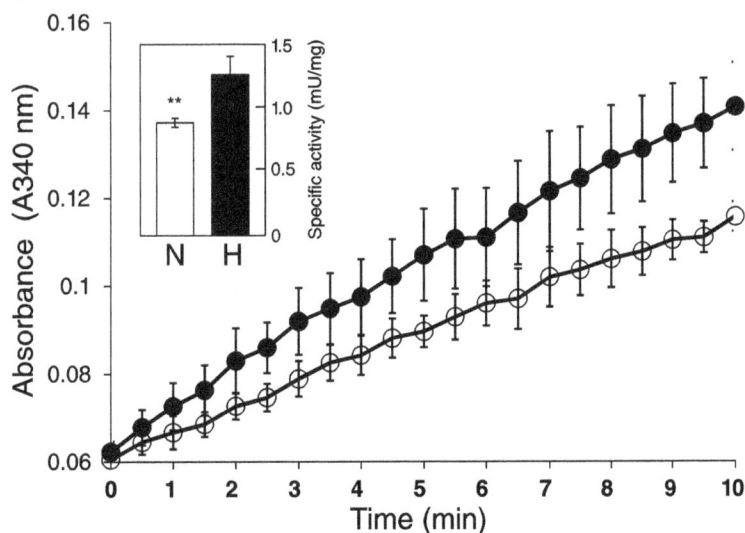

Figure 2 In vitro activities of DBT enzyme in normoxic and hypoxic mouse kidneys. Plots show time-dependent changes in absorbance at 340 nm. The histograms in the insets show the slopes (rate) of the changes in the linear portion of the curves. The y-axes in the histograms indicate the specific activity expressed as mU/mg mitochondrial protein. 1 mU is defined as the amount of enzyme needed to reduce 1 nmol of NAD^+ per min at 37°C. Assays were carried out with the mitochondrial fraction of homogenized tissue extract, and the reactions were run using sodium α-ketoisocaproate as the substrate. Data are expressed as mean values ± standard deviation (**p < 0.05). Open circles and histogram, normoxia; closed circles and histogram, hypoxia.

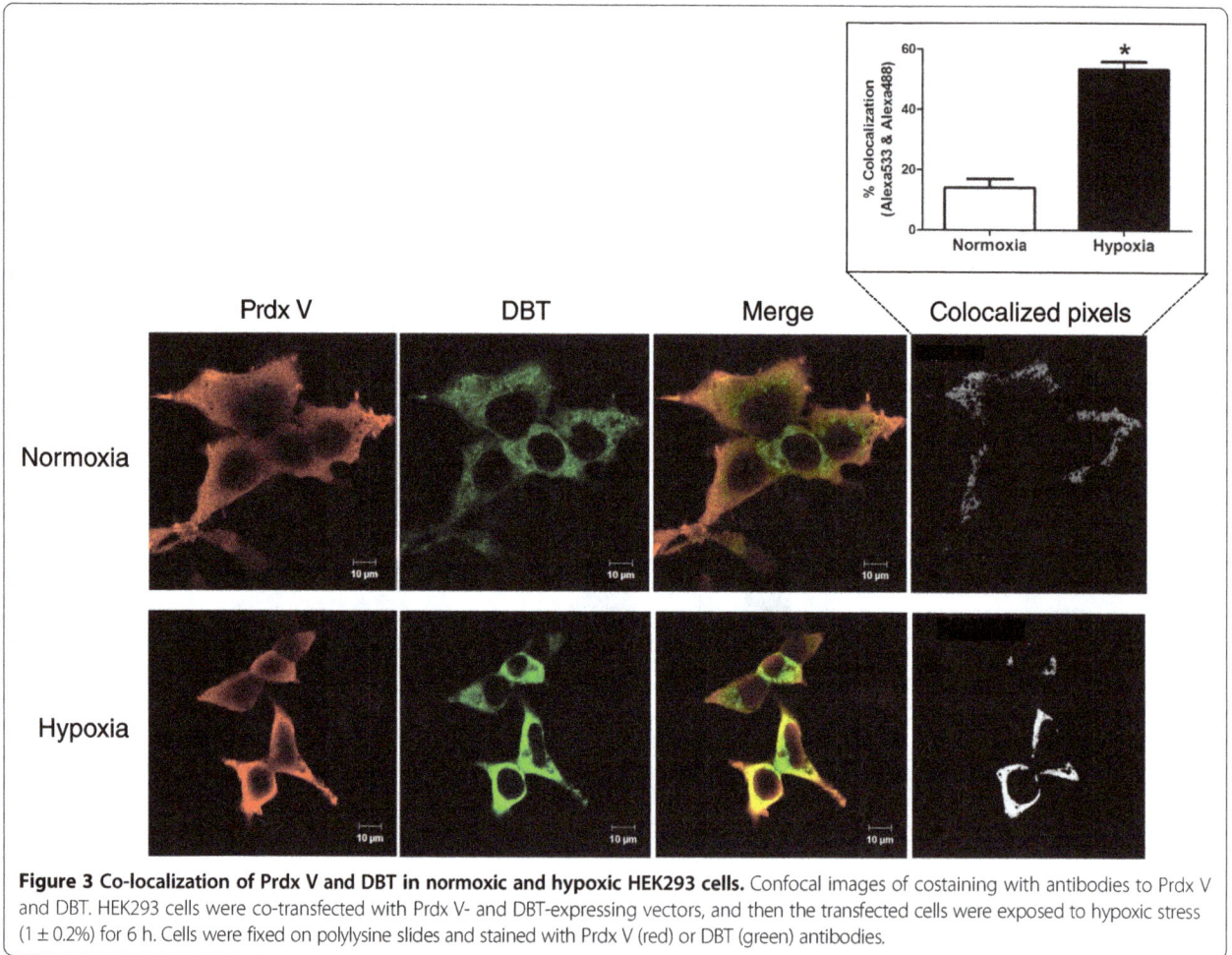

Figure 3 Co-localization of Prdx V and DBT in normoxic and hypoxic HEK293 cells. Confocal images of costaining with antibodies to Prdx V and DBT. HEK293 cells were co-transfected with Prdx V- and DBT-expressing vectors, and then the transfected cells were exposed to hypoxic stress (1 ± 0.2%) for 6 h. Cells were fixed on polylysine slides and stained with Prdx V (red) or DBT (green) antibodies.

interaction with DBT did not change compared with WT Prdx V. However, Prdx V C48S mutant and double mutant (C48/152S) showed diminished interactions with DBT under both normoxic and hypoxic conditions. This result is consistent with our previous finding of the cysteine 48 residue's role in the interaction of Prdx V and Jak2. Additionally, Prdx V C73S showed a notable increase in the interaction with DBT in both normoxic and hypoxic conditions. This needs further investigation. Using reverse immunoprecipitation, we also found that Prdx V and DBT interaction was increased in hypoxia. Although Prdx V 73S showed increased interaction with DBT, Prdx V C48S showed decreased interaction with DBT (Figure 4C). Thus, these data indicate that the Prdx V and DBT interaction during normoxic and hypoxic conditions requires the catalytic cysteine of Prdx V.

Discussion

Prdx V has a remarkably wide subcellular distribution compared with the other mammalian peroxiredoxins

[17]. Although Prdx V appears to be constitutively and ubiquitously expressed in most mammalian tissues, its expression is also upregulated in various pathophysiological situations in response to various stresses [15].

In this study, we employed a manageable and rapid protocol by combining Prdx V immunoprecipitation and shotgun proteomics (LC-MS/MS) to identify Prdx V and its interacting proteins in the hypoxic mouse kidney. Using this approach, we found several novel partners interacting with Prdx V from three independent replicates (Table 1). To confirm our proteomics analyses, co-immunoprecipitations were conducted using anti-target antibodies in normoxic and hypoxic mouse kidneys, and our candidate targets were validated (data not shown). Using functional annotation analysis, some Prdx V-interacting partners, such as DBT, Pcca, and Pccb, appeared to be related to various mitochondria-associated metabolism functions. This is consistent with a previous report regarding the targets in Prdx V knock-down mouse kidney under hypoxia [13]. Of note, Prdx V

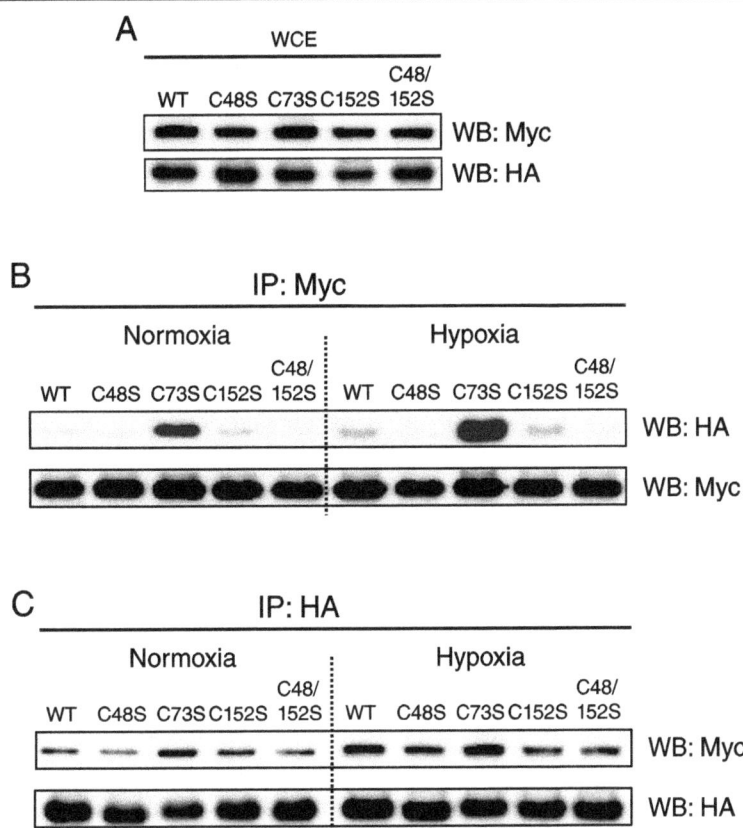

Figure 4 Comparative interactions of Prdx V WT or cysteine mutants with DBT in normoxic and hypoxic cells. (A) All WCE presented for every Prdx V mutant showed the same background of cellular expression for each transfected construct. **(B)** Coprecipitation of HA-tagged Prdx V with myc-tagged DBT. **(C)** Reverse immunoprecipitation of myc-tagged DBT with HA-tagged Prdx V. HEK293 cells were cotransfected with the HA-tagged Prdx V WT or cysteine mutant (C48S, C73S, C152S, C48/152S) and myc-tagged DBT expression vector, and then incubated under normoxic and hypoxic conditions. WCE, whole cell lysate of HEK293 cell; Mock, the empty vector tagged to myc or HA.

knockdown might cause an imbalance in the kidney under hypoxic stress by altering metabolism and causing mitochondrial dysfunction.

The BCKDH complex is an inner-mitochondrial enzyme complex involved in the breakdown of BCAAs, such as isoleucine, leucine, and valine, which is thought to be composed of a core of 24 transacylase (E2) subunits, and associated with decarboxylase (E1), dehydrogenase (E3), and regulatory subunits [18,19]. BCAAs are essential amino acids that serve as substrates for protein synthesis and also function in nutritional signals, regulating carbohydrate metabolism, energy balance and hormone secretion. Thus, the homeostasis of BCCAs is tightly regulated by the BCKDH complex. In particular, DBT belongs to the transacylase (E2) subunit and is a key enzyme in amino acid metabolism. Each E2 subunit consists of three independently functional domains: a lipoyl-bearing domain located in the N-terminal portion, an E1/E3-binding domain, and an inner-core domain at the C-terminal portion, with the three domains tethered by flexible linker regions [20]. The core domains of E2 subunit form a 24-meric scaffold, which is decorated

with multiple copies of E1 and E3 attached through the subunit-binding domain [19]. Although the basic enzyme activity of DBT is known, along with the protein sequence and structure, the physiological functions in amino acid metabolism under hypoxic stress are still unknown. Although BCAAs are essential amino acids, accumulation of BCAAs and their metabolites can be toxic to cells; thus, hypoxia-induced BCAA accumulation may promote kidney injury. The regulation of free amino acid levels is an important factor in understanding amino acid metabolism in acute hypoxia on tissues. Here, we identified that the reinforced interaction between Prdx V and DBT in mouse kidney was inducible by hypoxic stress. Additionally, we found that the enzyme activity of DBT was increased in hypoxia versus normoxia. Thus, we suggest that Prdx V may be an important molecule for the regulation of DBT activity through a direct interaction under hypoxic stress.

Unlike other members of Prdx family, human Prdx V does not contain a cysteine residue corresponding to the second conserved cysteine residue of the 2-Cys subgroup but possesses two additional cysteine residues (C73 and

C152) that are lacking in the 1-Cys subgroup; thus, Prdx V belongs to the atypical 2-Cys family [15]. The atypical 2-Cys human Prdx have side chains for all Cys residues (the peroxidatic Cys48, the resolving Cys152, and the additional Cys73) characterized by the presence of a free sulfhydryl group [21]. The peroxidatic residue Cys48 of Prdx V is a highly conserved Cys residue in all peroxiredoxins. This residue is located within the N-terminal part of the α2-helix, twisted at Ala60, in the reduced form, and lies in the positive charged active-site pocket built by surrounding factors such as Arg128, Thr45, and Val40 [15]. The resolving residue Cys152, located in the loop between β7 and α6, is associated with the proposed enzymatic mechanism of action of Prdx V. The proposed enzymatic mechanism requires the formation of an intramolecular disulfide bond between these two Cys residues. The additional Cys, Cys73, is found in the C-terminal part of the β4 stand of human Prdx V. However, this residue is not implicated in the enzymatic mechanism, because its mutation does not modify the activity [17]. It is thus surprising to find this Cys73 located close to the peroxidatic Cys48, at the bottom of the active-site pocket. Although the enzymatic mechanism involving the cysteine residues is well established, the physiological functions under specific stimulation conditions are still unknown. Other investigators have demonstrated that the catalytic cysteine of Prdx V has an important role in cytoprotection effects via its antioxidant activity [22]. Moreover, we recently reported a novel function of the peroxidatic Cys48 as a specific regulator of JAK2-Stat5 signaling triggered by lipopolysaccharide (LPS) [16]. Notably, it appeared that Prdx V played a role in the regulation of IL-6 expression through direct binding to Jak2 via its catalytic Cys48 residue, although this did not involve its peroxidase activity. Consistent with that, our current data show that hypoxia enhanced the direct interaction between Prdx V and DBT, and this enhancement did require the catalytic cysteine residue, Cys48, of Prdx V but not the other cysteine residues (Cys73 and Cys152).

In summary, this study provides new findings of novel Prdx V-interacting partners and showed alteration of DBT enzymatic activity in the kidney under hypoxia, which may be important for the kidney pathological conditions, such as hypoxic stress. Further studies are needed to explore the exact mechanism underlying Prdx V-DBT protein interaction associated with kidney homeostasis in chronic renal disease.

Conclusions

We have shown that Prdx V protein was interacting with other proteins in mouse kidney responding to hypoxic stress. One of interacting partner, DBT, showed enhanced interaction with Prdx V under hypoxia. Interestingly, DBT

enzymatic activity, known to be involved in mitochondrial metabolism, was increased under hypoxia as well. In addition, peroxidatic cysteine (Cys48) of Prdx V appeared to be important for interaction with DBT. This study proposes that Prdx V interacts with DBT and this interaction may be associated with mitochondrial metabolism in mouse kidney under hypoxia. Further functional analysis is required to confirm this hypothesis and to elucidate the underlying the role of Prdx V in biological mechanisms of hypoxic stress-induced mitochondrial metabolism.

Methods
Hypoxic conditions, protein extraction and immunoprecipitation
Mice (C57BL/6 J) were maintained under specific pathogen-free (SPF) conditions. All animal-related procedures were reviewed and approved under the Animal Care Regulations (ACR) of Chonnam National University (accession number: CNU IACUC-YB-2013-39).

Hypoxic stress was produced as described previously. Briefly, mice were placed in a chamber designed to regulate the flow of N_2 using a gas supply and the oxygen concentration was maintained at $8 \pm 0.5\%$ O_2 using an oxygen controller (Proox Model 110; BioSpherix, NY, USA). After 4 h of hypoxia, all mice (n = 10/each group, 8 weeks old) were induced with anesthesia under hypoxic conditions, and the kidneys were rapidly collected and frozen in liquid N_2 [13].

Total protein extraction was performed according to Yang et al. [13]. For immunoprecipitation assays, 500 μg of total extracted proteins was incubated with 10 μL of anti-Prdx V antibody at 4°C overnight. Then, immune complexes were pulled down by incubating with protein G agarose (Invitrogen, Carlsbad, CA) for 4 h at 4°C. The immunoprecipitated complex was eluted with 60 mM Tris–HCl (pH 6.8), 2.5% glycerol, 2% SDS, and 28.8 mM β-mercaptoethanol, and then the eluted complex was freeze-dried before being subjected to nano-UPLC-MS/MS analysis for comparative proteomics.

Interactome analysis by nano-UPLC-MS/MS
For gel-assisted digestion, the dried pellet was resuspended in 50 μl of 6 M urea, 5 mM EDTA, and 2% (w/v) SDS in 0.1 M tetra ethyl ammonium bicarbonate (TEABC) and incubated at 37°C for 30 min for complete dissolution. Proteins were reduced by adding 10 μL of 20 mM tris-2-carboxyethyl phosphine (TCEP) and alkylated followed by adding 20 μl of 20 mM iodoacetamide (IAM) at room temperature for 30 min. To incorporate proteins into a gel directly in the Eppendorf vial, 18.5 μl of acrylamide/bisacrylamide solution (40%, v/v, 29:1), 2.5 μl of 10% (w/v) ammonium persulfate, and 1 μl of 100% TEMED was then applied to the protein solution. The gel was cut into small pieces and then washed three

times with three volumes of TEABC containing 50% (v/v) ACN. The dehydrated gel samples were then digested with 15 μl trypsin (0.1 μg/μl) at 37°C for 18 hr. Then the digested peptides were recovered twice with a solution containing 50 mM ammonium bicarbonate, 50% acetonitrile, and 5% trifluoroacetic acid (TFA). The resulting peptide extracts were pooled, dried in a vacuum centrifuge, and then dissolved in 0.1% formic acid solution prior to MS or MS/MS analysis [14].

For nano-LC and tandem MS analysis, a nano-ACQUITY Ultra Performance LC Chromatography™ equipped Synapt™ G2-S System (Waters Corporation, MA, USA) used was previously described [23]. This step was performed on a 75 μm × 250 mm nano-ACQUITY UPLC 1.7 μm BEH300 C18 RP column and a 180 μm × 20 mm Symmetry C18 RP 5 μm enrichment column using a nano-ACQUITY Ultra Performance LC Chromatography™ System (Waters Corporation, MA, USA). Trypsinized peptides (5 μL) were loaded onto the enrichment column in mobile phase A (3% acetonitrile in water with 0.1% formic acid). A step gradient was then used at a flow rate of 300 nL/min. This included 3–40% mobile phase B (97% acetonitrile in water with 0.1% formic acid) run over 95 min, followed by 40–70% mobile phase B run over 20 min, and finally a sharp increase to 80% B over 10 min. Sodium formate (1 μmol/min) was used to calibrate the TOF analyzer in the range of m/z 50–2000, and [Glu1]-fibrinopeptide (m/z 785.8426) was run at 600 nL/min for lock mass correction. During data acquisition, the collision energies of low-energy mode (MS) and high-energy mode (MS^E) were set to 4 eV and 15–40 eV energy ramping, respectively. One cycle of the MS and MS^E modes of acquisition was performed every 3.2 s. In each cycle, MS spectra were acquired for 1.5 s with a 0.1 s interscan delay (m/z 300–1990), and the MS^E fragmentation (m/z 50–2000) data were collected in triplicate.

The continuum LC-MS^E data were processed and searched using the IDENTITYE algorithm in PLGS (ProteinLynx GlobalServer) version 2.5.2 (Waters Corporation). The data acquired by alternating low and high energy modes in the LC-MS^E were automatically smoothed, background subtracted, centered, deisotoped and charge state reduced, after which alignment of the precursor and fragmentation data were combined with retention time tolerance (±0.05 min) using PLGS software.

Processed ions were mapped against the IPI mouse database (version 3.87) using the following parameters: peptide tolerance, 10 ppm; fragment tolerance, 0.05 Da; missed cleavage, 1; and carbamidomethylation at C and oxidation at methionine and cysteine. Peptide identification was performed using the trypsin digestion rule with one missed cleavage. As a result, protein identification was completed with arrangement of at least two peptides.

All proteins identified on the basis the IDENTITYE algorithm are in keeping with > 95% probability. The false positive rate for protein identification was set at 5% in the databank search query option, based on the automatically generated reversed database in PLGS 2.5.2. Protein identification was also based on the assignment of at least two peptides comprised of seven fragments or more [23].

Cell culture and transfection

HEK293 cells were cultured in Eagle's Minimal Essential Medium (MEM) supplemented with 10% (v/v) fetal bovine serum (FBS), 1% glutamine, 100 mg/ml streptomycin, and 100 units/ml penicillin at 37°C in a humidified 5% CO_2 atmosphere. Lipofectamine 2000 (Invitrogen) reagents were used for transfection according to the manufacturer's instructions. At 48 hr after transfection, cells for hypoxic treatment were moved to a hypoxic incubator, and exposed to hypoxic stress, $1 \pm 0.2\%$ O_2, for 6 h.

Plasmid construction

Mouse Prdx V expression (WT, C48S, C73S, C152S, and C48/152S) vectors with N-terminal HA-tags were constructed as in our previous study [24]. The full-length DBT cDNA was purchased from Korea Human Gene Bank (clone ID: hMU001786). The human DBT was subcloned into the pCMV-Myc-N vector with N-terminal myc-tag (Clontech Laboratories, CA, USA).

DBT enzymatic activity

To measure the enzymatic activity of DBT, mitochondrial proteins from mouse hypoxic kidney (n = 10/each group) were isolated using the Qproteome Mitochondria Isolation Kit (Qiagen Sciences, Valencia, CA) according to the manufacturer's instructions. Using isolated mitochondrial proteins, branched-chain α-keto acid dehydrogenase (BCKDH) complex activity was determined spectrophotometrically at 37°C by measuring the rate of NADH generation from NAD^+ in the presence of substrate for BCKDH complex as described previously [25]. The reaction was started by adding the substrate and the change in the absorbance at 340 nm was recorded for 30 min. The BCKDH specific activity was expressed in mU/mg of mitochondrial protein: 1 mU was defined as the amount of enzyme that catalyzed the formation of NADH per min at 37°C.

Confocal fluorescence microscopy

HEK293 cells grown on poly L-lysine-treated glass coverslips were transiently transfected with Prdx V- and DBT-expressing vectors and then subjected to hypoxic stress, as described above. After hypoxic treatment, cells were fixed with 4% paraformaldehyde, permeabilized with 0.1% Triton X-100 and stained with anti-HA and anti-myc antibodies, followed by Alexa fluor555- and

Alexa fluor488-conjugated second antibodies. Then, cells were washed and mounted, and examined using a LSM 710 laser scanning confocal microscope (Carl Zeiss, Jena, Germany). Images were taken with a 63× objective at identical imaging settings.

Additional file

Additional file 1: Figure S1. Coprecipitation of endogenous Prdx V with Pccb, Rab43, and Alb proteins. Mouse kidney lysates were extracted under normoxic (N) and hypoxic (H) conditions, as indicated. The extracted kidney proteins were purified using Rab43, Pccb, or Alb antibodies, as indicated, and probed with Prdx V antibody to detect the interaction with Prdx V.

Abbreviations

Prdx V: Peroxiredoxin V; pO2: Pressure of oxygen; PHD: Prolyl-4-hydroxylase domain; HIF: Hypoxia-inducible factor; Prdx Vsi: Prdx V knock-down; Aco2: Aconitase 2; Acadm: Acyl-CoA dehydrogenase; Acox1: Acyl-CoA oxidase 1; BCAA: Branched-chain amino acid; WT: Wild-type; LPS: Lipopolysaccharide; SPF: Specific pathogen-free; ACR: Animal Care Regulations; TCEP: Tris-2-carboxyethyl phosphine; MEM: Minimal Essential Medium; FBS: Fetal bovine serum; BCKDH: Branched-chain α-keto acid dehydrogenase.

Competing interests

The authors declared that they have no competing interests.

Authors' contributions

T-H L proposed the project. S H A, H-Y Y, and G B T designed and performed the experiments and wrote the manuscript. J K, K-Y S, S K, Q T D, S J, and H-M L performed the data analysis. K-O C and T-H L provided guidance throughout the research and revised the manuscript. All authors read and approved the manuscript.

Acknowledgements

This work was supported by Basic Science Research Program through the National Research Foundation of Korea (NRF) funded by the Ministry of Science, ICT and future Planning (2014R1A2A2A01005448), and by the National Research Foundation of Korea (NRF) grant funded by the Korean government (MSIP) (2011-0030121). Sun Hee Ahn was also supported by Basic Science Research Program through the National Research Foundation of Korea (NRF) funded by the Ministry of Education (2013R1A1A2059721).

Author details

[1]Department of Oral Biochemistry, Dental Science Research Institute, Medical Research Center for Biomineralization Disorders, School of Dentistry, Chonnam National University, 300 Yongbong-Dong, Buk-Ku, Gwangju 500-757, Republic of Korea. [2]Department of Molecular Medicine, Graduate School, Chonnam National University, Gwangju, Republic of Korea. [3]Korea Basic Science Institute, Daejeon, Republic of Korea. [4]Laboratory of Veterinary Pathology, College of Veterinary Medicine, Chonnam National University, Gwangju, Republic of Korea. [5]Department of Oral and Maxillofacial Surgery, School of Dentistry, Chonnam National University, Gwangju, Republic of Korea.

References

1. Wood ZA, Schroder E, Robin Harris J, Poole LB. Structure, mechanism and regulation of peroxiredoxins. Trends Biochem Sci. 2003;28:32–40.
2. Chae HZ, Kang SW, Rhee SG. Isoforms of mammalian peroxiredoxin that reduce peroxides in presence of thioredoxin. Methods Enzymol. 1999;300:219–26.
3. Kropotov AV, Tomilin NV. A human B-box-binding protein downregulated in adenovirus 5-transformed human cells. FEBS Lett. 1996;386:43–6.
4. Knoops B, Clippe A, Bogard C, Arsalane K, Wattiez R, Hermans C, et al. Cloning and characterization of AOEB166, a novel mammalian antioxidant enzyme of the peroxiredoxin family. J Biol Chem. 1999;274:30451–8.
5. Shiota M, Izumi H, Miyamoto N, Onitsuka T, Kashiwagi E, Kidani A, et al. Ets regulates peroxiredoxin1 and 5 expressions through their interaction with the high-mobility group protein B1. Cancer Sci. 2008;99:1950–9.
6. Yamashita H, Avraham S, Jiang S, London R, Van Veldhoven PP, Subramani S, et al. Characterization of human and murine PMP20 peroxisomal proteins that exhibit antioxidant activity in vitro. J Biol Chem. 1999;274:29897–904.
7. Eckardt KU, Bernhardt WM, Weidemann A, Warnecke C, Rosenberger C, Wiesener MS, et al. Role of hypoxia in the pathogenesis of renal disease. Kidney international Supplement. 2005;68:S46–51.
8. Eckardt KU, Rosenberger C, Jurgensen JS, Wiesener MS. Role of hypoxia in the pathogenesis of renal disease. Blood Purification. 2003;21:253–7.
9. Fine LG, Norman JT. Chronic hypoxia as a mechanism of progression of chronic kidney diseases: from hypothesis to novel therapeutics. Kidney International. 2008;74:867–72.
10. Evans RG, Gardiner BS, Smith DW, O'Connor PM. Intrarenal oxygenation: unique challenges and the biophysical basis of homeostasis. Am J Physiol Renal Physiol. 2008;295:F1259–70.
11. Haase VH. Mechanisms of hypoxia responses in renal tissue. J Am Soc Nephrol. 2013;24:537–41.
12. Weinberg JM. Mitochondrial biogenesis in kidney disease. J Am Soc Nephrol. 2011;22:431–6.
13. Yang HY, Kwon J, Cho EJ, Choi HI, Park C, Park HR, et al. Proteomic analysis of protein expression affected by peroxiredoxin V knock-down in hypoxic kidney. J Proteome Res. 2010;9:4003–15.
14. Han CL, Chien CW, Chen WC, Chen YR, Wu CP, Li H, et al. A multiplexed quantitative strategy for membrane proteomics: opportunities for mining therapeutic targets for autosomal dominant polycystic kidney disease. Mol Cell Proteomics. 2008;7:1983–97.
15. Knoops B, Goemaere J, Van der Eecken V, Declercq JP. Peroxiredoxin 5: structure, mechanism, and function of the mammalian atypical 2-Cys peroxiredoxin. Antioxid Redox Signal. 2011;15:817–29.
16. Choi HI, Chung KJ, Yang HY, Ren L, Sohn S, Kim PR, et al. Peroxiredoxin V selectively regulates IL-6 production by modulating the Jak2-Stat5 pathway. Free Radic Biol Med. 2013;65C:270–9.
17. Seo MS, Kang SW, Kim K, Baines IC, Lee TH, Rhee SG. Identification of a new type of mammalian peroxiredoxin that forms an intramolecular disulfide as a reaction intermediate. J Biol Chem. 2000;275:20346–54.
18. Pettit FH, Yeaman SJ, Reed LJ. Purification and characterization of branched chain alpha-keto acid dehydrogenase complex of bovine kidney. Proc Natl Acad Sci U S A. 1978;75:4881–5.
19. Griffin TA, Lau KS, Chuang DT. Characterization and conservation of the inner E2 core domain structure of branched-chain alpha-keto acid dehydrogenase complex from bovine liver. Construction of a cDNA encoding the entire transacylase (E2b) precursor. J Biol Chem. 1988;263:14008–14.
20. Ono K, Hakozaki M, Suzuki T, Mori T, Hata H, Kochi H. cDNA cloning of the chicken branched-chain alpha-keto acid dehydrogenase complex. Chicken-specific residues of the acyltransferase affect the overall activity and the interaction with the dehydrogenase. Eur J Biochem. 2001;268:727–36.
21. Declercq JP, Evrard C, Clippe A, Stricht DV, Bernard A, Knoops B. Crystal structure of human peroxiredoxin 5, a novel type of mammalian peroxiredoxin at 1.5 A resolution. J Mol Biol. 2001;311:751–9.
22. Dubuisson M, Vander Stricht D, Clippe A, Etienne F, Nauser T, Kissner R, et al. Human peroxiredoxin 5 is a peroxynitrite reductase. FEBS Lett. 2004;571:161–5.
23. Yang HY, Kwon J, Park HR, Kwon SO, Park YK, Kim HS, et al. Comparative proteomic analysis for the insoluble fractions of colorectal cancer patients. Journal of Proteomics. 2012;75:3639–53.
24. Choi HI, Chung KJ, Yang HY, Ren L, Sohn S, Kim PR, et al. Peroxiredoxin V selectively regulates IL-6 production by modulating the Jak2-Stat5 pathway. Free Radic Biol Med. 2013;65:270–9.
25. Rodriguez-Bayona B, Peragon J. Stimulation of rat-liver branched-chain alpha-keto acid dehydrogenase activity by chronic metabolic acidosis. Int J Biochem Cell Biol. 1998;30:529–34.

16

Proteomics analysis of urine reveals acute phase response proteins as candidate diagnostic biomarkers for prostate cancer

Katarina Davalieva[1*], Sanja Kiprijanovska[1], Selim Komina[2], Gordana Petrusevska[2], Natasha Chokrevska Zografska[3] and Momir Polenakovic[1]

Abstract

Despite the overall success of prostate specific antigen (PSA) in screening and detection of prostate cancer (PCa), its use has been limited due to the lack of specificity. The principal driving goal currently within PCa research is to identify non-invasive biomarker(s) for early detection of aggressive tumors with greater sensitivity and specificity than PSA. In this study, we focused on identification of non-invasive biomarkers in urine with higher specificity than PSA. We tested urine samples from PCa and benign prostatic hyperplasia (BPH) patients by 2-D DIGE coupled with MS and bioinformatics analysis. Statistically significant (p < 0.05), 1.8 fold variation or more in abundance, showed 41 spots, corresponding to 23 proteins. The Ingenuity Pathway Analysis showed significant association with the Acute Phase Response Signaling pathway. Nine proteins with differential abundances were included in this pathway: AMBP, APOA1, FGA, FGG, HP, ITIH4, SERPINA1, TF and TTR. The expression pattern of 4 acute phase response proteins differed from the defined expression in the canonical pathway. The urine levels of TF, AMPB and HP were measured by immunoturbidimetry in an independent validation set. The concentration of AMPB in urine was significantly higher in PCa while levels of TF and HP were opposite (p < 0.05). The AUC for the individual proteins ranged from 0.723 to 0.754. The combination of HP and AMBP yielded the highest accuracy (AUC = 0.848), greater than PSA. The proposed biomarker set is quickly quantifiable and economical with potential to improve the sensitivity and specificity of PCa detection.

Keywords: Prostate cancer, Benign prostate hyperplasia, 2-D DIGE, MS, Urine analysis, Non-invasive biomarkers

Background

The introduction of prostate specific antigen (PSA) as a biomarker for prostate cancer (PCa) screening and detection has transformed the management of this disease [1-3]. Despite the overall success of the PSA blood test, its use has been limited due to the lack of specificity, especially in patients with total serum PSA levels in a range of 2–10 ng/ml. Various nonmalignant processes such as benign prostatic hyperplasia (BPH) and prostatitis, as well as manipulation and medical interventions of the prostate lead to serum PSA elevations and subsequently limit the specificity of PSA for cancer detection

[4]. Additionally, 15% of PCa cases occur in men with normal serum PSA levels [5]. These data have encouraged considerable investigation into the search for novel PCa biomarkers.

The rise of –omics technologies in recent years and their use in PCa research has delivered a number of new potential biomarkers for PCa [6-8]. These included proteins, fusing genes, RNA transcripts and epigenetic modifications of DNA. Among the available technologies, proteomics has shown large potential in identification of PCa biomarkers [9,10]. As a result of the found differences in protein expression profiles between BPH and PCa, and among different types and grades of cancer, a number of proteins in tissue and biological fluids (serum, plasma, urine, seminal plasma) were identified as potential diagnostic or prognostic markers for PCa. Prostate tissue, although a rich source of potential PCa biomarkers, has the

---

* Correspondence: katarina@manu.edu.mk
[1]Research Centre for Genetic Engineering and Biotechnology "Georgi D Efremov", Macedonian Academy of Sciences and Arts, Krste Misirkov 2, 1000 Skopje, Republic of Macedonia
Full list of author information is available at the end of the article

most invasive sampling, low tolerability and carries significant morbid risk [11]. On the other hand, a screening procedure based on biological fluid testing is highly desirable because of the minimally invasive and low-cost procedures for collecting samples. Hence, the current extensive investigation in PCa biomarkers is mostly oriented to the identification of highly specific non-invasive and easily accessible biomarkers in urine, seminal plasma and minimally invasive blood samples.

The number of newly identified diagnostic PCa biomarker candidates in biological fluids is rising over the time. Those found in serum include various precursor forms of PSA [12,13], α-2-macroglobulin [14,15], Zinc-α2-glycoprotein [14,16,17], Pigment epithelium-derived factor [16,18], Eukaryotic translation elongation factor 1 alpha 1, Fibronectin 1 [14], Chemokine ligand 16, Pentraxin 3, Spondin 2, Follistatine [19,20], panels of serum proteins [21] and many other soluble factors and intracellular proteins involved in structural or metabolic functions. Candidates for urine biomarkers include Annexin A3 [22], Inter-alpha-trypsin inhibitor heavy chain 4 [23], CD90 [24], Calgranulin/MRP-14 [25], Semenogelin 1, Uromodulin [26] and Engrailed-2 [27]. Some of these proteins have been identified in independent studies with different proteomics methods and their usefulness is yet to be validated in a large cohort within and across different ethnic populations. However, most of the data obtained until now is quite heterogeneous and there is a small percentage of overlap between independent studies. No single test or proposed biomarker to date can fulfill the requirements for the ideal PCa diagnostic biomarker and the next PCa screening tool. Furthermore, it is becoming clearer that the ideal PCa diagnostics test will most likely be based not on a single but multiple biomarkers, due to the clinical heterogeneity of the cancer and the need to distinguish the disease from greatly prevalent inflammatory and benign conditions [6,8]. This highlights the necessity of future extensive comparative analysis of well-defined samples for identification of a reliable diagnostic PCa biomarker or biomarker panel.

The aims of this study were to characterize the pattern of differential protein abundances in urine of PCa and BPH patients using two-dimensional difference in gel polyacrylamide gel electrophoresis (2-D DIGE) coupled with mass spectrometry (MS) and to identify a biomarker or marker panel for non-invasive PCa diagnosis preferentially with greater specificity and sensitivity from the ones that are currently in use. Another objective was to compare our results with those of other published studies and to assess the level of compatibility across different technological platforms used and different ethnic background of samples. The identified proteins with differential abundances between the PCa and BPH groups from this study were significantly associated with the Acute Phase Response Signaling pathway. We have been able to successfully validate these findings by confirming the differential abundance of TF, HP and AMBP in urine in an independent validation set. The results from this study may provide a clinically useful diagnostic set for the screening and detection of PCa.

Results

2-D DIGE analysis

The number of the detected spots in the DeCyder DIA workspaces in all gels ranged from 1308 to 1577. In the BVA module, the spots were matched among 4 gels with an average of 1063 matched spots. One hundred and thirty four spots showed statistically significant (p < 0.05) 1.8 fold variation or more in abundance. However, upon manual checking of the spot quality, the majority of the protein spots with differential abundance were either part of the most abundant protein in the urine, later identified as albumin [28], or were low quality spots. Therefore, only 41 spots were selected for further analysis (Figure 1). All of these spots fulfilled the criteria for presence in all spot maps. Among these, 22 spots showed higher abundance in the PCa group (up-regulated) and 19 spots had lower abundance in PCa (down-regulated).

Principle component analysis and hierarchical clustering analysis of the 41 spots with differential abundance were performed by EDA module, available within the DeCyder software (Figure 2). Two-dimensional scatter plots of the principal components of urine samples showed a good clear separation between samples from PCa and BPH patients (Figure 2A). Hierarchical clustering analysis (Figure 2B) showed that, based on the abundance pattern of the 41 spots, samples from PCa and BPH groups form two distinct separate clusters.

Identification and interpretation of proteins with differential abundance

All of the 41 spots with differential abundance were identified by MALDI-MS. Fold changes of the 41 identified protein spots in the two groups along with the detailed Mascot search results are given in Table 1. Several proteins were identified in multiple spots, most likely due to posttranslational modification leading to shifts in the 2-D. So, the identified spots with differential abundance equaled to 23 distinct proteins, among which 14 were up-regulated and 9 were down-regulated in PCa.

The 23 identified proteins were grouped into different classes based on sub cellular and functional information available (Additional file 1: Figure S1). The majority of the proteins were enzymes (30%), transporters (22%), enzyme inhibitors (13%) and proteins of unknown function (17%). Most of the proteins were secreted (74%), 17% were cytoplasmic and only 8% membrane or nucleus proteins. Binding was the major molecular function (50%), followed by catalytic activity (27%) and transport (15%). The GO

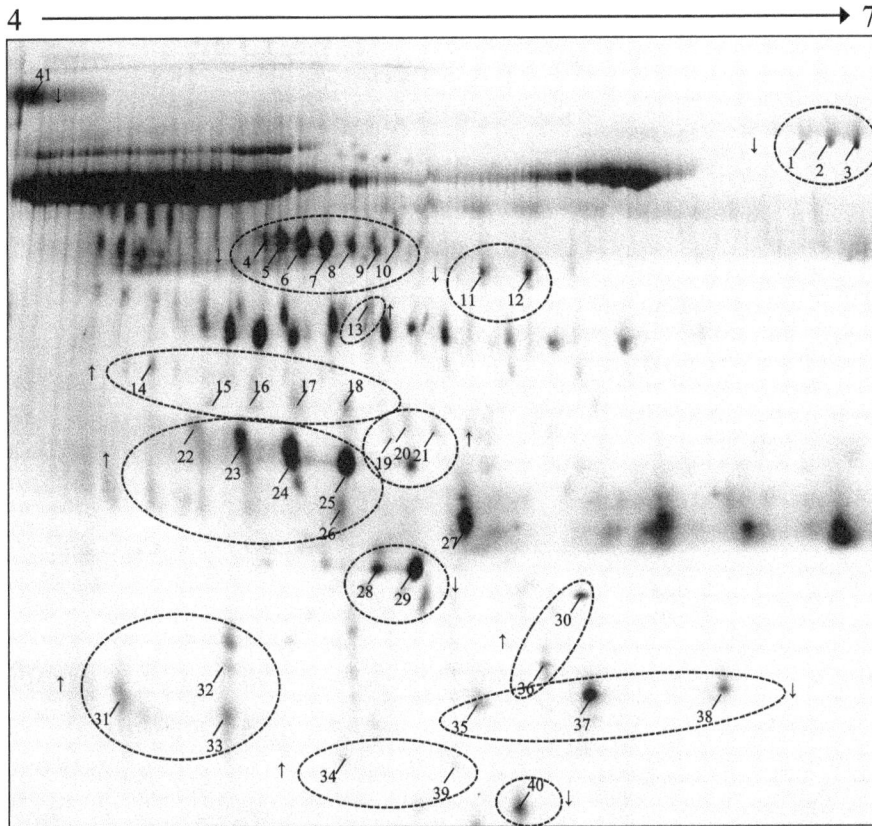

Figure 1 Representative 2-D map of the urine proteome obtained by using IEF on pH 4–7 IPG strip and 2-D gel electrophoresis on 12.5% SDS-PAGE. All proteins with differential abundance between studied groups are marked with numbered arrows. Details of these proteins identified by MALDI MS are tabulated in Table 1. Proteins with increased abundance in PCa are marked with red arrows.

based data of biological processes pointed to a variety of processes in which the identified proteins are putatively involved. Detailed information regarding biological function is given in Table 2.

The IPA analysis of associations between our set of proteins and known biological pathways showed significant association with the Acute Phase Response Signaling pathway ($p = 6,99E\text{-}14$). Nine of the 23 proteins with differential abundance are included in this pathway: α-1-microglobulin/bikunin (AMBP), apolipoprotein A-I (APOA1), fibrinogen alpha chain (FGA), fibrinogen gamma chain (FGG), haptoglobin (HP), inter-alpha-trypsin inhibitor (ITIH4), alpha-1-antitrypsin (SERPINA1), transferrin (TF) and transthyretin (TTR) (Figure 3A).

According to the IPA functional classification - Diseases and Disorders, 9 of our proteins (APOA1, CD59, FGG, GSN, HSPG2, MASP2, PTGDS, SERPINA1 and TYMP) are significantly associated with the category - Organism Injury and Abnormalities, sub category – Lesion formation ($p = 9,05E\text{-}10$) while in Molecular and Cellular Function category, 14 proteins (APOA1, CD59, FGA, GSN, HP, HSPG2, ILF2, ITIH4, PTGDS,

SERPINA1, TF, TTR, UMOD and TYMP) were found significantly associated with cell death ($p = 2,29E\text{-}5$).

The highest ranked protein network of functional associations between the differentially expressed proteins according to IPA was Cancer, Organism Injury and Abnormalities and Gastrointestinal Disease (score 46) (Figure 3B). Seventeen out of 23 proteins with differential abundance (TTR, HSPG2, CD59, FGA, FGG, GC, HP, APOA1, TF, ITIH4, AMBP, UMOD, SERPINA1, GSN, TYMP, PTGDS and LMAN2) are closely connected in the network through three major nodes: P38 Mitogen-activated protein kinase (P38 MARK), extracellular-signal-regulated kinase 1/2 (ERK1/2) and pro-inflamatory cytokines.

Our set of proteins was further analyzed using IPA biomarker filter which allows matching the input protein list with known disease profiles consisting of maps, networks and lists of biomarkers known for a disease. Results revealed that 12 of the proteins (AMBP, APOA1, FGA, GSN, HP, HSPG2, MASP2, PTGDS, SERPINA1, TF, TTR and TYMP) are predicted markers for prostate cancer (Figure 3C). The majority of these proteins (7/12) are also part of the Acute Phase Response Signaling pathway.

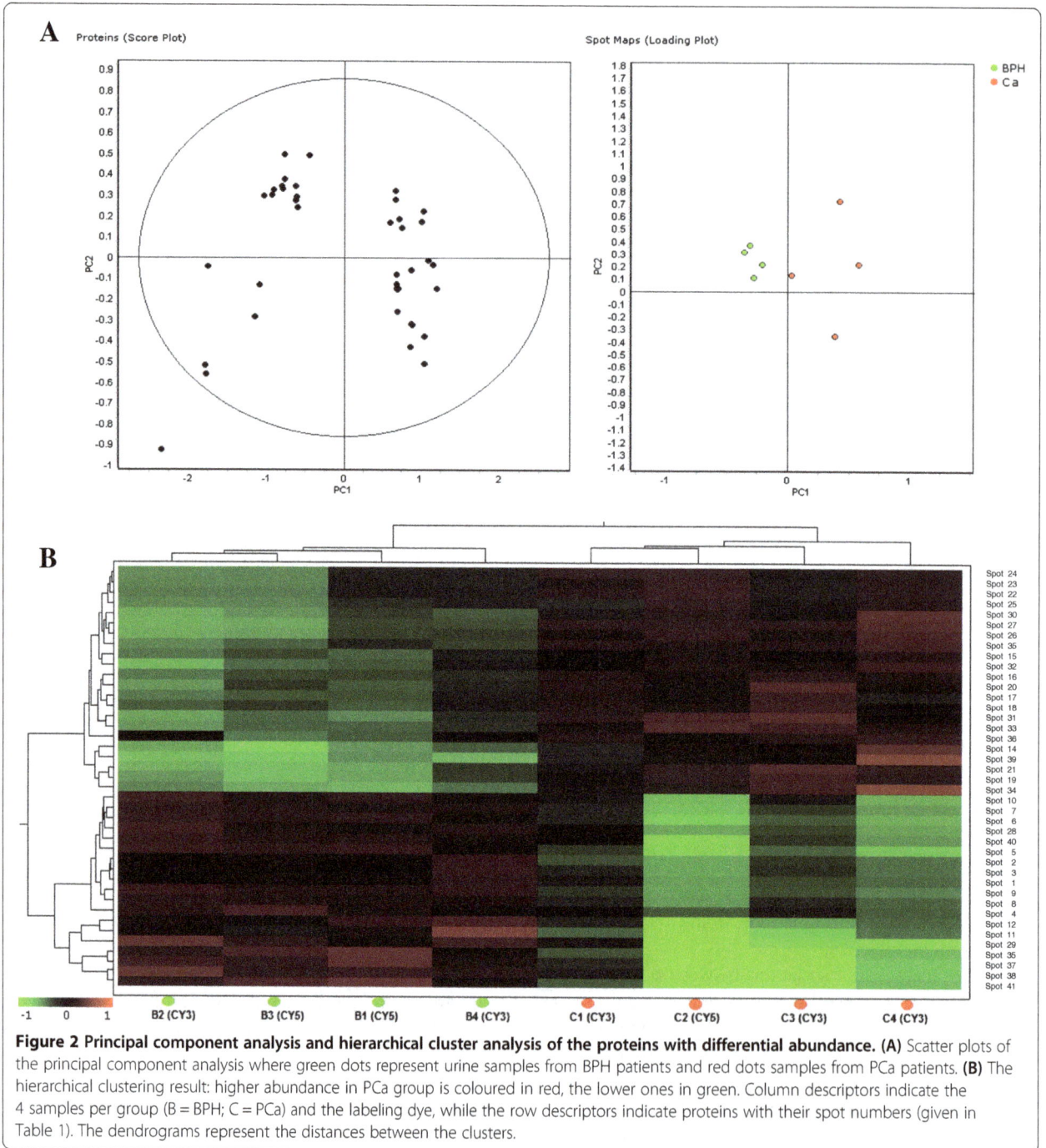

Figure 2 Principal component analysis and hierarchical cluster analysis of the proteins with differential abundance. (A) Scatter plots of the principal component analysis where green dots represent urine samples from BPH patients and red dots samples from PCa patients. **(B)** The hierarchical clustering result: higher abundance in PCa group is coloured in red, the lower ones in green. Column descriptors indicate the 4 samples per group (B = BPH; C = PCa) and the labeling dye, while the row descriptors indicate proteins with their spot numbers (given in Table 1). The dendrograms represent the distances between the clusters.

Validation of candidate biomarkers

Five proteins (TF, SERPINA1, APOA1, AMBP and HP) were further evaluated by immunoturbidimetry to test whether quantitative measurement in urine could be utilized as a diagnostic tool to distinguish patients with PCa and BPH. We manage to determine quantitatively the urine levels of three proteins (TF, AMPB and HP), while the concentrations of SERPINA1 and APOA1 were below the assays LOD. The measured concentrations of TF, AMPB and HP were normalized to urine creatinine to make correction for variations in urinary concentration. The results confirmed the abundance levels obtained by the DIGE experiment (Figure 4A). The concentration of AMPB in the PCa group showed a significantly higher level that in BPH group (Mann–Whitney U-test, p = 0.04). The levels of TF and HP were significantly higher in the BPH compared to the PCa group yielding p = 0.015 and p = 0.031, respectively (Figure 4B). Receiver operating curve (ROC) determined the diagnostic accuracy of the proteins in the validation set, using the histopathological

Table 1 List of proteins with differential abundance identified by MALDI-MS

Spot No.	T-test	Fold Ratio (PCa/ BPH)	Protein name	Swiss-Prot accession No.	Mw (kDa)	pI	Mascot protein score	RMS error (ppm)	p value	No. of matched peptides out of total	% of sequence coverage
1	0.010	−2.53	Serotransferrin	TRFE_HUMAN	79.29	6.81	98	80	3.1E-06	10/15	21
2	0.010	−2.64	Serotransferrin	TRFE_HUMAN	79.29	6.81	200	47	2.0E-16	23/43	38
3	0.010	−2.64	Serotransferrin	TRFE_HUMAN	79.29	6.81	243	57	1.0E-20	22/30	40
4	0.033	−2.17	Alpha-1-antitrypsin	A1AT_HUMAN	46.89	5.37	81	56	1.9E-04	12/31	31
5	0.002	−5.21	Alpha-1-antitrypsin	A1AT_HUMAN	46.89	5.37	128	82	3.2E-09	18/34	36
6	0.002	−4.34	Alpha-1-antitrypsin	A1AT_HUMAN	46.89	5.37	194	51	8.1E-16	23/43	53
7	0.002	−3.66	Alpha-1-antitrypsin	A1AT_HUMAN	46.89	5.37	74	109	7.7E-04	12/28	27
8	0.001	−3.05	Alpha-1-antitrypsin	A1AT_HUMAN	46.89	5.37	178	63	3.2E-14	20/31	41
9	0.000	−4.27	Alpha-1-antitrypsin	A1AT_HUMAN	46.89	5.37	66	102	4.9E-03	9/17	26
10	0.010	−3.18	Vitamin D-binding protein	VTDB_HUMAN	54.52	5.40	135	65	6.4E-10	12/24	50
11	0.011	−6.11	Fibrinogen gamma chain	FIBG_HUMAN	52.10	5.37	164	108	8.1E-13	17/30	47
12	0.019	−4.41	Thymidine phosphorylase	TYPH_HUMAN	50.32	5.36	64	104	7.4E-03	5/13	25
13	0.036	2.59	Gelsolin	GELS_HUMAN	86.04 (52.48)	5.90 (5.34)	80	81	2.0E-03	9/15	16
14	0.004	4.75	Fibrinogen alpha chain (fragment)	FIBA_HUMAN	95.65 (~50)	5.70 (4.65)	91	99	1.6E-05	10/18	15
15	0.003	2.47	Endonuclease domain-containing 1 protein	ENDD1_HUMAN	55.72	5.55	104	122	8.1E-07	9/13	20
16	0.000	3.07	E3 ubiquitin-protein ligase rififylin	RFFL_HUMAN	41.74	5.33	59	97	2.7E-02	5/12	26
17	0.009	2.72	Inter-alpha-trypsin inhibitor heavy chain H4 (fragment)	ITIH4_HUMAN	103,52 (~45)	6,51 (5,00)	64	51	9.1E-03	13/26	12
18	0.004	2.66	Inter-alpha-trypsin inhibitor heavy chain H4 (fragment)	ITIH4_HUMAN	103.52 (~45)	6.51 (5.15)	88	134	2.9E-05	11/26	13
19	0.001	4.75	Quinone oxidoreductase-like protein 1	QORL1_HUMAN	39.07	5.49	66	91	4.9E-02	4/12	12
20	0.005	3.17	Interleukin enhancer-binding factor 2	ILF2_HUMAN	43.26	5.19	62	82	3.0E-02	5/13	19
21	0.002	5.07	Vesicular integral-membrane protein VIP36	LMAN2_HUMAN	40.54	6.46	80	75	1.9E-04	7/14	21
22	0.041	2.26	Protein AMBP	AMBP_HUMAN	39.87	5.95	86	15	5.5E-05	8/15	22
23	0.048	2.25	Protein AMBP	AMBP_HUMAN	39.87	5.95	93	78	9.9E-06	10/24	23
24	0.044	2.26	Protein AMBP	AMBP_HUMAN	39.87	5.95	94	109	8.4E-06	10/19	24
25	0.035	2.35	Protein AMBP	AMBP_HUMAN	39.87	5.95	130	93	2.0E-09	15/35	37
26	0.023	3.33	Prostaglandin-H2 D-isomerase	PTGDS_HUMAN	21.24	7.66	61	52	1.8E-02	7/17	32
27	0.033	3.43	Prostaglandin-H2 D-isomerase	PTGDS_HUMAN	21.24	7.66	64	23	7.7E-03	8/36	33
28	0.005	−3.25	Apolipoprotein A-I	APOA1_HUMAN	30.75	5.56	116	57	5.1E-08	16/53	46
29	0.010	−6.82	Apolipoprotein A-I	APOA1_HUMAN	30.75	5.56	171	74	1.6E-13	20/55	57

Table 1 List of proteins with differential abundance identified by MALDI-MS (Continued)

30	0.009	3.55	Basement membrane specific heparan sulphate proteoglycan core protein (PERLECAN) (fragment)	PGBM_HUMAN	468.83 (20.65)	6.06 (5.47)	204	62	1.1E-15	15/33	88
31	0.001	4.60	CD59 glycoprotein	CD59_HUMAN	14.79	6.02	58	103	2.9E-02	4/7	21
32	0.006	3.11	CD59 glycoprotein	CD59_HUMAN	14.79	6.02	58	32	2.9E-02	4/7	21
33	0.000	4.17	CD59 glycoprotein	CD59_HUMAN	14.79	6.02	61	47	1.6E-02	5/15	28
34	0.001	6.56	Secreted and transmembrane protein 1	SCTM1_HUMAN	27.30	7.00	66	34	4.9E-03	6/13	22
35	0.019	−5.51	Haptoglobin (fragment, alpha 2 chain)	HPT_HUMAN	45.86 (~16)	6.13 (6.20)	73	79	9.3E-03	4/11	28
36	0.017	2.26	Mannan-binding lectin serine protease 2 (chainA, Human MBL associated protein 19)	MASP2_HUMAN	77.19 (19.53)	5.39 (5.44)	127	24	5.5E-08	7/14 (8/14)	37
37	0.015	−5.86	Haptoglobin (fragment, alpha 2 chain)	HPT_HUMAN	45.86 (~16)	6.13 (6.20)	88	32	7.5E-04	7/18	34
38	0.024	−6.58	Haptoglobin (fragment, alpha 2 chain)	HPT_HUMAN	45.86 (~16)	6.13 (6.20)	80	41	1.9E-04	11/34	31
39	0.004	6.21	Secreted and transmembrane protein 1	SCTM1_HUMAN	27.30	7.00	75	47	6.1E-04	6/9	22
40	0.012	−3.39	Transthyretin	TTHY_HUMAN	15.99	5.52	160	32	2.0E-12	10/27	73
41	0.010	−3.54	Uromodulin	UROM_HUMAN	72.64	5.05	158	57	3.2E-12	18/24	22

Table 2 Functional characterization of the proteins with differential abundance between PCa and BPH and association with urogenital cancers

Protein name	Symbol	Expression (PCa/BPH)	Biological function	Association with PCa[a]	Association with urogenital cancers[a]
Serotransferrin	TF	Down-regulated	Iron transport, acute-phase response, stimulation of cell proliferation	Down-regulated, serum [21]; Down-regulated, urine [29]	
Alpha-1-antitrypsin	SERPINA1	Down-regulated	Blood coagulation, acute phase response, hemostasis	Differentially expressed, tissue [30]; Up-regulated, serum [31]; Down-regulated, serum [21]	Up-regulated, urine, bladder cancer [32];
Vitamin D-binding protein	GC	Down-regulated	Vitamin D metabolic process, vitamin transport, transmembrane transport	Differentially expressed, tissue [33]	
Fibrinogen gamma chain	FGG	Down-regulated	Cell activation, protein complex assembly, response to stress, signal transduction, blood coagulation, hemostasis	Differentially expressed, tissue [30]	Degradation products in urine - markers for bladder cancer [34]; Up-regulated, urine, bladder cancer [35]
Thymidine phosphorylase	TYMP	Down-regulated	DNA replication, angiogenesis, response to stimulus	Potential marker for PCa [28]	Up-regulated, tissue, renal cell carcinoma [36]
Gelsolin	GSN	Up-regulated	Promote the assembly of monomers into filaments, transport, apoptosis, response to stress	Differentially expressed, tissue [33]; Down-regulated, tissue [37]	
Fibrinogen alpha chain	FGA	Up-regulated	Cell activation, protein complex assembly, response to stress, signal transduction, blood coagulation, hemostais		Degradation products in urine - markers for bladder cancer [34]; Up-regulated, urine, bladder cancer [35]
Endonuclease domain-containing 1 protein	ENDOD1	Up-regulated	Unknown, may act as DNase/RNase	Potential marker for PCa [28]	
E3 ubiquitin-protein ligase rififylin	RFFL	Up-regulated	Protein transport, proteolysis, apoptosis	Potential marker for PCa [28]	
Inter-alpha-trypsin inhibitor heavy chain H4	ITIH4	Up-regulated	Acute inflammatory response, carbohydrate metabolic process, nitrogen compound metabolic process	Up-regulated, urine [23]; Down-regulated, serum [31]	
Quinone oxidoreductase-like protein 1	CRYZL1	Up-regulated	Quinone cofactor metabolic process, cellular metabolic process	Potential marker for PCa [28]	
Interleukin enhancer-binding factor 2	ILF2	Up-regulated	Immune response, regulation of transcription, nucleobase-containing compound metabolic process	Potential marker for PCa [28]	
Vesicular integral-membrane protein VIP36	LMAN2	Up-regulated	Protein localization, protein transport, macromolecule localization		
Protein AMBP (Alpha-1-microglobulin/bikunin precursor)	AMBP	Up-regulated	Cell adhesion, immune response, regulator of biological processes, regulation of signal transduction	Up-regulated, urine [23]; Down-regulated, urine [38]; Down-regulated, serum [21]	
Prostaglandin-H2 D-isomerase	PTGDS	Up-regulated	Fatty acit metabolism, lipid biosynthesis, transport	Up-regulated, urine [38]	

Table 2 Functional characterization of the proteins with differential abundance between PCa and BPH and association with urogenital cancers *(Continued)*

Protein	Gene	Regulation	Function	Association	Association
Apolipoprotein A-I	APOA1	Down-regulated	Regulation of cytokine production, regulation of lipid transport, lipid metabolism	Differentially expressed, tissue [30]; Down-regulated, serum [31]; Down-regulated, serum [21]	Up-regulated, urine, bladder cancer [35]
Laminin G Like Domain 3 From Human Perlecan	HSPG2	Up-regulated	Developmental processes, cell adhesion, cell differentiation, localization		
CD59 glycoprotein	CD59	Up-regulated	Response to stress, cell surface receptor linked signal transduction, blood coagulation, hemostasis, response to external stimulus		
Secreted and transmembrane protein 1	SECTM1	Up-regulated	Immune response, regulation of signal transduction, multicellular organismal development		
Haptoglobin (α-chain)	HP	Down-regulated	Proteolysis, cellular iron ion homeostasis, response to stress, defense response, metabolic process	Down-regulated, urine [38]; Down-regulated, tissue [39]; Down-regulated, serum [21]	Up-regulated, urine, bladder cancer [35]
Mannan-binding lectin serine protease 2	MASP2	Up-regulated	Complement activation, lectin pathway, activation of immune response		
Transthyretin	TTR	Down-regulated	Transport, localization, establishment of localization	Up-regulated, serum [31]	
Uromodulin	UMOD	Down-regulated	Response to stress, cellular defense response, regulation of cell proliferation	Down-regulated, urine [26]	

aBased on published proteomics studies retrieved from PubMed.

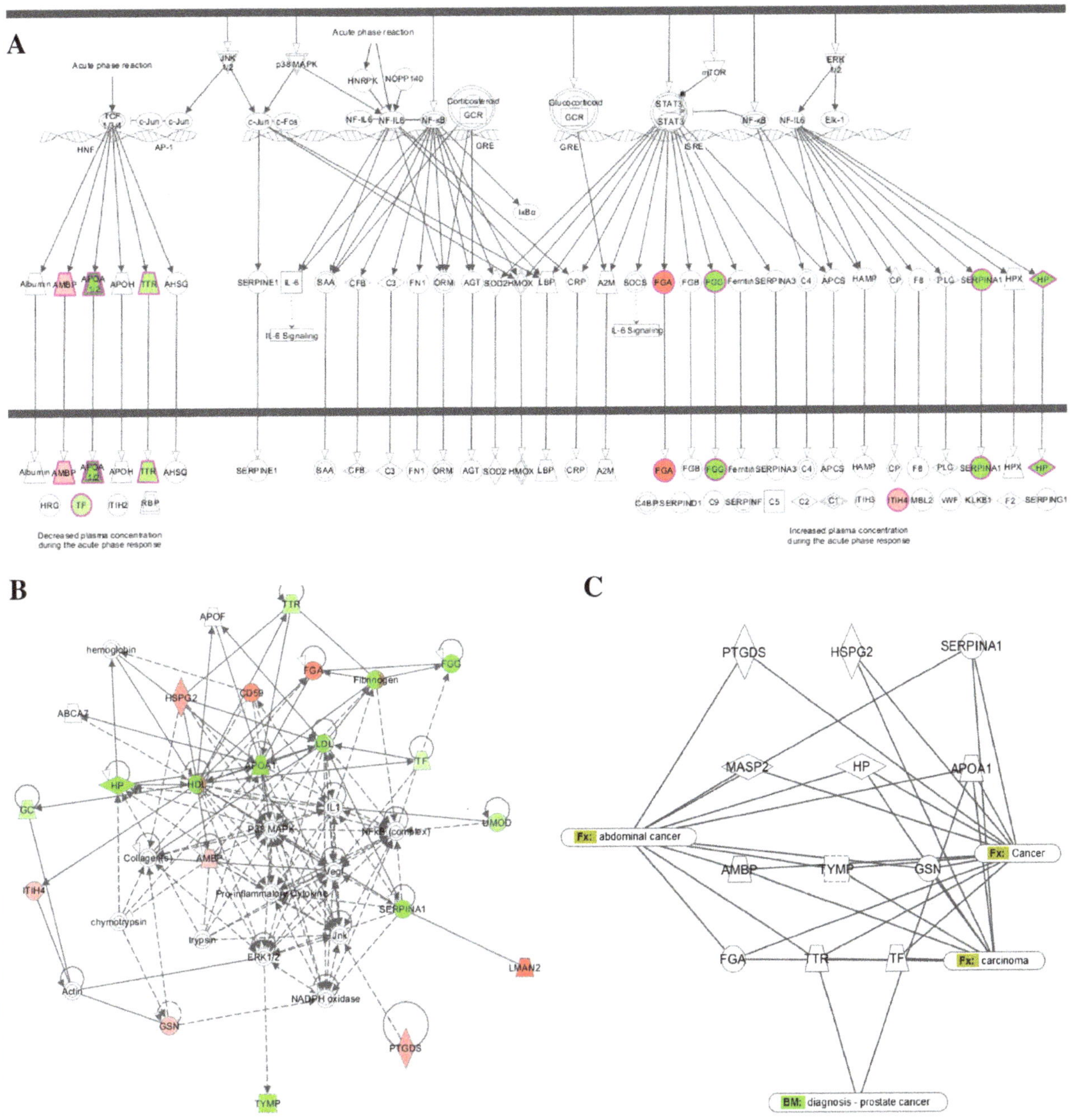

Figure 3 Pathways and networks associated with proteins with differential abundance according to IPA. (A) The top canonical pathway significantly associated with the differentially expressed proteins - Acute Phase Response Signaling (p = 6,99 e^{-14}). **(B)** Highest ranked protein network of functional associations between 23 proteins with differential abundance - Cancer, Organism injury and abnormalities and Gastrointestinal Disease. Most of the proteins with differential abundance are closely connected in the network through three major nodes: P38 MARK, Pro-inflamatory cytokine and ERK1/2. The network is graphically displayed with proteins as nodes and the biological relationships between the nodes as lines. The color of the shapes indicates the degree of over-expression (red) or under-expression (green) of the corresponding protein in PCa compared to BPH samples. Direct connection between molecules is represented by a solid line and indirect connection by broken line. The length of a line reflects published evidence supporting the node-to-node relationship concerned. **(C)** Selected subset of proteins with differential abundance associated with cancer in humans or cancer cell lines.

results as the gold standard for clinical diagnosis and classification (Figure 4C). The area under the curve (AUC) for TF was 0.754 (p = 0.002, 95% CI 0.596-0.912), for AMBP was 0.738 (p = 0.005, 95% CI 0.574-0.903) and for HP was 0.723 (p = 0.008, 95% CI 0.558-0.888). The optimal cutoffs for the proteins were: 12.81 mg TF/g creatinine (93.8% specificity, 56.3% sensitivity); 6.51 mg AMBP/g creatinine (50.0% specificity, 93.8% sensitivity); 2.40 mg HP/g

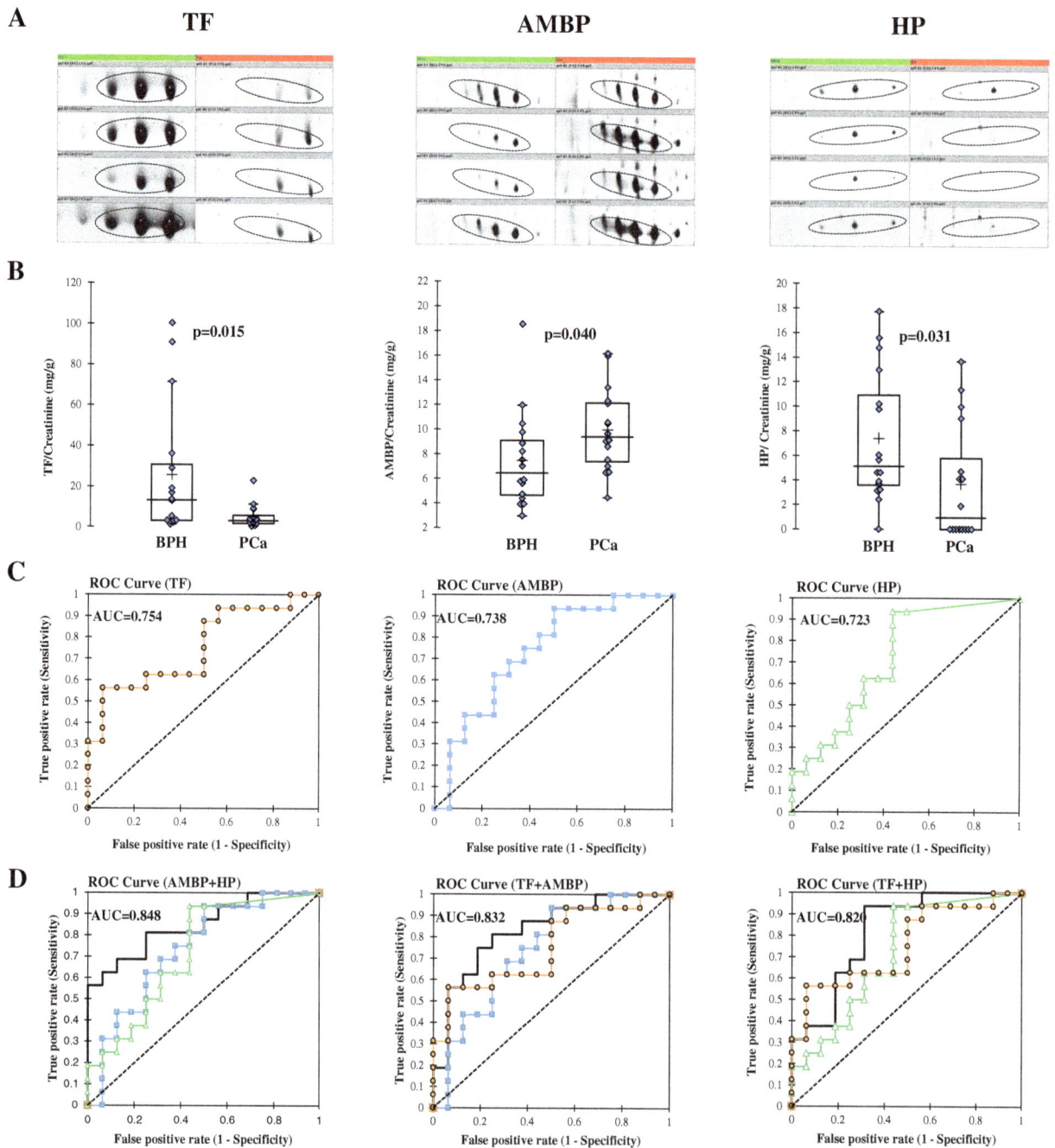

Figure 4 Validation of candidate biomarkers for the diagnosis of PCa. (A) 2-DE profiles of TF, AMBP and HP abundance in independent urine samples from BPH and PCa patients obtained by 2-D DIGE. Proteins with differential abundance were represented by clusters of 3–4 spots, highlighted with oval lines. Four gels corresponding to samples from each group were shown. **(B)** TF, AMBP and HP levels in urine of PCa and BPH patients, expressed as relative ratio to urine creatinine and obtained by immunoturbidimetry. AMBP level in PCa was significantly higher than that in BPH while TF and HP levels in PCa were significantly lower than in BPH (Mann–Whitney U-test, $P < 0.05$). In the combined dot/box plot graphs, concentration data (blue diamond), median (–), 25th and 75th percentiles and mean (+) are shown. **(C)** Urinary TF, AMBP and HP distinguish PCa on independent series of urine samples from patients with PCa and BPH. The optimal cutoffs for the proteins were: 12.81 mg TF/g creatinine (93.8% specificity, 56.3% sensitivity); 6.51 mg AMBP/g creatinine (50.0% specificity, 93.8% sensitivity); 2.40 mg HP/g creatinine (56.3% specificity, 93.8% sensitivity). ROC curves were based on series of 32 urine samples. **(D)** The diagnostic accuracy of TF, AMBP and HP combinations using logistic regression model. The combination AMBP and HP yielded highest diagnostic accuracy (AUC = 0.848).

creatinine (56.3% specificity, 93.8% sensitivity). The values of AUC and diagnostics cutoff for serum PSA in the validation set were 0.754 (p = 0.001, 95% CI 0.600-0.907) and 5.50 ng/ml (50.0% specificity, 93.8% sensitivity), respectively (data not shown).

Using a logistic regression model, the diagnostic accuracy of various combinations of TF, AMBP, HP and PSA were tested (Table 3 and Additional file 2). The combination of TF, AMBP and HP increased the individual diagnostic accuracy (Figure 4D). The highest AUC was obtained for the combination of HP and AMBP (AUC = 0.848). Inclusion of PSA or TF into the HP/AMBP test has not yielded an improved accuracy. The combination of each of the tested proteins with PSA also resulted in increased AUC, except for the AMBP where addition of PSA did not improve the diagnostic accuracy (0.773 (AMBP + PSA) vs. 0.738 (AMBP) vs 0.754 (PSA).

Discussion

The low sensitivity and specificity of current diagnostic methods for prostate cancer highlights the need for improvement in this area. In this study, we focused on identification of non-invasive biomarkers for PCa using urine samples from two age-matched groups of patients with histologically characterized diagnosis of PCa and BPH, respectively.

Urine is an attractive material in clinical proteomics because it can be sampled non-invasively in large quantities, contains generally soluble proteins which does not undergo significant proteolytic degradation and has lower dynamics range of protein concentrations compared to other biofluids [40]. Urine proteome represents modified ultrafiltrate of plasma combined with proteins derived from kidney and urinary tract [41]. Proteomic analysis of urine has suggested that it contains disease-specific information for a number of kidney diseases, cancers related to the urogenital system such as kidney, bladder and prostate cancer, as well as various non-nephrological/urogenital diseases [42]. The comparative proteomics studies for PCa biomarker identification in urine have reported a number of proteins with differential abundance between PCa and BPH, with some of them proposed as potential biomarkers for PCa diagnosis [22-27,43,44]. However, most of these

candidate biomarkers still lack an extensive validation and haven't been introduced into clinical practice. The high variability of the reported data so far highlights the need for more research in this area with consistent sample collection, storage methods and sample processing. The samples used in this study were collected and stored according to the standard protocol for urine collection with no addition of protease and phosphatase inhibitors and no pH adjustment [45]. Sample processing and manipulation were minimal without depletion of the highly abundant proteins, to exclude the possibility of losing low abundant proteins or low molecular weight proteins that exist in complexes with the highly abundant proteins. In addition, we have used the 2-D DIGE/MS platform who although has lover dynamic range of protein quantification than LC-MS based platforms is still an indispensable platform in proteomics, particularly for the visualization of the proteome and assessment of the individual posttranslational modifications [46].

Our proteomic data revealed 23 proteins with differential abundance in urine of PCa patients compared to BPH patients. In order to analyze the impact of coregulation of protein expression in context of biomarker identification, principal component analysis and hierarchical clustering has been applied on the differential protein expression data. Here, we have detected a clear separation of the two distinct groups (PCa and BPH) based on the abundance pattern of the 23 proteins. The clustering of samples showed formation of two separate clusters, clearly separating benign from tumor samples with general observation that PCa/BPH classification cannot be based on one individual protein, but on expression pattern of the selected subset of proteins.

Gene Ontology (GO) search for biological processes classified these proteins into proteins involved in the immune response and response to stimuli, regulators of different biological processes (transcription, proliferation, signal transduction, cytokine production), transport proteins involved in transport of ions, proteins and lipids and proteins involved into different metabolic processes.

Many of the proteins with differential abundance in this study have been associated with PCa or cancers of the urogenital system (Table 2). Overall, 17 of the identified proteins have been associated specifically with PCa in different proteomics studies. Our initial study of the protein components from urine of PCa patients, pointed out 11 proteins that might have some role in the pathogenesis of prostate cancer by comparison with other published studies analyzing normal urine proteome [28]. Five of these proteins (TYMP, ENDOD1, RFFL, CRYZL1 and ILF2) were detected with differential abundance when comparing to BPH group. In comparative studies using urine, the abundance levels of TF, ITIH4, AMPB, PTGDS, HP and UMOD were the same as in our study

Table 3 The diagnostic accuracy of various combinations of TF, AMBP, HP and PSA expressed as area under the ROC curve (AUC) using logistic regression model

Protein combinations tested								
TF	√	√			√	√	√	
HP	√	√	√	√		√		√	
AMBP	√	√	√	√	√				√
PSA	√		√				√	√	√
AUC	0.840	0.848	0.848	0.848	0.832	0.820	0.840	0.836	0.773

[23,26,29,38]. However, for ITIH4 and AMBP, there are also studies where opposite abundance levels were reported as well [31,38]. The abundance level of APOA1 in our study was the same as in a study using serum while the levels of SERPINA1 and TTR were found opposite to ours [31]. Some of the proteins such as SERPINA1, FGG, FGA, APOA1 and HP were also found with differential abundance in urine in proteomics studies of bladder cancer, and TYMP in tissue in renal cell carcinoma, but mostly with opposite abundance level compared to our study [32,34-36].

IPA analysis of our set of proteins revealed significant association with the Acute Phase Response Signaling pathway. Nine of our proteins (AMBP, APOA1, FGA, FGG, HP, ITIH4, SERPINA1, TF and TTR) were acute phase response proteins. The acute phase response is a rapid inflammatory response that provides protection against microorganisms using non-specific defense mechanisms [47]. The association of our proteins with this pathway complies with the generally accepted observation that inflammation is often observed in tumors and appears to play a dominant role in the pathogenesis of various cancer types [48]. Moreover, the highest ranked protein network of functional associations between the differentially expressed proteins revealed close connected in the network through three major nodes: P38 Mitogen activated protein kinase (P38 MARK), extracellular-signal-regulated kinase 1/2 (ERK1/2) and pro-inflamatory cytokines. P38 MARK and ERK1/2 are members of the mitogen activated protein kinase super family that can mediate cell proliferation and apoptosis [49]. Abnormal regulation of the MAPK pathways have been reported for a wide range of diseases including many cancers [50]. The pro-inflamatory cytokines interleukin-6 (IL-6), tumor necrosis factor alpha (TNFα) and interleukin-1 beta (IL-1β) are critical mediators of the systemic inflammatory response and main stimulators for the synthesis of an acute-phase response proteins [48]. These cytokines have been implicated in a variety of diseases including cancer where their role is more likely to contribute to tumour growth, progression and immunosuppression than to an effective host-antitumor response [51]. For this reason, cancer patients frequently present changes in various systemic parameters, comprising alterations in the level of serum inflammatory cytokines, acute-phase proteins and total albumin [52]. And finally, IPA biomarker filter revealed 12 proteins as candidate biomarkers for prostate cancer with majority being acute phase proteins.

A number of well characterized acute phase proteins have been linked to distinct cancer types and stages of malignancy [53]. The field of acute phase proteins as cancer biomarkers has vast potential. The identification of specific proteomic expression patterns in acute phase proteins related to cancer offers promise for novel diagnostic markers. Having in mind this and results from the IPA analysis, we considered the possibility of using acute phase response proteins found in urine as non-invasive biomarkers for PCa. This was additionally encouraged from the fact that the abundance pattern of 4 acute phase response proteins in our study differed from the defined expression (Figure 4A). Protein AMBP shows decreased plasma concentration during the acute phase response, but in our study increased expression in PCa was observed. Proteins HP, SERPINA1 and FGG showed decreased expression in PCa, opposite to the plasma concentration during the acute phase response. On the other hand, the expression levels of the acute phase response proteins in this study correlated with the observed levels in a number of independent studies on PCa, with minor exceptions where opposite abundance levels were also reported (reviewed in Table 2).

From the total set of acute phase response proteins detected with differential abundance, we highlighted TF, AMBP, HP, APOA1 and SERPINA1. These proteins were detected as clusters of 2-6 spots with the same abundance levels within clusters and therefore exhibited higher potential to be associated with PCa than proteins identified in only one spot. Notably, AMBP, HP and SERPINA1 were particularly interesting for us since they demonstrated the same abundance levels as in other proteomics studies and opposite abundances from the acute phase response. So, we chose to further evaluate the diagnostic potential of these 5 proteins in an independent set of patients by quantitative measurement in urine. Using immunoturbidimetry, we manage to determine quantitatively the urine levels of TF, AMBP and HP in all samples from the validation set. The results showed that there was a significant difference ($p<0.05$) in the urine levels of these proteins between PCa and BPH group, confirming the expression levels obtained by the DIGE experiment. The observed diagnostics accuracy of the proteins was moderate, with the lowest being found for HP (0.723 at 2.40 mg/g creatinine) and the highest for TF (0.754 at 12.81 mg/g creatinine). Similar diagnostics accuracy was found for serum PSA (0.754). A number of logistic regression models using combinations of TF, AMPB, HP and PSA were also tested in order to estimate if the diagnostic accuracy improves with the combination of proteins. In general, combination of proteins showed increased AUC with the highest value obtained for HP/AMBP. So, although the observed accuracy of the individual proteins were similar to the PSA, the combination of HP and AMBP yielded greater accuracy compared to individual tests. The results indicated that HP/AMBP combination could be potential biomarker set for the diagnosis of PCa with improved accuracy compared to the PSA. In addition, the proposed biomarker set complies with the requirements for diagnostics biomarker as it is easily accessible, non-invasive, quickly quantifiable and economical. Though maybe these proteins are not specific proteins for PCa, they could become an important check index and improve the sensitivity and specificity for early diagnosis.

Conclusions

Our study indicated that the 2-D DIGE/MS proteomic analysis of urinary proteins is feasible for the identification of non-invasive biomarkers for PCa diagnosis. As a result of this approach, a set of acute phase response proteins found in urine could serve as a diagnostics biomarkers for PCa. Moreover, our results confirmed the importance of previously identified proteins and highlighted new proteins that can add information regarding the pathophysiological mechanisms of PCa. However, further studies are needed to validate the proposed biomarkers in independent cohorts and to evaluate the diagnostic potential of the rest of the differentially expressed proteins found in this study which might further improve the diagnostics accuracy of the proposed set.

Methods

Samples

We analyzed 56 urine samples from patients with clinically and histological confirmed PCa and BPH obtained from the University Clinic for Urology, University Clinical Centre "Mother Theresa", Skopje, Republic of Macedonia. Informed consent for the use of these samples for research purposes was obtained from the patients in accordance with the Declaration of Helsinki. The study has been approved by the Ethics Committee of the Macedonian Academy of Sciences and Arts.

The patients referred to hospital because of clinical symptoms. The diagnosis was based on histological evaluation of tissues obtained by transurethral resection of the prostate (TURP) for BPH and whole prostate gland obtained by radical prostatectomy for PCa patients. None of the patients received preoperative therapy. Patient's clinical records including histology grading, tumor stage and pre-operative PSA were reviewed to preselect the urine samples used in this study (Additional file 3: Table S1). BPH patients were preselected to be without signs of inflammation (prostatitis). Urine samples for the 2D-DIGE analysis (screening set) consisted of 8 samples from PCa and 16 samples from BPH patients. Validation of the selected candidate biomarkers was done on an additional 16 PCa and 16 BPH urine samples. In the screening cohort, the mean values (± SD) for PSA, Gleason score and age were as follows: serum PSA was 9.0 ± 3.9 (ng/ml) for the PCa group and 6.4 ± 3.2 (ng/ml) for the BPH group; age was 69.0 ± 6.6 years for PCa and 65.3 ± 7.3 years for BPH; Gleason score was 7.4 ± 1.1. In the validation cohort, the mean values (± SD) for PSA, Gleason score and age were

as follows: serum PSA was 8.7 ± 3.2 (ng/ml) for PCa group and 5.6 ± 2.9 (ng/ml) for BPH group; age was 67.4 ± 5.0 years for PCa and 69.2 ± 8.2 years for BPH; Gleason score was 7.4 ± 1.1.

The first morning urine $(3-10)$ ml was collected from the patients prior to clinical intervention and stored on ice for short period (<1 h). Samples were centrifuged at 1000 g, for 10 min to remove cell debris and casts, aliquoted in 1.5 ml tubes and stored at $-80°C$ until use.

The stored urine samples were thawed and for each sample, proteins were isolated in triplicate from 100 μl urine using 2-D Clean-UP Kit (GE Healthcare) according to the manufacturer's instructions. The pellets from each replicate were dissolved in 10 μl of UTC buffer (8 M Urea, 2 M Thiourea, 4% CHAPS), pooled together for each sample, quantified by the Bradford method [54] in duplicate against a standard curve of Bovine Serum Albumin (BSA) and stored at $-80°C$ until use.

2-D DIGE, imaging and analysis

Equal amounts of protein extract from urine were pooled for the DIGE labeling: 2 PCA samples and 4 BPH samples per labeling reaction respectively. The pH of protein samples was adjusted to 8.5 with 1.5 M Tris–HCl. Proteins were labeled with the CyDye DIGE Fluor minimal dyes (GE Healthcare) following manufacturer's instructions. Forty five micrograms of protein per pool were minimally labeled with 400 pmol of Cy3 or Cy5, respectively. Cy2 was used to label an equivalent amount of internal standard containing equal amounts of all samples. Reactions were stopped with 10 mM L-lysine for 10 min. The samples were randomized between gels to ensure an even distribution between those labelled with Cy3 and Cy5 minimal dyes and to avoid repetitive linking of the same sample type with the same dye on multiple gels.

The first dimension of the 2-D DIGE analysis was performed using 24 cm Immobiline Drystrip gels (GE Healthcare) with linear pH 4–7 gradient. The separate CyDyes labeling reactions were combined, rehydration buffer (8 M Urea, 2 M Thiourea, 2% (w/v) CHAPS, 10 mM DTT, 1.2% (v/v) IPG-Buffer pH4-7, Trace of Bromophenol Blue) was added to a final volume of 450 μl and the gels were passively rehydratated overnight in IPG-Phor cassettes. Isoelectric focusing (IEF) was performed on the Ettan IPGphor 3 system (GE Heltchare) under the following conditions: 3 h at 300 V, 7 h gradient to 1000 V, 3 h gradient to 10000 V and 4 h 15 min at 10000 V, until total a of 64.5 kVh was reached. The focused proteins in the IPG strips were immediately equilibrated in two incubation steps, each lasting 15 min, at room temperature. In the first step, the equilibration buffer (6 M Urea, 2% SDS, 30% Glycerol, 50 mM Tris–HCl, pH 8.6) was supplemented with 1% (w/v) DTT for reduction, followed by alkylation in the same buffer containing 4.7% (w/v) iodoacetamide instead of DTT. The second dimension was carried out onto 12.5% homogeneous polyacrylamide gels using the Ettan DALTsix system (GE Healthcare), at 2.5 W per gel for 30 min, followed by 16 W/gel for 5 h.

The four 2-D DIGE gels were scanned on an Ettan DIGE imager (GE Healthcare). Gel images were normalized by adjusting the exposure time to obtain appropriate pixel value without any saturation. All gels were scanned at 100 dpi resolution. Images were cropped using Image QuantTL software v7.0 (GE Healthcare) to remove areas extraneous to the gel image.

DIGE images were analyzed using DeCyder 2-D Differential Analysis Software v7.1 (GE Healthcare). Spot detection and normalization was processed by the Differential Analysis (DIA) module using the estimated number of spots set to 3000 and spot volume < 30000 as exclusion filter. Gel-to-gel comparison and statistical analysis of the degree of difference in standardized protein abundance between PCa and BPH groups were performed with the Biological Variation Analysis (BVA) module. Matching was further improved by using landmarks and manually confirming potential spots of interest. Proteins with statistically significant differential abundance were selected based on two criteria: t-test < 0.05 and ratio > 1.8. Each spot was manually verified for an acceptable three dimensional characteristic protein profile and for adequate material for subsequent mass spectrometry identification. Spots not meeting these criteria were excluded from further analysis. The Extended Data Analysis (EDA) module of DeCyder and was used for principal component analysis and clustering studies. Selected protein spots with significant difference were excised from a preparative gel stained with CBB G-250 and identified by MALDI MS.

Setting up of preparative 2-D gels for spot picking

For preparative Coomassie brilliant blue (CBB) - stained gel, 60 μg of each of the 8 protein pools were combined, to give a total of 480 μg of protein. The volume was adjusted to 450 μl with the rehydration buffer. The IEF and second dimension SDS-PAGE were run according to standard procedures. Gel was fixed in 30% (v/v) ethanol, 2% (v/v) phosphoric acid for 30 min with two exchanges of the fixing solution, washed three times with 2% (v/v) phosphoric acid for 10 min each, balanced in pre-staining buffer (12% (w/v) $(NH_4)_2SO_4$, 2% (v/v) phosphoric acid, 18% (v/v) ethanol) for another 30 min and stained in staining solution (0.01% (w/v) CBB G-250, 12% (w/v) $(NH_4)_2SO_4$, 2% (v/v) phosphoric acid, 18% (v/v) ethanol) for 72 h. The gel was stored in the staining solution until the spots of interests were manually picked.

Mass spectrometry: in-gel tryptic digestion and identification

In-gel digestion was carried out manually with trypsin. Spots were first destained twice with a mixture of 50%

(v/v) ACN for 15 min each and than once with 100 mM NH_4HCO_3 and 50% (v/v) ACN for 15 min. Spots were dried in vacuum centrifuge and then reduced with 100 mM NH_4HCO_3 containing 10 mM DTT for 45 min at 56°C, and then alkylated with 54 mM iodoacetamide in 100 mM NH_4HCO_3 for 30 min in the dark, at room temperature. Gels pieces were washed with 100 mM NH_4HCO_3, shrunk with 50% ACN for 15 min and dried in a vacuum centrifuge. Gel particles were rehydrated with 20 μl of 0.01 μg/μl trypsin proteomics grade (Roche Diagnostics GmbH) in digestion buffer (95% 50 mM $NH_4HCO_3/5\%$ ACN) for 45 min at room temperature. The remaining enzyme supernatant was replaced with one gel volume of the digestion buffer and digestion was carried out at 37°C, overnight. After digestion, peptides were collected in a separate tube, extracted once with 20 μl of 50% ACN and twice with a mixture of 50% ACN/5% formic acid, dried in a vacuum centrifuge and reconstituted in 10 μl of 0.1% TFA.

For MS analysis, peptides were purified using ZipTip$_{C18}$ (Millipore Corporation) following the manufacturer's instructions and eluted in 2–3 μl of CHCA (4 mg/ml in 50% ACN/0.1% TFA) directly onto a MALDI target plate (Shimadzu Biotech Kratos Analytical). Droplets were allowed to dry at room temperature. Samples analysis was performed using AXIMA Performance MALDI-TOF-TOF mass spectrometer (Shimadzu Biotech Kratos Analytical). Spectra acquisition and processing was performed using the MALDI-MS software (Shimadzu Biotech Kratos Analytical) version 2.9.3.20110624 in positive reflectron mode at mass range 1–5000 Da with a low mass gate at 500 Da and pulsed extraction optimized at 2300 Da. External calibration was performed based on monoisotopic values of five well-defined peptides: Bradykinin fragment 1–5, Angiotensin II human, Glu1-Fibrinopeptide B human, Adrenocorticotropic Hormone Fragment 1–17 human and Adrenocorticotropic Hormone Fragment 7–38 human (Sigma-Aldrich). External calibration mix (500 fmol/μl) was diluted with the matrix in ratio 1:1 and applied onto MALDI target plate at final concentration of 250 fmol per spot. Each mass spectrum was acquired by 500 laser profiles (five pulses per profile) collected across the whole sample.

After filtering tryptic-, keratin- and matrix-contaminant peaks, the resulting monoisotopic list of m/z values was submitted to the search engine MASCOT (version 2.4.01, MatrixScience, UK) searching all human proteins and sequence information from Swiss-Prot (version 2014_05, 20265 sequences) and NCBInr (version 20140323, 276505 sequences). The following search parameters were applied: fixed modification-carbamidomethylation, variable modifications-methionine oxidation and N-terminal acetylation. Up to 1 missed tryptic cleavage was permitted and a peptide mass tolerance of ±0.40 Da was used for all mass searches. Positive identification was based on a Mascot score greater than 56, above the significance level ($p < 0.05$). The reported proteins were always those with the highest number of peptide matches.

Quantitative determination of selected proteins in urine

The concentrations of Serotransferrin, α1- antitrypsin, α1-microglobulin/bikunin, Apolipoprotein A-I and Haptoglobin in urine were measured by immunoturbidimetry using COBAS Integra 400 Plus (Roche Diagnostics). The kits used for measurement of these proteins in urine and detection limits were as follows: Serotransferrin, TQ Transferrin ver. 2 (Roche Diagnostics) with LOD = 0.013 g/L; α1- antitrypsin, TQ a-1 Antitrypsin ver. 2 (Roche Diagnostics) with LOD = 0.2 g/L; α1-microglobulin/ bikunin, TQ a-1 Microglobulin Gen. 2 (Roche Diagnostics) with LOD = 5 mg/L; Apolipoprotein A-I, TQ APO A-1 ver. 2 (Roche Diagnostics) with LOD = 0.2 g/L; Haptoglobin, TQ Haptoglobin ver. 2 (Roche Diagnostics) with LOD = 0.1 g/L. The level of creatinine in urine was determined using the Jaffé method by Crea Jaffe Gen. 2 Urine Kit (Roche Diagnostics) with 0.01 mmol/l limit of detection.

Bioinformatics and statistical analysis of the proteomics data

For an overview of the cellular localization, molecular function and biological processes in which identified proteins are included in, we used the UniProt Knowledgebase (UniProtKB) and Gene Ontology (GO) database. Pathway analysis was carried out for proteins found to be differently expressed between tumor and control samples using Ingenuity Pathway Analysis (IPA) (Ingenuity Systems, USA). Identified proteins were functionally assigned to canonical pathways and sub sequentially mapped to the most significant networks generated from previous publications and public protein interaction databases. A p value calculated with the right-tailed Fisher's exact test was used to yield a network's score and to rank networks according to their degree of association with our data set.

The Mann–Whitney U-test was used to analyze the correlation between the levels of the 3 selected proteins in urine and the pathological and clinical stage of prostate. The relative effectiveness of the diagnostic tests was illustrated by plotting the true-positive (sensitivity) versus the false-positive (1-specificity) results in receiver operating characteristic (ROC) curves. ROC curve analyses were used to define the most optimal diagnostic cutoff as well as the diagnostic performance given by areas under the curves (AUC). AUC were compared using a nonparametric method as described by Bamber [55]. In order to estimate the combined diagnostic potential of several candidate biomarkers, logistic regression analyses were performed with clinical diagnosis (PCa/BPH) as dependent variable and measurement of the protein concentrations in urine of selected proteins as independent variables. A confidence level of 95% ($p < 0.05$) was considered significant for all performed tests. Statistical analyses were performed using XLSTAT software (ver. 2014.4.06).

Additional files

Additional file 1: Figure S1. Classification of urine proteins with differential abundance between PCa and BPH. The molecular function, biological processes in which they are involved, subcellular location and type of the proteins were assessed by Gene Ontology search. The numbers represent percentages.

Additional file 2: Logistic regression models and ROC curves of combinations of TF, AMPB, HP and PSA. The summary of these results is presented in Table 3.

Additional file 3: Table S1. Clinical information of patients used to generate urine samples included in the study together with their PSA levels, histology grading and tumor stage.

Abbreviations

2-D: Two-dimensional; 2-D DIGE: Two-dimensional difference in gel polyacrylamide gel electrophoresis; ACN: Acetonitrile; AUC: Areas under the curves; BPH: Benign prostatic hyperplasia; BSA: Bovine serum albumin; BVA: Biological variation analysis; CHAPS: 3-[(3-Cholamidopropyl) dimethylammonio-1-propanesulfonate; CBB: Coomassie brilliant blue; Da: Dalton; DIA: Differential analysis; DTT: Dithiothreitol; GO: Gene ontology; IAA: 2-Iodoacetamide; IPA: Ingenuity pathways analysis; MALDI-TOF-TOF: Matrix-assisted laser desorption/ionization-time of flight-time of flight; MS: Mass spectrometry; PCa: Prostate cancer; PSA: Prostate-specific antigen; ROC: receiver operating characteristic; SDS: Sodium dodecyl sulfate; SDS-PAGE: Sodium dodecyl sulfate polyacrylamide gel electrophoresis; TFA: Trifluoroacetic acid; Tris: Tris(hydroxymethyl)amino.

Competing interests

The authors declare that they have no competing interests.

Authors' contributions

KD designed the study, participated in the sample selection and preparation, 2-D DIGE experiment, MS identifications, carried out bioinformatics analysis of the data and wrote the manuscript. SKiprijanovska participated in the sample preparation, 2-D DIGE experiment and MS identifications. SKomina and GP carried out the histological evaluation of the prostate tissue samples from the patients. NCZ carried out the immunoturbidimetry. MP conceived the study and participated in the design of the experiments. All authors have read and approved the final manuscript.

Acknowledgements

This work was supported by the funds for Science of the Macedonian Academy of Sciences and Arts (grant no. 09-114/1, Biomarker detection in prostate cancer with the use of 2-D DIGE/MALDI-MS technology). We thank patients for their participation in the study, the medical personal at the University Clinic for Urology at the University Clinical Centre "Mother Theresa", Skopje, Republic of Macedonia, for the collection of urine samples and Katerina Markovska for technical assistance.

Author details

[1]Research Centre for Genetic Engineering and Biotechnology "Georgi D Efremov", Macedonian Academy of Sciences and Arts, Krste Misirkov 2, 1000 Skopje, Republic of Macedonia. [2]Institute of Pathology, Medical Faculty, University "St. Cyril and Methodius", Skopje, Republic of Macedonia. [3]Biochemical laboratory, Clinical Hospital "Acibadem Sistina", Skopje, Republic of Macedonia.

References

1. Catalona WJ, Smith DS, Ratliff TL, Dodds KM, Coplen DE, Yuan JJ, et al. Measurement of prostate-specific antigen in serum as a screening test for prostate cancer. N Engl J Med. 1991;324(17):1156–61.

2. Oberaigner W, Horninger W, Klocker H, Schonitzer D, Stuhlinger W, Bartsch G. Reduction of prostate cancer mortality in Tyrol, Austria, after introduction of prostate-specific antigen testing. Am J Epidemiol. 2006;164(4):376–84.

3. Stamey TA, Yang N, Hay AR, McNeal JE, Freiha FS, Redwine E. Prostate-specific antigen as a serum marker for adenocarcinoma of the prostate. N Engl J Med. 1987;317(15):909–16.

4. Nadler RB, Humphrey PA, Smith DS, Catalona WJ, Ratliff TL. Effect of inflammation and benign prostatic hyperplasia on elevated serum prostate specific antigen levels. The Journal of urology. 1995;154(2 Pt 1):407–13.

5. Thompson IM, Pauler DK, Goodman PJ, Tangen CM, Lucia MS, Parnes HL, et al. Prevalence of prostate cancer among men with a prostate-specific antigen level < or =4.0 ng per milliliter. N Engl J Med. 2004;350(22):2239–46.

6. Makarov DV, Loeb S, Getzenberg RH, Partin AW. Biomarkers for prostate cancer. Annu Rev Med. 2009;60:139–51.

7. Prensner JR, Rubin MA, Wei JT, Chinnaiyan AM. Beyond PSA: the next generation of prostate cancer biomarkers. Sci Transl Med. 2012;4 (127):127rv123.

8. Velonas VM, Woo HH, Remedios CG, Assinder SJ. Current status of biomarkers for prostate cancer. Int J Mol Sci. 2013;14(6):11034–60.

9. Goo YA, Goodlett DR. Advances in proteomic prostate cancer biomarker discovery. J Proteomics. 2010;73(10):1839–50.

10. Pin E, Fredolini C, Petricoin 3rd EF. The role of proteomics in prostate cancer research: biomarker discovery and validation. Clin Biochem. 2013;46(6):524–38.

11. Loeb S. Prostate biopsy: a risk-benefit analysis. The Journal of urology. 2010;183(3):852–3.

12. Hori S, Blanchet JS, McLoughlin J. From prostate-specific antigen (PSA) to precursor PSA (proPSA) isoforms: a review of the emerging role of proPSAs in the detection and management of early prostate cancer. BJU Int. 2013;112(6):717–28.

13. Lazzeri M, Haese A, de la Taille A, Palou Redorta J, McNicholas T, Lughezzani G, et al. Serum isoform [-2]proPSA derivatives significantly improve prediction of prostate cancer at initial biopsy in a total PSA range of 2-10 ng/ml: a multicentric European study. Eur Urol. 2013;63(6):986–94.

14. Rehman I, Evans CA, Glen A, Cross SS, Eaton CL, Down J, et al. iTRAQ identification of candidate serum biomarkers associated with metastatic progression of human prostate cancer. PLoS One. 2012;7(2):e30885.

15. Zhang WM, Finne P, Leinonen J, Salo J, Stenman UH. Determination of prostate-specific antigen complexed to alpha(2)-macroglobulin in serum increases the specificity of free to total PSA for prostate cancer. Urology. 2000;56(2):267–72.

16. Byrne JC, Downes MR, O'Donoghue N, O'Keane C, O'Neill A, Fan Y, et al. 2D-DIGE as a strategy to identify serum markers for the progression of prostate cancer. Journal of proteome research. 2009;8(2):942–57.

17. Cima I, Schiess R, Wild P, Kaelin M, Schuffler P, Lange V, et al. Cancer genetics-guided discovery of serum biomarker signatures for diagnosis and prognosis of prostate cancer. Proc Natl Acad Sci U S A. 2011;108(8):3342–7.

18. Qingyi Z, Lin Y, Junhong W, Jian S, Weizhou H, Long M, et al. Unfavorable prognostic value of human PEDF decreased in high-grade prostatic intraepithelial neoplasia: a differential proteomics approach. Cancer Invest. 2009;27(7):794–801.

19. Sardana G, Dowell B, Diamandis EP. Emerging biomarkers for the diagnosis and prognosis of prostate cancer. Clinical chemistry. 2008;54(12):1951–60.

20. Sardana G, Jung K, Stephan C, Diamandis EP. Proteomic analysis of conditioned media from the PC3, LNCaP, and 22Rv1 prostate cancer cell lines: discovery and validation of candidate prostate cancer biomarkers. Journal of proteome research. 2008;7(8):3329–38.

21. Fan Y, Murphy TB, Byrne JC, Brennan L, Fitzpatrick JM, Watson RW. Applying random forests to identify biomarker panels in serum 2D-DIGE data for the detection and staging of prostate cancer. Journal of proteome research. 2011;10(3):1361–73.

22. Schostak M, Schwall GP, Poznanovic S, Groebe K, Muller M, Messinger D, et al. Annexin A3 in urine: a highly specific noninvasive marker for prostate cancer early detection. The Journal of urology. 2009;181(1):343–53.

23. Jayapalan JJ, Ng KL, Shuib AS, Razack AH, Hashim OH. Urine of patients with early prostate cancer contains lower levels of light chain fragments of inter-alpha-trypsin inhibitor and saposin B but increased expression of an inter-alpha-trypsin inhibitor heavy chain 4 fragment. Electrophoresis. 2013;34(11):1663–9.

24. True LD, Zhang H, Ye M, Huang CY, Nelson PS, von Haller PD, et al. CD90/THY1 is overexpressed in prostate cancer-associated fibroblasts and could serve as a cancer biomarker. Mod Pathol. 2010;23(10):1346–56.

25. Rehman I, Azzouzi AR, Catto JW, Allen S, Cross SS, Feeley K, et al. Proteomic analysis of voided urine after prostatic massage from patients with prostate cancer: a pilot study. Urology. 2004;64(6):1238–43.

26. M'Koma AE, Blum DL, Norris JL, Koyama T, Billheimer D, Motley S, et al. Detection of pre-neoplastic and neoplastic prostate disease by MALDI profiling of urine. Biochem Biophys Res Commun. 2007;353(3):829–34.

27. Morgan R, Boxall A, Bhatt A, Bailey M, Hindley R, Langley S, et al. Engrailed-2 (EN2): a tumor specific urinary biomarker for the early diagnosis of prostate cancer. Clinical cancer research: an official journal of the American Association for Cancer Research. 2011;17(5):1090–8.

28. Kiprijanovska S, Stavridis S, Stankov O, Komina S, Petrusevska G, Polenakovic M, et al. Mapping and Identification of the Urine Proteome of Prostate Cancer Patients by 2D PAGE/MS. International Journal of Proteomics. 2014;2014:12.

29. van Dieijen-Visser MP, Hendriks MW, Delaere KP, Gijzen AH, Brombacher PJ. The diagnostic value of urinary transferrin compared to serum prostatic specific antigen (PSA) and prostatic acid phosphatase (PAP) in patients with prostatic cancer. Clin Chim Acta. 1988;177(1):77–80.

30. Alaiya AA, Al-Mohanna M, Aslam M, Shinwari Z, Al-Mansouri L, Al-Rodayan M, et al. Proteomics-based signature for human benign prostate hyperplasia and prostate adenocarcinoma. Int J Oncol. 2011;38(4):1047–57.

31. Jayapalan JJ, Ng KL, Razack AH, Hashim OH. Identification of potential complementary serum biomarkers to differentiate prostate cancer from benign prostatic hyperplasia using gel- and lectin-based proteomics analyses. Electrophoresis. 2012;33(12):1855–62.

32. Yang N, Feng S, Shedden K, Xie X, Liu Y, Rosser CJ, et al. Urinary glycoprotein biomarker discovery for bladder cancer detection using LC/MS-MS and label-free quantification. Clinical cancer research: an official journal of the American Association for Cancer Research. 2011;17(10):3349–59.

33. Ummanni R, Mundt F, Pospisil H, Venz S, Scharf C, Barett C, et al. Identification of clinically relevant protein targets in prostate cancer with 2D-DIGE coupled mass spectrometry and systems biology network platform. PLoS One. 2011;6(2):e16833.

34. Schmetter BS, Habicht KK, Lamm DL, Morales A, Bander NH, Grossman HB, et al. A multicenter trial evaluation of the fibrin/fibrinogen degradation products test for detection and monitoring of bladder cancer. The Journal of urology. 1997;158(3 Pt 1):801–5.

35. Li H, Li C, Wu H, Zhang T, Wang J, Wang S, et al. Identification of Apo-A1 as a biomarker for early diagnosis of bladder transitional cell carcinoma. Proteome Sci. 2011;9(1):21.

36. Unwin RD, Harnden P, Pappin D, Rahman D, Whelan P, Craven RA, et al. Serological and proteomic evaluation of antibody responses in the identification of tumor antigens in renal cell carcinoma. Proteomics. 2003;3(1):45–55.

37. Lin JF, Xu J, Tian HY, Gao X, Chen QX, Gu Q, et al. Identification of candidate prostate cancer biomarkers in prostate needle biopsy specimens using proteomic analysis. International journal of cancer Journal international du cancer. 2007;121(12):2596–605.

38. Haj-Ahmad TA, Abdalla MA, Haj-Ahmad Y. Potential urinary protein biomarker candidates for the accurate detection of prostate cancer among benign prostatic hyperplasia patients. J Cancer. 2014;5(2):103–14.

39. Meehan KL, Holland JW, Dawkins HJ. Proteomic analysis of normal and malignant prostate tissue to identify novel proteins lost in cancer. The Prostate. 2002;50(1):54–63.

40. Decramer S, Gonzalez de Peredo A, Breuil B, Mischak H, Monsarrat B, Bascands JL, et al. Urine in clinical proteomics. Molecular & cellular proteomics: MCP. 2008;7(10):1850–62.

41. Hortin GL, Sviridov D. Diagnostic potential for urinary proteomics. Pharmacogenomics. 2007;8(3):237–55.

42. Rodriguez-Suarez E, Siwy J, Zurbig P, Mischak H. Urine as a source for clinical proteome analysis: from discovery to clinical application. Biochimica et biophysica acta. 2014;1844(5):884–98.

43. Principe S, Kim Y, Fontana S, Ignatchenko V, Nyalwidhe JO, Lance RS, et al. Identification of prostate-enriched proteins by in-depth proteomic analyses of expressed prostatic secretions in urine. Journal of proteome research. 2012;11(4):2386–96.

44. Theodorescu D, Schiffer E, Bauer HW, Douwes F, Eichhorn F, Polley R, et al. Discovery and validation of urinary biomarkers for prostate cancer. Proteomics Clin Appl. 2008;2(4):556–70.

45. Mischak H, Kolch W, Aivaliotis M, Bouyssie D, Court M, Dihazi H, et al. Comprehensive human urine standards for comparability and standardization in clinical proteome analysis. Proteomics Clin Appl. 2010;4(4):464–78.

46. Magdeldin S, Enany S, Yoshida Y, Xu B, Zhang Y, Zureena Z, et al. Basics and recent advances of two dimensional- polyacrylamide gel electrophoresis. Clin Proteomics. 2014;11(1):16.

47. Kushner I. Regulation of the acute phase response by cytokines. Perspect Biol Med. 1993;36(4):611–22.

48. Gabay C, Kushner I. Acute-phase proteins and other systemic responses to inflammation. N Engl J Med. 1999;340(6):448–54.

49. Mebratu Y, Tesfaigzi Y. How ERK1/2 activation controls cell proliferation and cell death: Is subcellular localization the answer? Cell Cycle. 2009;8(8):1168–75.

50. Roberts PJ, Der CJ. Targeting the Raf-MEK-ERK mitogen-activated protein kinase cascade for the treatment of cancer. Oncogene. 2007;26(22):3291–310.

51. Balkwill F, Mantovani A. Inflammation and cancer: back to Virchow? Lancet. 2001;357(9255):539–45.

52. Chechlinska M, Kowalewska M, Nowak R. Systemic inflammation as a confounding factor in cancer biomarker discovery and validation. Nature reviews Cancer. 2010;10(1):2–3.

53. Pang WW, Abdul-Rahman PS, Wan-Ibrahim WI, Hashim OH. Can the acute-phase reactant proteins be used as cancer biomarkers? Int J Biol Markers. 2010;25(1):1–11.

54. Bradford MM. A rapid and sensitive method for the quantitation of microgram quantities of protein utilizing the principle of protein-dye binding. Analytical biochemistry. 1976;72:248–54.

55. Bamber D. The area above the ordinal dominance graph and the area below the receiver operating graph. J Math Psychol. 1975;12:387–415.

17

The proteome profiles of the olfactory bulb of juvenile, adult and aged rats - an ontogenetic study

Michael Wille[1], Antje Schümann[1], Michael Kreutzer[2], Michael O Glocker[2], Andreas Wree[1], Grit Mutzbauer[3] and Oliver Schmitt[1*]

Abstract

Background: In this study, we searched for proteins that change their expression in the olfactory bulb (oB) of rats during ontogenesis. Up to now, protein expression differences in the developing animal are not fully understood. Our investigation focused on the question whether specific proteins exist which are only expressed during different development stages. This might lead to a better characterization of the microenvironment and to a better determination of factors and candidates that influence the differentiation of neuronal progenitor cells.

Results: After analyzing the samples by two-dimensional polyacrylamide gel electrophoresis (2DE) and matrix-assisted laser desorption/ionization time-of-flight mass spectrometry (MALDI-TOF-MS), it could be shown that the number of expressed proteins differs depending on the developmental stages. Especially members of the functional classes, like proteins of biosynthesis, regulatory proteins and structural proteins, show the highest differential expression in the stages of development analyzed.

Conclusion: In this study, quantitative changes in the expression of proteins in the oB at different developmental stages (postnatal days (P) 7, 90 and 637) could be observed. Furthermore, the expression of many proteins was found at specific developmental stages. It was possible to identify these proteins which are involved in processes like support of cell migration and differentiation.

Keywords: Proteomics, Rat, Brain, Olfactory bulb, Development

Background

The aim of this study was to analyze the differential proteome of the rat olfactory bulb (oB) within different developmental stages (7-day-old juvenile rats (P7), 90-day-old adult rats (P90) and 637-day-old aged rats (P637)).

The postnatal development of the oB corresponds to a bulging of the cerebral hemispheres. It develops from the cerebral vesicle near the placode. At the time of birth, the oB is one of the most rostral regions of the rat brain [1].

In addition, the oB integrates different neuronal microcircuits which are of fundamental physiological importance for olfaction. It has a particular source of sensory input (axons from olfactory receptor neurons of the olfactory epithelium) and one output source (mitral cell axons) which work as a filter to enhance the sensitivity of odor detection and its information processing [2]. The sensory input is created by axons from olfactory receptor neurons of the olfactory epithelium; the output is organized by the mitral cell axons. Additionally, information can also be received from other parts of the brain (amygdala, neocortex, hippocampus, locus coeruleus and substantia nigra). The peripheral olfactory system has also been recognized as a model of postnatal neurogenesis, because it displays the most robust and functional postnatal neurogenesis among neuronal populations that maintain stem cells [3]. In addition, neurogenesis is increased after lesioning of either the oB (termed bulbectomy) or the epithelium itself [4]. The epithelium contains a mixed population of basal cells, basal daughter cells, immature and mature olfactory receptor neurons (ORNs),

* Correspondence: schmitt@med.uni-rostock.de
[1]Department of Anatomy, Gertrudenstr. 9, 18055 Rostock, Germany
Full list of author information is available at the end of the article

and sustentacular cells [5,6]. Basal cells and their daughter cells generate the new neurons that repopulate the olfactory epithelium (OE) throughout life [7-10]. The presence of stem cells in the adult brain which can develop into mature neurons in vitro suggests that postnatal neurogenesis could be used to repopulate certain neuronal lineages [11,12]. Therefore, different factors for neuronal precursor proliferation and maturation have been determined (including neurotrophins, neuropeptides, and cytokines [13,14]). Despite the usage of certain mitogens, the proliferation of neuronal stem cells can be induced in vitro. The identification of factors that regulate and modulate the maturation of neurons within the postnatal brain is critical to the application of stem cell therapy [15-17].

According to Bandeira et al. [1], a postnatal increase of the rat brain-mass occurs in the first 3 months after birth (by a factor of ~20.7 in all brain areas). The time course of changes in the postnatal brain can be divided into 4 phases, the first 2 phases (first phase [P0-P2]: "idle time", no significant mass increase of the brain; second phase [P2-P25]: heavy increase of the brain's mass) can be referred to all brain structures. Furthermore, from P25 on, the hippocampus, oB and the rest of the brain show a slow growth up to P90. The weight of body and brain increases till P275, but much slower with rising age [18].

For the postnatal changes in the number of neurons, the rat brain generates between 1 and over 5 million neurons per day during a period of 3–4 days. During the first postnatal week, then, the number of neurons increases by nearly 50% in the brain. At this stage, the number of neurons is followed by a marked net loss of neurons during the second postnatal week, when 60%–70% of the number of neurons in the brain is lost. In contrast, an addition of neurons proceeds in the oB until adulthood. An identifiable period of neuronal loss within adulthood is not detectable in this brain area. The generation and integration of new neurons is possible because the rostral migratory stream (RMS) of the subventricular zone (SVZ) at the anterior horn of the lateral ventricles progressively fills up and replaces neurons of the oB which degenerate [19-23]. It presents an active neurogenerative region in adulthood (schematic overview, see Figure 1c). Here, the progenitor cells proliferate continuously and differentiate to granular and periglomerular cells in the oB [23-27]. These progenitor cells migrate from the SVZ to their functional destination in the oB. Studies by Luskin [27], Lois and Alvarez-Buylla [23] and Luskin and Boone [28] identified the migratory route as a pathway where the velocity of migration constitutes 5 mm/7 days in the adult rat and mouse between the anterior horn of the lateral vesicle and the olfactory ventricle. The cells spread radially towards the granular and periglomerular layers, where they are thought to differentiate into neurons [23,27,29].

In adult rats, this subependymal layer is a remnant of the embryonic subventricular zone which originates after the occlusion of the primitive olfactory ventricle where cell proliferation can occur in the adulthood. This proliferation is organized in a well-defined pathway of tangential and radial migration. The migration is performed by a meshwork of astrocytic processes and cell bodies which form long tangentially oriented canals. These canals act as physical barriers, allowing the displacement of long chains of migrating cells from the lateral ventricle to the oB and prevent any dispersion along their tangential migration route. Interestingly, cells inside these glial tubes contain proteins like class 3 beta-tubulin [30]. This beta-tubulin shows a diffuse staining which signals a not yet organized form of mature finally assembled microtubules [26]. With this mechanism, it is possible for neurons as well as for progenitor cells (differentiated in the stem cell niche) to migrate within the oB. The progenitor cells can differentiate to dopamineric interneurons. The main period of origin of interneurons is located between P1 and P10, after this period their development decreases. This means the chosen period of P7 in this study can be seen as one of the main peaks of generating interneurons.

Another major amount of olfactory interneurons are the granular cells. This GABAergic cell type is important for generating inhibition to regulate the processing of sensory output. These cells are essential for learning and for the manipulation of sensory experiences early in the postnatal period. They are able to modify the structure and function of the oB [31-34]. Regarding their morphology, these cells are axonless interneurons and their dendrites span several bulbar layers and have distinct layer-specific features [35]. Furthermore, granular cells run through considerable changes of their formation of spines and filopodias. During development, three major changes are known: decrease of stubby spines, accompanied by an increase in the frequency of typical spines, then a distinct pattern of change in spine, filopodium density in the different dendritic domains and an overshoot in the number of spinesor filopodia, respectively, in the basal dendrites (peak at P28).

A main aspect during growth and development of the rat brain are the expression- and activity changes of glycolytic enzymes. For example, the activity of phosphofructokinase (a key enzyme which catalyzes the conversion of fructose-6-phosphate to fructose-1, 6-bisphosphate and which also represents the rate-limiting step in the glycolysis) is detectable in fetal brains at the twelfth day of gestation. It shows little change in enzyme activity in the brain from 5 days before birth to 8 day after birth. During this period, activity was 30-40% of the normal value found in an adult brain. After 12 days, a rapid increase in activity to 21 days occurred [36]. Furthermore, the activities of additional glycolytic enzymes (hexokinase [37], aldolase [38],

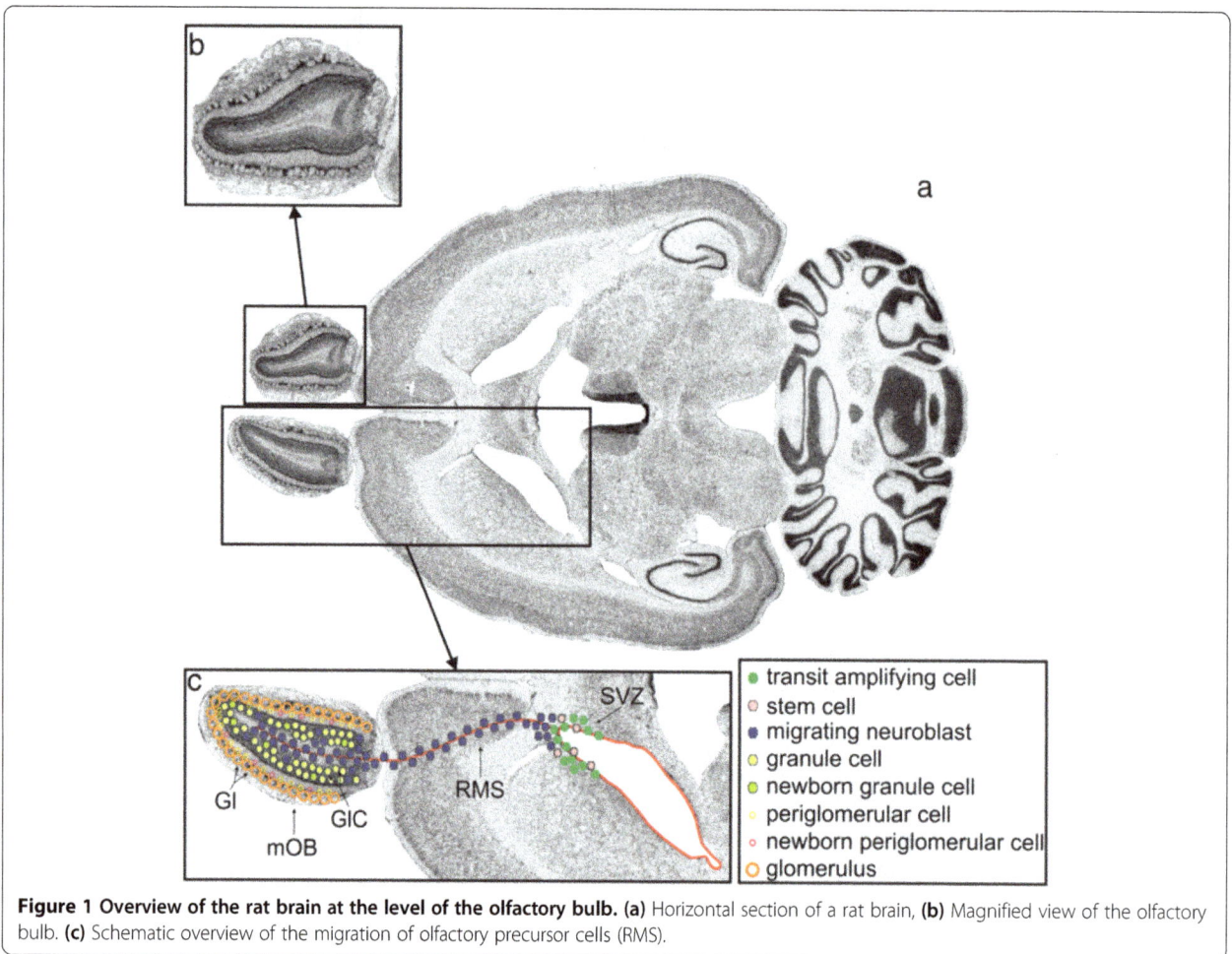

Figure 1 Overview of the rat brain at the level of the olfactory bulb. (a) Horizontal section of a rat brain, **(b)** Magnified view of the olfactory bulb. **(c)** Schematic overview of the migration of olfactory precursor cells (RMS).

lactate-dehydrogenase [39]) do show substantial increases during maturation of the rat brain.

A main feature of Parkinson's disease is the loss of dopaminergic neurons in the substantia nigra [40]. Current therapeutic options of this neurodegenerative disease are symptomatic and temporary only. The experimental investigation of optimizing therapeutic strategies like stem cell transplantation in well described experimental models of Parkinson's disease, both in vitro and in vivo are still essential [41]. From previous studies [42,43], we know that transplanted progenitor cells in the neonatal and adult striata develop differently in neural and glial cell types. A better characterization of the development of the microenvironment at the proteome level may support a better understanding of survival and differentiation of transplanted progenitor cells.

In this study, proteins from the oB of juvenile, adult and aged rats were separated by 2DE and identified by MALDI-TOF-MS. Proteins which change their expression during ontogenesis were identified. As yet, protein expressions in the developing proteome are not fully

understood. As mentioned before, stem cells of the oB in the adult brain can develop into mature neurons in vitro which suggests that postnatal neurogenesis could be used to repopulate certain neuronal lineages [11,12]. Additionally, with the usage of certain mitogens and cytokines, the proliferation of neuronal stem cells can be induced [15-17]. Based on this, the oB seems an interesting system to analyze differences in the expression of proteins depending of its development.

Results

To identify the differences in the protein expression between the groups of different ages (P7, P90 and P637), 2DE with subsequent gel-matching, spot-warping and differential spot analysis combined with protein identification by MALDI-TOF-MS was performed. A precise, manual labeling of the spots was realized by using the software Progenesis PG200.

First, in Figure 1a, a horizontal section of the oB of a rat brain is shown. In Figure 1b, a magnified view of the analyzed brain area (oB) is given. In Figure 1c, a schematic overview of the migration of olfactory precursor cells is presented. Here, the olfactory precursor cells (stem cells, transit amplifying cells) proliferate mainly in the SVZ where the differentiation in immature neuroblasts also happens. These neuroblasts then migrate through the RMS to the oB. After a period of 5–7 days (after birth), these neuroblasts shift towards the granular (granular cell layer, GlC), periglomerular and external plexiform cell layers of the oB. At a time of 15–30 days after birth, matured neuroblasts generate interneurons in the oB.

In Figure 2, the images of the reference gels of the different developmental stages are presented ((a) P7, (b) P90 and (c) P637). The spot compositions turn out to be comparable. On average, 828 (±71) spots were detected in the six gels of P7, 775 (±152) spots in the gels of P90 and 785 (±56) spots in the gels of P637. Furthermore, in Figure 3, an example of the manual spot editing (segmentation, delineation) by using a gel image of the olfactory bulb in Progenesis PG200 is shown.

The differentially expressed proteins were classified into 13 functional protein groups. In the following, only those proteins which show the most differential expression are described in detail (Figures 4(a) and 5(a)). A complete overview of the expression changes of all analyzed proteins as well as a description of the remaining protein categories is listed in Additional file 1.

In the following, a more detailed analysis of the up- and down-regulation of the proteins from the different categories and developmental stages in comparison to P90 is described (4(b), 5(b)).

For P7, the carbohydrate metabolism consists of 14 proteins which are down-regulated in this stage towards P90. For example, the protein 2-oxoglutarate dehydrogenase (Ogdh) which catalyzes the overall

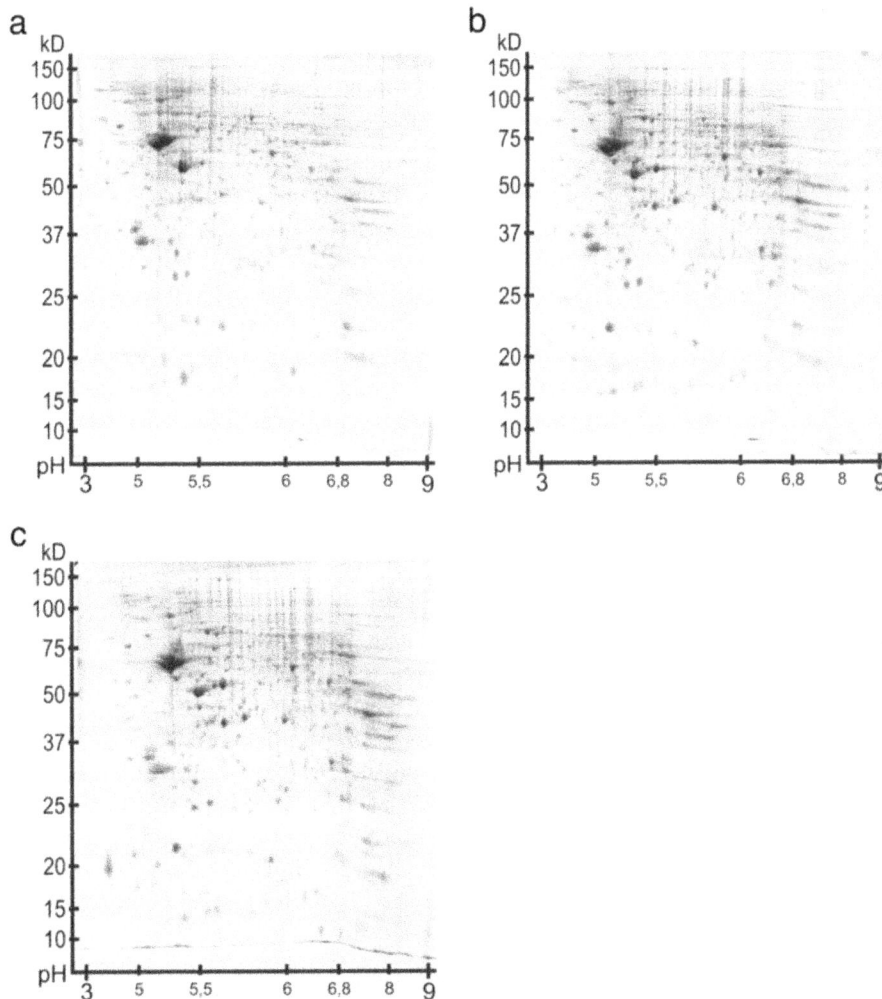

Figure 2 Overview of the reference gel images from the olfactory bulb. The coomassie blue stainings of the gels have similiar intensities. The 2DE-gel image of a P7 animal is shown in **(a)**. In **(b)** the 2DE-gel image of a P90 animal is presented. **(c)** shows the 2DE-gel image of an oldest P637 rat.

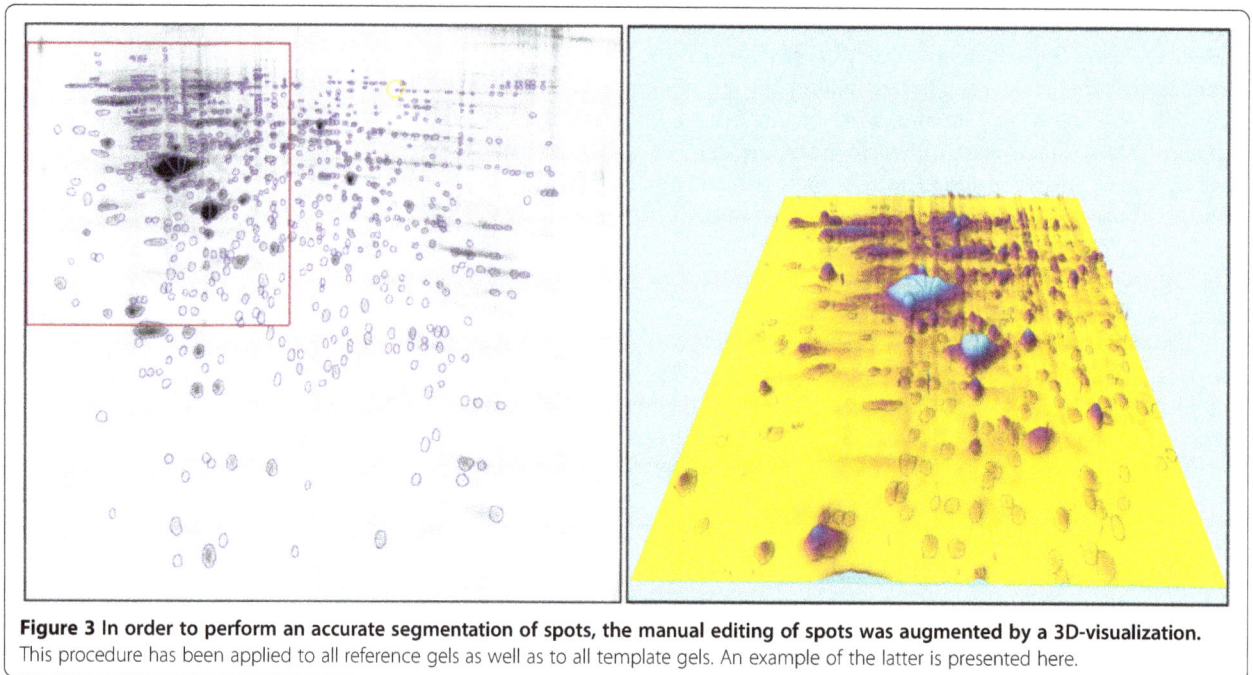

Figure 3 In order to perform an accurate segmentation of spots, the manual editing of spots was augmented by a 3D-visualization. This procedure has been applied to all reference gels as well as to all template gels. An example of the latter is presented here.

conversion of alpha-ketoglutarate to succinyl-CoA and CO_2 during the Krebs-cycle and which also participates in mitochondrial degradation of glutamate shows a down-regulation. Aconitate hydratase (Aco2) that catalyzes the interconversion of citrate to isocitrate via cis-aconitate in the second step of the TCA-cycle is down-regulated as well. Another down-regulated protein towards P90 is the gamma-enolase (Eno2) which has neurotrophic and neuroprotective properties on a broad spectrum of neurons. This expression pattern could also be validated by Western Blot analysis of the same protein homogenates (Figure 6). It binds in a calcium-dependent manner to cultured neocortical neurons and promotes the survival of the cells. Furthermore, proteins like glucose-6-phosphate 1-dehydrogenase (G6pdx), which produces pentose sugars for nucleic acid synthesis and which is a secondary antioxidant enzyme that plays an important role in detoxifying ROS/RNS by maintaining a supply of intermediates such as glutathione and NADPH are differentially expressed. Glyceraldehyde-3-phosphate dehydrogenase (Gapdh), a protein involved in the important energy-yielding step in carbohydrate metabolism, catalyzes the reversible oxidative phosphorylation of glyceraldehyde-3-phosphate and is also participating in the organization and assembly of the cytoskeleton. Altogether 5 proteins including alpha glucosidase (Ganab) with a hydrolytic function as well as transaldolase (Taldo1) are up-regulated at this stage. Taldo1 presents an important regulator of metabolites in the pentose-phosphate pathway, for example. Furthermore, 4 proteins are absent at this stage. For example, the protein 3-hydroxyisobutyrate dehydrogenase

(Hibadh) plays a critical role in the catabolism of L-valine by catalyzing the oxidation of 3-hydroxyisobutyrate to methylmalonate semialdehyde. Fructose-bisphosphate aldolase C (Aldoc) which gets expressed specifically in the hippocampus and in Purkinje cells of the brain, catalyzes the reversible aldol cleavage of fructose-1.6-biphosphate and fructose 1-phosphate to dihydroxyacetone phosphate and either glyceraldehyde-3-phosphate or glyceraldehydes. Glycerol-3-phosphate dehydrogenase 1-like protein (Gpd1l) catalyzes the conversion of glycerol 3-phosphate to glycerone phosphate. The protein isocitrate dehydrogenase subunit alpha (Idh3a) catalyzes the oxidative decarboxylation of isocitrate to 2-oxoglutarate.

As for the proteins of the biosynthesis, 13 proteins are up-regulated. For example, the protein serine tRNA ligase (Sars) catalyzes the attachment of serine to tRNA. Additionally, different subunits needed for splicing (splicing factor 3a, subunit 3 (Sf3a3)), elongation (elongation factor 1-gamma (Eef1g)) and translation (eukaryotic translation initiation factor 4A1 (Eif4a1)) display an up-regulation. Furthermore, different members of the family of the heterogeneous nuclear ribonucleoproteins (hnRNPs) are up-regulated towards P90 (heterogeneous nuclear ribonucleoprotein (Hnrnpf), heterogeneous nuclear ribonucleoprotein C (Hnrnpc), heterogeneous nuclear ribonucleoprotein K (Hnrnpk)). These members are essential for mRNA metabolism, DNA-related functions and microRNA biogenesis. Four proteins are down-regulated towards P90. These proteins are involved in the metabolism of biosynthesis like the elongation factor Tu (Tufm) or serine/threonine-protein

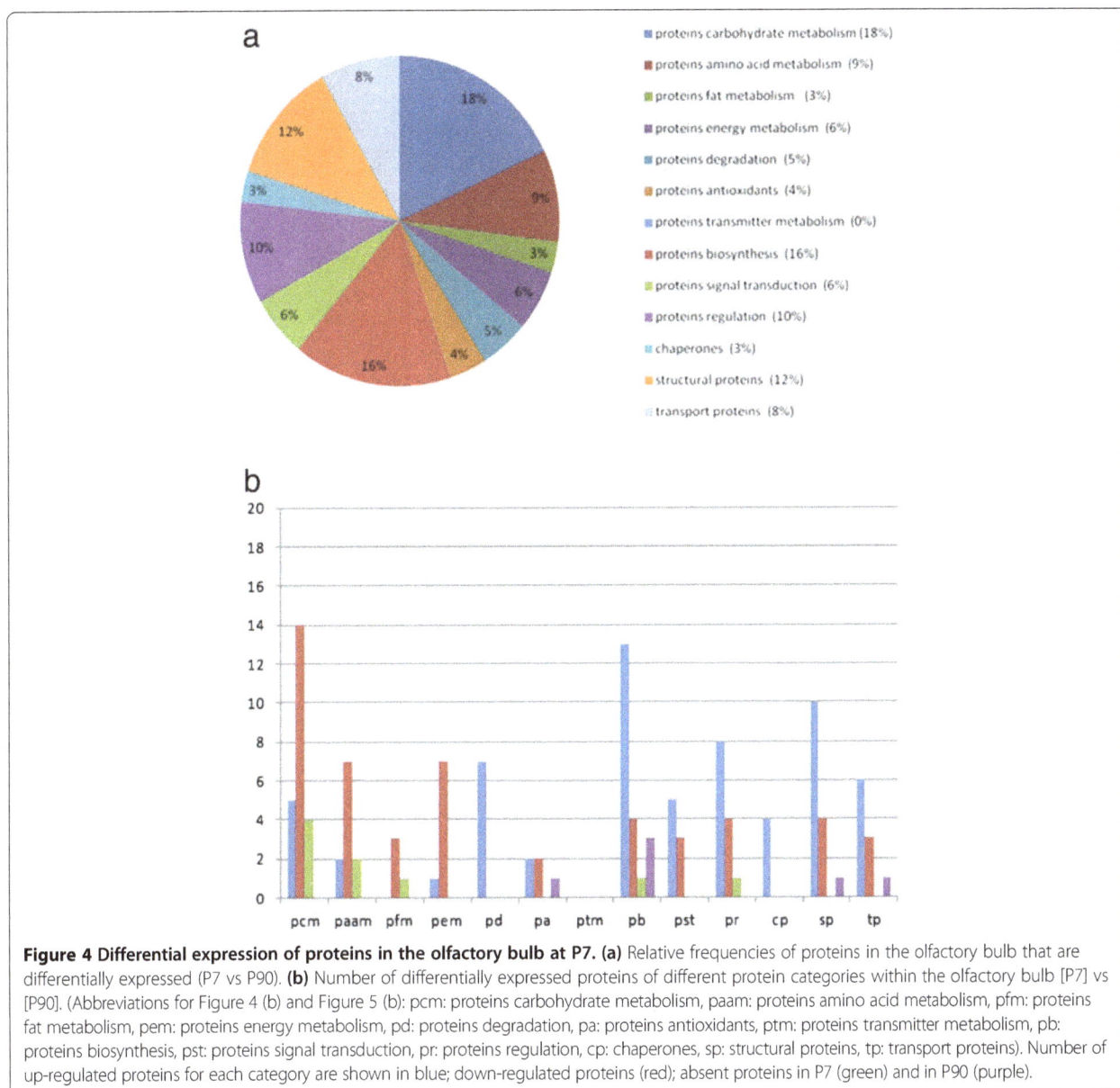

Figure 4 Differential expression of proteins in the olfactory bulb at P7. (a) Relative frequencies of proteins in the olfactory bulb that are differentially expressed (P7 vs P90). **(b)** Number of differentially expressed proteins of different protein categories within the olfactory bulb [P7] vs [P90]. (Abbreviations for Figure 4 (b) and Figure 5 (b): pcm: proteins carbohydrate metabolism, paam: proteins amino acid metabolism, pfm: proteins fat metabolism, pem: proteins energy metabolism, pd: proteins degradation, pa: proteins antioxidants, ptm: proteins transmitter metabolism, pb: proteins biosynthesis, pst: proteins signal transduction, pr: proteins regulation, cp: chaperones, sp: structural proteins, tp: transport proteins). Number of up-regulated proteins for each category are shown in blue; down-regulated proteins (red); absent proteins in P7 (green) and in P90 (purple).

phosphatase 2A 55 kDa regulatory subunit B alpha (Ppp2r2a). One protein is absent in the gels from the P7 protein extract (ribose-phosphate pyrophosphokinase 1 (Prps1)) which catalyzes the phosphoribosylation of ribose 5-phosphate to 5-phosphoribosyl-1-pyrophosphate, necessary for purine metabolism and nucleotide biosynthesis. Three proteins are absent in the gels from the P90 protein extract: The serine/arginine-rich splicing factor 3 (Sfrs3), the RNA- binding motif protein, X-linked-like-1 (Rbmxl1) and the putative RNA-binding protein 3 (Rbm3), a protein which is expressed widely during early brain development in glutamatergic and GABAergic cells as well as in newly formed and migrating neurons, peaking in the first to second postnatal weeks.

The amount of up- and down-regulation of the structural proteins is more or less balanced. Thirteen proteins are up-regulated at this stage. These proteins are mainly involved in the dynamic organization of the cytoskeleton, like actin-related protein 2 (Actr2), member of the Arp2/3 complex, dihydropyrimidinase-related protein 2 (Dpysl2) or F-actin-capping protein subunit alpha-1 (Capza1). Other up-regulated proteins are actin beta (Actb) which can support the acceleration of axonal outgrowth, for example, or stathmin (Stmn1) involved in the regulation of the microtubule filament system by destabilizing microtubules. Six proteins are down-regulated towards P90. These are mainly members of the neurofilament family (neurofilament light polypeptide (Nefl), neurofilament medium polypeptide

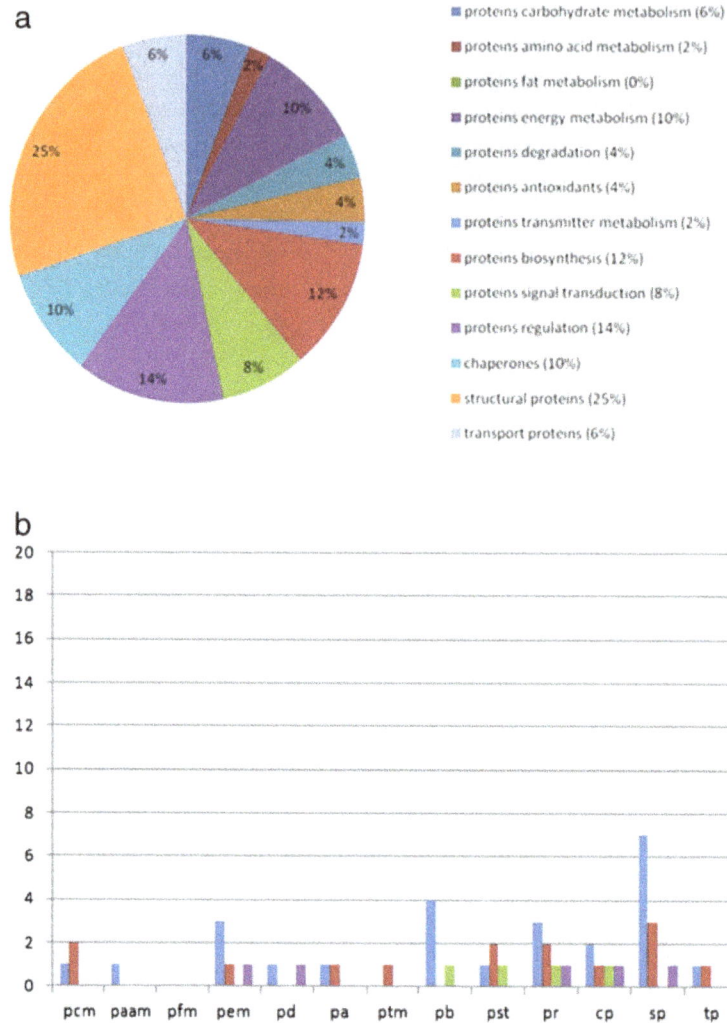

Figure 5 Differential expression of proteins in the olfactory bulb at P637. (a) Relative frequencies of proteins in the olfactory bulb that are differentially expressed (P637 vs P90). **(b)** Number of differentially expressed proteins of different protein categories within the olfactory bulb [P637] vs [P90]. Same abbreviations and labeling performed as in Figure 4 (b); absent proteins in P637 (green) and in P90 (purple).

(Nefm)) which comprise the axoskeleton and functionally maintain the neuronal caliber. They may also participate in intracellular transport to axons and dendrites. This pattern of expression could also be validated by Western blot analysis for the neurofilament medium polypeptide as well as for the neurofilament light polypeptide (Figure 6). The tropomyosin alpha-3 chain is the only protein which is absent at P90 in this category. This protein has the ability to regulate the molecular composition of microfilaments, which in turn regulates the dynamic and functional properties of the resulting actin filament population.

All in all, at developmental stage P7, a total of 129 proteins is expressed differentially towards P90, of which 63 proteins are up-regulated, 51 proteins are down-regulated, 9 proteins are absent at P7 and 6 proteins are absent at P90 (Table 1).

For P637 in comparison to P90, differences in the expression are also remarkable. Beginning with the functional group with the most differentially expressed proteins, the structural proteins; this category consists of 7 proteins which are up-regulated, 4 proteins which are down-regulated. The up-regulated protein faction contains actin beta (Actb) and myristoylated alanine-rich C-kinase substrate (Marcks) which also shows an up-regulated expression at P7 towards P90. Other up-regulated proteins are members of the tubulin family (tubulin alpha-1A chain (Tuba1a), tubulin beta-3 chain (Tubb3), tubulin beta 4A (Tubb4)). These proteins are generally required for the organization of the cytoskeleton as well as migration and differentiation of neurons, for example, by assembling microtubules into a highly organized mitotic spindle. The down-regulated proteins contain lamin-B1 (Lmnb1) and septin-11 (Sept11), for example. The protein Lmnb1

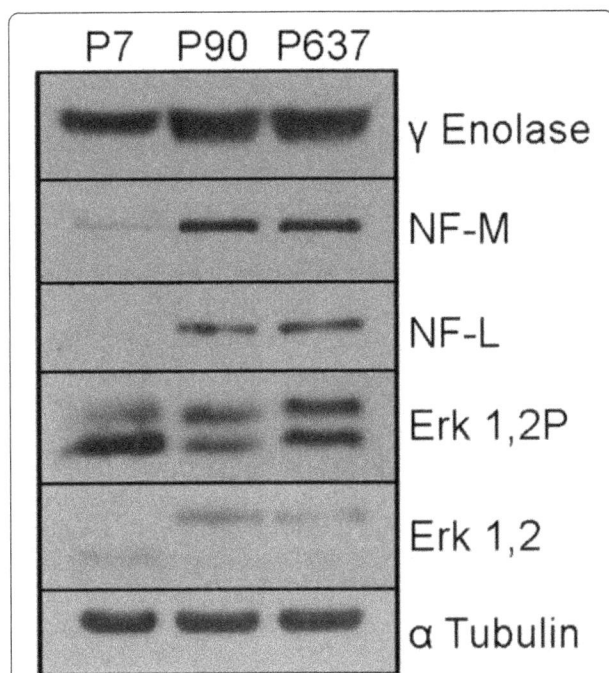

Figure 6 The validation of some differially expressed proteins was performed by Western Blots (Mitogen-activated protein kinase 1, 2 (Erk1,2), Phospho Mitogen-activated protein kinase 1, 2 (Erk1,2P), Neurofilament Low (NF-L), Neurofilament Medium (NF-M) and Gamma Enolase (γ Enolase) in the olfactory bulb between the different development stages (P7, P90, P637). α-tubulin staining was used to ensure the equal loading of proteins. This protein is expressed also differentially.

(involved in nuclear stability, chromatin structure and gene expression) shows an up-regulation at P7 towards P90 and Sept11 (filament-forming cytoskeletal GTPase, involved in cytokinesis and vesicle trafficking) displays a down-regulation at P7.

As for the regulatory proteins, 3 proteins are up-regulated. These contain the Rab GDP dissociation inhibitor alpha (Gdi1) which is also up-regulated at P7 in comparison to P90. Another protein is the Rap2-interacting protein × (Rufy3) which is implicated in the formation of a single axon of developing neurons. A

total of 2 proteins (Rho GDP-dissociation inhibitor 1 (Arhgdia), annexin A5 (Anxa5)) is down-regulated. Arhgdia is involved in the organization of the actin cytoskeleton in response to extracellular signals for proper cell morphology, growth, proliferation, differentiation, motility and adhesion. The protein Anxa5, whose functional meaning was described earlier, is also down-regulated at P7 towards P90. One protein is absent at P637 (calcium binding protein 39 (Cab39)). This protein presents a necessary element in cell metabolism that is required for maintaining energy homeostasis. At P90 the protein myosin regulatory light chain 12B (Myl12b) is absent.

The proteins which are involved in pathways of the protein biosynthesis present 4 proteins which are up-regulated and 1 protein which is absent at P637 (transcriptional activator protein Pur-beta (Purb)). The up-regulated group contains members of the heterogeneous nuclear ribo-nucleoprotein family (heterogeneous nuclear ribonucleoprotein F (Hnrnpf), heterogeneous nuclear ribonucleoprotein K (Hnrnpk)). These proteins are relevant for mRNA metabolism, DNA-related functions and microRNA biogenesis.

At P637, all in all 47 proteins are expressed differentially towards P90, of which 25 proteins show an up-regulation, 14 proteins show a down-regulation, 4 proteins are absent at P7 and 4 proteins are absent at P90 (Table 1).

Discussion

In the following, the identification and characterization of differentially expressed proteins in the oB during the postnatal day of development (P7 and P637) in comparison to P90 will be discussed.

In general, the oB does not consist of a homogeneous cell population. Therefore, a differential proteome analysis is a representation of proteome changes of the whole oB concerning all cell populations and the neuropile. The separation technique used in this study allows protein separation within the range of approximately 10–100 kDa and between a pH-range of 3–10. For this reason, it is possible to analyze a substantial part, but not the entire proteome of the oB [44]. Moreover, some parts of the gel may show a lower resolution which can lead to a misinterpretation of the regulation analysis of some protein spots. Furthermore, it is known that proteins may be represented by multiple spots with different locations in the gel which can result from different post-translational modifications and isoforms of the same protein. Therefore, several proteins may show specific regulation changes for one spot only and not for the entire protein. By applying other mass spectrometric detection methods like the stable isotope ratio mass spectrometry (SIRMS), it would be possible to unambiguously identify both the protein and its variants and to overcome the sample-to-sample recovery variabilities associated with

Table 1 Overview of the differential expression of the proteins in P7 and P637 vs. P90

Categories	Amount [P7] (absolute numbers)	Amount [P637] (absolute numbers)
Total	129	47
Upregulated	63	25
Downregulated	51	14
Absent	9	4
Absent P90	6	4

non-SIRMS MS-proteomic methods. Moreover, the usage of affinity-enrichment-MS methods (e.g. IDB-EST™, iTRAQ™) would enhance the possibility for targeted biomarker discovery applications to drill down to lower-abundance proteins and to improve its analysis [45]. Hence, the results of the presented proteome analysis of the rat oB should be considered as a first step to elucidate the patterns of proteins which are differentially expressed in this region of the brain.

This study focuses on proteins that differed in abundance and give indications for developmental changes in the oB. Expression levels of proteins were determined and their changes were interpreted by comparison with regard to the literature. Specific changes of cytoplasmatic proteins of the postnatal developing oB were found.

At P7 in comparison to P90, those proteins which are involved in carbohydrate metabolism, biosynthesis, amino acid metabolism and regulatory as well as structural proteins show an increased expression. For the protein expression at P637 versus P90, proteins of the energy metabolism, proteins involved in biosynthesis, chaperones, regulatory and structural proteins show various changes in their expression.

The majority of differentially expressed proteins of the carbohydrate metabolism are down-regulated at P7 with comparison to P90. Most of these proteins are involved in the tricarboxylic acid cycle and glycolysis. For example, the protein 2-oxoglutarate dehydrogenase (Ogdh) which catalyzes the conversion of alpha-ketoglutarate to succinyl-CoA and CO_2 during the Krebs-cycle and which also participates in mitochondrial degradation of glutamate [46] shows a down-regulation at P7. Other proteins like aconitate hydratase (Aco2) participating in the interconversion of citrate to isocitrate or the protein gamma-enolase (Eno2) which catalyzes the conversion of 2-phosphoglycerate to phosphorenolpyruvate and which also has neuroprotective properties [47] are involved in these pathways. Already Wilbur et al. [48] found that the activities of several proteins involved in tricarboxylic acid cycle as well as of the pyruvate metabolism show low expression levels in newborn rats, whereas they are increased markedly to adult levels during 10–30 postnatal days but do not decrease in adulthood.

The majority of proteins which are involved in the processes of protein biosynthesis display an up-regulation at P7 in comparison to P90. Especially proteins for splicing (splicing factor 3a, subunit 3 (Sf3a3)), elongation (elongation factor 1-gamma (Eef1g)) and translation (eukaryotic translation initiation factor 4A1 (Eif4a1)) are up-regulated. Of particular relevance in the mammalian nervous system are transcripts exhibiting multiple splicing patterns, particularly important growth processes such as axon guidance, synaptogenesis, and the regulation of membrane physiology [49-51]. Furthermore, different members of the

family of the heterogeneous nuclear ribonucleoproteins (hnRNPs) are up-regulated towards P90 (heterogeneous nuclear ribonucleoprotein (Hnrnpf), heterogeneous nuclear ribonucleoprotein C (Hnrnpc), heterogeneous nuclear ribonucleoprotein K (Hnrnpk)). These members generally play key roles in mRNA metabolism, DNA-related functions and microRNA biogenesis. In general, regulation of stability, localization and translation of a number of transcripts are important in the establishment and maintenance of nerve cell functions [52-56]. The regulation of the expression of many RNAs depends on the activity of RNA-binding proteins, often acting in concert and organized in complexes, probably regulating the activity and the expression of the other components of the complex itself [57]. These complexes contain elements involved in the control of different aspects of RNA metabolism, including the interactions established by heterogeneous nuclear proteins (hnRNPs) during the transport of the mRNAs from nucleus to cytoplasm and to the site of translation. They also contain mRNA–proteins and pre-mRNP complexes which can be actually found in association with the cytoskeleton. Models have been proposed about the role of cytoskeleton-associated hnRNPs in mRNA biogenesis-control at the post-transcriptional level [58,59]. For example, hnRNPK and the Abelson-interacting protein 1 (Abi-1) can act in a synergistic manner in a multiprotein complex that regulates the crucial balance between filopodia formation and synaptic maturation in neurons [60].

Especially proteins for the dynamic organization of the cytoskeleton are up-regulated at the early developmental stage of P7, as, for example, the actin-related protein 2 (Actr2) which is a member of the Arp2/3 complex. This complex is a stable assembly of two actin-related proteins (Arp2 and Arp3) and subsequently five other proteins. This nucleator acts by serving as a template for monomer addition by mimicking the barbed end of a growing filament [61]. This complex gives rise to branched actin filaments, as the complex binds to the sides of actin mother filaments and remains associated with the pointed end leaving the barbed end free to elongate [62,63]. Other up-regulated proteins (F-actin-capping protein subunit alpha-1 (Capza1), actin beta (Actb), stathmin (Stmn1)) are also involved in the dynamic organization which can support neuronal differentiation, including axonal growth and branching, or dendritic development [64]. In addition, Kaltwasser et al. [65] and Stühler et al. [66] reported that with increasing age a down-regulation of Stmn1 could be observed in the rat brain. Some members of the neurofilament family are down-regulated (neurofilament light polypeptide (Nefl), neurofilament medium polypeptide (Nefm)). These are involved in the promotion of axonal growth, neuronal polarity through direct binding to tubulin and known to be required for signaling in axon guidance [67].

As McAllister [68] described, dendritic growth and its arborization are crucial for the proper postnatal development of the nervous system. It suggests that levels of activity at individual synapses can locally modify the structure of the postsynaptic neuron. Furthermore, most of the extra- and intracellular signals that influence dendritic growth also alter synapse formation and axon guidance. This means that especially in the first postnatal days, the development of neurons and formation of synapses e.g. is essential, therefore, a dynamic expression of structural proteins is important at this stage of development.

A comparable development expression is also detectable for regulatory proteins. This category includes 8 up-regulated proteins and 4 down-regulated proteins. Cofilin-1 (Cfl1) and neuromodulin (GAP-43) among others belong to the up-regulated proteins at P7. Cfl1 has the property to directly regulate actin dynamics, severing and depolymerizing actin filaments to generate new barbed ends for an initiation of the actin polymerization. One of the best characterized ways of regulation is the phosphorylation on serine 3 that inhibits its F-actin activity [69]. Moreover, Sparrow et al. [70] showed, that cofilin plays an essential role during myelination in combination with neuregulin-1. This presents a highly specialized form of cell motility in which protrusive expansion of the leading edge of the inner mesaxon, accompanied by high rates of membrane synthesis, drives the glial membrane repeatedly around the axon to generate the myelin sheath. As stated by Philpot et al. [71], a proper formation of myelin has important consequences for the physiological function. They were able to show that an increased myelination occurs in the oB beginning about P11 and continues through P30. These results are corroborated by the appearance and rapid increase in expression of myelin-specific mRNAs in the rat bulb from P10-P15 [72]. The protein neuromodulin (GAP-43) presents a neuron-specific phosphoprotein that appears to be participating in the development and functional modulation of synaptic relationships [73]. During development, GAP-43 is one of a small number of proteins that is expressed selectively in association with process outgrowth [74]; its synthesis and axonal transport persist at high levels throughout axogenesis and synaptogenesis, and then decline precipitously with the establishment of stable synaptic relationships [75-81]. Jacobson et al. [82] observed that throughout the rat brain as a whole, levels of GAP-43 are highest during the first postnatal week, a period in which synaptic organization is still taking place, then fall by more than 90% over the next few weeks. For the down-regulated proteins, e.g., members of the annexin family (annexin A3 (Anxa3), annexin A5 (Anxa5)) show a lower amount of expression compared to P90. An accumulation of annexin 5 in cultured glioma cells during differentiation rather than proliferation could be shown by Giambanco et al. [83]. It is known that brain

development and maturation in the rat are postnatal events, consisting of glial cell proliferation and differentiation after the tenth postnatal day and in synaptogenesis during the third and fourth postnatal week [84-86]. Furthermore, Giambanco et al. [83] provided evidence that the accumulation of annexin 5 shows a sharp increase after the first postnatal week which appears to be in line with the possibility that the expression of this protein might be related to brain development.

The proteins of the amino acid metabolism show a higher amount of down-regulated proteins (7 proteins) than up-regulated proteins (2 proteins). The protein GMP synthase (GMPS) is an example for an up-regulated protein in comparison to P90 in this category. GMP synthetase is a key enzyme in the de novo synthesis of guanine nucleotides [87]. In the guanine nucleotide pathway, two enzymes are involved in converting IMP (inosinmonophosphat) to GMP. Here, GMP synthetase catalyzes the amination of XMP (Xanthosinmonophosphat) to GMP. The accumulation of guanine nucleotides is not only essential for DNA and RNA synthesis, but it also provides GTP, which is involved in a number of cellular processes important for cell division. GTP hydrolysis is also required for protein glycosylation [88], synthesis of adenine nucleotides [89], protein translation [90], activation of G-proteins [91] and microtubule assembly [92] which presents a possible explanation for the up-regulated expression of the protein at P7 compared to P90. A dynamic organization of the cytoskeleton is essential for neuronal differentiation (axonal growth and branching, dendritic development) in the early stages of development. Furthermore, a higher amount of down-regulated proteins was detectable at this postnatal developmental stage for the regulatory proteins. As mentioned earlier, the protein glutamine synthetase (Glul) displays a lower expression at P7 than at P90. Also Patel et al. [93] presented a differential pattern within development for this protein in different brain regions. At birth, the protein activity was the highest in the oB followed by forebrain and cerebellum (these data are not shown here). The developmental increase in enzyme activity was more or less linear in the forebrain and oB up to 20 days and in the cerebellum up to 50 days. This suggests an increase in glutamine synthetase activity which is associated with the maturation process rather than the proliferation of astrocytes. Another protein which is down-regulated is glutamate dehydrogenase 1 (Glud1) which catalyzes the reversible interconversion between glutamate and α-ketoglutarate. Glutamate serves as an important metabolite in the central nervous system [94] and is present in high concentrations in the brain [95-97]. The development of glutamate dehydrogenase activity in the whole brain has been shown to be low immediately post partum, however, its activity increases towards the adult brain [98-100]. Furthermore, as

Leong et al. [101] described, the general pattern of glutamate dehydrogenase activity development correlates inversely with the decreasing ammonia concentrations observed in the rat brain as the animal gets older, suggesting a role for this enzyme in ammonia detoxication. Glutamate dehydrogenase may also be potentially required in the maintenance of glutamate and, by further metabolism of γ-amino butyric acid, in neurotransmitter function, whereas this property varies not only with the age of the animal but also with the region of the brain.

Differential protein expression was also observed in aged rat brains (P637). The main differences in the expression concerned the categories of proteins of the energy metabolism, proteins of the biosynthesis, regulatory proteins, chaperones and structural proteins.

As described in the results, the expression of 3 proteins involved in the energy metabolism displays an up-regulation (e.g. creatine kinase B chain (Ckb) and cytochrome c oxidase subunit 5B (Cox5b)). The protein ATP synthase subunit alpha (Atp5a1) is down-regulated towards P90. As stated by Villa et al. [102], there is clear evidence that the oxidative energy metabolism gradually decreases with aging and the capacity of the brain to meet stress conditions is reduced with advancing age. Also Turpeenoja et al. [103] described that mitochondria from old animals contain less respiratory units than those from young animals and that the energy transduction capability is impaired by aging. They could observe two inner membrane proteins (cytochrome b and subunit II of cytochrome oxidase) which showed an age-related increase and they assumed the abundance of these proteins may be related to a decreased degradation rate leading to the accumulation of some polypeptides, including abnormal and inactive forms of proteins. This might cause partial inactivation of some enzymes with a concomitant decline in their specific activity. It has been postulated that senescent cells have reduced capacity to selectively identify and degrade defective proteins [104] with the consequence of accumulating functionally inactive polypeptides.

The proteins involved in protein biosynthesis reveal a higher expression (members of the heterogeneous nuclear ribo-nucleoprotein family (heterogeneous nuclear ribonucleoprotein F (Hnrnpf), heterogeneous nuclear ribonucleoprotein K (Hnrnpk)) at P637 in comparison to P90. As stated above, these proteins are involved in mRNA metabolism, DNA-related functions and micro-RNA biogenesis. This also includes processing in alternative splicing during neuronal differentiation [52]. As stated by Tollervey et al. [105], especially genes with metabolic functions show expression changes during aging which is linked with alternative splicing. This indicates that alternative splicing complements the transcriptional regulation in modifying the molecular machinery for the repair of oxidative DNA damage in the brain [106,107].

Also regulatory proteins exhibit major changes in the expression at P637 versus P90; the protein Rab GDP dissociation inhibitor alpha (Gdi1), for example, is up-regulated at this stage as well as at P7 compared to P90. This protein is a regulator of transport processes such as vesicular trafficking, docking off transport vesicles with their corresponding acceptor membranes [108] which is important at early stages of development. The higher expression of this protein at later stages, in this case at P637, could be inferred as an effect leading to endoplasmic reticulum-associated protein degradation by either the proteasome or macroautophagy pathway at higher age [109]. However, this expression pattern could not be shown for Rho GDP-dissociation inhibitor 1 (Arhgdia) and annexin A5 (Anxa5)) which show a down-regulation at P637. The calcium-binding protein 39 (Cab39) could not be detected at this developmental stage. As Iacopino et al. [110] described with the analysis of decreased expression of 28 kDa calbindin-D, another calcium-binding protein, in 27-month-old rat brain tissue, aging in both the rat and human central nervous system is accompanied by increased intracellular levels of free calcium, reduced activity of Ca^{2+}/Mg^{2+}-ATPase, and a decreased ability of the mitochondria to sequester calcium [111,112]. This altered calcium homeostasis is affected by both normal aging and various neurodegenerative diseases.

Moreover, the group of chaperones contains proteins which have a differential expression at the analyzed developmental stages. For example, two members of the heat shock protein 90 family (HSP 90-alpha (Hsp90aa1), heat shock protein HSP 90-beta (Hsp90ab1)) show an up-regulation towards P90. As described by Proctor et al. [113], neurodegeneration is an age-related disorder which is characterized by the accumulation of aggregated protein, loss of protein homeostasis and neuronal cell death. Two central systems in protein homeostasis are the chaperone system (includes Hsp90), which promotes correct protein folding, and the cellular proteolytic system, which degrades misfolded or damaged proteins. A stochastic model of Hsp90 [114] showed that under conditions of low or transient stress, chaperone capacity is sufficient to maintain protein homeostasis. Even under conditions of increasing stress with age, normal chaperone capacity is able to process the increasing overload due to the mechanism of up-regulation of HSPs after stress. This observation could also be the reason for the up-regulation of the proteins in our analysis. However, other proteins (T-complex protein 1 subunit beta (Cct2)) which are also involved in chaperone activity show a down-regulation towards P90 or are absent at this stage (protein disulfide-isomerase A3 (Pdia3)). Also Vanguilder and Freeman [115] mentioned a decreased expression of these proteins as an example for age-related dysregulation, which indicates a loss of protein quality regulation that may contribute to a buildup of cytoskeletal proteins (e.g., increased neurofilament light chain and tubulin) with increasing age.

The category of structural proteins showed 7 up-regulated proteins (actin beta (Actb), myristoylated alanine-rich C-kinase substrate (Marcks) as well as several members of the tubulin family (tubulin alpha-1A chain (Tuba1a), tubulin beta-3 chain (Tubb3), tubulin beta 4A (Tubb4)). It could be hypothesized that the up-regulation of these structural proteins might be due to the cytoskeletal dynamics in age-related neuronal dysfunction which was mentioned above. An indicator for this circumstance could also be the down-regulation of other structural proteins (lamin-B1 (Lmnb1) and septin-11 (Sept11)) towards P90. This means, by up-regulating the expression of established structural proteins, the organism tries to counteract the loss of essential neuronal key elements during aging by maintaining the functional organization of the cytoskeleton in general, for example.

Five differentially expressed proteins were validated in Western Blots (Figure 6). In comparison to the 2DE-analysis (Figure 7), the same developmental expression changes of these proteins were detectable. All of these proteins are involved in generating and regulating cell processes, axons, dendrites and also processes of glial cells.

Beyond that it could be established that additional differentially expressed proteins of the three developmental stages are involved in proliferation, migration and differentiation. Protein-members of the energy and amino acid metabolism, for example, as well as the structural proteins show that the maintenance of the oB with substrates differs during the development of the rat brain. Some of these proteins could be candidates for further investigation regarding the survival and differentiation of transplanted progenitor cells in neonatal and adult striata. This also gives the option to connect the differentially expressed proteins to their specific metabolic pathways in future studies

and to analyze factors and candidates which influence the differentiation of neuronal progenitor cells.

Furthermore, a differential expression of proteins was also detectable in brains of adult and aged rats (P90, P637). This could be an indicator for the capability of the oB to still generate and integrate cells of the nervous system by the RMS of the SVZ in adulthood. In general, the oB is of special interest as it reveals spontaneous neurogenesis throughout the entire lifetime, suggesting that it plays a functional role in physiological cell replacement in aging, learning and cognition, as well as proposing a therapeutic potential in neurological diseases [13,116], including neurodegenerative disorders like Alzheimer's and Parkinson's disease.

Conclusions

In conclusion, the proteomic analysis of the oB at three different developmental stages (P7, P90, P637) indicates several changes of protein expression including up- and down-regulation of single proteins or the appearance of proteins at specific developmental stages only, supporting the concept of an extraordinarily high plasticity of this part of the rat brain.

Yet, the performed differential 2D gel spot abundance analysis does not exclude the simultaneous presence of proteins/isoforms with no detectable abundance differences.

However, the analysis may contain the same proteins other than that identified as differential with regards to its abundance. As listed in the results, some abundant proteins also show an absence at single developmental stages which may result from different posttranslational modifications. However, the aim of the present study was not to characterize structural differences of the same protein in different spots but rather to detect protein spots that differed in abundance and provide indications for the changes in protein expression during development. For the precise analysis of expression changes of single proteins, further

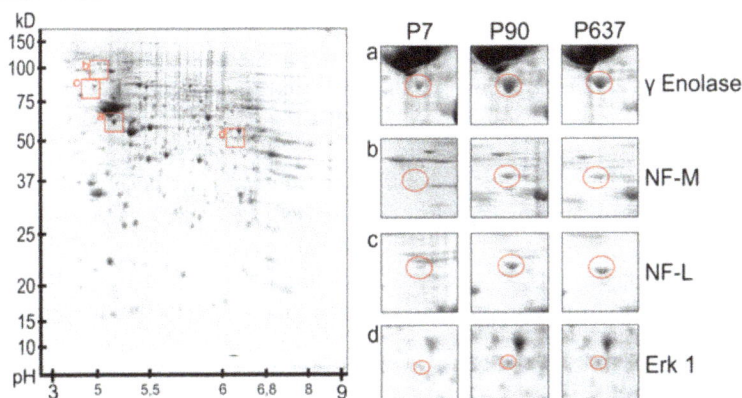

Figure 7 Left side: overview of a control gel (P90) with rectangles (a-d) around regions of spots with differentially regulated proteins.
Right side: Magnification of the marked regions (a-d) for each developmental stage (P7, P90, P637). Each spot is marked by a circle.

studies are needed to verify if all individual spots of this protein were detected to determine its exact differential regulation. This also includes spots which are located in regions of the gel that show a lower resolution, for example.

In this study, a relative small amount of proteins is considered to be expressed differentially by applying a restricted SVQ of greater 1.67 and smaller 0.6. An additional analysis of extensive SVQ ranges (e.g. P0, P20, P40, P200) could give a more detailed view about the development of the oB. Furthermore, usage of additional methods (e.g. difference gel electrophoresis, DIGE) could increase the systems sensitivity, and reproducibility. Also an additional usage of purification protocols could provide a more detailed view on small proteins (e.g. neurotransmitters) and plasma membrane proteins for further analysis.

Methods
Treatment
Male Wistar rats (Rattus norvegicus, Charles River, Sulzfeld, Germany) of different ages (7, 90, 637 postnatal days) with 6 animals in each group were used [117,118]. The animals were housed at $22 \pm 2°C$ under an artificial day and night rhythm with a 12 h light–dark cycle with free access to water and standard nutrition. All animal treatment and experimental procedures were conducted in compliance with the regulations and licensing of the local authorities (Landesamt für Landwirtschaft, Lebensmittelsicherheit and Fischerei Mecklenburg Vorpommern, Germany) and the Animal Care and Use Committee of the University of Rostock. According to the European Communities Council Directive of 24th November 1986 (86/609/EEC) and in accordance with the above-mentioned local authorities adequate measures were taken to minimize pain or discomfort.

Perfusion and dissection
Perfusion was performed at defined dates (7, 90, 637 days postnatal). The animals were anesthetized with ether and killed by intraperitoneal Pentobarbital-Na$^+$-injection (60 mg/kg BW). Transcardial perfusion was performed with 100 – 400 ml (bodyweight depending) cooled (4°C) 0.9% NaCl- solution. After decapitation and brain dissection, the dissected brain regions were weighed and stored at –80°C until homogenization. From pentobarbital injection to –80°C storage it took less than 5 min.

Homogenization
Protein extraction was performed according to published standardized protocols [119,120]. The brain sections were incubated with (9 × probe mass [mg]) μL lysis buffer consisting of 7 M urea (Sigma, Steinheim, Germany), 2 M thiourea (Sigma), 4% CHAPS (Sigma), 70 mM DTT (Sigma), 0.5% Bio-Lyte Ampholytes pH 3–10 (Fluka,

Buchs, Switzerland) and a mixture of protease inhibitors (Roche, Basel Switzerland) additionally enriched with (0.1× probe mass [mg]) μL Pepstatin A and PMSF (Fluka) and snap-frozen at –150°C. The samples were quickly thawed and transferred into a 2 mL Wheaton potter (neo-lab, Heidelberg, Germany) for homogenization. Glass beads (Roth, Karlsruhe, Germany) were added to the suspension, following a 15 s sonication, 15 s vortexing, which was repeated six times and finished by shock freezing the suspension at 150°C. The samples were quickly thawed. They were put in a beaker on a magnetic stirrer that was filled with ice water for 15 min. Finally, the samples were centrifuged at 17.860 × g for 20 min at 4°C. The supernatant was very carefully removed using a 2 ml syringe (Becton Dickinson, Heidelberg, Germany) with a 0.5 × 25 mm needle (Becton Dickinson), because of a thick lipid coverage derived from myelinated nerve fibers. The protein concentration of the supernatant was determined by the Bradford assay.

Two dimensional polyacrylamide gel electrophoresis (2DE)
Rehydration
The first dimension was performed in a PROTEAN IEF cell system (Bio-Rad, CA, USA). Protein extracts of 1 mg protein were loaded on immobilized pH 3–10 nonlinear gradient strips with a length of 17 cm (GE-Healthcare, Art.: 17-1235-01) and actively rehydrated with 300 μL rehydration buffer consisting of 6 M urea (Sigma), 2 M thiourea (Sigma), 2% CHAPS (Sigma), 16 mM DTT (Sigma), 0.5% Bio-Lyte Ampholytes pH 3 10 (Fluka) at 50 V for 12 h at 20°C.

First dimension: isoelectric focussing
After rehydration, electrode wicks (Bio-Rad) were added to reduce artifacts. Focusing started with the "conditioning step" (2 h) which subdivides in 2 sub-steps: (a) linear voltage rise to 500 V, step-hold 30 min, (b) linear voltage rise to 2500 V, step-hold 1 h. After that, the "slow voltage ramping" (2.5 h): quadratic voltage rise to 8000 V and the "final focusing": actual process of focusing (duration: 50.000 Vhrs) were performed. During the whole IEF the temperature was constantly kept at 20°C. After focusing the strips were stored at –80°C.

Second dimension: polyacrylamide gel electrophoresis
Focused IPG-strips were equilibrated in two steps of 30 min each in 5 mL of freshly prepared SDS equilibration solution consisting of 1.5 M Tris–HCl pH 8.8 (Roth), 6 M urea (Sigma), 30% Glycerol (Sigma), 2% SDS (Sigma), trace of bromophenol blue (Roth) supplemented with 10 mg/mL DTT and 40 mg/mL iodoacetamide.

The strips were transferred on 12% homogeneous selfcast sodium dodecyl sulfate polyacrylamide gels (200 mm × 250 mm × 1.5 mm). They were run at 125 V per gel (Power Pac 1000, Bio-Rad) in the PROTEAN Plus Dodeca Cell (Bio-Rad). A cooling device (Julabo F10, Julabo Labortechnik, Seelbach, Germany) was used to ensure a constant buffer-temperature of 10°C.

Fixation and staining
Fixation was performed with acetic acid-methanol-solution (45% methanol, 1% acetic acid overnight. Staining of the gels was performed in a colloidal CBB G250 solution (1 g/1000 ml) (Roth) as described previously [121]. After 24 h, the gels were destained with ultra pure water and were held in cold storage (4°C) with ultra pure water until digitization.

Gel analysis
Digitization
The stained gels (n = 6) were scanned as 12 bit gray scale tif-images with a F4100 scanner (Heidelberg, Heidelberg, Germany) at 300 dpi resolution. Gels were rinsed in 0.02% sodium azide (Aldrich Chemie, Steinheim, Germany), shrink-wrapped in plastic and stored at 4°C until picking for MALDI-TOF-MS.

Digital gel processing
For 2DE-gel image analysis, the software package Progenesis PG200 Version 2006 (Nonlinear Dynamics Ltd., Newcastle upon Tyne, U.K.) was used. Each developmental stage (P7, P90, P637) consisted of six gel images. For P90 and P637, each gel image presents a single dissected brain region, whereas in P7 the samples were pooled for 2DE. The gels were registered to a reference gel (the gel which contained most spots with highest separation and staining quality and least artifacts) and manually edited spots were matched to allow comparability of all gels.

Determination of differentially abundant protein spots
Protein spots in 2DE were quantified by normalizing spot volumes using the Progenesis PG200 and spot volume differences were calculated. After the comparison of the normalized gray value-spot volumes of all generated spot pairs with Access and Excel (Windows, Microsoft), the comparison of the spot volumes was determined by calculation of the spot volume quotient (SVQ, P90/P7 or P90/P637). Only those spots were considered to be significantly up- or down-regulated that showed a SVQ (Spot Volume Quotient) of 0.6 or less and 1.67 or greater. The differences were evaluated significantly if differentially expressed spots in at least four gel images (correlated spots) were detected which belong to one test group. If multiple spots of one protein were detected by mass spectrometric analysis, the differential expression was determined by the mean of their individual expression levels. If the differential expression tendency was ambiguous, the

protein expression is marked as "up-down" (see Additional file 1: Table S1). The classification of the differentially expressed proteins in their respective functional protein groups itself was generated by a comparison of the proteins properties; in addition, the basic function of each protein is briefly described (according to Entrez Gene, GeneCards, UniProtKB/Swiss-Prot, and/or UniProtKB/ TrEMBL).

Mass spectrometric analysis of protein spots
Protein spots were excised from the gels with a spot picker (Flexys Proteomics picker, Genomic Solutions, Ann Arbor, MI, USA), transferred into 96-well plates, and subjected to in-gel digestion with trypsin. The gel plugs were washed twice with 30% acetonitrile (ACN) in 25 mM ammonium bicarbonate and 50% ACN in 10 mM ammonium bicarbonate, respectively, shrunk with ACN, and dried at 37°C. The dried gel plugs were re-swollen with 5 μl protease solution (sequencing grade trypsin, 10 ng/μl in 3 mM Tris–HCl, pH 8.5, Promega, Madison, WI, USA) and incubated for 8 h at 37°C. Thereafter, 5 μl of extraction solution (0.3% trifluoroacetic acid, 50% ACN) were added and the samples were agitated at room temperature for 30–60 min before the peptide extracts were transferred into the 96-well collection plates. The resulting peptide-containing solution was prepared for MALDI analysis by spotting 0.6 μl of the tryptic digest and 0.45 μl of matrix solution consisting of 9 mg/ml α-cyano-4-hydroxy-cinnamic acid (CHCA) in 50% ACN, 0.1% trifluoroacetic acid on standard stainless steel MALDI plates. MALDI-MS analysis was performed on a 4700 Proteomics Analyzer MALDI-TOF/ TOF mass spectrometer (Applied Biosystems, Foster City, CA, USA). All acquired spectra were processed using 4700 Explore™ software (Applied Biosystems). For protein identification, spectra were submitted to MASCOT (version 2.4.0, Matrix Science, London, UK) via the MASCOT Deamon. Searches were performed against the subset of rat proteins of the UniProtKB protein sequence database (2012_01; 42755 sequences from Rattus). A mass tolerance of 60 ppm and 1 missing cleavage site were set, oxidation of methionine residues was considered as variable modification, and carbamidomethylation of cysteines as fixed modification. Peptide masses of trypsin autoproteolysis products and matrix-derived peaks were excluded. Identifications with Mascot scores greater than 59 were considered significant ($p < 0.05$). All results were examined carefully for reliability and occurrence of multiple proteins in the same sample.

Immunoblot analysis
The homogenated protein samples of the oB were dissolved in sample buffer containing 1% SDS and boiled for 10 min. Protein contents in the samples were determined according to Neuhoff et al. [122]. For

immunoblotting, total cellular extracts (10–50 μg per lane) were separated by SDS-Page using 7,5% polycryamide gels and transferred to PVDF membranes (0.2 mm, Bio-Rad, Munich, Germany). The blots were blocked with 5% non-fat dry milk powder in TBS for 30 min and incubated with the individual primary antibodies. The following antibodies were used (dilutions are given in the brackets): rabbit polyclonal anti-Gamma-enolase (1:500), mouse monoclonal anti –α Tubulin (1:1000), goat monoclonal anti-neurofilament medium (1:2000), mouse monoclonal anti-neurofilament low (1:500), rabbit polyclonal anti-Erk 1,2 (1:500) and mouse monoclonal anti-Erk 1,2 P (1:500). After washing, membranes were incubated with secondary HRP-conjugated anti-mouse (1:2000), anti-goat (1:2000) and anti-rabbit (1:2000) IgG, and visualized by the enhanced chemiluminescence (ECL) procedure as described by the manufacturer (Thermo scientific, Pierce, Rockford, USA).

Additional file

Additional file 1: Some differentially expressed proteins of the categories of P7 P90 and P637 which are not listed in the results are presented and described. In Table S1. Remaining categories of differentially expressed proteins of P7 in comparison to P90. Table S2. Remaining categories of differentially expressed proteins of P637 in comparison to P90.

Abbreviations

A: Axon; CBB: Coomassie blue; Cpg: Cytoplasmic granule; Cts: Centrosome; Cr: Chromosome; Csk: Cytoskeleton; cp: Chaperones; Cpm: CYtoplasm; Cpv: Cytoplasmic vesicle; Cts: Cytosol; ER: Endoplasmic reticulum; Eds: Endosome; Exr: Extracellular region; Gc: Growth cone; Gj: Gap junction; Golgi: Golgi apparatus; hGc: Heterotrimeric G-protein complex; La: Lipid-anchor; Lyso: Lysosome; M: Membrane; Micro: Microsome; Mmt: Mitochondrial matrix; Mm: Mitochondrial membrane; Mim: Mitochondrion inner membrane; MimS: Mitochondrion intermembrane space; Mom: Mitochondrion outer membrane; Mel: Melanosome; Mito: Mitochondrion; MT: Microtubule; Nc: Nucleus; Nf: Neurofilament; Nm: Nucleus matrix; Np: Nucleoplasm; P: Proteasome; pa: Proteins antioxidants; paam: Proteins amino acid metabolism; pb: Proteins biosynthesis; pcm: Proteins carbohydrate metabolism; pd: Proteins degradation; Per: Peroxisome; pem: Proteins energy metabolism; Pema: Proteinaceous extracellular matrix; pfm: Proteins fat metabolism; PMp: Peripheral Membrane protein; pr: Proteins regulation; pst: Proteins signal transduction; ptm: Proteins transmitter metabolism; Rs: Ribosome; S: Synapse; Sc: Secreted; Scc: Spliceosomal complex; sER: Smooth endoplasmic reticulum; sp: Structural proteins; Sv: Synaptic vesicle; Ss: Synaptosome; SR: Sarcoplasmic reticulum; TCA: Trichloroacetic acid; tp: Transport proteins; Ulc: Ubiquitin ligase complex.

Competing interests

The authors declare that they have no competing interests.

Authors' contributions

MW carried out the Western Blots, spot segmentation and drafted the manuscript. AS carried out the two-dimensional polyacrylamide gel electrophoresis. GM optimized two-dimensional polyacrylamide gel electrophoresis for brain tissue. MK participated in the database management of the spot lists and performed the picking of spots. MOG, AW and OS participated in the design of the study. OS performed the dissection of the material and the sample preparation. OS conceived of the study, and participated in its design and coordination and helped to draft the manuscript. All authors read and approved the final manuscript.

Acknowledgements
We appreciate the outstanding work of Dr. Stefan Mikkat by identifying the proteins with the technique of MALDI-TOF-MS.

Author details
[1]Department of Anatomy, Gertrudenstr. 9, 18055 Rostock, Germany.
[2]Proteome Center Rostock, Schillingallee 69, 18055 Rostock, Germany.
[3]Department of Pathology, Josef-Schneider-Str. 2, 97080 Würzburg, Germany.

References

1. Bandeira F, Lent R, Herculano-Houzel S. Changing numbers of neuronal and non-neuronal cells underlie postnatal brain growth in the rat. Proc Natl Acad Sci U S A. 2009;106:14108–13.
2. Imamura F, Albert E, Ayoub A, Greer C. Timing of neurogenesis is a determinant of olfactory circuitry. Nat Neurosci. 2011;14:331–7.
3. Graziadei PP, Graziadei GA. Neurogenesis and neuron regeneration in the olfactory system of mammals. I. Morphological aspects of differentiation and structural organization of the olfactory sensory neurons. J Neurocytol. 1979;8:1–18.
4. Costanzo RM. Comparison of neurogenesis and cell replacement in the hamster olfactory system with and without a target (olfactory bulb). Brain Res. 1984;307:295–301.
5. Morrison EE, Costanzo RM. Morphology of the human olfactory epithelium. J Comp Neurol. 1990;297:1–13.
6. Uraih LC, Maronpot RR. Normal histology of the nasal cavity and application of special techniques. Environ Health Perspect. 1990;85:187–208.
7. Schultz E. Repair of the olfactory mucosa. Am J Pathol. 1960;37:1–19.
8. Graziadei PP, Monti Graziadei AG. Regeneration in the olfactory system of vertebrates. Am J Otolaryngol. 1983;4:228–33.
9. Calof AL, Chikaraishi DM. Analysis of neurogenesis in a mammalian neuroepithelium: proliferation and differentiation of an olfactory neuron precursor in vitro. Neuron. 1989;3:115–27.
10. Huard JM, Youngentob SL, Goldstein BJ, Luskin MB, Schwob JE. Adult olfactory epithelium contains multipotent progenitors that give rise to neurons and non-neural cells. J Comp Neurol. 1998;400:469–86.
11. Shetty AK, Turner DA. In vitro survival and differentiation of neurons derived from epidermal growth factor-responsive postnatal hippocampal stem cells: inducing effects of brain-derived neurotrophic factor. J Neurobiol. 1998;35:395–425.
12. Shetty AK, Turner DA, Carolina N. Neurite outgrowth from progeny of epidermal growth factor – responsive hippocampal stem cells Is significantly less robust than from fetal hippocampal cells following grafting onto organotypic hippocampal slice cultures: effect of brain-derived neurotrophic factor. J Neurobiol. 1999;38:391–413.
13. Abe K. Therapeutic potential of neurotrophic factors and neural stem cells against ischemic brain injury. J Cereb Blood Flow Metab. 2000;20:1393–408.
14. Rosser AE, Tyers P, Dunnett SB. The morphological development of neurons derived from EGF- and FGF-2-driven human CNS precursors depends on their site of integration in the neonatal rat brain. Eur J Neurosci. 2000;12:2405–13.
15. Thomson JA. Embryonic stem cell lines derived from human blastocysts. Science. 1998;282:1145–7.
16. Vogel G. Harnessing the power of stem cells. Science. 1999;283(5407):1432–4.
17. Edwards BE, Gearhart JD, Wallach EE. The human pluripotent stem cell: impact on medicine and society. Fertil Steril. 2000;74:1–7.
18. Kishimoto Y, Davies E, Radin N. Developing rat brain: changes in cholesterol, galactolipids, and the individual fatty acids of gangliosides and glycerophosphatides. J Lipid Res. 1965;4:532–6.
19. Smart I. The subependymal layer of the mouse brain and its cell production as shown by radioautography after thymidine-H3 injection. J Comp Neurol. 1961;116:325–47.
20. Altman J. Autoradiographic and histological studies of postnatal neurogenesis. IV. Cell proliferation and migration in the anterior forebrain, with special reference to persisting neurogenesis in the olfactory bulb. J Comp Neurol. 1969;137(4):433–57.
21. Reynolds BA, Weiss S. Generation of neurons and astrocytes from isolated cells of the adult mammalian central nervous system. Science. 1992;255:1707–10.

22. Okano J, Pfaff W. RB and Cdc2 expression in brain: correlations incorporation and neurogenesis with 3H-thymidine. J Neurosci. 1993;13(7):2930–8.

23. Lois C, Alvarez-Buylla A. Proliferating subventricular zone cells in the adult mammalian forebrain can differentiate into neurons and glia. Proc Natl Acad Sci U S A. 1993;90:2074–7.

24. Shimada M. Cytokinetics and histogenesis of early postnatal mouse brain as studied by 3H-thymidine autoradiography. Arch Histol Jpn. 1966;26(4):413–37.

25. Bayer SA. 3H-thymidine-radiographic studies of neurogenesis in the rat olfactory bulb. Exp Brain Res. 1983;50(2–3):329–40.

26. Kishi K. Golgi studies on the development of granule cells of the rat olfactory bulb with reference to migration in the subependymal layer. J Comp Neurol. 1987;258(1):112–24.

27. Luskin MB. Restricted proliferation and migration of postnatally generated neurons derived from the forebrain subventricular zone. Neuron. 1993;11:173–89.

28. Luskin MB, Boone MS. Rate and pattern of migration of lineally-related olfactory bulb interneurons generated postnatally in the subventricular zone of the rat. Chem Senses. 1994;19:695–714.

29. Lois C, Alvarez-Buylla A. Neuronal stem cells in the brain of adult vertebrates. Stem Cells. 1995;13(3):263–72.

30. Peretto P, Merighi A, Fasolo A, Bonfanti L. Glial tubes in the rostral migratory stream of the adult rat. Brain Res Bull. 1997;42:9–21.

31. Brunjes PC, Frazier LL. Maturation and plasticity in the olfactory system of vertebrates. Brain Res. 1986;396(1):1–45.

32. Sullivan RM, Leon M. One-trial olfactory learning enhances olfactory bulb responses to an appetitive conditioned odor in 7-day-old rats. Brain Res. 1987;432(2):307–11.

33. Brunjes PC. Unilateral naris closure and olfactory system development. Brain Res Rev. 1994;19(1):146–60.

34. Wilson DA, Sullivan RM. Neurobiology of associative learning in the neonate: early olfactory learning. Behav Neural Biol. 1994;61(1):1–18.

35. Matsutani S, Yamamoto N. Postnatal development of dendritic spines on olfactory bulb granule cells in rats. J Comp Neurol. 2004;473(4):553–61.

36. Adlard BP, Dobbing J. Phosphofructokinase and fumarate hydratase in developing rat brain. J Neurochem. 1971;18:1299–303.

37. Teichgräber P, Biesold D. Properties of membrane-bound hexokinase in rat brain. J Neurochem. 1968;15(9):979–89.

38. Hamburgh M, Flexner LB. Biochemical and physiological differentiation during morphogenesis. XXI. Effect of hypothyroidism and hormone therapy on enzyme activities of the developing cerebral cortex of the rat. J Neurochem. 1957;1(3):279–88.

39. De Vellis J, Schjeide OA, Clemente CD. Protein synthesis and enzymic patterns in the developing brain following head x-irradiation of newborn rats. J Neurochem. 1967;14(5):499–511.

40. Datta I, Bhonde R. Can mesenchymal stem cells reduce vulnerability of dopaminergic neurons in the substantia nigra to oxidative insult in individuals at risk to Parkinson's disease? Cell Biol Int. 2012;36(7):617–24.

41. Ganser C, Papazoglou A, Just L, Nikkhah G. Neuroprotective effects of erythropoietin on 6-hydroxydopamine-treated ventral mesencephalic dopamine-rich cultures. Exp Cell Res. 2010;316(5):737–46.

42. Hovakimyan M, Haas SJ, Schmitt O, Gerber B, Wree A, Andressen C. Mesencephalic human neural progenitor cells transplanted into the neonatal hemiparkinsonian rat striatum differentiate into neurons and improve motor behaviour. J Anat. 2006;209(6):721–32.

43. Haas SJ, Petrov S, Kronenberg G, Schmitt O, Wree A. Orthotopic transplantation of immortalized mesencephalic progenitors (CSM14.1 cells) into the substantia nigra of hemiparkinsonian rats induces neuronal differentiation and motoric improvement. J Anat. 2008;212(1):19–30.

44. Becker M, Schindler J, Nothwang HG. Neuroproteomics - the tasks lying ahead. Electrophoresis. 2006;27(13):2819–29.

45. Schneider LV, Hall MP. Stable isotope methods for high-precision proteomics. Drug Discov Today. 2005;10(5):353–63.

46. Bunik VI, Kabysheva MS, Klimuk EI, Storozhevykh TP, Pinelis VG. Phosphono analogues of 2-oxoglutarate protect cerebellar granule neurons upon glutamate excitotoxicity. Ann N Y Acad Sci. 2009;1171:521–9.

47. Takei N, Kondo J, Nagaike K, Ohsawa K, Kato K, Kohsaka S. Neuronal survival factor from bovine brain is identical to neuron-specific enolase. J Neurochem. 1991;57(4):1178–84.

48. Wilbur DO, Patel MS. Development of mitochondrial pyruvate metabolism in rat brain. J Neurochem. 1974;22(5):709–15.

49. Black DL, Grabowski PJ. Alternative pre-mRNA splicing and neuronal function. Prog Mol Subcell Biol. 2003;31:187–216.

50. Lipscombe D. Neuronal proteins custom designed by alternative splicing. Curr Opin Neurobiol. 2005;15(3):358–63.

51. Ule J, Darnell RB. RNA binding proteins and the regulation of neuronal synaptic plasticity. Curr Opin Neurobiol. 2006;16(1):102–10.

52. Cao W, Razanau A, Feng D, Lobo VG, Xie J. Control of alternative splicing by forskolin through hnRNP K during neuronal differentiation. Nucleic Acids Res. 2012;40(16):8059–71.

53. Kuhl D, Skehel P. Dendritic localization of mRNAs. Curr Opin Neurobiol. 1998;8(5):600–6.

54. Derrigo M, Cestelli A, Savettieri G, Di Liegro I. RNA-protein interactions in the control of stability and localization of messenger RNA (review). Int J Mol Med. 2000;5(2):111–23.

55. Roegiers F, Jan YN. Staufen: a common component of mRNA transport in oocytes and neurons? Trends Cell Biol. 2000;10(6):220–4.

56. Raimondi L, Cannino G, D'Asaro M, Sala A, Savettieri G, Di Liegro I. Regulation of RNA metabolism in the nervous system. Recent Res Dev Neurochem. 2002;5:39–48.

57. Keene JD, Tenenbaum SA. Eukaryotic mRNPs may represent posttranscriptional operons. Mol Cell. 2002;9(6):1161–7.

58. Percipalle P, Raju CS, Fukuda N. Actin-associated hnRNP proteins as transacting factors in the control of mRNA transport and localization. RNA Biol. 2009;6(2):171–4.

59. Maher-Laporte M, Berthiaume F, Moreau M, Julien LA, Lapointe G, Mourez M, et al. Molecular composition of staufen2-containing ribonucleoproteins in embryonic rat brain. PLoS One. 2010;5(6):e11350.

60. Proepper C, Steinestel K, Schmeisser MJ, Heinrich J, Steinestel J, Bockmann J, et al. Heterogeneous nuclear ribonucleoprotein k interacts with Abi-1 at postsynaptic sites and modulates dendritic spine morphology. PLoS One. 2011;6(11):e27045.

61. Kessels MM, Schwintzer L, Schlobinski D, Qualmann B. Controlling actin cytoskeletal organization and dynamics during neuronal morphogenesis. Eur J Cell Biol. 2011;90(11):926–33.

62. Pollard TD. Regulation of actin filament assembly by Arp2/3 complex and formins. Annu Rev Biophys Biomol Struct. 2007;36:451–77.

63. Campellone KG, Welch MD. A nucleator arms race: cellular control of actin assembly. Nat Rev Mol Cell Biol. 2010;11(4):237–51.

64. Devaux S, Poulain FE, Devignot V, Lachkar S, Irinopoulou T, Sobel A. Specific serine-proline phosphorylation and glycogen synthase kinase 3β-directed subcellular targeting of stathmin 3/Sclip in neurons. J Biol Chem. 2012;287(26):22341–53.

65. Kaltwaßer B, Schulenborg T, Beck F, Klotz M, Schäfer KH, Schmitt M, et al. Developmental changes of the protein repertoire in the rat auditory brainstem: a comparative proteomics approach in the superior olivary complex and the inferior colliculus with DIGE and iTRAQ. J Proteomics. 2013;79:43–59.

66. Stühler K, Pfeiffer K, Joppich C, Stephan C, Jung K, Müller M, et al. Pilot study of the Human Proteome Organisation Brain Proteome Project: applying different 2-DE techniques to monitor proteomic changes during murine brain development. Proteomics. 2006;6(18):4899–913.

67. Maurya DK, Sundaram CS, Bhargava P. Proteome profile of the mature rat olfactory bulb. Proteomics. 2009;9(9):2593–9.

68. McAllister AK. Cellular and molecular mechanisms of dendrite growth. Cereb Cortex. 2000;10(10):963–73.

69. Huang TY, DerMardirossian C, Bokoch GM. Cofilin phosphatases and regulation of actin dynamics. Curr Opin Cell Biol. 2006;18(1):26–31.

70. Sparrow N, Manetti ME, Bott M, Fabianac T, Petrilli A, Bates ML, et al. The actin-severing protein cofilin is downstream of neuregulin signaling and is essential for Schwann cell myelination. J Neurosci. 2012;32(15):5284–97.

71. Philpot BD, Klintsova AY, Brunjes PC. Oligodendrocyte/myelin-immunoreactivity in the developing olfactory system. Neuroscience. 1995;67(4):1009–19.

72. Kanfer J, Parenty M, Goujet-Zalc C, Monge M, Bernier L, Campagnoni AT, et al. Developmental expression of myelin proteolipid, basic protein, and 2′,3′-cyclic nucleotide 3′-phosphodiesterase transcripts in different rat brain regions. J Mol Neurosci. 1989;1(1):39–46.

73. Benowitz LI, Apostolides PJ, Perrone-Bizzozero N, Finklestein SP, Zwiers H. Anatomical distribution of the growth-associated protein GAP-43/B-50 in the adult rat brain. J Mol Neurosci. 1988;8(1):339–52.

74. Perrone-Bizzozero NI, Finklestein SP, Benowitz LI. Synthesis of a growth-associated protein by embryonic rat cerebrocortical neurons in vitro. J Neurosci. 1986;6(12):3721–30.

75. Benowitz LI, Shashoua VE, Yoon MG. Specific changes in rapidly transported proteins during regeneration of the goldfish optic nerve. J Neurosci. 1981;1(3):300–7.

76. Skene JH, Willard M. Axonally transported proteins associated with axon growth in rabbit central and peripheral nervous systems. J Cell Biol. 1981;89(1):96–103.

77. Benowitz LI, Lewis ER. Increased transport of 44,000- to 49,000-dalton acidic proteins during regeneration of the goldfish optic nerve: a two-dimensional gel analysis. J Neurosci. 1983;3(11):2153–63.

78. Freeman JA, Bock S, Deaton M, McGuire B, Norden JJ, Snipes GJ. Axonal and glial proteins associated with development and response to injury in the rat and goldfish optic nerve. Exp Brain Res. 1986;13:34–47.

79. Kalil K, Skene JH. Elevated synthesis of an axonally transported protein correlates with axon outgrowth in normal and injured pyramidal tracts. J Neurosci. 1986;6(9):2563–70.

80. Moya KL, Benowitz LI, Jhaveri S, Schneider GE. Enhanced visualization of axonally transported proteins in the immature CNS by suppression of systemic labeling. Brain Res. 1987;428(2):183–91.

81. Perry GW, Burmeister DW, Grafstein B. Fast axonally transported proteins in regenerating goldfish optic axons. J Neurosci. 1987;7(3):792–806.

82. Jacobson RD, Virág I, Skene JH. A protein associated with axon growth, GAP-43, is widely distributed and developmentally regulated in rat CNS. J Neurosci. 1986;6(6):1843–55.

83. Giambanco I, Verzini M, Donato R. Annexins V and VI in rat tissues during post-natal development: immunochemical measurements. Biochem Biophys Res Commun. 1993;196(3):1221–6.

84. Eayrs JT, Goodhead B. Postnatal development of the cerebral cortex in the rat. J Anat. 1959;93:385–402.

85. Aghajanian GK, Bloom FE. The formation of synaptic junctions in developing rat brain: a quantitative electron microscopic study. Brain Res. 1967;6(4):716–27.

86. Caley DW, Maxwell DS. An electron microscopic study of neurons during postnatal development of the rat cerebral cortex. J Comp Neurol. 1968;133(1):17–44.

87. Hirst M, Haliday E, Nakamura J, Lou L. Human GMP synthetase. Protein purification, cloning, and functional expression of cDNA. J Biol Chem. 1994;269(38):23830–7.

88. Antonino LC, Wu JC. Human IMP dehydrogenase catalyzes the dehalogenation of 2-fluoro- and 2-chloroinosine 5'-monophosphate in the absence of NAD. Biochemistry. 1994;33(7):1753–9.

89. Stayton MM, Rudolph FB, Fromm HJ. Regulation, genetics, and properties of adenylosuccinate synthetase: a review. Curr Top Cell Regul. 1983;22:103–41.

90. Simmer JP, Kelly RE, Rinker Jr AG, Scully JL, Evans DR. Mammalian carbamyl phosphate synthetase (CPS) DNA sequence and evolution of the CPS domain of the Syrian hamster multifunctional protein CAD. J Biol Chem. 1990;265(18):10395–402.

91. Hepler JR, Gilman AG. G proteins. Trends Biochem Sci. 1992;17(10):383–7.

92. Gelfand VI, Bershadsky AD. Microtubule dynamics: mechanism, regulation, and function. Annu Rev Cell Biol. 1991;7:93–116.

93. Patel AJ, Hunt A, Tahourdin CSM. Regional development of glutamine synthetase activity in the rat brain and its association with the differentiation of astrocytes. Dev Brain Res. 1983;8(1):31.

94. Berl S. Drugs, catecholamines and Krebs cycle interaction (CNS). Psychopharmacol Bull. 1975;11(2):40.

95. Agrawal HC, Himwich WA. Amino acids, proteins, and monoamines of developing brain. In: Himwich WA, editor. Developmental Neurobiology, Thomas, Illinois. 1970. p. 287–310.

96. Van den Berg CJ, Matheson DF, Ronda G, Reijnierse GLA, Blokhuis GGD, Kroon MC, et al. A model of glutamate metabolism in brain: a biochemical analysis of a heterogeneous structure. Metab Compart Neurotransmission. 1975;31:515–43.

97. Maker HS, Clarke DD, Lajthab AL. Intermediary metabolism of carbohydrates and amino acids. In: Siegel GJ, Albers RW, Katzman R, Agranoff BW, editors. Basic Neurochem. 2nd ed. p. 279–307.

98. Kuhlman RE, Lowry OH. Quantitative histochemical changes during the development of the rat cerebral cortex. J Neurochem. 1956;1(2):173–80.

99. Page MA, Krebs HA, Williamson DH. Activities of enzymes of ketone-body utilization in brain and other tissues of suckling rats. Biochem J. 1971;121(1):49–53.

100. Baquer NZ, McLean P, Greenbaum AL. Systems relationships and the control of metabolic pathways in developing brain. In: Homines FA, Van DenBerg CJ, editors. Normal and Pathological Development of Energy Metabolism. 1975. p. 109–32.

101. Leong SF, Clark JB. Regional development of glutamate dehydrogenase in the rat brain. J Neurochem. 1984;43(1):106–11.

102. Villa RF, Gorini A, Ferrari F, Hoyer S. Energy metabolism of cerebral mitochondria during aging, ischemia and post-ischemic recovery assessed by functional proteomics of enzymes. Neurochem Int. 2013;63(8):765–81.

103. Turpeenoja L, Villa RF, Magri G, Giuffrida Stella AM. Changes of mitochondrial membrane proteins in rat cerebellum during aging. Neurochem Res. 1988;13(9):859–65.

104. Bradley MO, Hayflick L, Schimke RT. Transformation, and amino acid analogs (WI-38). Effects of aging, viral protein degradation in human fibroblasts. J Biol Chem. 1976;251:3521–9.

105. Tollervey JR, Wang Z, Hortobágyi T, Witten JT, Zarnack K, Kayikci M, et al. Analysis of alternative splicing associated with aging and neurodegeneration in the human brain. Genome Res. 2011;21(10):1572–82.

106. Lu T, Pan Y, Kao SY, Li C, Kohane I, Chan J, et al. Gene regulation and DNA damage in the ageing human brain. Nature. 2004;429(6994):883–91.

107. Su B, Wang X, Nunomura A, Moreira PI, Lee HG, Perry G, et al. Oxidative stress signaling in Alzheimer's disease. Curr Alzheimer Res. 2008;5(6):525–32.

108. Yang C, Slepnev VI, Goud B. Rab proteins form in vivo complexes with two isoforms of the GDP-dissociation inhibitor protein (GDI). J Biol Chem. 1994;269(50):31891–9.

109. Nishtala K, Phong TQ, Steil L, Sauter M, Salazar MG, Kandolf R, et al. Proteomic analyses of age related changes in A.BY/SnJ mouse hearts. Proteome Sci. 2013;11:29.

110. Iacopino AM, Christakos S. Specific reduction of calcium-binding protein (28-kilodalton calbindin-D) gene expression in aging and neurodegenerative diseases. Proc Natl Acad Sci U S A. 1990;87(11):4078–82.

111. Martinez A, Vitórica J, Satrústegui J. Cytosolic free calcium levels increase with age in rat brain synaptosomes. Neurosci Lett. 1988;88(3):336–42.

112. Whitehouse PJ. Clinical and neurochemical consequences of neuronal loss in the nucleus basalis of Meynert in Parkinson's disease and Alzheimer's disease. Adv Neurol. 1987;45:393–7.

113. Proctor CJ, Lorimer IA. Modelling the role of the Hsp70/Hsp90 system in the maintenance of protein homeostasis. PLoS One. 2011;6(7):e22038.

114. Proctor CJ, Soti C, Boys RJ, Gillespie CS, Shanley DP, Wilkinson DJ, et al. Modelling the actions of chaperones and their role in ageing. Mech Ageing Dev. 2005;126(1):119–31.

115. Vanguilder HD, Freeman WM. The hippocampal neuroproteome with aging and cognitive decline: past progress and future directions. Front Aging Neurosci. 2011;8:8.

116. Brüstle O, McKay RD. Neuronal progenitors as tools for cell replacement in the nervous system. Curr Opin Neurobiol. 1996;6(5):688–95.

117. International Genetic Standardization (IGS) Program. [www.criver.com/files/pdfs/rms/rm_rm_r_igs.aspx]

118. Pass D, Freeth G. The Rat. Anzccart News. 1993;6(4):1–4.

119. Klose J, Kobalz U. Two-dimensional electrophoresis of proteins: an updated protocol and implications for a functional analysis of the genome. Electrophoresis. 1995;16:1034–59.

120. Lorenz P, Bantscheff M, Ibrahim SM, Thiesen HJ, Glocker MO. Proteome analysis of diseased joints from mice suffering from collagen-induced arthritis. Clin Chem Lab Med. 2003;41:1622–32.

121. Westermeier R. Sensitive, quantitative, and fast modifications for Coomassie Blue staining of polyacrylamide gels. Proteomics. 2006;6(2):61–4.

122. Neuhoff V, Philipp K, Zimmer HG, Mesecke S. A simple, versatile, sensitive and volume-independent method for quantitative protein determination which is independent of other external influences. Hoppe Seylers Z Physiol Chem. 1979;360(11):1657–70.

Permissions

List of Contributors

Shi Qinghong, Gao Shen, Song Lina, He Chengyan and Li Hongjun
Jilin University China-Japan Union Hospital, Changchun 130033, China

Zhao Yueming
Jilin University China-Japan Union Hospital, Changchun 130033, China
Tumor Hospital of Jilin Province, Changchun 130021, China

Zhao Haifeng
Jiamusi University, Jiamusi 154002, China

Li Xiaoou
Tumor Hospital of Jilin Province, Changchun 130021, China

Wu Jianlin
State Key Laboratory for Quality Research in Chinese Medicines, Macau University of Science and Technology, Avenida Wai Long, Taipa, Macau, China

Gry H Dihazi, Gerhard A Mueller, Marwa Eltoweissy, Johannes T Wessels and Hassan Dihazi
Department of Nephrology and Rheumatology, University Medical Center Goettingen, Georg-August University Goettingen, Robert-Koch-Strasse 40, D-37075 Goettingen, Germany

Abdul R Asif
Department of Clinical Chemistry, Georg-August University Goettingen, Robert-Koch-Strasse 40, D-37075 Goettingen, Germany

Peng Zhang, Yang Zhang, Zhixian Gao, Nan Ji and Liwei Zhang
Department of Neurosurgery, Beijing Tiantan Hospital, Capital Medical University, No. 6 TiantanXili, Dongcheng District, Beijing 100050, China

Zhengguang Guo, Danqi Wang, Lili Zou and Wei Sun
Core Facility of Instrument, Institute of Basic Medical Sciences, Chinese Academy of Medical Science/ School of Basic Medicine, Peking Union Medical College, No. 5 Dongdan Santiao, Dongcheng District, Beijing 100005, China

Chun-Yan Wu, Chun-Dong Liu, Yu-Xiang Wang, Wei Na, Ning Wang and Hui Li
Key Laboratory of Chicken Genetics and Breeding of Agriculture Ministry, Key Laboratory of Animal Genetics, Breeding and Reproduction of EducationDepartment of Heilongjiang Province, College of Animal Science and Technology, Northeast Agricultural University, Harbin 150030, Heilongjiang, China

Yuan-Yuan Wu
Key Laboratory of Chicken Genetics and Breeding of Agriculture Ministry, Key Laboratory of Animal Genetics, Breeding and Reproduction of Education Department of Heilongjiang Province, College of Animal Science and Technology, Northeast Agricultural University, Harbin 150030, Heilongjiang, China
Weifang Academy of Agricultural Sciences, Weifang 261071, Shandong, China

Bing Zhang, Shao-Jun Du, Jue Hu, Di Miao and Jin-Yuan Liu
Laboratory of Plant Molecular Biology, Center for Plant Biology, School of Life Sciences, Tsinghua University, Beijing 100084, People's Republic of China

Caiyun He Guori Gao and Aiguo Duan
State Key Laboratory of Tree Genetics and Breeding, Key Laboratory of Tree Breeding and Cultivation of the State Forestry Administration, Research Institute of Forestry, Chinese Academy of Forestry, Beijing, People's Republic of China

Jianguo Zhang
State Key Laboratory of Tree Genetics and Breeding, Key Laboratory of Tree Breeding and Cultivation of the State Forestry Administration, Research Institute of Forestry, Chinese Academy of Forestry, Beijing, People's Republic of China
Collaborative Innovation Center of Sustainable Forestry in Southern China, Nanjing Forestry University, Nanjing, People's Republic of China

Hongmei Luo
Experimental Center of Desert Forestry, Chinese Academy of Forestry, InnerMonglia, People's Republic of China

Shan-Feng Jiang, Kai Dang, Hui-Ping Wang and Shen-Hui Xu
Key Laboratory of Resource Biology and Biotechnology in Western China (College of Life Sciences, Northwest University), Ministry of Education, Xi'an 710069, People's Republic of China

Hui Chang and Yun-Fang Gao
Key Laboratory of Resource Biology and Biotechnology in Western China(College of Life Sciences, Northwest University), Ministry of Education, Xi'an 710069, People's Republic of China
Shaanxi Key Laboratory for Animal Conservation, Northwest University, Xi'an 710069, People's Republic of China

Naveena B. Maheswarappa, K. Usha Rani, Y. Praveen Kumar and Vinayak V. Kulkarni
National Research Centre on Meat, Chengicherla, Hyderabad, Telangana 500092, India

Srikanth Rapole
Proteomics Lab, National Centre for Cell Science, Pune 411007, India

Jize Zhang
Institute of Grassland Research, Chinese Academy of Agricultural Sciences, Hohhot 010010, People's Republic of China
State Key Laboratory of Animal Nutrition, Institute of Animal Sciences, Chinese Academy of Agricultural
Sciences, Beijing 100193, People's Republic of China

Qingping Lu, Renna Sa and Hongfu Zhang
State Key Laboratory of Animal Nutrition, Institute of Animal Sciences, Chinese Academy of Agricultural
Sciences, Beijing 100193, People's Republic of China

Yang Gao
College of Animal Science and Technology, Jilin Agricultural University, Changchun 130118, People's Republic of China

Lina Zhang and Kasper Hettinga
Dairy Science and Technology, Food Quality and Design Group, Wageningen University, Postbox 8129, 6700EV Wageningen, The Netherlands.

Aalt D. J. van Dijk
Biometris, Wageningen University and Research Centre, P.O. Box 166700 AA Wageningen, The Netherlands

Bioinformatics Group, Wageningen University Droevendaalsesteeg 1, 6708 PB Wageningen, The Netherlands
Bioscience, cluster Applied Bioinformatics, Wageningen University and Research, Droevendaalsesteeg 1, 6708 PB Wageningen, The Netherlands

Weibo Ren, Xiangyang Hou, Yuqing Wang, Xiliang Li, Yong Ding, Zinian Wu, Ningning Hu, Lingqi Kong, Chun Chang, Chao Jiang and Jize Zhang
Key Laboratory of Forage Grass, Ministry of Agriculture, Institute of Grassl and Research, Chinese Academy of Agricultural Sciences, Hohhot 010010, Inner Mongolia, China

Warwick Badgery
NSW Department of Primary Industries, Orange Agricultural Institute, Orange, NSW 2800, Australia

Huiqin Guo
College of Life Sciences, Inner Mongolia Agricultural University, Hohhot 010019, Inner Mongolia, China

N. C. Rath
USDA/Agricultural Research Service, Poultry Production and Product Safety Research Unit, Fayetteville, AR 72701, USA

R. Liyanage and J. O. Lay Jr
Statewide Mass Spectrometry Facility, University of Arkansas, Fayetteville, AR 72701, USA

S. K. Makkar
Department of Poultry Science, University of Arkansas, Fayetteville, AR 72701, USA

Samuel Antwi-Baffour, Ransford Kyeremeh and George Awuku Asare
Department of Medical Laboratory Sciences, School of Biomedical and Allied Health Sciences, College of Health Sciences, University of Ghana, P. O. Box KB 143Korle-Bu, Accra, Ghana

Jonathan Kofi Adjei
Department of Medical Laboratory Sciences, School of Biomedical and Allied Health Sciences, College of Health Sciences, University of Ghana, P. O. Box KB 143Korle-Bu, Accra, Ghana
Department of Molecular Medicine, School of Medical Sciences Kwame Nkrumah University of Science and Technology, Kumasi, Ghana

Francis Agyemang-Yeboah and Max Annani-Akollor
Department of Molecular Medicine, School of Medical Sciences Kwame Nkrumah University of Science and Technology, Kumasi, Ghana

Ben Gyan
Noguchi Memorial Institute of Medical Research, University of Ghana, Legon, Ghana

Seyma Katrinli and Gizem Dinler Doğanay
Molecular Biology Biotechnology and Genetics Research Center (MOBGAM), Istanbul Technical University, Sariyer, Istanbul, Turkey

Kamil Ozdil, Abdurrahman Sahin, Oguzhan Ozturk, Mehmet Sokmen and H. Levent Doğanay
Gastroenterology,Umraniye Teaching and Research Hospital, Umraniye, Istanbul, Turkey

Gozde Kir
Pathology, Umraniye Teaching and Research Hospital, Umraniye, Istanbul, Turkey

Ahmet Tarik Baykal, Emel Akgun
Department of Medical Biochemistry, School of Medicine, Acibadem University, Istanbul, Turkey

Omer Sinan Sarac
Computer Engineering, Istanbul Technical University, Sarıyer, Istanbul, Turkey

Sun Hee Ahn, Hee-Young Yang and Suhee Kim
Department of Oral Biochemistry, Dental Science Research Institute, Medical
Research Center for Biomineralization Disorders, School of Dentistry, Chonnam National University, 300 Yongbong-Dong, Buk-Ku, Gwangju 500-757, Republic of Korea

Tae-Hoon Lee
Department of Oral Biochemistry, Dental Science Research Institute, MedicalResearch Center for Biomineralization Disorders, School of Dentistry, Chonnam National University, 300 Yongbong-Dong, Buk-Ku, Gwangju 500-757, Republic of Korea Department of Molecular Medicine, Graduate School, Chonnam National University, Gwangju, Republic of Korea

Gia Buu Tran and Quoc Thuong Dinh
Department of Molecular Medicine, Graduate School, Chonnam National University, Gwangju, Republic of Korea

Joseph Kwon
Korea Basic Science Institute, Daejeon, Republic of Korea.

Ha-Mi Lee, Kyoung-Oh Cho and Kyu-Yeol Son
Laboratory of Veterinary Pathology, College of Veterinary Medicine, Chonnam National University, Gwangju, Republic of Korea.

Seunggon Jung
Department of Oral and Maxillofacial Surgery, School of Dentistry, Chonnam National University, Gwangju, Republic of Korea

Katarina Davalieva, Sanja Kiprijanovska and Momir Polenakovic
Research Centre for Genetic Engineering and Biotechnology "Georgi D Efremov", Macedonian Academy of Sciences and Arts, Krste Misirkov 2, 1000 Skopje, Republic of Macedonia

Selim Komina and Gordana Petrusevska
Institute of Pathology, Medical Faculty, University "St. Cyril and Methodius", Skopje, Republic of Macedonia

Natasha Chokrevska Zografska
Biochemical laboratory, Clinical Hospital "Acibadem Sistina", Skopje, Republic of Macedonia

Michael Wille, Antje Schümann, Andreas Wree and Oliver Schmitt
Department of Anatomy, Gertrudenstr. 9, 18055 Rostock, Germany

Michael Kreutzer and Michael O Glocker
Proteome Center Rostock, Schillingallee 69, 18055 Rostock, Germany

Grit Mutzbauer
Department of Pathology, Josef-Schneider-Str. 2, 97080 Würzburg, Germany

Index

www.ingramcontent.com/pod-product-compliance
Lightning Source LLC
Chambersburg PA
CBHW082044190326
41458CB00010B/3454

* 9 7 8 1 6 4 1 1 6 1 5 7 2 *